BIOCATALYSIS FOR GREEN CHEMISTRY AND CHEMICAL PROCESS DEVELOPMENT

BIOCATALYSIS FOR GREEN CHEMISTRY AND CHEMICAL PROCESS DEVELOPMENT

Edited by

JUNHUA (ALEX) TAO
ROMAS KAZLAUSKAS

WILEY

A JOHN WILEY & SONS, INC., PUBLICATION

Published by John Wiley & Sons, Inc., Hoboken, New Jersey.
Published simultaneously in Canada.

Library of Congress Cataloging-in-Publication Data:

Biocatalysis for green chemistry and chemical process development / edited by Junhua (Alex) Tao, Romas Kazlauskas.
p. cm.
Includes index.
ISBN 978-0-470-43778-0 (cloth)
1. Environmental chemistry–Industrial applications. 2. Enzymes–Biotechnology. 3. Green technology. I. Tao, Junhua. II. Kazlauskas, R. J. (Romas J.), 1956-
TP155.2.E58B56 2011
660.6′3–dc22

2010046369

Printed in Singapore

oBook ISBN: 978-1-118-02830-8
ePDF ISBN: 978-1-118-02828-5
ePub ISBN: 978-1-118-02829-2

10 9 8 7 6 5 4 3 2 1

CONTENTS

CONTRIBUTORS

BANNER, TODD (Ph.D.) Cargill Biotechnology Development Center (BioTDC), Minneapolis, MN

BARRETT, JOHN S. F. (Graduate Research Assistant) Department of Chemical Engineering and Material Science, University of Minnesota, Minneapolis, MN

CHEN ZHENMING (Ph.D.) Institute of Sustainability, Hangzhou Normal University, Hangzhou, China

DE SOUZA, MERVYN (Ph.D.) Cargill Specialty Canola Oils, Fort Collins, CO

DIETRICH, JEFFREY A. (Ph.D.) UCB-UCSF Joint Graduate Group in Bioengineering, UC-Berkeley, UC-San Francisco, Berkeley, CA

FARID, SUZANNE S. (Professor) Advanced Centre for Biochemical Engineering, Dept. of Biochemical Engineering, University College London, Torrington Place, London, UK

FORTMAN, J.L. (Ph.D.) Fuels Synthesis Division, Joint BioEnergy Institute, Lawrence Berkeley, National Lab, Berkeley, CA

FOSMER, ARLENE (Ph.D.) Cargill Biotechnology Development Center (BioTDC), Minneapolis, MN

GORDON, JOHN (D.Phil.) LES R&D Evaluations and Business Process Excellence, Lonza AG, Visp., Switzerland

HO, SA (Ph.D.) Pfizer Biotherapeutics Pharmaceutical Sciences, Chesterfield, MO

HU, SEAN (Ph.D.) Novozymes, DAVIS, CA

JACKSON, DAVID A. (Ph.D.) Syngenta Crop Protection Muenchwilen AG., Muenchwilen, Switzerland

JESSEN, HOLLY (Ph.D.) Cargill Biotechnology Development Center (BioTDC), Minneapolis, MN

JUMINAGA, DARMAWI (Ph.D.) California Institute of Quantitative Biomedical Research (QB3), University of California-Berkeley, Berkeley, CA

KAZLAUSKAS, ROMAS (Professor) Department of Biochemistry, Molecular Biology and Biophysics and The Biotechnology Institute, University of Minnesota, Saint Paul, MN

KEASLING, JAY D. (Professor) Fuels Synthesis Division, Joint BioEnergy Institute, Lawrence Berkeley National Laboratory, Berkeley, CA

KIM, BYUNG GEE (Professor) Department of Chemical and Biological Engineering, Seoul National University, Seoul, South Korea

MARASCO, ERIN (Ph.D.) Cargill Biotechnology Development Center (BioTDC), Minneapolis, MN

MCLAUGHLIN, JOSEPH M. (Ph.D.) Pfizer Biotherapeutics Pharmaceutical Sciences, Chesterfield, MO

MEYER, HANS-PETER (Ph.D.) Lonza Innovation for Future Technologies (LIFT), Lonza AG, Visp, Switzerland

PATEL, RAMESH (Ph.D.) SLRP Associates, Biotechnology Consultation, Bridgewater, NJ

POLLOCK, JAMES (Professor) Department of Biochemical Engineering, University College London, London, UK

ROBINS, KAREN (MSc. Applied Science) Senior Research Associate, Lonza AG, Switzerland

RUSH, BRIAN (Ph.D.) Cargill Biotechnology Development Center (BioTDC), Minneapolis, MN

SCHNEIDER, MANFRED P. (Professor) FB C–Organische Chemie, Bergische Universitaet Wuppertal, Wuppertal, Germany

SERRA, STEFANO (Professor) C.N.R., Istituto di Chimica del Riconoscimento Molecolare, Milano, Italy

SHELDON, ROGER A. (Professor) Department of Biotechnology, Delft University of Technology, The Netherlands

SRIENC, FRIEDRICH (Professor) Department of Chemical Engineering and Material Science, University of Minnesota, Minneapolis, MN

TAO, JUNHUA (ALEX) (Ph.D.) Institute of Sustainability, Hangzhou Normal University, Hangzhou, China

VELDHOUSE, JON (Ph.D.) Cargill Biotechnology Development Center (BioTDC), Minneapolis, MN

WERBITZKY, OLEG (Ph.D.) Lonza Innovation for Future Technologies (LIFT), Lonza AG, Visp, Switzerland

WOHLGEMUTH, ROLAND (Ph.D.) Sigma-Aldrich, Buchs, Switzerland

XU, FENG (Ph.D.) Novozymes, Davis, CA

YIN, YIFENG (Ph.D.) Trinity Biosystems, Inc., Menlo Park, CA

PREFACE

Green chemistry is a goal and an approach to design safer, more efficient, and less expensive chemical processes. Many disciplines can contribute to green chemistry: organometallic chemists design new catalytic reactions and invent better ligands; others discover organocatalysts or safer solvents derived from biomass. Process chemists and engineers scale up reactions and find ways to carry out several steps in the same solvent, often water. Although this wide-ranging and multidisciplinary approach is essential for progress in green chemistry, it makes it hard for green chemists to exchange ideas and the tips and tricks that lead to progress in the field.

This book focuses on one important tool of green chemistry—biocatalysis. Though the focus is only on this one tool, this book explores more of the details, and exchanges some of these tips and tricks with others to spur further progress in this area. It shows how biocatalysis contributes to a wide range of industrial applications.

Biocatalysis is one of the most important tools for green chemistry. Biocatalysis is environmentally benign (often even edible!), and, because it can catalyze otherwise difficult transformations it can eliminate multiple steps involved in complex chemical syntheses. Eliminating the steps reduces waste and hazards, improves yields, and cuts costs.

This book describes recent progress in biocatalysis-driven green syntheses of industrially important molecules. The first three chapters introduce recent technological advances in enzymes and cell-based transformations, and green chemistry metrics for synthetic efficiency. The remaining chapters are case studies of biotechnological production of pharmaceuticals, including small molecules, natural products and biologics, flavor, fragrance and cosmetics, fine chemicals, value-added chemicals from glucose and biomass, and polymeric materials.

Currently, there are no books specifically devoted to a comprehensive overview of green chemistry applications of enzyme-driven transformations for a wide range of industries. There are a number of books that discuss green chemistry or biocatalysis, which mention these applications, but not in a comprehensive manner.

Saint Paul, Minnesota
Hangzhou, China

ROMAS KAZLAUSKAS
JUNHUA (ALEX) TAO

PART I

INTRODUCTION CHAPTERS

CHAPTER 1

BIOTECHNOLOGY TOOLS FOR GREEN SYNTHESIS: ENZYMES, METABOLIC PATHWAYS, AND THEIR IMPROVEMENT BY ENGINEERING

ROMAS J. KAZLAUSKAS
Department of Biochemistry, Molecular Biology and Biophysics and
The Biotechnology Institute, University of Minnesota, Saint Paul, Minnesota

BYUNG-GEE KIM
Department of Chemical and Biological Engineering, Seoul National University, Seoul, Korea

1.1 INTRODUCTION

Green chemistry is the design of products and processes that eliminate or reduce waste, toxic, and hazardous materials. Green chemistry is not a cleanup approach, but a prevention approach. Preventing problems is inevitably easier and less expensive than contending with difficulties after they occur.

The risk associated with a chemical depends both on how dangerous it is (hazard) and on one's contact with it (exposure) (Figure 1.1). In the past, governments and industry focused on reducing risk by minimizing exposure. Rules limit the exposure of workers to hazardous chemicals and the release of these chemicals into the environment. This approach is expensive; it is difficult to establish a safe level of hazardous chemicals, and currently, only a small fraction of the chemicals manufactured are regulated.

The green chemistry approach focuses on reducing risk by reducing or eliminating the hazard. Hazardous materials are eliminated by, for example, replacing them with nonhazardous ones. Hazardous materials are eliminated also by increasing the yield of a reaction, as higher yield eliminates some of the waste from that reaction. In addition, higher yield allows any preceding reactions to

Biocatalysis for Green Chemistry and Chemical Process Development, First Edition.
Edited by Junhua (Alex) Tao and Romas Kazlauskas.

risk = hazard • exposure

Figure 1.1 The risk associated with a chemical depends on both how dangerous it is (hazard) and one's contact with it (exposure). In the past, the focus was on minimizing exposure by rules that limit the amounts of hazardous chemicals in air and water. The green chemistry approach is to eliminate or reduce hazardous materials. This change requires redesigning of synthetic approaches.

be carried out on a smaller scale, thus eliminating some of the waste from these steps as well. This prevention approach saves money, as fewer raw materials are needed and the cost of treatment of waste is reduced.

Reducing costs and being environmentally friendly are goals that everyone agrees on. Why has this not been done before? One reason is that environmental costs were ignored in the early days of the chemical industry. Now that more of the cleanup cost falls on the manufacturer, there is a big financial incentive to be greener. Another reason is that chemists in research and design laboratories did not view environmental hazards as their problems. It was something to be fixed later in the scale-up stage. The green chemistry approach changes this thinking. By thinking about hazards and environmental consequences at the research and design stage, many problems are prevented and do not need a fix later. The principles of green chemistry outlined by Anastas and Warner [1] provide specific guidelines for what to look for at the research and design stage to make a greener process. These principles are discussed below in the context of biocatalysis.

The first use of biochemical reactions for organic synthesis was probably in 1858, when Louis Pasteur resolved tartaric acid by using a microorganism to destroy one enantiomer [2]. In spite of this early demonstration, chemists have used biocatalysis only sporadically. Chemists gradually recognized the potential of biochemical reactions, but there were both practical and conceptual hurdles. Practical problems were how to get the enzymes and how to stabilize them. The conceptual problems were beliefs that enzymes accept only a narrow range of biochemical intermediates as substrates, and that enzymes are too complex to consider engineering them for key properties like stability, stereoselectivity, substrate range, and even reaction type. The recent advances in biotechnology have solved many of the practical problems, and the increased understanding of biochemical structures and mechanisms has made biocatalysts more understandable to chemists. This chapter surveys the state of the art for engineering biocatalysts for chemistry applications. If you find an enzyme that catalyzes your desired reaction, regardless of how poor the enzyme is, it is highly likely that it can be engineered into an enzyme suitable for industrial and large-scale use.

1.2 THE NATURAL FIT OF BIOCATALYSIS WITH GREEN CHEMISTRY

Biotechnology methods fit naturally to the goals and principles of green chemistry. Green chemistry, or sustainable chemistry, seeks to integrate industrial manufacturing practice with the natural world. This natural world is the biological world, where sustainability and recycling are integral parts. Use of the biological methods for industrial manufacturing is an excellent starting point to create a green process. In some cases, biotechnology tools, unlike chemical tools, are even edible. Baker's yeast, used to make bread, also catalyzes the reduction of various carbonyl compounds. Lipases are the most commonly used enzymes for biocatalysis. These enzymes are also eaten in multigram amounts by patients with pancreatic insufficiency and in smaller amounts when food-grade lipases are used in the manufacture of cheese.

Although the "bio" part of biocatalysis makes it environmentally friendly, it is the "catalysis" part that provides the green chemistry advantage. Catalysis, in place of reagents, converts many substrate molecules to products and eliminates the need for stoichiometric reagents. Catalysis is fast, so the reaction may not need to be heated. This saves energy and may eliminate side reactions that occur at higher temperatures. Catalysis is selective, eliminating the need to add and remove protective groups or use auxiliaries to control reactivity. Catalysis can enable complex and otherwise difficult reactions. This ability can eliminate steps and simplify syntheses.

The 12 principles of green chemistry outline the design goals for synthesis. Making progress toward any one of these goals will make a synthesis greener; progress toward several goals is of course better. Table 1.1 lists some suggestions on how biocatalysis can help a synthesis toward these goals.

Given the wide-ranging ways in which biocatalysis can contribute to green chemistry, it is no surprise that many recent winners of the US Environmental Protection Agency's Presidential Award in Green Chemistry have used biocatalysis as the key improvement technique (Table 1.2). Many other winners have also used biocatalysis; these examples were selected to show the range of techniques used.

The 2010 winner of the Greener Reaction Conditions Award was an improved synthesis of sitagliptin, the active ingredient in an oral type 2 diabetes drug. Merck and Codexis collaborated to develop the key step, which is a transaminase-catalyzed formation of a chiral amine (R form) [3]. The starting transaminase required 27 amino acid substitutions to fit the large substrate in the active site, to increase the reaction rate and to stabilize the enzyme to the reaction conditions. The previous synthesis of sitagliptin was already a good synthesis, which won a green chemistry award in 2006 [4]. The new synthesis, which added a biocatalysis step, eliminates four steps, including an asymmetric hydrogenation using a rhodium catalyst that requires a high-pressure reactor.

LS9, Inc., winners of the 2010 prize for a small business, engineered new metabolic pathways into microorganisms to make biofuels [5]. For example, to make *Escherichia coli* bacteria produce biodiesel, they added the genes for plant thioesterases to divert normal fatty acid biosynthesis into synthesis of several fatty acids suitable for biodiesel. Next, they added genes for enzymes to make ethanol and an enzyme to couple the ethanol and fatty acids to make fatty acid ethyl esters, which can be used for biodiesel. Finally, the researchers added genes for

TABLE 1.1 How Biocatalysis Follows the Twelve Principles of Green Chemistry

Prevent waste

The high selectivity of enzymes and their ability to carry out difficult chemical reactions eliminate synthesis steps and the associated waste in multistep reactions. Reagents and solvents for eliminated steps are not needed; higher yield and selectivity in the remaining steps also eliminate waste. Biocatalysis usually uses water as the solvent, which eliminates organic solvent waste

Design safer chemicals and products

Biocatalysts are typically used only for manufacture and are not the product themselves. In some cases, a biocatalyst can be a product, such as an enzyme-based drain cleaner that replaces a drain cleaner based on a strong acid or base. Biocatalysis can enable the manufacture of new products, such as biodegradable polyesters, which is not practical to manufacture chemically

Design less hazardous chemical syntheses

Eliminating hazardous steps or replacing hazardous reagents with biocatalysts makes the manufacturing process safer

Use renewable feedstock

Biocatalysts are typically manufactured by growing microorganisms that secrete the biocatalysts. In other cases, the whole cells can be used as the catalysts. The feedstock to grow microorganisms are sugars and amino acids, which are renewable

Use catalysts, not stoichiometric reagents

Biocatalysts are highly efficient catalysts, that is, each enzyme molecule converts thousands to millions of substrate molecules to product. Thus, reactions are fast and the turnover number is high

Avoid chemical derivatives

The high selectivity of enzymes usually eliminates the need for protective groups and for resolutions involving chemical derivatives

Maximize atom economy

The high selectivity of enzymes and their ability to carry out difficult chemical reactions allow most of the starting material (including all reagents) atoms to be converted to the product. Eliminating synthesis steps, reagents, and derivatives reduces the number of atoms of the starting materials needed

Use safer solvents and reaction conditions

Biocatalysis typically uses water as the solvent, neutral pH, and room temperature. If needed for substrate solubility or faster reaction, biocatalysis can tolerate a wide range of reaction conditions, including organic solvents

Increase energy efficiency

Because biocatalysts are fast catalysts, one does not need to heat a reaction, thus saving energy. One also rarely needs to cool a reaction, which also requires energy, since the reaction rate can be reduced by adding less catalyst. The selectivity of biocatalysts is high, so there is no need to reduce the temperature to increase selectivity as there is with chemical reagents and catalysts

TABLE 1.1 ***(Continued)***

Design for degradation
Biocatalysts are biodegradable; some catalysts (e.g., baker's yeast and food-grade proteases) are even edible
Analyze in real time to prevent pollution
No special advantage of biocatalysis. Biocatalytic reactions can be monitored in real time by analyzing pH, oxygen, cofactor concentrations, and other parameters
Minimize the potential for accidents
Mild reaction conditions for biocatalysis typically minimize the potential for explosions, fire, and accidental release. Typical reaction conditions are near room temperature, atmospheric pressure, nonflammable solvent, and dilute solutions

cellulases and xylanases to this *E. coli* bacteria so that instead of using glucose as the starting material for this synthesis, the *E. coli* could use inexpensive biomass. The amount of biodiesel produced is at least 10-fold too low for commercial use, but further engineering will likely increase this amount.

The 2009 prize winner, Eastman Kodak Company, used lipase B from *Candida antarctica*, the most widely used lipase for synthesis. This lipase is highly active in organic solvents and shows high stereoselectivity. The high activity in organic solvents was the key characteristic for this work. It allowed researchers to eliminate the solvent and directly react the starting acid (or an ethyl ester) and alcohol together [6]. The mild reaction conditions allow researchers to use delicate unsaturated fatty acids that would undergo side reactions if an acid or a base is used to catalyze the ester formation. In diols, the enzyme's regioselectivity ensured that only one alcohol group reacted.

The 2006 prize was for a synthesis of a key fragment of atorvastatin (active ingredient of Lipitor), a cholesterol-lowering drug [7]. Codexis engineered 180 variations of the ketoreductase. All variants are stable and highly active, but differ in their substrate specificity. Screening these ketoreductases toward a target substrate identifies which ones have the correct substrate specificity, and further protein engineering to improve the ketoreductases is possible. Codexis demonstrated this approach for the atorvastatin fragment, but a similar approach should work for other problems.

The 2005 prize was for engineering a metabolic pathway. Organisms that naturally produce bacterial polyesters grow slowly and are inconvenient to work with. Researchers at Metabolix transferred the entire pathway into *E. coli* to enable it to produce polyesters [8]. Optimization of the metabolism increased the yield to practical levels for manufacture.

The last example is a plant cell fermentation for a multistep synthesis of paclitaxel (Taxol), a complex anticancer drug containing eight stereocenters [9]. Chemical synthesis is impractical (40 steps, ~2% overall yield) as is its isolation from the yew tree, since the amounts are so low. The existing process isolated a precursor of paclitaxel from leaves and twigs followed by 11 chemical

TABLE 1.2 Select Recent Winners of the US Presidential Green Chemistry Awards That Use Biocatalysis

Winner	Discovery	Green Features	Other Advantages	Biotechnology Tools
Merck & Codexis (2010)	Improved synthesis of sitagliptin, an oral diabetes drug.	Eliminates several steps, including an asymmetric hydrogenation using a rhodium catalyst that required a high-pressure reactor 10–13% higher yield and 19% less waste.	Can use existing equipment for synthesis New transaminase may be useful in other syntheses	Combined rational design and directed evolution methods to improve the starting transaminase activity 25,000-fold
LS9 (2010)	Engineered a new metabolic pathway in *Escherichia coli* to make biodiesel	Diverts fatty acid biosynthesis to make fatty acid ethyl esters	Can grow on inexpensive biomass	Cloning enzymes from other organisms into *E. coli*
Eastman Chemical Company (2009)	A solvent-free biocatalytic synthesis of esters as ingredients for cosmetics and personal care products	Eliminates >10 L of organic solvent/kg of ester Purification is often not needed	Allows use of delicate raw materials such as unsaturated fatty acids Regioselective monoesterification of diols such as 4-hydroxy-benzyl alcohol	Immobilized lipase catalyzes esterification and transesterification

Codexis (2006)	A simplified manufacturing process for a key intermediate in Liptor.	Reduces the formation of by-products Reduces generation of waste Eliminates the need for hydrogen gas Reduces solvent use Increases yield	Fewer unit operations, including no fractional distillation of the product Less purification equipment needed	Genetic engineering methods to create up to 4000-fold improved enzymes that catalyze the 3-step synthesis
Metabolix (2005)	Bioplastics (polyhydroxyalkanoates) made by genetically engineered organisms	Made from renewable resources (sugars) instead of petroleum Biodegradable Saves energy	Reduces greenhouse gas emissions	Add new biochemical pathways to microorganisms and optimize metabolism
Bristol Myers Squibb (2004)	Synthesis of paclitaxel using plant cell fermentation technology	Eliminates an estimated 14,000 pounds of hazardous chemicals and materials per year Eliminates 10 solvents and six drying steps Saves energy	Eliminates harvest of yew trees Increases the amount of available drug	Growth of specific paclitaxel-producing cells in fermentation tanks. Optimize metabolism

http://www.epa.gov/greenchemistry/pubs/pgcc/technology.html (accessed 21 September 2010).

transformations to the product. The plant cell fermentation eliminates all of these steps because the specific cell line yields paclitaxel directly.

1.3 WHY BIOCATALYSTS NEED TO BE ENGINEERED

Three of the six examples in Table 1.2 involve unnatural substrates. The ability of the enzymes to accept these synthetic intermediates is due to good luck, not evolutionary pressure in nature. One reason to engineer enzymes is to better accommodate unnatural substrates. Enzymes involved in the uptake of nutrients (lipases, proteases) often have broad substrate specificity because they must act on a broad range of possible food sources. Similarly, enzymes involved in detoxification (*P*450 monooxygenases, glutathione *S*-transferases) also have a broad substrate range to accommodate many possible natural substrates. In contrast, enzymes involved in primary metabolism, such as glycolysis, typically have a narrow substrate range. The broad substrate range of many enzymes is critical to their usefulness because it allows chemists to use enzymes to catalyze reactions on their synthetic intermediates, and not just on biochemical intermediates.

It is also due to luck that these enzymes often show high enantioselectivity toward these synthetic intermediates. The ability of enzymes to both accept a wide range of substrates, which suggests an open active site, and show high selectivity, which suggests a restricted active site, is somewhat surprising. Protein engineering can reshape the active site to increase the selectivity of the enzyme.

Another common reason to engineer enzymes is to increase their stability under the reaction conditions. Reaction conditions can differ dramatically from those present in a cell. Reaction conditions may involve high temperatures, extremes of pH, high substrate and product concentrations, oxidants, and organic cosolvents. Sometimes an enzyme must tolerate these conditions for only a few minutes or hours, but in a continuous manufacturing process, an enzyme may need to tolerate them for months.

A third reason to engineer enzymes and metabolic pathways is to create new reactions or new biochemical pathways. For example, Ran and Frost expanded the substrate range of an aldolase and thus created a new metabolic pathway to make shikimic acid for an influenza drug synthesis (Section 1.5).

Combining enzymes from different organisms and biochemical pathways can create a new biochemical pathway. The artemisinin synthesis discussed in Chapter 7 is an excellent example of this type of protein engineering and several other examples are discussed later in this chapter (Section 1.5).

1.4 STRATEGIES TO ENGINEER ENZYMES

The strategy chosen to engineer an enzyme depends on both the property being engineered and how much information is known about the enzyme, Table 1.3. Some engineering approaches require that the structure of the enzyme known,

TABLE 1.3 Examples of Protein Engineering of Biocatalysts

Strategy	Enzyme	Details	Improvement
Rational design	Phosphodiesterase	Decreases the size of one pocket to increase enantioselectivity; reverses the sizes of two pockets to reverse enantioselectivity	Increase or reverse enantioselectivity
Rational design	T4-lysozyme	Introduces proline residues and/or removed glycine residues in loop regions, introduces disulfide bonds	Higher thermostability
Rational design		Designs an enzyme that catalyzes a retro-aldol cleavage; not all designs gave catalytic activity	New catalytic activity
Directed evolution	Subtilisin	Random mutagenesis using error-prone PCR and screening to identify more stable variants	>250-fold higher activity in 60% DMF
Directed evolution	Phosphite dehydrogenase for NAD(P)H regeneration	Random mutagenesis using error-prone PCR and screening to identify more stable variants; combines the 12 best substitutions using site-directed mutagenesis	>7000-fold higher thermal stability
Directed evolution	Nitrilase for synthesis of Lipitor side chain	Saturation mutagenesis at each position and screening to find the best amino acid substitution	Increases enantioselectivity to >100
Directed evolution		Saturation mutagenesis at selected locations in the active site	Increases E from 12 to >50
Bioinformatics	Phytase	Amino acid substitutions to bring the amino acid sequence closer to the consensus sequence	Higher thermostability
Bioinformatics	Thermolysin	Replaces amino acids with those found at equivalent positions in naturally occurring, more thermostable variants	Stable to boiling water >2 h
Statistical correlation	Halohydrin dehalogenase for synthesis of Lipitor side chain	Identifies beneficial or detrimental effects of various amino acid substitutions (ProSAR) and combines 35 beneficial mutations	>4000-fold higher volumetric productivity
Statistical correlation	Humulene synthase	Correlates amino acid substitution in the active site to which product is formed	Changes major product from humulene to sibirene
Statistical correlation	Cellulase	Recombines eight blocks from three parents; identifies those blocks that increase thermostability	Higher thermostability

while others do not require any structural information, but require larger libraries and more screening. In some cases, typically for thermostability, screening is easy, while in other cases, for example, less product inhibition under process conditions, the screening can be slow and limited to several hundred variants.

1.4.1 Rational Design

Rational design requires structural information and a molecular level understanding of how the protein structure influences the property to be engineered. This requires answers to questions such as "What is the molecular basis of catalysis?" and "What determines efficient protein expression in different hosts?" These questions cannot be completely answered. This inability limits the success of rational protein engineering.

In spite of these difficulties, there are a number of examples where rational design has worked. For example, changing the pocket sizes in an organophosphorus hydrolase predictably altered the enantioselectivity (E) [10]. The wild-type enzyme favors the S_P enantiomer of ethyl phenyl 4-nitrophenyl phosphate ($E = 21$). Decreasing the site of the small site using a Gly60Ala mutation increased enantioselectivity to $E > 100$. Four other amino acid substitutions to reverse the relative sizes of the subsites reversed the enantiopreference.

Protein stability can also be engineered rationally in many cases. Protein stability depends on the equilibrium between the folded and unfolded states. The hydrophobic effect and inter-residue interactions stabilize the folded state, while the entropy associated with main chain flexibility favors the unfolded state. Rational strategies that reduce the flexibility of the unfolded state increase protein stability. In particular, introducing disulfide cross-links or proline residues (a less flexible residue) or removing glycine residues (the most flexible residue) all stabilized enzymes [11].

The most advanced rational design uses quantum mechanical models of the transition state and search strategies that test many possible conformations of the enzyme. Baker and collaborators designed enzymes that catalyzed an unnatural model reaction, the Kemp elimination [12], a retro-aldol cleavage [13], and a Diels–Alder cycloaddition [14] using this approach. These calculations modeled hundreds of thousands of possibilities and narrowed them to several dozen candidates. Synthesis of these predicted variants revealed a number with detectable catalytic activity.

1.4.2 Directed Evolution

Directed evolution needs no structure. One makes random changes in an enzyme to create many enzymes, and then screens these enzymes for the ones that show improvements. This has to be repeated until the enzyme is sufficiently improved. Chen and Arnold were the first to demonstrate that directed evolution can improve a biocatalyst [15]. They improved the stability of subtilisin in organic solvents.

This problem was well suited to directed evolution because the molecular basis for enzyme stability in organic solvents is not well understood and therefore cannot be engineered rationally. Chen and Arnold created amino acid substitutions throughout the protein and screened to find the more solvent-stable variants. After three rounds of mutagenesis and screening, they had a subtilisin variant with 10 amino acid substitutions that was 250-fold more active in 60% dimethylformamide (DMF) than the wild type. Using a similar approach, Zhao and coworkers increased the stability of a dehydrogenase >7000-fold [16] and DeSantis and coworkers increased the enantioselectivity of a nitrilase [17]. In the last example, researchers did not have structural information for the nitrilase, so they could not use rational design.

If some structure information is available, it can speed up directed evolution by eliminating regions in the protein that are less likely to contain a solution. For example, residues close to substrate (either direct contact or the next sphere of residues) are more likely to yield big improvements in enantioselectivity than distant residues [18]. For example, Horsman *et al.* [19] increased the enantioselectivity of a *Pseudomonas fluorescens* esterase by targeting only four amino acid residues in the active site ($4 \times 19 = 76$ possibilities). They found two variants with an enantioselectivity of ~60 toward 3-bromo-2-methyl propanoate esters, which was a dramatic increase over the wild-type enantioselectivity of 12. The focus on the active site is a good approach to changing enantioselectivity or substrate specificity, but mutations that improve thermal stability, catalytic activity, and probably stability toward organic solvents are scattered throughout the protein.

Amino acid substitutions are the most common way to create enzyme variants (review [20]). The most common methods to make these substitutions are error-prone polymerase chain reaction (epPCR) and saturation mutagenesis. The epPCR approach is simple and makes random single amino acid substitutions throughout the protein, but likely misses some substitutions because they require the statistically unlikely two nucleotide substitutions in a single codon [21]. Site saturation mutagenesis gives all possible substitutions at selected locations. Saturation mutagenesis at every position (one position at a time) is also possible, but is expensive and requires a large screening effort. Another approach to making protein variants is DNA shuffling, which exchanges longer sections of proteins between two or more parents. The sections will differ in several amino acid substitutions, but may also include insertions and deletions.

1.4.3 Bioinformatics Approaches

Bioinformatics compares the amino acid or DNA sequence of the biocatalyst with other known sequences. Genome sequencing projects, enabled by the rapid advances in DNA sequencing methods, have created vast amounts of sequence information. The amount of this information is much larger and broader than the amount of functional or structural information. Extracting information for protein engineering from these sequences relies on the hypothesis that biocatalyst with similar sequences are related in evolutionary history. This relatedness means that they will have similar function and similar structure.

One straightforward application of bioinformatics is to search for additional variants of an enzyme in the sequence databases. Starting from one or a few enzymes that catalyze the desired reaction, one searches for similar amino acid sequences. These sequences likely correspond to enzymes that also catalyze the desired reaction, but may have altered substrate specificity, stability to reaction conditions, or thermostability. For example, Fraaije *et al*. identified a thermostable Baeyer–Villiger monooxygenase enzymes by searching for a sequence characteristic of Baeyer–Villiger monooxygenases in the genome of a thermophile [22].

When the current variant of an enzyme already has many desirable characteristics, it is preferable to engineer the remaining characteristics instead of discovering new enzymes that may be lacking in other ways. Bioinformatics can also guide the engineering of more stable enzymes in two ways. First, sequence comparison can identify similar enzymes in thermophilic organisms. Some of the sequence differences account for the different stability. If one can identify which differences are most likely to lead to increase stability, then making these changes in the less stable enzyme can increase its stability. For example, Bae and Phillips stabilized adenylate kinase (T_m increased by 5°C.) by adding three salt bridges that they identified in the adenylate kinase from a thermophile [23].

The second approach to using bioinformatics to engineer more stable enzymes is the consensus sequence concept. The underlying hypothesis is that conserved amino acids contribute most to stability. If an amino acid substitution yields an unstable protein, the organism will not survive, so the conserved amino acids are more likely to contribute to stability than the nonconserved amino acids. The stabilization strategy is to compare the related sequences and to engineer the target protein to resemble the consensus sequence, that is, the most abundant amino acid at each position. Lehman and coworkers dramatically increased the thermostability of phytase using this approach; T_m increased up to 35°C from 55 to 90°C. The new enzyme contained >10 amino acid substitutions, where each substitution contributed slightly to the stability [24].

Bioinformatics is also the basis for homology modeling—the extrapolation of three-dimensional structures to proteins that have similar amino acid sequences (>35% identical amino acids). These three-dimensional structures are less accurate than experimentally derived structures, but can be a good starting point for rational design. A homology model can identify the active site of an enzyme, so saturation mutagenesis to change the shape of the active site can be attempted. For example, Keasling and coworkers made a homology model of γ-humulene synthase, and then targeted the residues predicted to be in the active site for mutagenesis [25]. More details about this experiment are in the next section.

1.4.4 Statistical Correlation Approaches

Unlike rational design, which uses molecular design principles to improve a desired property, statistical correlation approaches take a more empirical approach. A set of enzyme variants are tested, and then statistics are used to correlate the changes with the improvements. Replacement of changes that degrade the property with neutral or beneficial changes improves the enzyme.

The best example is the ProSAR (protein structure activity relationship) approach used by Codexis researchers to improve the reaction rate of a halohydrin dehalogenase >4000-fold [4]. Researchers made random amino acid substitutions (an average of 10) in the dehalogenase and measured catalysis by the variants. Then, they made a statistical correlation whether a particular substitution was beneficial. For example, variants that contained a F186Y substitution were, on average, better than those that did not. Some variants that contained F186Y were poor because of the detrimental effects of other mutations, but the statistical analysis identified that, on average, it was a beneficial mutation. The final improved enzyme contained 35 amino acid substitutions among its 254 amino acids. Combining beneficial mutations and removing deleterious ones yields an improved enzyme.

γ-Humulene synthase catalyzes the cyclization of farnesyl diphosphate via cationic intermediates to γ-humulene in 45% yield, but forms 51 other sesquiterpenes in smaller amounts. Keasling and coworkers substituted amino acid residues in the active site and identified the contribution of each one to the product distribution [25]. Substitutions were combined to favor formation of one of the other sesquiterpenes. For example, a triple substitution created an enzyme that formed 78% sibirene. The contribution of each substitution was additive.

In another example, Arnold and coworkers improved the thermal stability of cellulases [26]. Instead of single amino acid substitutions, they recombined peptide fragments from three parents to make the variants that contained several substitutions as a group. They divided three parent cellulases into eight blocks and then recombined these blocks to make new cellulases. They made a set of 24 variants, measured their thermal stability, and identified the contribution of each block. For example, block 6 from parent 3 tended to stabilize the cellulases. Combining the stabilizing blocks and removing the destabilizing blocks yielded more stable cellulases.

Although this is an empirical approach that focuses on what works rather than why it works, it does not preclude the use of molecular design principles in choosing which changes to make. For example, the Codexis researchers hypothesized that changes nearest the active site would likely change the substrate fit for the halohydrin dehalogenase and therefore directed many of the amino acid substitutions to this region. Finally, by carefully examining the changes in protein structure that work, researchers may learn the molecular basis for the improvements.

1.4.5 Multiple Criteria

Practical applications require that the biocatalysts should meet multiple criteria: they must be stable at the temperature for the reaction, they must tolerate the pH and solvent of the reaction, they must tolerate high concentration of the starting material and product, they must show high selectivity, and they must react quickly with the substrate even when it is an unnatural compound. Often, one starts with a catalyst that meets some of these criteria and tries to engineer in the missing properties. These multiple requirements make engineering more difficult than engineering a single property because engineering to improve one property must not destroy other properties in the process. For rational engineering, this means that you must understand the molecular basis of all of them, which is currently not possible. For methods that rely on screening, this means that you must screen for all the important properties.

For example, Schmidt and coworkers identified an esterase variant with three amino acid substitutions that showed higher enantioselectivity (E) toward the synthetic building block (S)-but-3-yn-2-ol ($E = 89$ as against $E = 3$ in wild type). Unfortunately, these substitutions degraded protein expression. Bacteria produced high amounts of the soluble starting esterase, but only small amounts of the variant, and most of it was in an unfolded insoluble form [27]. Additional experimentation revealed a variant with only two amino acid substitutions from wild type that was both highly enantioselective and efficiently produced in soluble form.

1.5 ENGINEERING OF METABOLIC PATHWAYS

Multistep metabolic pathways offer the opportunity for complex syntheses, but require use of whole cells. Examples of multistep processes on commercial scales are fermentation of glucose to ethanol by yeast, fermentation of glucose to citric acid by *Aspergillus* fungi, and production of penicillin G by *Penicillium* fungi. Although isolated enzymes can be used much like chemical catalysts in a wide range of temperatures, solvents, and reactors, whole cells are more limited in the reaction conditions they tolerate. In addition, they require the ability to work with microorganisms and typically use dilute aqueous solutions. The key advantages of whole cells are that they can stabilize enzymes that are difficult to isolate and that they can contain multiple enzymes. An isolated enzyme approach with multiple enzymes would require multiple enzyme isolations and optimization, which would eliminate the advantages of isolated enzymes.

Metabolic pathways are not limited to those existing in nature; the sections below highlight current strategies to create new metabolic pathways [28]. These pathways may be more efficient routes to existing products, or they may be new routes to create new products, Table 1.4.

1.5.1 New Pathways To Increase Yields

Combining existing pathways from different organisms created a pathway that yielded high concentrations of 1,3-propanediol used in polymers on a multi-ton scale. Several microorganisms convert glycerol to 1,3-propandiol, but Dupont and Genencor engineered a new pathway that allows use of glucose, a less expensive carbon source [29], Figure 1.2a. The pathway combines three different pathways from three different organisms in an *E. coli* strain. The first pathway is the naturally occurring glycolysis pathway in *E. coli*, which converts glucose to dihydroxyacetone phosphate. The second pathway consists of two enzymes added from the

TABLE 1.4 Examples of Engineering Metabolic Pathways

Product/Substrate	Microorganism	Details
1,3-propanediol from glucose	Recombinant *E. coli*	The natural glycerol to 1,3-propandiol pathway is combined with a glucose to glycerol pathway. Glucose as a carbon source is less expensive than fermentation grade glycerol. 1,3-Propanediol is used for synthesis of a polyester (poly(trimethylene terephthalate)). This biocatalytic process uses 40% less energy than conventional processes, and reduces greenhouse gas emissions by 20%
Shikimate from glucose	Recombinant *E. coli*	Shikimate, used for the synthesis of oseltamivir phosphate (Tamiflu®), is made by the aromatic amino acid biosynthetic pathway. The natural pathway uses phosphoenolpyruvate as a starting compound, but the engineered pathway uses the more abundant pyruvate and gives larger amounts of product
Copolymer of β-hydroxybutyrate and β-hydroxyvalerate from glucose and propionic acid	*Rastonia eutrophus*	Excess glucose is converted to poly(β-hydroxybutyrate) via acetyl-CoA. Co-feeding with propionic acid makes propionyl-CoA, which results in a copolymer that is more flexible and useful than the natural homopolymer
Higher alcohols from glucose	Recombinant *E. coli*	An added decarboxylase diverts 2-keto acids from the normal amino acid biosynthesis to form aldehydes. Reduction of the aldehydes yields alcohols, which may be useful as biofuels

yeast *Saccharomyces cerevisiae*, which convert the dihydroxyacetone phosphate to glycerol. The final pathway consists of several enzymes from *Klebsiella pneumonia*, which convert the glycerol to 1,3-propandiol. Surprisingly, researchers found that an uncharacterized oxidoreductase in *E. coli* worked better than 1,3-propandiol dehydrogenase from *Klebsiella*, so the production strain uses this *E. coli* oxidoreductase in this pathway. The production strain also incorporates gene deletions that eliminate nonproductive reactions. Another reason to move a pathway from one organism to another is to circumvent native regulation and thus make higher concentrations of product.

Ran and Frost designed a new pathway to shikimic acid by inventing a new step [30], Figure 1.2b. Shikimic acid is a precursor of oseltamivir phosphate (Tamiflu®), an influenza drug. The normal pathway to shikimic acid uses phosphoenol pyruvate as a starting compound. Unfortunately, low levels of

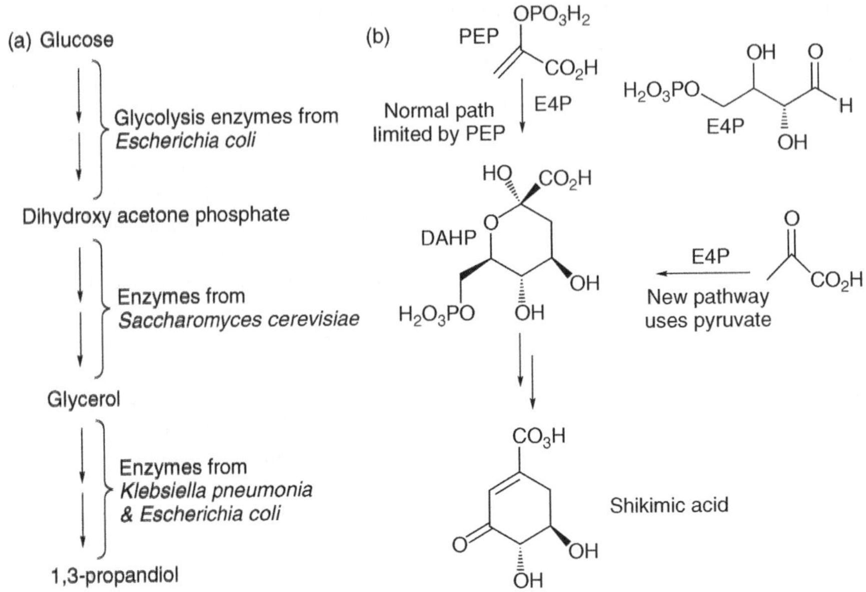

Figure 1.2 Two examples of strategies to engineer more efficient metabolic pathways to existing products. (a) Combining existing pathways from different organisms creates new pathways. Three pathways—conversion of glucose to dihydroxyacetone phosphate, conversion of dihydroxyacetone phosphate to glycerol, and glycerol to 1,3-propandiol—are combined in a strain of *E. coli* for the manufacture of 1,3-propandiol. (b) The amount of phosphoenolpyruvate limits the yield of shikimic acid from the natural pathway. A new pathway, created by expanding the substrate range of an aldolase, uses the more abundant pyruvate and generates higher amounts of shikimic acid.

phosphoenol pyruvate limit the amount of shikimic acid that can be produced. The solution was introducing a new enzyme, an aldolase that can use the more abundant pyruvate as the starting compound. The new enzyme was created by changing the substrate specificity of an existing aldolase by directed evolution. The new pathway gave higher yields of the desired compound on glucose.

1.5.2 New Pathways To Make Different Products

One way to get a new product from a metabolic pathway is to add a new intermediate that can enter the pathway. One example is the production of a mixed bacterial polyester, Figure 1.3a. Many bacteria, such as *Rastonia eutropha*, store excess carbon as granules of poly(β-hydroxybutyrate). This polyester is a potential biodegradable, renewable replacement for polypropylene. However, this natural polyester is crystalline and too brittle for most applications. A solution is to feed propionic acid along with the normal carbon source [31]. The acetyl-CoA intermediate forms from the normal carbon source, while propionyl-CoA forms from the propionic acid. Both enter the polyester synthesis pathway resulting in

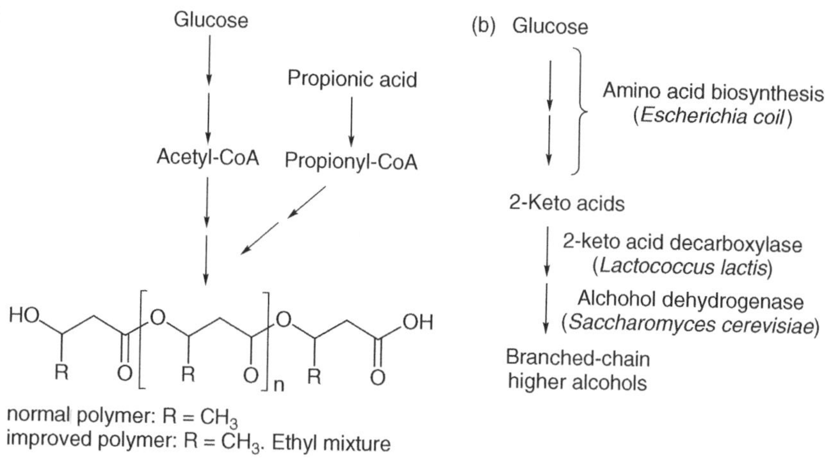

Figure 1.3 Two examples of strategies to engineer new metabolic pathways that create new products. (a) Adding propionic acid to the bacteria (*Rastonia eutropha*) making poly(β-hydroxybutyrate) leads to a mixed copolymer of β-hydroxyvalerate and β-hydroxybutyrate. This polymer is more flexible and useful than the natural poly(β-hydroxybutyrate). (b) Adding two new enzymes to *E. coli* diverts the 2-keto acids from amino acid biosynthesis to the synthesis of alcohols. The 2-keto acids are decarboxylated to the aldehyde and then reduced to the alcohol. The alcohols with five to eight carbons are potential replacements for ethanol as a biofuel

a copolymer of 4-carbon (β-hydroxybutyrate) and 5-carbon (β-hydroxyvalerate) subunits. This mixed polymer is less crystalline, more flexible, and more useful for application. This unnatural polymer does not form naturally because acyl-CoA intermediates with an odd number of carbon atoms are rare in natural biosynthetic pathways.

Another way to get new products from a metabolic pathway is to add new steps that divert an intermediate to the newly added pathway. One example is the diversion of 2-ketoacids from amino acid biosynthesis to the synthesis of alcohols for biofuels [32], Figure 1.3b. Adding a 2-keto acid decarboxylase converts these acids to aldehydes, which are reduced to alcohols. The higher alcohols (five to eight carbon atoms) are potential second generation biofuels to replace ethanol.

1.6 OUTLOOK

Biocatalysis tools have improved dramatically in the last 10 years and continuing advances in biology indicate that there will be additional improvements. Biocatalysis will be an increasingly important strategy and will become one of the key core technologies for chemical manufacturing in the next decade. It is a green chemistry technology that works perfectly with the emerging trend of bio-based sustainable feedstocks (biorefinery). The subsequent chapters detail additional examples where biocatalysis enables greener organic syntheses.

ACKNOWLEDGMENT

The authors thank the Korean Ministry of Education, Science and Technology, for funding this work (WCU program grant R32-10213).

REFERENCES

1. Anastas PT, Warner JC. Green chemistry: theory and practice. New York: Oxford University Press; 1998.
2. Pasteur L. Mé'moire sur la fermentation de l'acide tartrique. C R Séances Acad Sci 1858;46:615–618.
3. Savile CK, Janey JM, Mundorff EC, Moore JC, Tam S, Jarvis WR, Colbeck JC, Krebber A, Fleitz FJ, Brands J, Devine PN, Huisman GW, Hughes GJ. Biocatalytic asymmetric synthesis of chiral amines from ketones applied to sitagliptin manufacture. Science 2010;329:305–309.
4. Hansen KB, Hsiao Y, Xu F, Rivera N, Clausen A, Kubryk M, Krska S, Rosner T, Simmons B, Balsells J, Ikemoto N, Sun Y, Spindler F, Malan C, Grabowski EJJ, Armstrong JD III. Highly efficient asymmetric synthesis of sitagliptin. J Am Chem Soc 2009;131:8798–8804.
5. Steen EJ, Kang Y, Bokinsky G, Hu Z, Schirmer A, McClure A, del Cardayre SB, Keasling JD. Microbial production of fatty-acid-derived fuels and chemicals from plant biomass. Nature 2010;463:559–562.
6. Clendennen SK, Boaz NW. Cosmetic emulsifiers, United States Patent Eastman Chemical Company. (Kingsport, TN, US 7667067). 2010.
7. Fox RJ, Davis SC, Mundorff EC, Newman LM, Gavrilovic V, Ma SK, Chung LM, Ching C, Tam S, Muley S, Grate J, Gruber J, Whitman JC, Sheldon RA, Huisman GW. Improving catalytic function by ProSAR-driven enzyme evolution. Nat Biotechnol 2007;25:338–344.
8. Madison LL, Huisman GW. Metabolic engineering of poly(3-hydroxyalkanoates): from DNA to plastic. Microbiol Mol Biol Rev 1999;63:21–53.
9. Patel RN. Tour de paclixatel: biocatalysis for semisynthesis. Annu Rev Microbiol 1998;52:361–395.
10. Chen-Goodspeed M, Sogorb MA, Wu F, Raushel FM. Enhancement, relaxation, and reversal of the stereoselectivity for phosphotriesterase by rational evolution of active site residues. Biochemistry 2001;40:1332–1339.
11. (a) Matthews BW, Nicholson H, Becktel WJ. Enhanced protein thermostability from site-directed mutations that decrease the entropy of unfolding. Proc Natl Acad Sci USA 1987;84:6663–6667; (b) Wetzel R. Harnessing disulfide bonds using protein engineering. Trends Biochem Sci 1987;12:478–482; (c) Matsumura M, Signor G, Matthews BW. Substantial increase of protein stability by multiple disulphide bonds. Nature 1989;342:291–293; (d) Suzuki Y. The proline rule. A strategy for protein thermal stablization. Proc Jpn Acad Ser B Phys Biol Sci 1999;75:133–137.
12. Röthlisberger D, Khersonsky O, Wollacott AM, Jiang L, DeChancie J, Betker J, Gallaher JL, Althoff EA, Zanghellini A, Dym O, Albeck S, Houk KN, Tawfik DS, Baker D. Kemp elimination catalysts by computational enzyme design. Nature 2008;453:190–195.

13. Jiang L, Althoff EA, Clemente FR, Doyle L, Röthlisberger D, Zanghellini A, Gallaher JL, Betker JL, Tanaka F, Barbas CF III, Hilvert D, Houk KN, Stoddard BL, Baker D. De novo computational design of retro-aldol enzymes. Science 2008;319:1387–1391.
14. Siegel JB, Zanghellini A, Lovick HM, Kiss G, Lambert AR, St. Clair JL, Gallaher JL, Hilvert D, Gelb MH, Stoddard BL, Houk KN, Michael FE, Baker D. Computational design of an enzyme catalyst for a stereoselective bimolecular Diels-Alder reaction. Science 2010;329:309–313.
15. Chen K, Arnold FH. Tuning the activity of an enzyme for unusual environments: Sequential random mutagenesis of subtilisin E for catalysis in dimethylformamide. Proc Natl Acad Sci USA 1993;90:5618–5622.
16. Johannes TW, Woodyer RD, Zhao H. Directed evolution of a thermostable phosphite dehydrogenase for NAD(P)H regeneration. Appl Environ Microbiol 2005; 71:5728–5734.
17. DeSantis G, Wong K, Farwell B, Chatman K, Zhu Z, Tomlinson G, Huang H, Tan X, Bibbs L, Chen P, Kretz K, Burk MJ. Creation of a productive, highly enantioselective nitrilase through Gene Site Saturation Mutagenesis (GSSM). J Am Chem Soc 2003;125:11476–11477.
18. Morley KL, Kazlauskas RJ. Improving enzyme properties: when are closer mutations better? Trends Biotechnol 2005;23:231–237.
19. Park S, Morley KL, Horsman GP, Holmquist M, Hult K, Kazlauskas RJ. Focusing mutations into the *P. fluorescens* esterase binding site increases enantioselectivity more effectively than distant mutations. Chem Biol 2005;12:45–52.
20. Neylon C. Chemical and biochemical strategies for the randomization of protein encoding DNA sequences: library construction methods for directed evolution. Nucleic Acids Res 2004;32:1448–1459.
21. Wong TS, Roccatano D, Zacharias M, Schwaneberg U. A statistical analysis of random mutagenesis methods used for directed protein evolution. J Mol Biol 2006;355:858–871.
22. Fraaije MW, Wu J, Heuts DPHM, van Hellemond EW, Lutje Spelberg JH, Janssen DB. Discovery of a thermostable Baeyer–Villiger monooxygenase by genome mining. Appl Microbiol Biotechnol 2005;66:393–400.
23. Bae E, Phillips GN Jr. Identifying and engineering ion pairs in adenylate kinases: insights from molecular dynamics simulations of thermophilic and mesophilic homologues. J Biol Chem 2005;280:30943–30948.
24. (a) Lehmann M, Loch C, Middendorf A, Studer D, Lassen SF, Pasamontes L, van Loon APGM, Wyss M. The consensus concept for thermostability engineering of proteins: further proof of concept. Protein Eng 2002;15:403–411; (b) Lehmann M, Wyss M. Engineering proteins for thermostability: the use of sequence alignments versus rational design and directed evolution. Curr Opin Biotechnol 2001;12:371–375.
25. Yoshikuni Y, Ferrin TE, Keasling JD. Designed divergent evolution of enzyme function. Nature 2006;440:1078–1082.
26. Heinzelman P, Snow CD, Wu I, Nguyen C, Villalobos A, Govindarajan S, Minshull J, Arnold FH. A family of thermostable fungal cellulases created by structure-guided recombination. Proc Natl Acad Sci USA 2009;106:5610–5615.
27. Schmidt M, Hasenpusch D, Kähler M, Kirchner U, Wiggenhorn K, Langel W, Bornscheuer UT. Directed evolution of an esterase from Pseudomonas fluorescens yields a mutant with excellent enantioselectivity and activity for the kinetic resolution of a chiral building block. ChemBioChem 2006;7:805–809.

28. Prather KLJ, Martin CH. De novo biosynthetic pathways: rational design of microbial chemical factories. Curr Opin Biotechnol 2008;19:468–474.
29. Nakamura CE, Whited GM. Metabolic engineering for the microbial production of 1,3-propanediol. Curr Opin Biotechnol 2003;14:454–459.
30. Ran N, Frost JW. Directed evolution of 2-keto-3-deoxy-6-phosphogalactonate aldolase to replace 3-deoxy-D-arabino-heptulosonic acid 7-phosphate synthase. J Am Chem Soc 2007;129:6130–6139.
31. Rehm BHA. Bacterial polymers: biosynthesis, modifications and applications. Nat Rev Microbiol 2010;8:578–592.
32. Atsumi S, Hanai T, Liao JC. Non-fermentative pathways for synthesis of branched-chain higher alcohols as biofuels. Nature 2008;451:86–89.

CHAPTER 2

HOW GREEN CAN THE INDUSTRY BECOME WITH BIOTECHNOLOGY?

HANS-PETER MEYER and OLEG WERBITZKY
Lonza Innovation for Future Technologies (LIFT), Lonza AG, Visp, Switzerland

2.1 INTRODUCTION

There is a general consensus that the perspectives of white or industrial biotechnology are very good. It represents one of the largest, if not the largest, business opportunities in biotechnology for the future. New developments in technology and methods, changing ecological and economic conditions, recent policies and political commitments have led to a renaissance in industrial biotechnology. It is not only believed that the share of biotechnologically produced fine chemicals could increase several fold over the next 10–20 years, but even a switch from a petro-based economy to a biobased economy is expected to be complete within the next 30–80 years.

The global chemistry market is estimated at \$2.292 billion and is expected to grow to \$3.235 billion by 2015 and to \$4.012 billion by 2020. Of the \$2.292 billion, the global industrial biotechnology market is believed to represent only about \$50 billion [1] (not including biofuels). The global white (or industrial) biotechnology market of approximately \$50 billion is smaller than the red biotechnology (large-molecule biopharmaceuticals) market (Table 2.1). Needless to say there is a lot of overlap, for example, between the white or industrial biotechnology and the pharmaceutical or red biotechnology. However, industrial biotechnology represents a larger long-term business potential than red biotechnology. Of the \$50 billion biotechnology products mentioned above, about 25% are fine chemicals and the share of biotechnologically produced fine chemicals is expected to grow rapidly. It is estimated that >20% of all chemicals could be produced by biotechnological means in 2020! This potential is spread over diverse markets and various product/molecule classes; moreover, in order to meet the expectations

Biocatalysis for Green Chemistry and Chemical Process Development, First Edition.
Edited by Junhua (Alex) Tao and Romas Kazlauskas.

TABLE 2.1 Summary of the Different Biotechnology Markets and the Market Volumes in Billion US Dollars, the Compounded Annual Growth Rate (CAGR), and the Approximate Number of Companies Globally Active in the Area

	Industrial Biotechnology	Pharmaceutical Biotechnology	Environmental Biotechnology	Agricultural Biotechnology	Marine Biotechnology
Markets served	Many different markets such as small-molecule pharma and fine chemicals, and flavor and fragrance, bulk chemicals	Large-molecule pharma products such as therapeutic proteins and monoclonal antibodies	Service and solution for bioremediation and waste treatment	Transgenic or genetically modified (GM) seeds and plants	Products and lead substances from the marine environment
Color code	White	Red	Gray	Green	Blue
Market size US$	>50 bio without biofuels	>70 bio	NA	>7 bio	NA
CAGR	15%	> 20%	NA	15–20%	NA
Companies	4000	6000	NA	>50	NA

Abbreviation: NA, not available.

of the industry and investors, we have to overcome many technical challenges, some of which are described in this chapter.

Why are we not able to better exploit nature's toolbox as our ancestors in Mesopotamia, Egypt, Mexico, and Sudan were when using, for example, *Acetobacter* and yeast concomitantly for the production of chemicals? We are still not able to cultivate most microorganisms nor are we able to exploit them for synthetic or biocatalytic purposes. It is clear that, in order to realize the expectations of society, investors, and other stakeholders, not only may the industry concerned have to realize enormous structural changes to introduce more and more biological processes into production but it may also have to develop the necessary toolbox.

2.2 HOW GREEN (SUSTAINABLE) CAN THE INDUSTRY BECOME?

Green chemistry is a concept and guiding principle to encourage the development of manufacturing processes and products with the lowest possible environmental impact or footprint. The goal of the green chemistry concept and principles is to prevent pollution and reduce resource consumption, instead of merely eliminating them. The 12 principles of green chemistry formulated by Anastas and Warner [2] in 1998 are summarized in Table 2.2. These principles have been followed consciously or subconsciously by synthetic and process organic chemists for a long time for one simple reason: these principles also make a lot of sense from an economical point of view.

Considering these 12 principles, it is clear that biotechnology should be able to make a much bigger contribution to green chemistry than is presently the case. The potential contribution of biotechnology is discussed also under synonymously used terms, such as third-wave biotechnology, suschem, greenchem, or white

TABLE 2.2 The 12 Principles of Green Chemistry Are Summarized Below, Published by Anastas and Warner

1. Prevent waste
2. Design safer chemicals and products
3. Design less hazardous chemical syntheses
4. Use renewable feedstocks: Use raw materials and feedstocks
5. Use catalysts, not stoichiometric reagents
6. Avoid chemical derivatives
7. Maximize atom economy
8. Use safer solvents and reaction conditions
9. Increase energy efficiency
10. Minimize the potential for accidents
11. Design chemicals and products to degrade after use
12. Analyze in real time to prevent pollution

biotechnology. The common denominator is sustainability, which always includes three dimensions: ecological sustainability, economic sustainability, and social sustainability [3].

"Green chemistry" principles should be considered during the very early design phase of building blocks, intermediates, and new chemical entities. As the structures of small-molecule active ingredients and drugs become more and more complex, chemical synthesis becomes more and more tedious and wasteful. Biotransformation and biosynthesis will become an ecological and economic necessity for their production. Table 2.3 reminds us why biotransformation should have a preeminent role for synthetic organic chemists [2].

Unfortunately, the commercially available enzyme toolbox for organic chemical synthesis does not correspond to the commercial potential of the fine chemical and pharmaceutical industry. Considering the theoretical options and today's reality, one can say that the commercial enzyme toolbox is more or less still empty. This situation is aggravated by the prevailing interest and flow of funds toward biofuels. This is regrettable, as the fine chemical and small-molecule chemical industry could drastically and immediately increase its sustainability by applying biotechnological principles [4].

There are several *in vitro* and *in vivo* methods available for the manufacture of chemicals and proteins on an industrial scale by biotechnological means [5] (Table 2.4). These include reactions with enzymes in chemical reactors, microbial fermentation, plant cell culture, or manufacturing products with transgenic plants ("pharming").

TABLE 2.3 Why will Biotransformation Play a Preeminent Role in Organic Chemistry?

Specificity and yield

- High chemo-, regio-, and stereoselectivity
- Usually high yields
- Enzymes can be adapted to organic solvents

Working safety and ecology

- Mild reaction conditions and biodegradable catalyst
- No runaway reactions possible
- Reduced health hazards for technical staff

Economic advantages

- No protection deprotection groups needed
- Heterologous protein expression allows competitive production of enzymes
- Low corrosive reaction media result in longer life-span of multipurpose installations

There are three main reasons to use enzymes. However, they must be commercially available for large-scale applications.

TABLE 2.4 Summary of Existing *In Vivo* Biotechnological Manufacturing Options

- Enzymatic conversion for small molecules
- Bacterial, yeast, and fungal fermentation for proteins and small molecules
- Microalgae fermentation for proteins and small molecules
- Plant cell culture for proteins and small molecules
- Protozoa fermentation for proteins
- Transgenic plants proteins and small molecules
- Mammalian cell culture for glycosylated proteins
- Transgenic animals for glycosylated proteins
- Insect cell culture (transient) for proteins

The buzzword "biotechnology" is connected with high expectations of being able to provide green and sustainable solutions for manufacturing and greenhouse gas reduction. But how realistic are these expectations regarding biotechnology? Biofuels are used to give a green touch to different administrations without, however, considering socioeconomic costs. We believe that with respect to energy, biotechnology will play only a minor role in very specific fields because of the limited availability of arable land and water, and because of the very low efficiency of natural photosynthesis.

On the other hand, biotechnology should have an especially high impact in the production of complex chemicals used for pharmaceuticals, fine chemicals and specialties (Table 2.5). In addition, the so-called "bioplastics," derived from renewable agricultural feedstocks are another meaningful application of biotechnology.

However, with respect to renewable feedstocks from agriculture, forestry, or marine sources, we have to consider that ours is a feedstock-limited world. Although ample renewable energy resources are available, we remain focused on cheap oil, of which we use over 80 million barrels a day. Yet, only about 5%–7% of all that oil is used for chemical manufacturing such as plastics, chemicals, pharmaceuticals, and other products. The rest is burned! Thus, the sooner we are able to replace oil as energy carrier by alternatives such as solar, wind, hydroelectric, and geothermal, the more oil will be left for use as a

TABLE 2.5 Waste Generated per Kilogram of Product Produced in the Chemical Industry

Bulk chemicals (kg)	> 0.1
Fine chemicals (kg)	5–50
Pharmaceuticals (kg)	25–100
Oligonucleotides (kg)	<1000
Peptides (kg)	<1000
Oligosaccharides (kg)	<1000

Source: Adapted from J Woodley.

feedstock for chemistry. While we must expect radical changes with an uncertain outcome in many areas, there is too much focus on bioconversion to biofuels and commodity products, leaving the toolbox for biotechnological options for more complex chemicals largely empty. Lonza, for example, has started a global corporate initiative called *Lonza Innovation for Future Technologies* (LIFT), which is funding long-term, innovative ideas with great potential. This program also includes the search for radically new approaches to improve chemical manufacturing and products, and the LIFT program is open to any institution or organization with interesting new ideas.

The trend to ever more complex molecules makes the situation even worse than described by John M. Woodley, who did not include "tides" and oligosaccharide synthesis [6]. Why this is a problem is described in the sections below.

2.3 FINE CHEMICALS AND SMALL-MOLECULE PHARMACEUTICALS

The pharmaceutical market remains the most important driver for innovation, also for industrial biotechnology. Global pharmaceutical sales more than doubled between 1998 and 2006. Of the 252 drugs that entered clinical trials in 2007, 135 were still small molecules, but 61 were already monoclonal antibodies. We will also see a change from a block-buster mentality to a chemically more differentiated therapeutic approach. Chemical diversity will follow the genetic diversity and genetic profiles of patients. Thus, novel small molecules will be of much greater diversity, almost exclusively chiral and highly functionalized. Examples of novel bioreaction tools needed include peptide biosynthesis, glycosylation of small molecules and peptides, functionalization of biopolymers as well, and the production and derivatization of highly active compounds. The challenge is that biotransformation processes are far more diverse than therapeutic protein production processes. The desired products are often toxic, poorly soluble, or instable novel small molecules. Timeline compressions in the development cycle of pharmaceuticals, in combination with a missing broad strain and enzyme choice, results in the fact that bioprocesses typically represent the second-generation process choice in the manufacture of a small-molecule pharmaceutical or intermediate. To make things worse, academic research and development does not yet follow these needs, as the majority of the academic work is still on hydrolytic bioprocesses (Table 2.6).

Table 2.6 summarizes an analysis of the last four "International Symposium on Biocatalysis and Biotransformations" [7]. Of the 365 lectures and posters that were analyzed in 2007, less than 70 papers dealt with new enzymes. Lipases and esterases still represent 64% of all papers dealing with hydrolytic enzymes, and 26% of all papers. On the other hand, there are "biotransformation" territories to be conquered and radical innovation is needed. In our opinion, both aspects can be best dealt with in an academic environment where "risky" and more long-term projects can be performed better than in industry. However, hydrolytic enzymes no longer really represent a prime academic topic, in particular, as there are small companies specialized in improving and adapting biocatalysts to industrial needs.

TABLE 2.6 An Overview of the Enzyme Classes Presented as Oral or Poster Presentations at the Last Three Biotrans Symposia

Enzyme Class	2009	2007	2005	2003
1. Oxidoreductases (%)	32	34	24	28
2. Transferases (%)	10	8	6	3
3. Hydrolases (%)	46	41	55	58
4. Lyases (%)	9	12	12	10
5. Isomerases (%)	3	2	2	1
6. Ligases (%)	0	1	0	0

An example of the percentage calculation: 34% of all presentations in 2007 dealt with oxidoreductases. The sum does not add up to 100% because some of the presented papers did not deal with biotransformations.

2.4 OLIGOPEPTIDES

The name peptide corresponds to a broad group of very different compounds of various sizes and various derivations, as it is a general term for a polymeric chain of two or more amino acids linked by an amide bond. This definition is pretty clear for short, synthetic peptides, for example, aspartame, which is the methyl ester of a dipeptide, but becomes more ambiguous for longer peptides, for example, insulin, which also might be considered a short protein. Taking into account the recent progress achieved in the development of new methods for the chemical synthesis of longer peptides [8], let us consider amino acid chains of <100 acids as peptides. Peptides are ubiquitous and widely represented in nature, they are involved in numerous biochemical and physiological processes, and therefore of high relevance for the pharmaceutical industry, but they also have a great potential in other attractive applications such as food additives [9], cosmetics (e.g., Argireline), or antimicrobial peptides.

Pharma applications still represent the largest group of industrially relevant peptides, with about 50 peptide drugs approved currently on the market; in addition, 400 compounds are in different phases of clinical development [10]. The size of the pharmaceutical peptide market in 2008 was estimated at $8.5 billion (excluding insulin), with an annual demand approaching 1500 kg. These numbers illustrate the importance of economically and ecologically sound peptide synthesis processes! In principle, therapeutic peptides can be made using a number of different technologies. Besides solid- and liquid-phase chemical synthesis, recombinant and semisynthetic methodologies are also utilized. In the meantime, all these strategies have been successfully scaled up to industrial scale and can in principle be regarded as viable options. However, as each of the current methods still has its particular strengths and limitations, there is no single technology that offers a universal and competitive solution for the manufacturing of all the different peptide drug candidates in clinical development. The selection of the route of choice for a specific product is generally made considering not only the structural characteristics of the target compound, but also a number of other important aspects. Product specifications, targeted peak quantities and related manufacturing costs, time to market, potential investment needed for the implementation of specific technologies in relation to the overall risk of a clinical failure of the corresponding product—all these different factors build the framework for the decision process during route selection (Table 2.7).

Figure 2.1 gives an overview of technologies currently applied in the manufacturing of the commercial peptide Active Pharmaceutical Ingredients (APIs) grouped according to the length of the peptide chain.

Historically, liquid-phase peptide synthesis was the first technique to have been developed (Nobel prize for V. du Vigneaud in 1955) for the synthesis of Oxytocin [11], and remained for a long time the only solution for the synthesis of peptides. The concept of liquid-phase synthesis is typically convergent; single amino acids, bearing protecting groups, are reacted with other amino acids to give dipeptides, and then smaller peptides are extended by attaching additional amino acids. Finally, the different building blocks are combined to the desired product. Even highly complex targets such as calcitonin (32 amino acids) used to be produced in liquid phase. The main characteristic of this approach is its similarity to other classical organic synthesis. Liquid-phase chemistry can be run in standard equipment and on a large scale. On the other hand, there are also a number of drawbacks: the technology is very laborious since many different steps have to run separately leading to a great number of intermediates, which have to be isolated, dried, characterized, and stored. The process development times and the cycle times during production are generally rather long. Owing to the large number of isolation steps, it is also clear that the ecological footprint of such a liquid-phase peptide process is considerable, as it corresponds to the footprint of a standard organic synthesis, with the corresponding number of chemical steps.

It is also important to mention that the overall peptide manufacturing process can be rather complex and always includes other critical steps in addition to the elongation of the peptide chain. These steps can be additional chemical reactions (e.g., removal of potential protecting groups after chemical syntheses), or separation steps, such as the separation of different cell components (membranes, DNA,

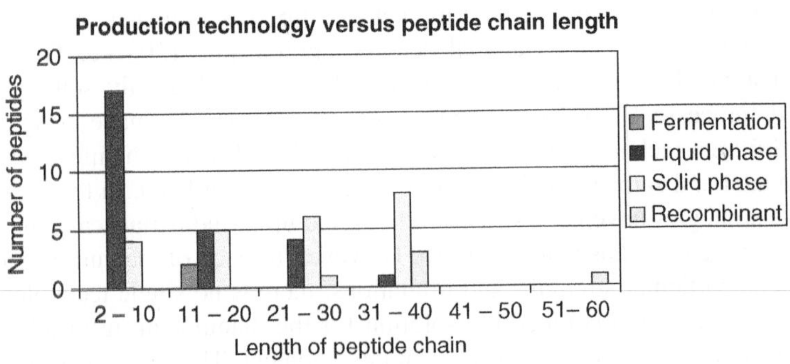

Figure 2.1 Technologies currently applied in the manufacturing of the commercial products grouped according to the length of the peptide chain.

host cell proteins, etc.) in case of recombinant processes. Generally, a peptide manufacturing process also includes final purification and isolation steps. The purification is usually a chromatographic process, whereas the isolation of the target compound, is generally accomplished using lyophilization, spray drying, or in single cases, precipitation/filtration/drying sequences.

One of the biggest challenges in chemical peptide synthesis remains the enantiomerization of activated amino acids via formation of oxazolones during the coupling reactions. One possibility to overcome this issue is the design of strategies in which the fragments are joined at glycine, which is achiral, or proline, which does not form oxazolones. Depending on the desired sequence, this is not always possible, and other fragment condensations should be carefully optimized. During a solid-phase synthesis, the oxazolone issue is less critical, as *N*-urethane-protected amino acids only form oxazolones when they are strongly activated. In addition, they are quite resistant to deprotonation, so that racemization during Solid-Phase Peptide Synthesis (SPPS) remains typically low.

Chemical solid-phase synthesis using the Fmoc/tBu strategy is currently the most utilized approach for the synthesis of mid-sized and longer peptides. In a typical process, the desired peptide is elongated stepwise on a solid support, to which all intermediates up to the desired product remain covalently attached until the end of the elongation. The elongation of the desired peptide chain starts with the attachment of the α-*N*- and side-chain-protected C-terminal amino acids to the solid support (initial loading), and then each of the following amino acids is attached successively to the growing chain, using a sequence of two reactions: (i) cleavage of the α-*N*-protecting group from the product on the solid support and (ii) coupling the next protected amino acid to the product on the solid support. The usage of the base-labile Fmoc group for amino-protecting group was first proposed by Carpino [12] in the early 1970s, as an extension of Merrifield's Nobel price winning SPPS methodology [13]. Characterized by milder reaction conditions in general, this methodology has a number of additional advantages. Owing to the orthogonal cleavage properties of the side-chain-protecting groups, it is possible to prepare intermediates with protected side-chain functionalities. This allows hybrid approaches in convergent synthesis routes, using solid-phase synthesis for making protected peptide fragments, and liquid-phase peptide chemistries for coupling two of these fragments to give longer peptides [14]. The major advantages of SPPS and hybrid concepts are the straightforward approach to longer peptides, and the excellent scalability up to ton scale. However, a considerable drawback is the high material flow, and as a consequence, the environmental impact. Table 2.8 shows an example of the material flow for 20 kg of a 10-mer and 100 kg of a 35-mer peptide manufactured by SPPS.

Manufacture of peptides by recombinant technologies is a highly promising approach to the environmental problem. However, despite a successful history for specific cases (e.g., insulin) and a growing number of new processes, the development of a recombinant process for peptides remains a difficult task, which is considered viable only when high volumes are required. One reason is

TABLE 2.7 Overview of the Different Technologies for Peptide Production

	Liquid Phase	Solid-Phase and Hybrid Approach	Recombinant and Semisynthesis
Process development	Long	Rapid	Very long
Raw material costs	Medium	High	Low
Advantages	Similar to standard organic chemistry	Straightforward approach to longer peptides	Practically no limitations in scale and amounts
	Established technology	Excellent scalability	
Limitations	Very laborious, will remain expensive	High material flows at large scale	Unnatural sequences potentially problematic
Environmental impact	Medium–high	High	Low–medium

TABLE 2.8 Two Examples of Material Flow for 20 kg of a 10-mer and 100 kg of a 35-mer Peptide Manufactured by SPPS Subdivided in API Demand (top) and Subsequent Needs for Amino Acids, Reagents and Solvents

	Peptide/API versus Raw Materials	
Raw Material	20 kg 10-mer	100 kg 35-mer
Amino acids	0.4 to	3.5 to
Coupling reagents	0.3 to	4.0 to
Piperidine	6.5 to	11.0 to
NMP/DMF	100 to	300 to
DCM	80 to	60 to
ACN	35 to	110 to
United States Pharmacopeia (USP) water	300 to	375 to

certainly the low productivity of the established recombinant systems when used for peptides. Owing to their smaller size and also to the lack of tertiary structure, peptides generally remain dissolved in the cytoplasm of the host cell, and are therefore rapidly degraded. Several approaches have been proposed to overcome this productivity issue: (i) synthesis of peptides as parts of fusion proteins, (ii) synthesis of large multicopy constructs (concatemers), and (iii) use of secretion processes. However, each of theses concepts still has considerable drawbacks. The fusion protein approach, in which the target peptide is expressed as a synthetic construct with a larger protein, overcomes the degradation issue; however, the cell has also to produce large amounts of an undesired product (protein part of the fusion protein, which is often significantly larger than the desired peptide);

in this case, the overall metabolic burden creates another type of limitation for productivity. The second setback of the fusion protein approach is related to the liberation of the desired peptide from the rest of the fusion construct. Depending on the sequence, this additional step can be performed chemically or enzymatically. Both methods can show selectivity problems; in addition, the enzymatic method requires the corresponding enzyme, which might be expensive or even not available on a commercial scale.

Another related technology for recombinant peptide production is the concatemer approach, in which the concatemer corresponds to an artificial protein constructed from multiple copies of the desired peptide separated from one another by a specific spacer. With regard to expression, the concatemer approach is certainly more elegant than the fusion protein concept, as most of the overexpressed product corresponds to the desired product. However, this approach usually requires two different enzymes, one for cutting the concatemer into single pieces and one for trimming the C-terminus in order to obtain the desired product. Here again the availability and the cost of the enzymes become critical.

The secretion technology is, in principle, free of these problems; however, the productivity for secreted peptides reported until today remains on the low side [15,16].

Two additional limitations, valid for all recombinant approaches, should be mentioned here. The first is that, currently peptides with an amide function at the C-terminus, as required for many APIs, cannot be obtained directly by recombinant synthesis. This difficulty is generally overcome, either by using an enzymatic approach, starting from a recombinant product extended by an additional C-terminal glycine [15] or by chemical synthesis [17]. The second constraint is that until now only peptides containing natural peptides can be made efficiently in recombinant systems. All efforts to include unnatural or modified amino acids in recombinant products are still in early development [18].

2.5 OLIGONUCLEOTIDES

Oligonucleotides are a class of biomacromolecules with an enormous potential as therapeutic agents for a number of different applications. The era of antisense research started in 1978, when Paul Zamecnik and Mary Stephenson reported on the inhibition of virus replication in cell culture by a DNA oligonucleotide complementary to the target RNA [19]. This was the beginning of a remarkable development in nucleotide research, leading to the discovery of several other mechanisms of action for oligonucleotides (e.g., the 1989 Nobel Prize awarded to Sidney Altman and Thomas Cech for the discovery of Ribozymes and the 2006 Nobel Prize awarded to Andrew Fire and Craig Mello for the discovery of siRNA) and to the development of a whole range of interesting candidate compounds.

Therapeutic oligonucleotides are typically a linear chain composed of around 20 DNA or RNA monomers or chemically modified analogs thereof. For some of the applications, oligonucleotides are used as single-chain products (antisense compounds, immunomodulators, aptamers, enzyme inhibitors, and ribozymes), whereas for others double-stranded oligonucleotides are needed (siRNA, molecular decoy DNA). In order to overcome the general problem of the enzymatic degradation of natural oligonucleotides, numerous concepts for modification of the natural oligonucleotide structure have been developed, mainly to the sugar-phosphate backbone of the oligonucleotide chain (e.g., phosphorthioates, introduction of modified substituents in the 2′-position of the ribose, phosphorthioamidates, LNAs (locked nucleic acid), morpholino compounds, and gapmers).

Despite the significant differences in their respective chemical structures, all different types of oligonucleotides are currently produced by rather similar processes based on the same generic scheme. This standard manufacturing process for oligonucleotides comprises four steps: (i) synthesis, (ii) cleavage/deprotection, (iii) purification, and (iv) isolation.

1. Today, all oligonucleotides are synthesized chemically on a solid support, using the same type of nucleoside phosphoramidite chemistry for the elongation of oligonucleotide chain. During this step, monomeric phosphoramidite derivatives of the corresponding nucleosides are sequentially coupled to the elongating oligonucleotide chain, which is covalently bound to a solid support.
2. The elongation process is followed by a cleavage/deprotection step, in which the product is cleaved from the solid support, and at the same time, the protecting groups are also cleaved from the molecule (backbone, bases, 2′-deprotection). Depending on the type of 2′-protecting group for RNA, the deprotection of the product is performed in two separated reactions in some cases (e.g., 2′-*tert*-Butyldimethylsilyl (2′-TBDMS), or 2′-Triisopropylsilyloxymethyl (2′-TOM)).
3. The previous cleavage/deprotection step leads to a solution of the crude oligonucleotide, which is then purified by chromatography, either HPLC (high performance liquid chromatography) on reverse phase resin or ion exchange chromatography. After the chromatographic purification, the solution of the desired purified product undergoes some additional operations depending on the specific requirements of the product and the process. This can be desalting, removal of 5′-DMT (*N,N*-dimethyltryptamine) protecting group, removal of organic solvent, concentration, precipitation, or annealing. The result of this step is typically an aqueous solution of the desired final product in purified form.
4. Finally, the desired oligonucleotide product is generally isolated as a solid, which is normally obtained via direct lyophilization of the product solution from the previous step.

The key part of this process is certainly the chemical synthesis of the oligonucleotide chain. The concept of the synthesis process is still based on the four-step solid-phase phosphoramidite approach first described by Beaucage and

Caruthers [20] in 1981. Optimized in different aspects (reagents, solid supports, and synthesizer design) over the years, this remarkable chemistry turns out to be highly efficient in terms of overall and step yields and process times (about 80 chemical reactions in less than 24 h), and it is very robust and highly generic to all different types of target oligonucleotides. In the meantime, two additional important characteristics have underlined the value of this technology: the degree of automation and the scalability of the process. Currently, this synthesis process can be run fully automatically under cGMP-compliant conditions on scales delivering between 1 g and 3 kg (corresponds to 900 mMol initial loading) of isolated oligonucleotide per batch. The present process technology has been demonstrated for the manufacture of oligonucleotides on a multi-100 kg scale. On the basis of the present market requirements, with only two minor oligonucleotide-based drugs on the market and a limited number of late stage candidates in the clinical pipeline, these synthesis capabilities are certainly sufficient.

However, if the expected breakthrough in drug delivery technologies can be achieved, solving the related issue of efficient delivery of oligonucleotide drugs, the required quantities of oligonucleotides could increase dramatically. This would immediately lead to a major sustainability issue. Table 2.9 outlines the mass-balance for the synthesis step for the production of an oligonucleotide. Without considering the requirements for the purification of the product, the presented figures indicate a E-factor of about 4700 kg/kg.

From this perspective, it would be interesting to understand whether biological methods for the production of large quantities of oligonucleotides could be envisaged in the future. In principle, concepts for enzymatic amplification of DNA exist and are widely utilized in analytical applications such as PCR. A limitation of the PCR approach for preparative applications is that in addition to the template sequence, a primer is required. An interesting technology working on the same principle as PCR, but without the requirement of a primer, has recently been reported by the German start-up company Riboxx, which claims to be able to amplify RNA using a viral RNA-dependant RNA polymerase. According to the inventors, the technology even tolerates some modifications in the structures of the sugar moieties [21].

TABLE 2.9 E-Factor in Oligonucleotide Manufacture: Yogesh S. Sanghvi, Presentation Eurotides Conference 2006, Hamburg, Germany

Deprotection reagent and solvent (L)	8000
Amidites (L)	315
Coupling reagent (L)	577
Thiolation reagent (L)	448
Capping reagent solutions (L)	780
Acetonitrile (L)	18,250
Total	28,370
Oligonucleotide product (kg)	6

2.6 OLIGOSACCHARIDES

"Ordinary" sugar, first used in the seventh century AD, and for which the first (industrial) sugar mills were established in the early seventeenth century, is a commodity with an annual consumption of ~150 million tons per annum. Besides use in the kitchen and in the fermentation industry, sugars have numerous biological roles and play a vital function in various fundamental metabolic and physiological processes. It is no wonder that there is an increasing interest in sugar-based products, as they play a role in almost all areas of human activity. In addition, a wide range of very potent and bioactive natural products, such as anticancer compounds or antimicrobial compounds, are often glycosylated peptides, polyketides, indolocarbazols, and other chemical structures. The potential applications of glycocompounds range from the better known pharmaceuticals to industrial applications of which the world is much less aware. To give just one example: for 50 years, it has been known that the opportunistic human pathogen *Pseudomonas aeruginosa* produces potent glycolipid biosurfactants, so-called rhamnolipids [22]. The University of Arizona developed and patented an industrial production process, and a small company (Jeneil Biosurfactants Co, Milwaukee, Wisconsin USA) produces the so-called "Zonix." Table 2.10 gives an overview of some applications of sugar-based products.

Although glycocomponds have many potential applications, they are rarely used in the pharmaceutical and other industries; mainly, it is difficult to produce adequate quantities at competitive prices. Moreover, the number of possible structures for oligosaccharides is orders of magnitude higher than that for peptides and oligonucleotides. This problem is somewhat alleviated as mammalian or plant glycosides seem to be derived from a rather small pool of monomers [23,24].

TABLE 2.10 Examples of Sugar-Based Products and Applications

Pharma, fine chemicals, and building blocks: Oligosaccharide antibiotics (everninomycins, ziracins, dalbavancin, vancomycin, and teicoplanin), antiparasitic compounds (natamycin and avermectins), antitumor compounds (calicheamycin, rebeccamyn, staurosporin, mithramycin, and aureolic acids), antiviral compounds (deoxysugars), Alzheimers disease (scyllo-inositol), thiosugars (thiolactomin, 1,6-thioanhydrosugars, and mycothiol), iminosugars (deoxynojirimycin and derivates, and imminocylcitols), vaccines (gylcoconjugates), and excipients and drug formulations (D-tagatose, D-gulose).
Food additifs: Sweeteners (tagatose, stevioside, D-glucose, palatinose, and sugar alcohols), health food (β-glucans), prebiotics (fructooligosaccharides, lactosucros, inulin, galactooligosacharides, and arabinogalactans), and additives (D-ribose)
Agrochemicals: Plant growth regulators. blight control (validoxylamine)
Cosmetics: Skin products (hyaluronic acid)
Industrial applications: Surfactants (rhamnilipid, mannosylerythritollipid, and glycosphingolipid)

2.6.1 Production of Glycocompounds

Well-developed synthetic methods are available for all the four major classes of biological compounds (proteins, oligonucleotides, lipids, and carbohydrates) except for the last one. Chemical synthesis of peptides and nucleotides has been well established for several decades as shown in the previous two chapters. Although glycobiology is a rapidly expanding field, with a growing need for access to large quantities of oligosaccharides, their production remains unresolved, although several approaches are possible as summarized in Table 2.11.

2.6.1.1 Chemical Synthesis of Oligosaccharides. As in many other areas where chemical and biotechnological production methods are possible in principle, chemistry has delivered the first-generation process, while biotechnology often provides only a second-generation option, since the necessary toolbox is not available. Almost any oligosaccharide can be prepared by classic organic chemical methods, but the methods are tedious, have low yields, and are wasteful and expensive. Synthesis of the monosaccharide building blocks, if they are not available from natural sources, requires extensive protecting group manipulations.

Nevertheless, a first fully automated oligosaccharide synthesis has been described by the group of Peter H Seeberger at the Swiss Federal Institute in Zürich, Switzerland [23]. The procedure as summarized in Table 2.12 gives an idea of the unparalleled complications encountered in carbohydrate synthesis.

TABLE 2.11 Options for the Synthesis of Oligosaccharides

Chemical synthesis: Solid-phase and liquid-phase chemical synthesis
Biotransformation: Using enzymes such as glycosyltransferases, gylcosidases, and glycosynthases to convert biologically or chemically derived precursors
Biosynthesis: *De novo* synthesis of glycocompound or their building blocks by fermentation
Extraction: Extraction from natural sources, for example, plant material

Methods can of course be combined.

TABLE 2.12 Steps in the Automated Chemical Solid-Phase Oligosaccharide Synthesis of Oligosaccharide Developed by the Group of Peter H Seeberger, Switzerland

1. Attachment of first building block
2. Coupling and deprotection
3. Capping of nonreacted–OH groups
4. Cleavage of linker
5. Removal of protecting groups
6. Purification

The table assumes that the stable sugar building blocks, the resin with a linker, and the needed hardware (synthesizer) including real-time monitoring are readily available. The process is an iterative one, the number of steps depending on the length of the oligosaccharide or monomers to be coupled.

This short description shows that a large-scale chemical synthesis of an oligosaccharide would be prohibitively expensive and an ecological disaster!

At this place, a short note on an interesting observation called "glycation": So-called "glycation" can improve the thermostability of nonglycosylated proteases such as trypsin and chymotrypsin [25]. It is interesting that the nonenzymatic attachment of carbohydrates to the amino groups of proteins, called *glycation*, was carried out *in vacuo* on lyophilized proteins.

2.6.1.2 Enzymatic Synthesis. As no tedious protection/deprotection steps are necessary, enzymatic oligosaccharide synthesis is an attractive alternative to chemical oligosaccharide synthesis. Moreover, because of the high selectivity of the enzyme-catalyzed reactions, side products typically need not be removed after the reaction. However, the donor substrates (sugar nucleotides, sugar phosphates) and the mono-, di-, or oligosaccahride acceptors must be produced separately and cost efficiently. Table 2.13 summarizes the options offered by enzymes for oligosaccharide synthesis.

Leloir glycosyltransferases require expensive nucleotide-activated sugar donors. The advantage of non-Leloir glycosyltransferases are the cheaper substrates and starting materials such as sucrose [26] and sucrose analogs. Unfortunately, oligosaccharide synthesis in general is still very restricted owing to a lack of commercially available glycosyltransferases of both types as well as to the high cost of some of their substrates. Fortunately, interest in this enzyme class has been increasing as shown in Table 2.6, but much more effort is needed to make the oligosaccharide toolbox available to the synthetic chemist.

Glycosidases, which are used in reversed reactions (equilibrium or kinetically controlled), can also be used for the synthesis of carbohydrates. Table 2.6 shows that hydrolytic enzymes are the best investigated enzyme class, although this is mainly due to lipases and esterases. The number of available glycosidase enzymes for catalysis is higher and more scientific and technical data can be obtained than on glycosyltransferases.

Glycosidases are also robust biocatalysts, and several are commercially available. The same can be said about proteases, which can be applied in a similar manner as hydrolytic enzymes for peptide synthesis.

TABLE 2.13 Strategies for the Biocatalytic Oligosaccharide Synthesis

1. Leloir glycosyltransferases: stereo- and regio-slective transfer of activated monosaccharides (glucose, galactose, and xylose nucleotides as donors)
2. Non-Leloir glycosyltransferases: transfer of sugar phosphates as donors
3. Glycosidases: glycosidation or transglycosylation reaction (reverse reaction of hydrolysis) using glycosydases
4. Glycosynthases: mutant glycosidases forming glycosidic linkages

There are two basic methods for enzymatic oligosaccharide synthesis, glycosylatransferases or the reverse reaction of glycosyl hydrolases. All the methods should can of course be combined.

Glycosynthases are specifically engineered glycosidases that catalyze the formation of glycosidic bonds from a glycosyl donor and an acceptor alcohol. This powerful method, first reported in 1998, uses site-directed mutation of exoglucosidase to generate an enzyme that is hydrolytically inert. New and improved glycosynthases act more rapidly and transfer to a much wider array of acceptors [27]. Thanks to this development, the above-mentioned lack of a wide range of glycosyltransferases can be addressed. Isomerases are an important addition to the enzyme arsenal, as they allow access to the needed monomer.

Solid-phase synthesis can be combined with enzymatic methods in which substrates are bound (immobilized) to solid surfaces for the synthesis of natural biopolymers: peptides, oligonucleotides, and oligosaccharides [28].

In conclusion, the enzymatic toolbox for oligosaccharide synthesis is far from being well established, and chemical synthesis remains the method of choice for oligosaccharide synthesis. Thus, carbohydrate biotechnology remains a true frontier for biocatalysis with a lot of research work remaining to be done.

2.6.1.3 Extraction. Many products in nature are glycosylated. For example, bacterial secondary metabolites contain a vast repertoire of unusual sugars. These products represent a natural pool for the sourcing of complex mono-, di-, or oligosaccharides.

2.6.2 Examples of Glycosynthesis

2.6.2.1 Fondaparinux—A Synthetic Heparin Substitute. Heparin (Figure 2.2) is a heavily *O*- and *N*-sulfated glycocompound. Isolated from animal organs, for example, porcine mucosal tissue, it has been used as an anticoagulant drug since the 1940s.

It was found that specific pentasaccharide sequence (antithrombin-binding domain) was responsible for the anticoagulation activity of heparin. Fully synthetic pentasaccharide analogs have been developed and tested to overcome the clinical problems of heparin as well as the issue of animal-derived injectable pentasaccharides (heparin is injected subcutaneously). These efforts led to the pentasaccharides Fondaparinux and Idraparinux, both of which display higher activity and bioavailability than heparin.

Fondaparinux is the active ingredient of Arixtra® marketed by GSK (Glaxo Smith Kline). Patent protection has expired, and the Australian company Alchemia has developed a generic product, which will be marketed with the

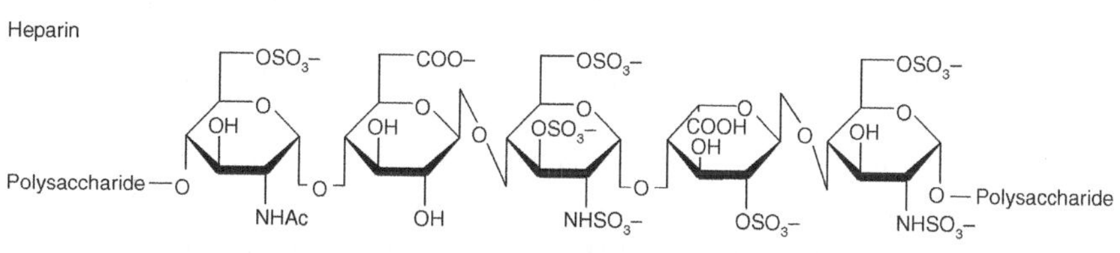

Figure 2.2 The chemical structure of heparin.

Indian partner Dr. Reddy's. Fondaparinux is a typical example of how chemistry, with a broad and applicable toolbox developed over the last 200 years, was able to deliver the first-generation process. Heparin and its derivatives could possibly be synthesized enzymatically [29]. Biotechnology has, however, failed to deliver a cheaper and more ecological alternative, although the development of alternative pentasaccharides began in the early 1980s.

The first chemical synthesis of the Fondaparinux consisted of over 60 steps, but has been reduced to 55 steps from naturally occurring carbohydrates. The pentasaccharide idraparinux needs only 25 chemical synthetic steps from naturally occurring carbohydrates [30]. Consideration of the yields after 55 or 25 chemicals steps, the raw material requirements, and waste production show that oligosaccharide synthesis is in real need of more ecological biotechnology approaches.

2.6.2.2 *Phosphomannopentaose.* PI-88 (phosphomannopentaose sulfate) is an example of a combination of fermentation and biocatalysis [31]. PI-88 is a mix of sulfated oligosaccharides and heparanase inhibitor, prepared by the hydrolysis of the extracellular phosphomannan produced with the yeast *Pichia holstii*.

2.6.2.3 *NANA.* *N*-acetyl-D-neuraminic acid (NANA) is a ubiquitous sialic acid, which is an important building block for antiviral drugs. NANA can be produced by a number of methods, for example, by recombinant fermentation using *Escherichia coli* via colominic acid, or by a variety of enzymatic processes, or starting from the polymer chitin, an unbranched β(1–4) polymer of NAG (*N*-acetyl-glucosamine). The unbranched biopolymer chitin is degraded to NAG. In the next step, NAG can be chemically epimerized under alkaline conditions to NAM (*N*-acetyl-D-mannosamine) and finally, NAM can be further converted to NANA.

2.6.2.4 *Hyaluronic Acid, Chondrotin Sulphate, and Rhamnan Sulphate.* All the three compounds are polysaccharide compounds that can nowadays be produced by fermentation. Hyaluronic acid, which is used as an ingredient not only in skin care, but also in ophthalmology, has been traditionally isolated from rooster combs. Animal-derived material is being replaced gradually by material produced by microbial fermentation. The molecular weight distribution of hyaluronic acid for pharmaceutical products should fulfill stringent quality requirements, which is not the case for cosmetic applications. This is an opportunity for hyaluronic acid fermentation, to avoid fractionation of the wasteful animal products.

Chondroitin sulfate is a mucopolysaccharide that is used in cartilage repair in arthritis. The material is mostly extracted but fermentative options are in development. Rhamnan sulphate is presently extracted from seaweeds, and alternatives are being sought.

2.6.2.5 Conjugated Vaccines. Conjugated vaccines represent 25% of the current $4.5 billion vaccine market. These therapeutic glycoconjugates are highly complex and expensive, and require biological as well as chemical steps. GlycoVaxyn [GlycoVaxyn AG, 8008 Zürich, Switzerland. www.glycovaxy.com], a small spin-off of the Swiss Federal Institute of Technology in Zürich, has developed technologies that enable the manufacture of bioconjugate vaccines and other complex polysaccharide structures, using the genetic information that is located on an operon in *Campylobacter jejuni*, which encodes the metabolic apparatus capable of carrying out glycosylation of recombinant proteins in *E. coli*. These bioconjugates are directly synthesized *in vivo* using these appropriately engineered bacterial cells.

This work signifies another interesting development. Until recently, biology textbooks taught that prokaryotes are not able to *N*-glycosylate proteins. Today we know not only that there are bacterial glycosylation processes, but also that glycosylation in bacteria leads to a much greater diversity of glycan composition, linkage units, and glycosylation sequences on polypeptides than in eukaryotes [32,33]. Bacterial *N*- and *O*-linkages are formed with a wider range of sugars than those observed in eukaryotic proteins.

Japanese scientists have been at the forefront of developing biotechnological strategies for rare sugars. The strategy named "Izumoring" lays down the principle of producing all of the 59 different terose, pentose, and hexose sugars. In principle, the discovery of some key isomerases, epimerases, and microbial oxidative pathways now makes it possible to manufacture any of these sugars from cheap starting materials such as starch and lignocellulose. The International Society of Rare Sugars (ISRS) [34] has published several excellent reviews that describe the production of rare sugars and oligosaccharides.

The European Project NEPSA, which focuses on the production of novel monosaccharides, is another valuable initiative [35] for the production of carbohydrate-based building blocks.

2.7 SUMMARY AND OUTLOOK

The potential of biotechnology in industry is far from being fully realized, even in established areas such as biocatalysis for fine chemicals. With the argument of sustainability, various biofuel projects have triggered an interest, which by far exceeds their real economic and ecological potential. Industrial biotechnology covers a broad range of products and services ranging from a biobased economy with biomass-derived basic chemicals and enzymes in the food, paper, and textile industries to high-value fine chemicals and pharmaceuticals. It is in fine chemicals and pharmaceuticals that sustainable solutions are badly needed. These, however, are not low-hanging fruits, as shown by the examples of oligopeptide, oligonucleotide, and oligosaccharide synthesis.

Development of a toolbox containing commercially available biocatalysts of the six different enzyme classes is required to meet the ever increasing degree of sophistication of the molecules to be produced. What is most important, however, is to reach an agreement between chemists and biologists on a roadmap and on priorities, that is, where to focus development, where to reduce priorities, and where to choose which feedstocks for which value chain.

ACKNOWLEDGMENTS

The authors thank their Lonza colleague Gareth Griffiths, who made sure that the text in English can be well understood by the esteemed readers.

REFERENCES

1. Meyer H-P. We need a different roadmap. CHEManager Eur 2009;5:8.
2. Anastas PT, Warner JC. Green chemistry: theory and practice. New York: Oxford University Press; 1998.
3. Leresche J, Meyer H-P. Chemocatalysis and biocatalysis (biotransformation): some thoughts of a chemist and a biotechnologist. Org Proc Res Dev 2006;10:572–580.
4. Meyer H-P, Ghisalba O, Leresche JE. Biotransformation and the pharma industry. In: Crabtree RH, editor. Volume 3, Handbook of green chemistry, biocatalysis. Weinheim: Wiley-VCH GmbH; 2009. pp. 171–212.
5. Meyer H-P, Turner NJ. Biotechnological manufacturing options for organic chemistry. Min Rev Org Chem 2009;6:300–306.
6. Woodley JM. New opportunities for biocatalysis: making pharmaceutical processes greener. TIBTEC 2008;26:321–327.
7. Biotrans Oviedo 2007. Conference proceedings of the 8th International Symposium on Biocatalysis and Biotransformation; 2007 July 8–13; Oviedo Spain, http://www.uniovi.es/biotrans2007.
8. Kent SBH. Total chemical synthesis of proteins. Chem Soc Rev 2009;38:338–351.
9. Ager DJ, Pantaleone DP, Henderson SA, Katritzky AR, Prakash I, Walters DE. Commercial, synthetic nonnutritive sweeteners. Angew Chem Int Ed 1998;37:1802–1817.
10. Moorcroft M. Peptide manufacturing—the rising star. sp2; 2009, Mar 16–18.
11. du Vigneaud V, Ressler C, Swan JM, Roberts CW, Katsoyannis PG, Gordon S. J Am Chem Soc 1953;75:4879–4880.
12. Carpino LA, Han GY. The 9-fluorenylmethoxycarbonyl function, a new base-sensitive amino-protecting group. J Am Chem Soc 1970;92:5748–5749.
13. Merrifield RB. Solid phase peptide synthesis I: the synthesis of a tetrapeptide. J Am Chem Soc 1963;85:2149–2154.
14. Bray BL. Large-scale manufacturing of peptide therapeutics by chemical synthesis. Nat Rev Drug Discov 2003;2:587–593.
15. Metha NM. Oral delivery and recombinant production of peptide hormones Part II: recombinant production of therapeutic peptides. BioPharm Inter 2004;17(7):44–46.
16. Leonhartsberger S. *E. coli* expression system efficiently secretes recombinant proteins into culture broth. Bioprocess Int 2006; April 64–66.

17. Giraud M, Williner M, Werbitzky O. patname Method for the synthesis of peptide amides. patno WO 2006008050.
18. Ryu Y, Schultz PG. Efficient incorporation of unnatural amino acids into proteins in *Escherichia coli*. Nat Methods 2006;3:263–265.
19. Zarnechnik PC, Stephenson ML. Inhibition of *Rous sarcoma* virus replication and transformation by a specific oligonucleotide. Proc Natl Acad Sci USA 1978; 75:285–288.
20. Beaucage SL, Caruthers MH. Deoxynucleoside phosphoramidites—A new class of key intermediates for deoxypolynucleotide synthesis. Tetrahedron Lett 1981;22:1859.
21. Rohayem J, Rudolph W, Jaeger K, Jacobs E. WO 2007/012329.
22. Soberon-Chavez G. Production of rhamnolipids by *Pseudomonas aeruginosa*. Appl Microbiol Biotechnol 2005;68:718–725.
23. Seebeger PH. Automated oligosaccharide synthesis. Chem Soc Rev 2008;37:19–28.
24. Thorson JS, Vogt T. Glycosylated natural products. In: Wong C-H, editor. Carbohydrate-based drug discovery. Weinheim: Wiley-VCH Verlag GmbH; 2003. pp. 685–711.
25. Pham VT, Ewing E, Kaplan H, Choma C, Hefford MA. Glycation improves the thermostability of trypsin and chymotrypsin. Biotechnol Bioeng 2008;101(3):452–459.
26. Seibel A, Beine R, Behringer C, Buchholz KA. New pathway for the synthesis of oligosaccharides by the use of non-Leloir glycosyltransferases. Biocatal Biotransformation 2006;24(1/2):157–165.
27. Perugino G, Trincone A, Rossi M, Moracci M. Oligosaccharide synthesis by glycosynthases. TIBTECH 2004;22(1):31–37.
28. Laurent N, Haddoub N, Flitch S. Enzyme catalysis on solid surfaces. TIBTECH 2008;26(6):328–337.
29. Linhard RJ, Dordick KS. Enzymatic synthesis of glycosaminoglycan heparin. Semin Thromb Hemost 2007;33(5):453–465.
30. Petitou M, van Boeckel CAA. A synthetic antithrombin III binding pentasaccharide is now a drug! What comes next? Angew Chem Int Ed 2004;43:3118–3133.
31. Scott EN, Thomas AA. Phosphomannopentaose sulphate—antiangiogenic heparanase inhibitor. Drugs Future 2008;33(1). Published on-line. DOI: 10.1358/dof.2008.033.01.1165464 PI-88.
32. Benz I, Schmidt AA. Never say never again: protein glycosylation in pathogenic bacteria. Mol Biol 2002;45:267–276.
33. Wacker M, Linton D, Hitchen PG, Nita-Lazar M, Mihai H, North S, Panico M, Morris HR, Wren B, Aebi M. N-linked glycosylation in *Campylobacter jejuni* and its functional transfer into *Escherichia coli*. Science 2002;298:1790–1793.
34. The International Society of Rare Sugars (ISRS), Kagawa, Japan. http://isrs.kagawa-u.ac.jp/indexe.html.
35. 5th Framework Programme (FP5) of the European Commission Research & Innovation, NEPSA (NEw Products from Starch-derived 1,5-Anhydro-D-fructose). http://www.eurice.de/NEPSA/ F Giffhorn—Co-ordinator.

CHAPTER 3

EMERGING ENZYMES AND THEIR SYNTHETIC APPLICATIONS

ZHENMING CHEN and JUNHUA (ALEX) TAO
Institute of Sustainability, Hangzhou Normal University, Hangzhou, China

3.1 EMERGING ENZYMES

Nature essentially makes all types of molecules by enzymes. As a result of recent technology breakthrough in large-scale DNA sequencing, there is an exponential growth in GenBank. This section focuses on biocatalyst libraries developed recently.

3.1.1 Enzymes for Reduction

3.1.1.1 Ketoreductases. As a result of the commercial availability of a large library of diversified ketoreductase (KREDs), these enzymes have been demonstrated on an industrial scale to catalyze regio- and stereoselective conversion of a wide range of ketones to chiral alcohols through cost-effective cofactor regeneration using either formate dehydrogenases or glucose dehydrogenases (Scheme 3.1) [1–3].

3.1.1.2 Enoate (Ene) Reductases. Enoate (Ene) reductase catalyzes reduction of carbon–carbon double bonds of α,β-unsaturated aldehydes, ketones, nitros, and nitriles (Scheme 3.2, EWG = electron withdrawing group). Therefore, compounds with up to two chiral centers can be generated in one step [4].

Biocatalysis for Green Chemistry and Chemical Process Development, First Edition.
Edited by Junhua (Alex) Tao and Romas Kazlauskas.

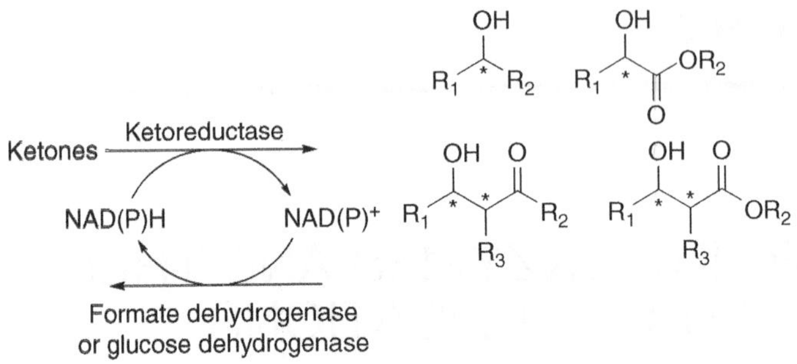

Scheme 3.1 Enzymatic reduction of ketones.

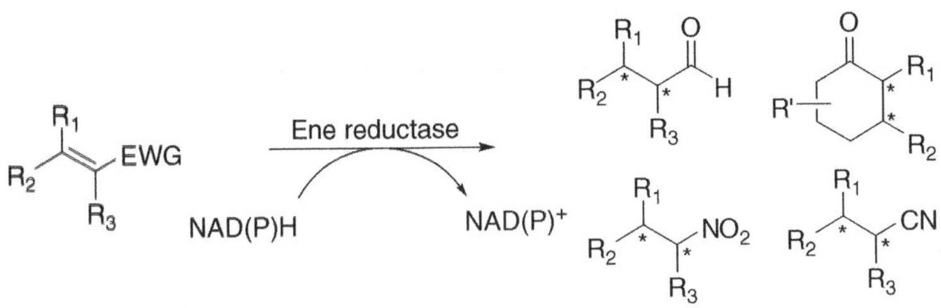

Scheme 3.2 Enzymatic reduction of double bonds.

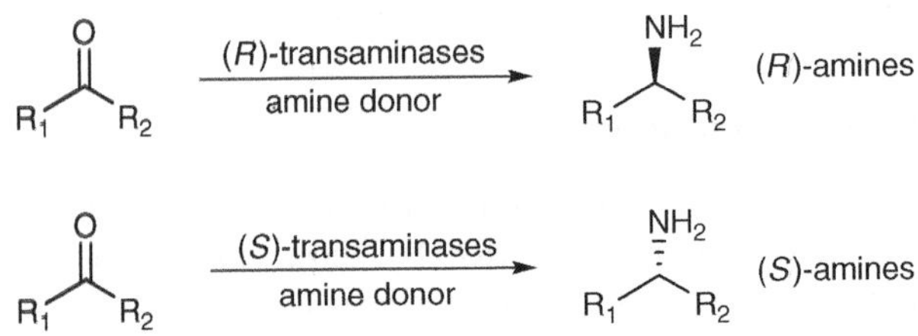

Scheme 3.3 Enzymatic amination of ketones.

3.1.1.3 Transaminases. ω-Transaminases catalyze direct conversion of ketones to chiral amines (Scheme 3.3). Both the (*R*)- and (*S*)- enantiomers can be obtained in one step [5–8]. The potential of the transformation is still underexplored because of the limited commercial availability of these enzymes.

3.1.2 Enzymes for Oxidation

3.1.2.1 Alcohol Oxidases. Alcohol oxidases catalyze oxidation of alcohols to aldehydes or ketones in the presence of oxygen. In contrast to chemical oxidation, enzymatic oxidation does not require toxic solvents or metal oxidants. Some of the industrialized alcohol oxidases include sugar oxidases, aromatic alcohol oxidases, and cholesterol oxidases (Scheme 3.4) [9–11].

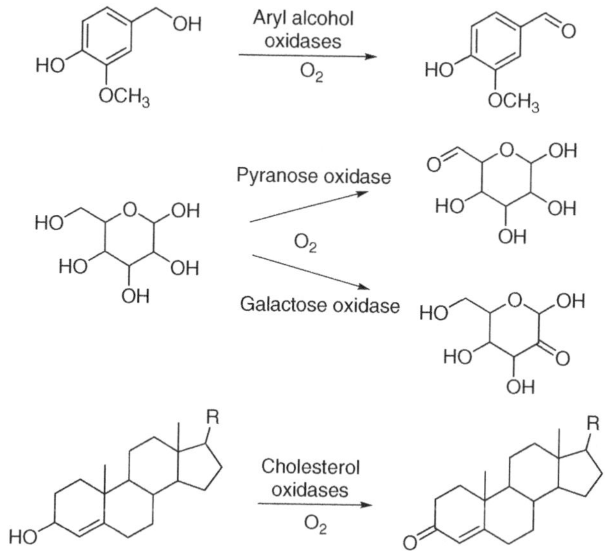

Scheme 3.4 Enzymatic oxidation by alcohol oxidases.

3.1.2.2 Halogenases. For electron-rich substrates, nature usually uses flavin-dependent halogenases or haloperoxidases for chlorination, bromination, or iodination [(1 and 2), Scheme 3.5]. For electron-deficient molecules such as alkanes, mononuclear iron halogenases are often applied for halogenation [(3), Scheme 3.5] [12–15]. On the other hand, fluorination relies on a nucleophilic substitution mechanism [16,17].

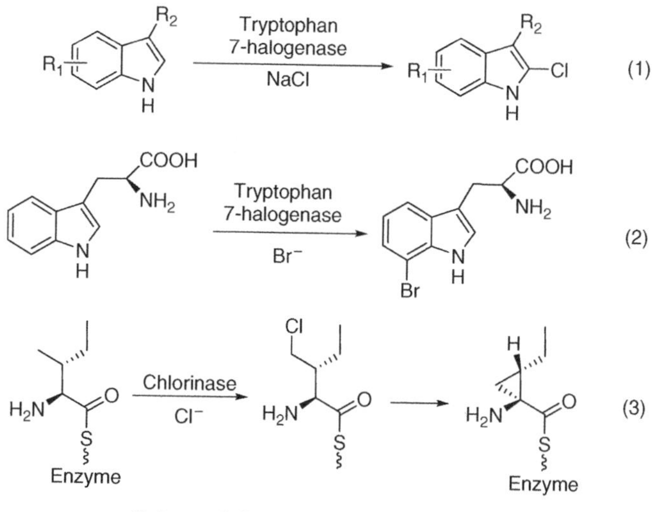

Scheme 3.5 Enzymatic halogenation.

3.1.2.3 *Oxidation by Oxygenases.* Oxygenases are categorized as either monooxygenases or dioxygenases, depending upon the insertion of one or both atoms of dioxygen into a substrate (Scheme 3.1) [18–20]. For example, aromatic dioxygenases are important pathways in the degradation of aromatics into *cis*-dihydrodiols (Scheme 3.6) [21,22]. However, dioxygenation may also take place in the absence of cofactors [23–25].

3.1.2.4 *Peroxidases.* The active site of peroxidases may involve a heme unit, selenium, vanadium, and manganese. These enzymes catalyze a wide range of oxidations using H_2O_2 as the oxidant for hydroxylation of arenes, oligomerization of phenols and aromatic amines, epoxidation of olefins, and so on (Scheme 3.7) [26–28].

3.1.3 Enzymes for C–C Formation

3.1.3.1 *Aldolases.* Aldolases catalyze C–C bond forming reactions with simultaneous creation of multiple functional groups and chiral centers, which

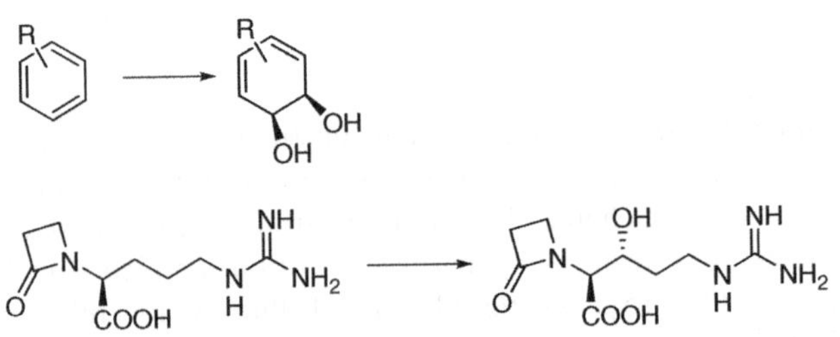

Scheme 3.6 C—H activation by oxygenases.

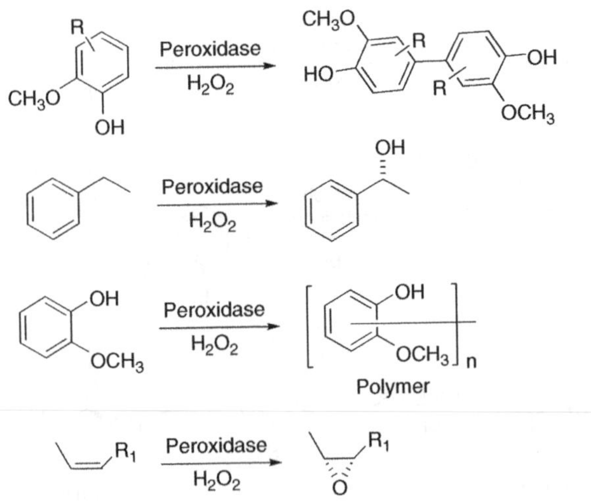

Scheme 3.7 Enzymatic oxidation by H_2O_2.

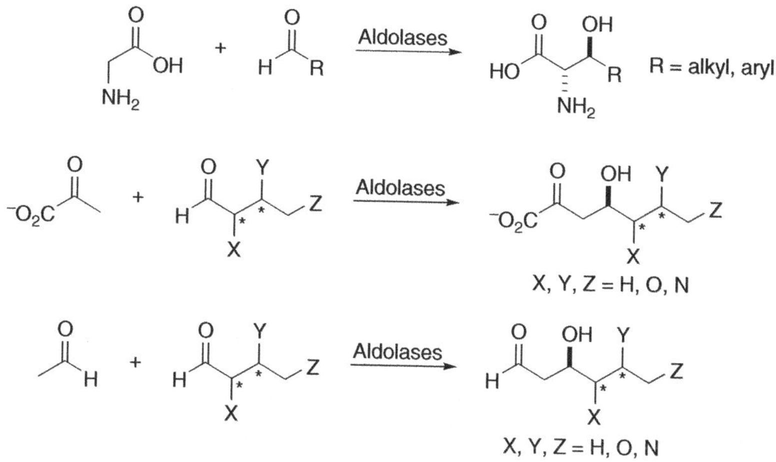

Scheme 3.8 C—C bond formation catalyzed by aldoalses.

have been used to prepare a wide range of carbohydrates such as azasugars, cyclitols, and densely functionalized chiral molecules (Scheme 3.8) [29].

3.1.3.2 Oxynitrilases (Hydroxynitrile Lyases). Hydroxynitrile lyases or oxynitrilases catalyze C–C bond formation between aldehydes and a cyanide source to produce optically active (*R*)- or (*S*)-cyanohydrins (Scheme 3.9), versatile building blocks for chiral compounds [30,31].

3.1.4 C–X (X = O, N, etc.) Bond Formation or Cleavage

3.1.4.1 Glycosyltransferases. Glycosyltransferases accept activated sugars such as uridine diphosphate (UDP) or glycosyl phosphates as monosaccharide donors under either retention or inversion of the configuration at the anomeric center (Scheme 3.10, only inversion scenario shown). Many promiscuous glycosyltransferases have been discovered to prepare oligosaccharides and glycosylated natural products, which are otherwise difficult to obtain [32–35]. For large-scale production, several methods have been developed to recycle released nucleotide monophosphate (NMP) or nucleotide diphosphate (NDP) [36,37].

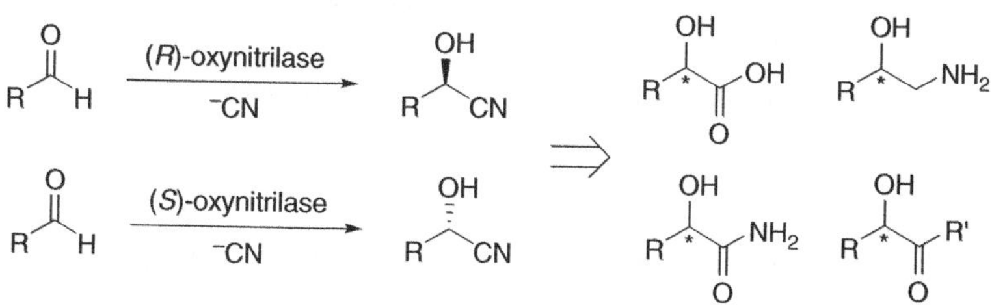

Scheme 3.9 Enzymatic synthesis of chiral cyanohydrins.

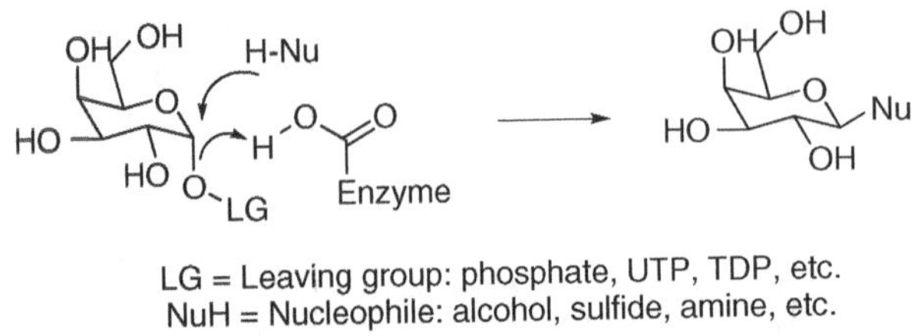

Scheme 3.10 Enzymatic glycosylation.

3.1.4.2 Glycosidases. Glycosidases catalyze the cleavage of glycosidic linkages via an oxonium intermediate or transition state similar to acid-catalyzed hydrolysis of glycosides under either retention or inversion of the configuration at the anomeric center (Scheme 3.11) [38]. Glycosidases catalyze synthesis of glycosides under either equilibrium-controlled conditions or kinetic-controlled conditions [39–41]. In the former case, the reaction is designed to shift the equilibrium toward the product, for example, by adding organic solvents. In the latter case, activated glycosyl donors, such as di- or oligosaccharides, aryl glycosides, or glycosyl fluorides are used.

3.1.4.3 Nitrilases. Nitrilases catalyze the conversion of nitriles to carboxylic acids and NH_3 through a cysteine residue in the active site [42]. Nitrilases are "green" catalysts for the synthesis of a variety of carboxylic acids and derivatives (Scheme 3.12) [43–45].

3.1.4.4 Nitrile Hydratases. Nitrile hydratase (NHase) catalyzes the hydration of nitriles to produce amides (Scheme 3.13) and have been used in the industrial production of acrylamide and nicotinamide. They are roughly classified into iron and cobalt types on the basis of the metal ions in the active site [46–48].

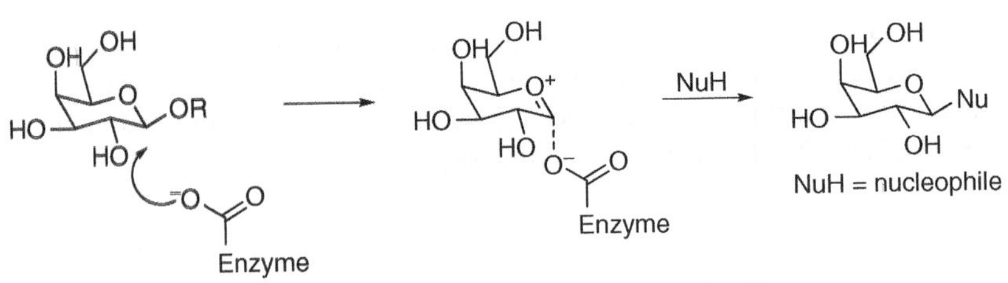

Scheme 3.11 Hydrolysis of carbohydrates.

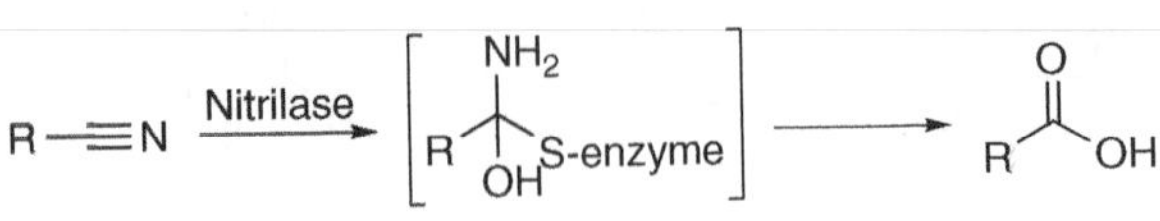

Scheme 3.12 Enzymatic hydrolysis of nitriles.

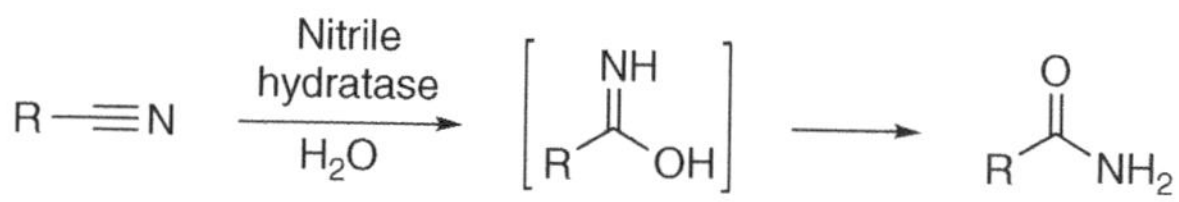

Scheme 3.13 Hydration of nitriles to give amides.

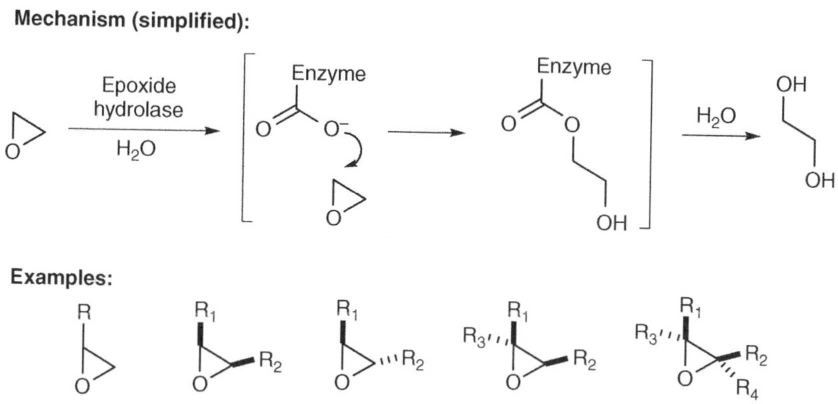

Scheme 3.14 Enzymatic hydrolysis of epoxides.

3.1.4.5 Epoxide Hydrolases. Epoxide hydrolases catalyze the hydrolysis of epoxides to give chiral diols or triols (Scheme 3.14) [49–52].

3.1.4.6 Cyclases (Acyltransferases). Macrocyclic motifs are often essential for a natural product's biological properties. Macrocyclization is usually carried out by cyclases toward the end of chain elongation. For example, in the biosynthesis of antibiotic tyrocidine A, a linear enzyme-bound decapeptide is cyclized via an intramolecular S_N2 reaction between the N-terminal amine nucleophile and the C-terminal thioester [(1), Scheme 3.15] [53]. This cyclase has great versatility. Not only does it catalyze the formation of macrolactams of ring sizes from 18 to 42 atoms with excellent catalytic efficiency, but also four amino acids [Asn, Gln, Tyr, Val, (2), Scheme 3.15] can be replaced by polyketide (PK) moieties, leading to the production of hybrid peptide/polyketide (PP/PK) cyclic molecules [54,55]. In contrast to chemical macrocyclization, no protection was required for the linear precursors.

Cyclases are also responsible for formation of macrocyclic depsipeptides and lactones scyclization. For example, a di-domain excised from fengycin synthase was able to catalyze the formation of a macrolactone through the formation of a C–O bond [56]. Several cyclases have also been characterized to be functional including the epothilone D cyclase [57,58] (Scheme 3.16).

3.1.4.7 Enzymatic Heterocyclization. Heterocycles are common in natural products and may occur as single, tandem, and multiple moieties (Scheme 3.17)

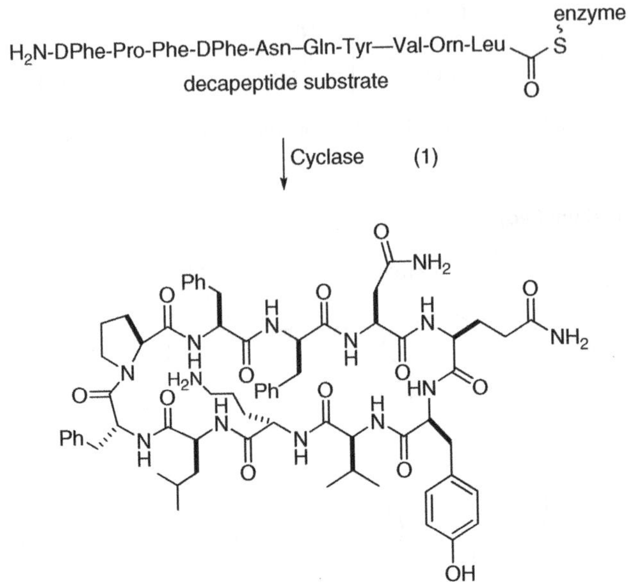

Tyrocidine A

Scheme 3.15 Enzymatic macrocyclization.

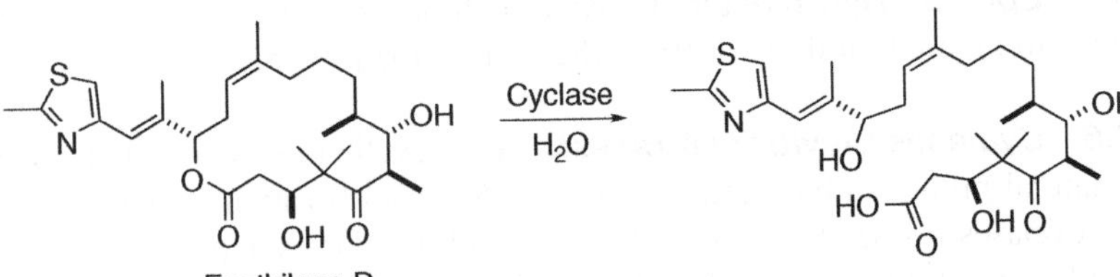

Epothilone D

Scheme 3.16 Enzymatic hydrolysis of epothilone D.

[59–61]. An oxazoline was usually formed from a dipeptide containing serine in the second position upon dehydration (Scheme 3.17) [62]. Biosynthesis of a thiazoline from cysteine and a 2-methyl oxazoline from threonine follows a similar mechanism. These heterocycles can be further modified to form thiazolidines/oxazolidines upon reduction or thiazoles/oxazoles upon oxidation.

A different mechanism was adopted in the biosynthesis of cyclic polyethers such as momensin, where the cascade polyether formation is triggered by epoxidation of a polyene template (Scheme 3.18) [63]. Similar mechanisms are probably used to make other polyether antibiotics containing tetrahydrofurans and tetrahydropyrans [64].

3.1.4.8 Methyl Transferases. Methyl transferases accept a wide range of nucleophiles such as halides, amines, hydroxyls, and enolates [(1 and 2),

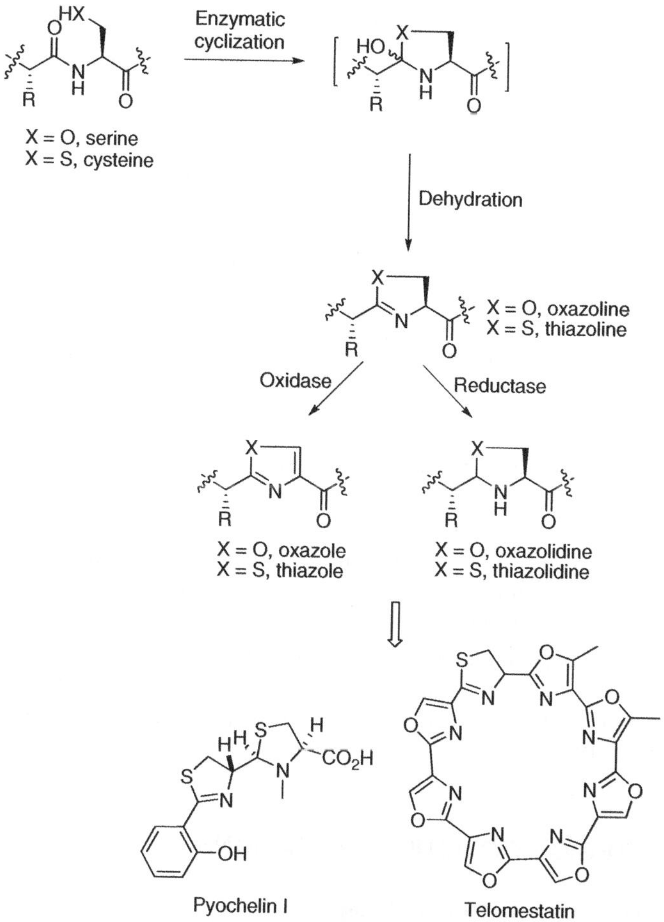

Scheme 3.17 Biosynthesis of N-heterocycles.

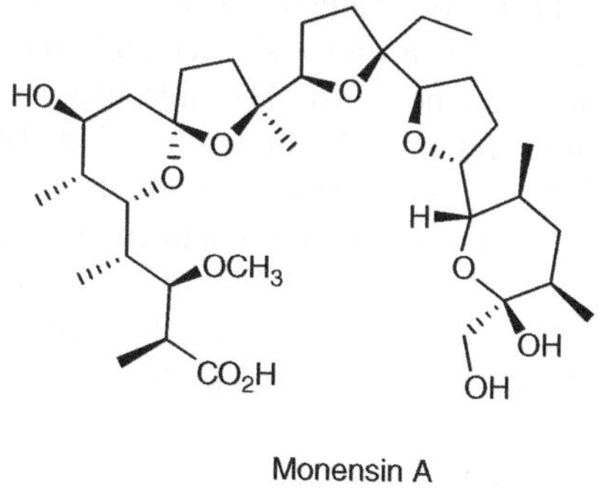

Scheme 3.18 Biosynthesis of heterocyclic polyethers.

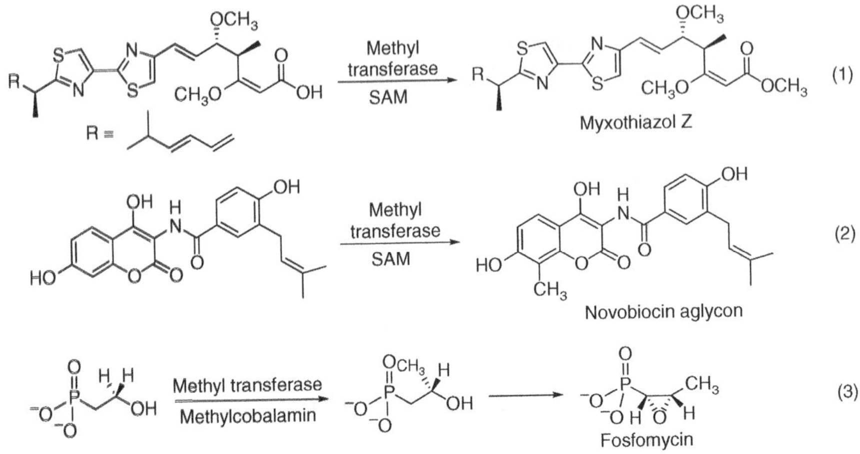

Scheme 3.19 Enzymatic methylation.

Scheme 3.17] [65–67]. For example, in the biosynthesis of novobiocin, methylation takes place at only one phenolic carbon and not at the remaining three hydroxyl groups [68,69]. On the other hand, methyl transfer to electron-deficient substrates often occurs under radical mechanisms requiring methylcobalamin as the cofactor as shown in the biosynthesis of fosfomycin, where only one of the two enantiotopic hydrogen atoms was replaced by the methyl group (Scheme 3.19) [70].

3.2 STRATEGIC SYNTHETIC APPLICATIONS

3.2.1 Retrosynthetic Biocatalysis

In order to claim new intellectual properties, cut steps, and increase greenness, it is essential to integrate biotransformations into chemical transformations at the retrosynthetic level [71]. For example, chemically, atorvastatin, the API of Lipitor®, can be prepared from a diol ("side chain") and a diketone precursor by Paal–Knorr reaction (Scheme 3.20). The side chain is the most expensive piece because of two chiral centers. The first-generation chemical manufacturing process is lengthy requiring two cryogenic steps. On the other hand, this piece can be prepared by either an aldoalse-catalyzed sequential aldol reaction from two aldehydes (pathway a, Scheme 3.20) [72–74] or a potentially KRED-catalyzed reduction of a 3-ketoester (pathway b, Scheme 3.06) obtainable chemically from a 3-hydroxyester, which can be synthesized from a bis-nitrile under nitrilase or 3-ketoester under KREDs [75–77].

In another example, pregabalin can be prepared from a number of biocatalytic routes (Scheme 3.21) [78–80]. In pathway a and b, the amino group is derived from a nitrile precursor, which can be prepared highly stereoselectively from

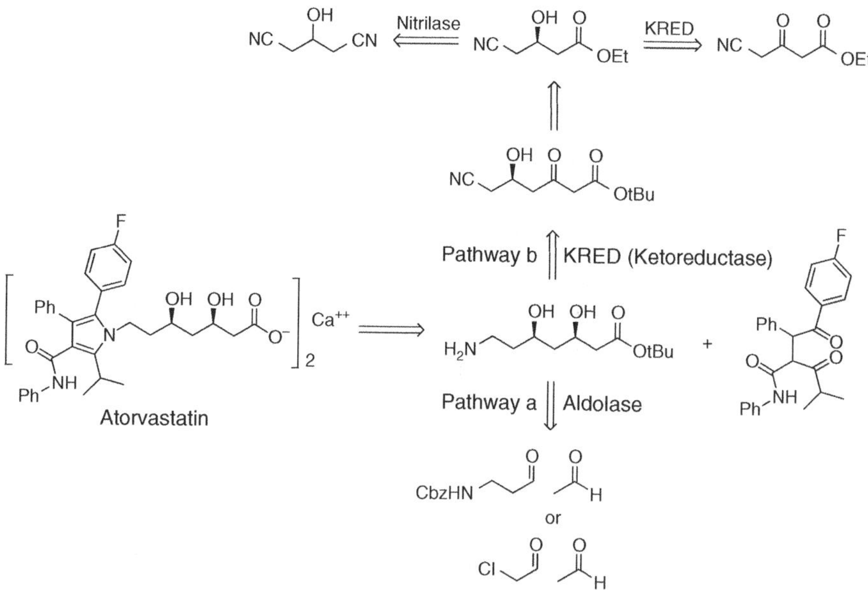

Scheme 3.20 Retrosynthetic analysis of atorvastatin.

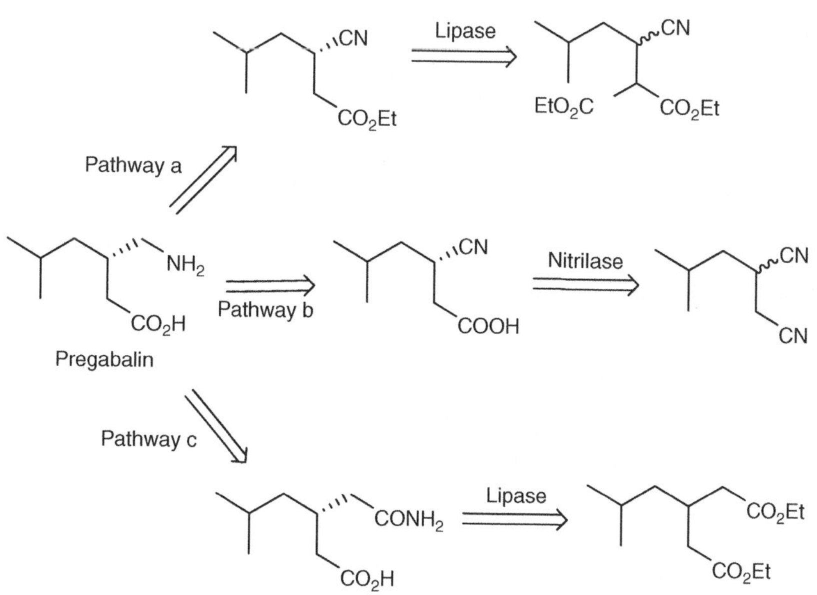

Scheme 3.21 Retrosynthetic analysis of pregabalin.

either a diester using a lipase or a bis-nitrile using a nitrilase. Alternatively, the amino group can be obtained under Hoffman degradation from an amide, which can be obtained enzymatically from a diester by enzymatic desymmetrization.

3.2.2 Optimization of Biotransformations

Once a biocatalytic route is designed and leading biocatalysts are selected from comprehensive screening of a diverse enzyme library, it is often essential to optimize the biotransformation step, where reactivity, selectivity, inhibition, and deactivation have to be thoroughly studied [81]. In some cases, substrates can be redesigned to accommodate the function of enzymes without sacrificing process efficiency, while in other reactions, engineering can be applied or enzymes need to be immobilized. In many cases, the biocatalysts have to be evolved to improve reactivity, selectivity, or stability.

3.2.2.1 Substrate Modulation. One strategy to improve process efficiency is to modify substrates not ideal for biocatalysis without sacrificing the overall process efficiency. For example, a different protecting group may significantly change the reactivity, solubility, and lipophilicity of a molecule leading to high selectivity or reactivity during biotransformations. To optimize biocatalytic reactions, substrate modulation can be a much faster and economic approach to the enzyme-directed evolution approach. For example, in the synthesis of 3,3-difluoro-4,4-dimethylproline, a key precursor for an HIV protease inhibitor, only Subtilisin carlsberg was identified to be active from a library of over 150 hydrolases if the nitrogen is protected by a Boc group (Scheme 3.22) [82]. The reaction was slow requiring 4–5 days even under a low substrate loading of 10 g/L. However, by replacing the Boc group with benzyl, pig liver esterase was selected from the same enzyme library, which catalyzes the kinetic resolution with excellent resolution efficiency: substrate loading 100 g/L, 44% yield, >99% *ee*, 24 h.

3.2.2.2 Solvent Tuning. For enzymatic catalysis, reaction media including pH, solvent types and contents, and other additives often have a profound effect

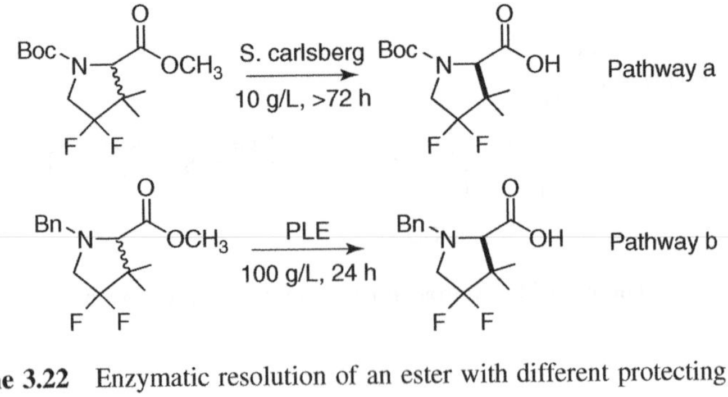

Scheme 3.22 Enzymatic resolution of an ester with different protecting groups.

on protein conformation. Therefore, it is often crucial to apply this principle for biotransformation process optimization. For example, in the resolution of a racemic mixture of the *cis*- and *trans*-diastereomers, initial screening of a large library of hydrolases produced only one hit: *Candida antarctica* lipase B with a poor E-value of 10 [83]. Through extensive solvent screening and pH optimization, the E-value was improved to >100 in the presence of 40% acetone (Scheme 3.23). Preliminary 2D-TROSY NMR study using N^{15}-labeled *C. antarctica* lipase B indicated that the dramatic solvent effect was probably caused by a global conformational change of the enzyme [83].

3.2.2.3 *Enzyme Inhibition and Deactivation.* At high substrate or product concentration, enzymes can often be inhibited or deactivated. In some cases, substrate inhibition can be overcome by substrate absorption or a continuous process, where a substrate is converted to its corresponding product under a short residence time [84–86]. On the other hand, product inhibition is often more challenging. For example, a careful study showed that a catalytic amount of calcium acetate can be added to remove product inhibition in a process to produce pregabalin [79].

3.2.2.4 *Directed Evolution.* Directed evolution is an important part of the integrated solution when the best wild-type enzyme cannot deliver process metrics after comprehensive screening and optimization. Over the past five years, directed evolution has proved to be a robust technology to improve enzyme function. As more and more practical and high-throughput enzymatic assays are established, the time required to implement this technology has been shortened significantly and often, more than 10–30-fold improvement in reactivity and enantioselectivity can be achieved within 6–12 months. While this timeline may not fit well for drug molecules at preclinical or early clinical stages where time is often a constraint, directed evolution can be integrated efficiently to process development for compounds in late stages or on the market. For example, in the development of a nitrilase-catalyzed synthesis of (*R*)-4-cyano-3-hydroxybutyric acid, a key intermediate in the current manufacturing process of Atorvastatin (Scheme 3.24), the wild-type nitrilase identified from the initial screening has

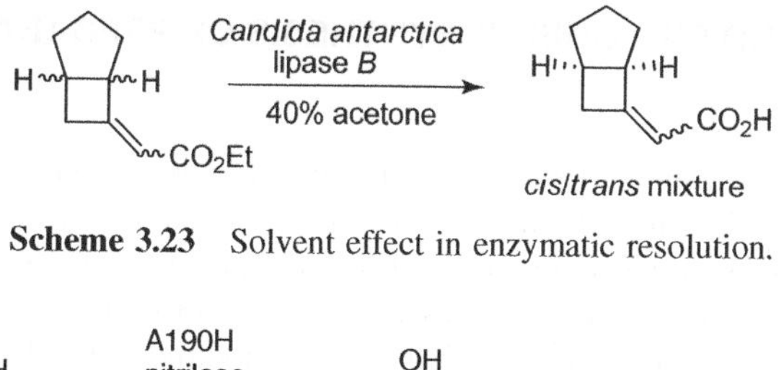

Scheme 3.23 Solvent effect in enzymatic resolution.

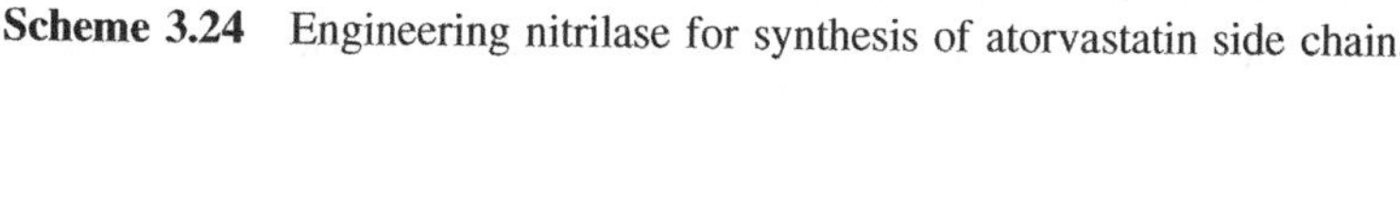

Scheme 3.24 Engineering nitrilase for synthesis of atorvastatin side chain.

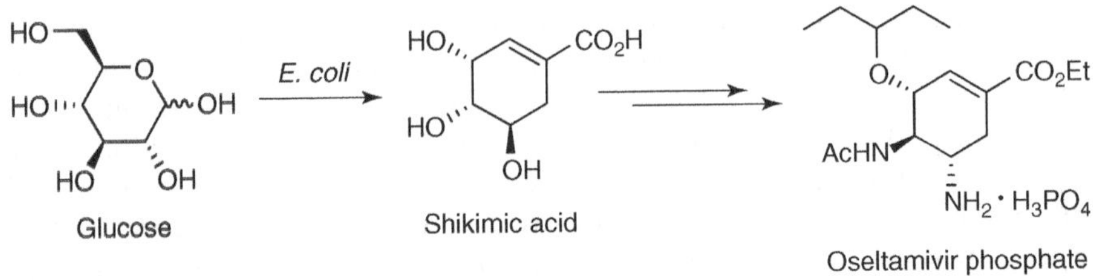

Scheme 3.25 Synthesis of shikimic acid through metabolic engineering.

an enantioselectivity (*ee*) of 88% under a loading of 3 M of the substrate [87]. Directed evolution using gene site saturation mutagenesis (GSSM) was used to improve the stereoselectivity of the lead enzyme and a Ala190His single mutation resulted in an excellent enantioselectivity of 99% under the same substrate loading [76].

In addition to random mutagenesis, where it is often time consuming to screen a large library of enzyme mutants generated from directed evolution, semirational approaches can also be used to build a high-quality, focused library. If two enzymes have high sequence homology, a homologous 3D model for an enzyme, whose structure is unknown, can be constructed from protein structures solved by either crystallography or NMR. Docking programs can then be used to guide a subset of amino acids for mutations [28,88,89].

3.2.2.5 Metabolic Engineering. Recent advances in metabolic engineering, combinatorial biosynthesis, system biology, and synthetic biology has led to the development of cells suitable, with engineered pathways, for industrial-scale applications in the energy, chemical, and pharmaceutical industries. For example, shikimic acid is a key intermediate for the production oseltamivir phosphate, the API of Tamiflu® (Scheme 3.25), used as part of the prevention plan against an influenza pandemic. Through metabolic engineering, a genetically engineered *Escherichia coli* strain was developed to produce shikimic acid in 33% yield with a titer of 84 g/L from glucose [90,91].

3.3 CASE STUDIES: "GREEN" SYNTHESIS OF PREGABALIN, API OF LYRICA

Enzymatic catalysis can achieve high regio- and stereoselectivity under mild conditions in water, allowing green processes to be developed with significant cost advantages over chemical approaches. For example, pregabalin was developed by Pfizer to be a more potent successor to gabapentin and was launched first under the brand name Lyrica® in 2004. The compound is used in the treatment of peripheral neuropathic pain as adjunctive therapy for partial seizures resulting from epilepsy and fibromyalgia. It is one of the fastest growing drugs with a worldwide sale of over $2.0 billion in 2008. The first-generation route for the production of pregabalin involves diastereomeric resolution of a racemic γ-amino acid, which was prepared from a racemic cyanodiester (CNDE) after hydrolytive decarboxylation followed by hydrogenation under Raney-Ni (Scheme 3.26) [92]. In order to achieve high optical purity (>99.5%) for the final API, (*S*)-mandelic acid resolution was followed by another step of recrystallization in THF/H_2O, which has a combined two-step yield of only 25–29%. As a result, the total yield for this process is only 18–21%. Since the undesired enantiomer could not be recycled, poor process efficiency leads to high usage of raw materials, solvents, and energy. For example, over 70% of all process materials before the resolution were ultimately turned into wastes.

To overcome these issues, several alternative rotues were developed. For example, a two metal-catalyzed process was developed where the key transformation involves asymmetric hydrogenation of a *t*-butylamine salt, which was prepared from a Baylis–Hillman adduct followed by Pd-catlyzed carbonylation under high pressure (Scheme 3.27) [93]. Though the overall yield is much higher, the process requires special high-pressure reactor. Moreover, since the homogeneous catalyst is expensive, its loading usually cannot be higher than 0.1%, which often leads to issues related to process robustness as a result of catalyst poisoning due to impurities in the substrates, reactor, or air (oxygen, water, etc.).

Another process was developed using enzymatic resolution, which took place in the first step and the undesired enantiomer could be easily recycled (Scheme 3.28) [79,94]. Regio- and stereospecific hydrolysis of CNDE led to a mono acid with 45% conversion and >98% *ee*. After decarboxylation, the resulting mono ester was telescopically hydrolyzed to give the final product upon hydrogenation.

In comparison with the first chemical resolution process, it is projected that, between 2007 and 2020, the enzymatic synthesis is expected to eliminate [94,95]:

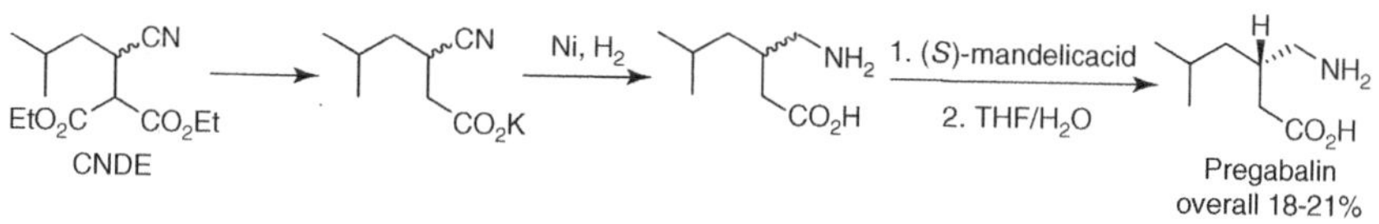

Scheme 3.26 Synthesis of pregabalin by chemicla resolution.

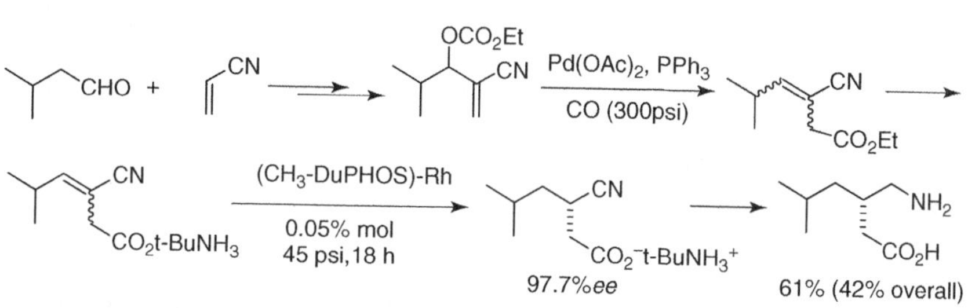

Scheme 3.27 Synthesis of pregabalin under asymmetric hydrogenation.

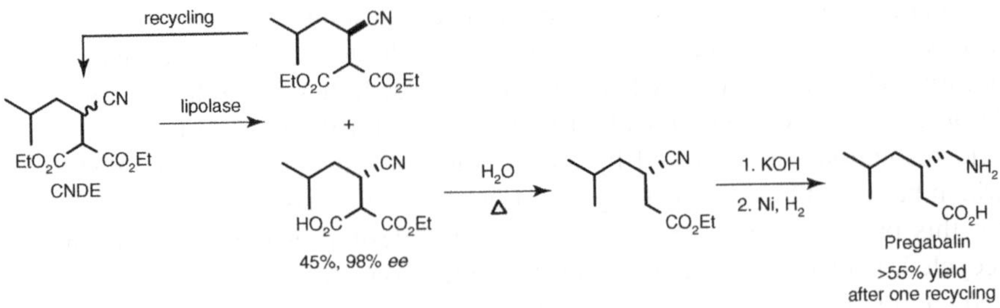

Scheme 3.28 Biocatalytic synthesis of pregabalin.

- 185,000 tons of solvent, an 88% reduction
- 4800 tons of mandelic acid, a 100% reduction
- 1890 tons of Raney nickel catalyst, an 85% reduction
- 10,000 tons of starting material, a 39% reduction

3.4 CONCLUSION

Biocatalysis is well suited for the development of green chemical processes as a result of ready access to GenBank and enzymes capable of catalyzing a wide range of reactions, and a wide range of strategies for imporving enzyme functions. The key is to integrate biotransformations into the toolboxes of chemical transformations at the retrosynthetic level to reduce the consumption of solvents, raw materials, and energy.

REFERENCES

1. Goldberg K, *et al.* Biocatalytic ketone reduction—a powerful tool for the production of chiral alcohols—Part I: processes with isolated enzymes. Appl Microbiol Biotechnol 2007;76(2):237–248.
2. Husain M, Husain Q. Applications of redox mediators in the treatment of organic pollutants by using oxidoreductive enzymes: A review. Crit Rev Environ Sci Technol 2008;38(1):1–42.
3. Moore J, *et al.* Advances in the enzymatic reduction of ketones. Acc Chem Res 2007;40(12):1412–1419.
4. Chaparro-Riggers JF, *et al.* Comparison of three enoate reductases and their potential use for biotransformations. Adv Synth Catal 2007;349(89):1521–1531.
5. Hwang B, Kim B. High-throughput screening method for the identification of active and enantioselective [omega]-transaminases. Enzyme Microb Technol 2004; 34(5):429–436.
6. Rozzell J, *et al.* Hydrolysis and formation of C-N bonds—transaminations. In: Drauz K, Waldmann H, editors. Volume 2, Enzmes catalysis on organic synthesis. 2nd ed. Weinheim: Wiley -VCH Verlag GmbH & Co. KGaA; 2002. pp. 873–893.

7. Stewart JD. Dehydrogenases and transaminases in asymmetric synthesis. Curr Opin Chem Biol 2001;5(2):120–129.
8. Matcham G, *et al.* Enzyme and reaction engineering in biocatalysis: synthesis of (S)-methoxyisopropylamine (=(S)-1-Methoxypropan-2-amine). CHIMIA Int J Chem 1999;53(12):584–589.
9. Kroutil W, *et al.* Biocatalytic oxidation of primary and secondary alcohols. Adv Synth Catal 2004;346(2–3):125–142.
10. Kumar AK, Goswami P. Functional characterization of alcohol oxidases from Aspergillus terreus MTCC 6324. Appl Microbiol Biotechnol 2006;72(5):906–911.
11. Overhage J, Steinbüchel A, Priefert H. Harnessing eugenol as a substrate for production of aromatic compounds with recombinant strains of Amycolatopsis sp. HR167. J Biotechnol 2006;125(3):369–376.
12. Burd V, Van Pée KH. Halogenating enzymes in the biosynthesis of antibiotics. Biochemistry (Moscow) 2003;68(10):1132–1135.
13. Vaillancourt FH, *et al.* Cryptic chlorination by a non-haem iron enzyme during cyclopropyl amino acid biosynthesis. Nature 2005;436(7054):1191–1194.
14. Galoni DP, Vaillancourt FH, Walsh CT. Halogenation of unactivated carbon centers in natural product biosynthesis: trichlorination of leucine during barbamide biosynthesis. J Am Chem Soc 2006;128(12):3900–3901.
15. Vaillancourt FH, Vosburg DA, Walsh CT. Dichlorination and bromination of a threonyl-S-carrier protein by the non-heme FeII halogenase SyrB2. Chembiochem 2006;7(5):748–752.
16. O'Hagan D, *et al.* Biochemistry: biosynthesis of an organofluorine molecule. Nature 2002;416: 279.
17. O'Hagan D. Recent developments on the fluorinase from Streptomyces cattleya. J Fluor Chem 2006;127(11):1479–1483.
18. Fessner WD, Helaine V. Biocatalytic synthesis of hydroxylated natural products using aldolases and related enzymes. Curr Opin Biotechnol 2001;12(6):574–586.
19. Chen H, Walsh CT. Coumarin formation in novobiocin biosynthesis:[beta]-hydroxylation of the aminoacyl enzyme tyrosyl-S-NovH by a cytochrome P450 NovI. Chem Biol 2001;8(4):301–312.
20. Que L. One motif—many different reactions. Nat Struct Mol Biol 2000;7(3): 182–184.
21. Gibson DT, Parales RE. Aromatic hydrocarbon dioxygenases in environmental biotechnology. Curr Opin Biotechnol 2000;11(3):236–243.
22. Boyd DR, Sharma ND, Allen CCR. Aromatic dioxygenases: molecular biocatalysis and applications. Curr Opin Biotechnol 2001;12(6):564–573.
23. Fetzner S. Oxygenases without requirement for cofactors or metal ions. Appl Microbiol Biotechnol 2002;60(3):243–257.
24. Kendrew SG, Hopwood DA, Marsh E. Identification of a monooxygenase from Streptomyces coelicolor A3 (2) involved in biosynthesis of actinorhodin: purification and characterization of the recombinant enzyme. J Bacteriol 1997;179(13): 4305.
25. Bauer I, *et al.* 2,4-dioxygenases catalyzing N-heterocyclic-ring cleavage and formation of carbon monoxide: purification and some properties of 1H-3-Hydroxy-4-Oxoquinaldine 2,4-dioxygenase from arthrobacter sp. Ru61a and comparison with 1H-3-Hydroxy-4-Oxoquinoline4-Dioxygenase from *Pseudomonas Putida* 33/1. FEBS J 1996;240(3):576–583.

26. Adam W, *et al.* Biotransformations with peroxidases, biotransformations. In: Faber K, editor. Volume 63, Advances in biochemical engineering/biotechnology. Berlin: Springer-Verlag; 1999. pp. 73–108.
27. Costa-Ferreira M, *et al.* An overview on the use of microbial and enzymatic systems for dye bitransformations. J Nat Fibers 2006;3:69–80.
28. Liu HL, *et al.* Biotransformation of ferulic acid into two new dihydrotrimers by Momordica charantia peroxidase. Chem Pharm Bull 2005;53(7):816–819.
29. Fessner WD. Enzyme catalyzed aldol additions. In: Mahrwald R, editor. Volume 1, Modern aldol reactions. Germany: Wiley-VCH Verlag; 2004. pp. 201–272.
30. Purkarthofer T, *et al.* Potential and capabilities of hydroxynitrile lyases as biocatalysts in the chemical industry. Appl Microbiol Biotechnol 2007;76(2):309–320.
31. Sharma M, Sharma NN, Bhalla TC. Hydroxynitrile lyases: at the interface of biology and chemistry. Enzyme Microb Technol 2005;37(3):279–294.
32. Rowan AS, Hamilton CJ. Recent developments in preparative enzymatic syntheses of carbohydrates. Nat Prod Rep 2006;23(3):412–443.
33. Brik A, Wong CH. Sugar-Assisted ligation for the synthesis of glycopeptides. Chem Eur J 2007;13(20):5670–5675.
34. Trincone A, Giordano A. Glycosyl hydrolases and glycosyltransferases in the synthesis of oligosaccharides. Curr Org Chem 2006;10(10):1163–1193.
35. Rabbani S, Schwardt O, Ernst B. Glycosyltransferases: an efficient tool for the enzymatic synthesis of oligosaccharides and derivatives as well as mimetics thereof. CHIMIA Int J Chem 2006;1(2):23–27.
36. Rupprath C, *et al.* An enzyme module system for *in situ* regeneration of deoxythymidine 5O-Diphosphate (dTDP)-activated deoxy sugars. Adv Synth Catal 2007;349:1489–1496.
37. Chen X, *et al.* Sugar nucleotide regeneration beads (superbeads): a versatile tool for the practical synthesis of oligosaccharides. J Am Chem Soc 2001;123(9):2081–2082.
38. Faijes M, Planas A. In vitro synthesis of artificial polysaccharides by glycosidases and glycosynthases. Carbohydr Res 2007;342(12–13):1581–1594.
39. Lu WY, *et al.* Facile synthesis of alkyl [beta]-d-glucopyranosides from d-glucose and the corresponding alcohols using fruit seed meals. J Mol Catal B Enzym 2007;44(2):72–77.
40. Yu HL, *et al.* Environmentally benign synthesis of natural glycosides using apple seed meal as green and robust biocatalyst. J Biotechnol 2008;133(4):469–477.
41. Yu HL, *et al.* Assembly of a three-dimensional array of glycoconjugates by combinatorial biocatalysis in nonaqueous media. J Comb Chem 2007;10(1):79–87.
42. O'Reilly C, Turner P. The nitrilase family of CN hydrolysing enzymes—a comparative study. J Appl Microbiol 2003;95(6):1161–1174.
43. Singh R, Sharma R, Tewari N. Nitrilase and its application as a green catalyst. Chem Biodivers 2006;3(12):1279–1287.
44. He YC, *et al.* Biocatalytic synthesis of (R)-(-)-mandelic acid from racemic mandelonitrile by a newly isolated nitrilase-producer Alcaligenes sp. ECU0401. Chin Chem Lett 2007;18(6):677–680.
45. Mylerova V, Martinkova L. Synthetic applications of nitrile-converting enzymes. Curr Org Chem 2003;7(13):1279–1295.

46. Kobayashi M, Shimizu S. Metalloenzyme nitrile hydratase: structure, regulation, and application to biotechnology. Nat Biotechnol 1998;16(8):733–736.
47. Harrop TC, Mascharak PK. Fe (III) and Co (III) centers with carboxamido nitrogen and modified sulfur coordination: lessons learned from nitrile hydratase. Acc Chem Res 2004;37(4):253–260.
48. DiCosimo R. Nitrilases and nitrile hydratases. In: Patel RN, editor. Biocatalysis in the pharmaceutical and biotechnology industries. Boca Raton (FL): CRC Press LLC; 2007. pp. 1–26.
49. Tang Y, *et al.* Biocatalytic preparation of (S)-glycidyl phenyl ether using newly isolated Bacillus sp. ECU1001. J Mol Catal B Enzym 2001;13:61–68.
50. Smit M, Labuschagne M. Diversity of epoxide hydrolase biocatalysts. Curr Org Chem 2006;10(10):1145–1161.
51. Lee EY, Shuler ML. Molecular engineering of epoxide hydrolase and its application to asymmetric and enantioconvergent hydrolysis. Biotechnol Bioeng 2007;98(2):318–327.
52. Arand M, *et al.* Epoxide hydrolases: structure, function, mechanism, and assay. Methods in enzymology. Orlando, FL: Academic Press, Inc; 2005. pp. 569–588.
53. Kohli RM, Walsh CT, Burkart MD. Biomimetic synthesis and optimization of cyclic peptide antibiotics. Nature 2002;418(6898):658–661.
54. Kohli RM, *et al.* Chemoenzymatic route to macrocyclic hybrid peptide/polyketide-like molecules. J Am Chem Soc 2003;125(24):7160–7161.
55. Lin H, Walsh CT. A chemoenzymatic approach to glycopeptide antibiotics. J Am Chem Soc 2004;126(43):13998–14003.
56. Sieber SA, *et al.* Peptidyl thiophenols as substrates for nonribosomal peptide cyclases. Angew Chem Int Ed Engl 2004;43:493–498.
57. Lu H, *et al.* Expression, site-directed mutagenesis, and steady state kinetic analysis of the terminal thioesterase domain of the methymycin/picromycin polyketide synthase. Biochemistry 2002;41(42):12590–12597.
58. Boddy CN, *et al.* Epothilone C macrolactonization and hydrolysis are catalyzed by the isolated thioesterase domain of epothilone polyketide synthase. J Am Chem Soc 2003;125(12):3428–3429.
59. Schneider TL, Walsh CT, O'Connor SE. Utilization of alternate substrates by the first three modules of the epothilone synthetase assembly line. J Am Chem Soc 2002;124(38):11272–11273.
60. Patel HM, Tao J, Walsh CT. Epimerization of an l-cysteinyl to a d-cysteinyl residue during thiazoline ring formation in siderophore chain elongation by pyochelin synthetase from pseudomonas aeruginosa. Biochemistry 2003;42(35):10514–10527.
61. Shin-ya K, *et al.* Telomestatin, a novel telomerase inhibitor from Streptomyces anulatus. J Am Chem Soc 2001;123(6):1262–1263.
62. Roy RS, *et al.* Thiazole and oxazole peptides: biosynthesis and molecular machinery. Nat Prod Rep 1999;16(2):249–263.
63. Hughes-Thomas ZA, *et al.* Intermediates released from a polyether-producing polyketide synthase provide insight into the mechanism of oxidative cyclization. Angew Chem Int Ed Engl 2003;42(37):4475–4478.
64. Sun Y, *et al.* A complete gene cluster from streptomyces nanchangensis NS3226 encoding biosynthesis of the polyether ionophore nanchangmycin. Chem Biol 2003; 10(5):431–441.

65. Wuosmaa AM, Hager LP. Methyl chloride transferase: a carbocation route for biosynthesis of halometabolites. Science 1990;249(4965): 160.

66. Blackburn GM, Gamblin SJ, Wilson JR. Mechanism and control in biological amine methylation. Helv Chim Acta 2003;86(12):4000–4006.

67. Weinig S, *et al.* Melithiazol biosynthesis: further insights into myxobacterial PKS/NRPS systems and evidence for a new subclass of methyl transferases. Chem Biol 2003;10(10):939–952.

68. Pacholec M, Tao J, Walsh CT. CouO and NovO: C-methyltransferases for tailoring the aminocoumarin scaffold in coumermycin and novobiocin antibiotic biosynthesis. Biochemistry 2005;44(45):14969–14976.

69. Tao J, *et al.* Synthesis of proposed oxidation- cyclization- methylation intermediates of the coumarin antibiotic biosynthetic pathway. Org Lett 2003;5(18):3233–3236.

70. Woodyer RD, *et al.* New insight into the mechanism of methyl transfer during the biosynthesis of fosfomycin. Chem Commun 2007;43(4):359–361.

71. Tao J, *et al.* Biocatalysis retrosynthesis. In: Liese A, Seelbach K, Wandrey C, editors. Industrial biotransformation. 2nd ed. Weinheim: Wiley-VCH Verlag GmbH & Co. KGaA; 2006. pp. 63–91.

72. Greenberg WA, *et al.* Development of an efficient, scalable, aldolase-catalyzed process for enantioselective synthesis of statin intermediates. Proc Natl Acad Sci U S A 2004;101(16):5788–5793.

73. Jennewein S, *et al.* Directed evolution of an industrial biocatalyst: 2-deoxy-D-ribose 5-phosphate aldolase. Biotechnol J 2006;1(5):537–548.

74. Hu S, Tao J. Xie Z. Process for producing atorvastatin, pharmaceutically acceptable salts thereof and intermediates thereof. WO patent WO/2006/134,482. 2006.

75. Davis CS, Fox RJ, Huisman GW, Gavrilovic V, Newman LM. Halohydrin dehalogenases and related polynucleotides. US patent 20050272064. 2005.

76. DeSantis G, *et al.* Creation of a productive, highly enantioselective nitrilase through gene site saturation mutagenesis (GSSM). J Am Chem Soc 2003;125(38): 11476–11477.

77. Bergeron S, *et al.* Nitrilase-catalysed desymmetrisation of 3-hydroxyglutaronitrile: preparation of a statin side-chain intermediate. Org Process Res Dev 2006;10(3): 661–665.

78. Martinez CA, *et al.* Development of a chemoenzymatic manufacturing process for pregabalin. Org Process Res Dev 2008;12(3):392–398.

79. Hu S, Martinez C, Tao S, Tully W, Kelleher P, Dumond Y. US20051222. 2005.

80. Xie Z, *et al.* Cloning and optimization of a nitrilase for the synthesis of (3S)-3-cyano-5-methyl hexanoic acid. J Mol Catal B Enzym 2006;41(3–4):75–80.

81. Tao J, Zhao L, Ran N. Recent advances in developing chemoenzymatic processes for active pharmaceutical ingredients. Org Process Res Dev 2007;11(2):259–267.

82. Hu S, *et al.* Efficient enzymatic process for the production of (2S)-4,4-difluoro-3,3-dimethyl-N-Boc-proline, a key intermediate in the synthesis of HIV protease inhibitors. Org Process Res Dev 2006;10(3):650–654.

83. Yazbeck D, *et al.* Enzymatic process for the synthesis of Z/E-(2R, 5R)-bicyclo [3.2. 0] hept-6-ylidene-acetate: solvent effect and NMR study. Org Process Res Dev 2006;10(3):655–660.

84. Anderson BA, *et al.* Application of a practical biocatalytic reduction to an enantioselective synthesis of the 5H-2, 3-benzodiazepine LY300164. J Am Chem Soc 1995; 117(49):12358–12359.

85. Liese A, Seelbach K, Wandrey C, editors. Industrial biotransformations. 2nd ed. Weinheime: Wiley-VCH, Verlag GmbH & Co. KGaA; 2006.

86. Tao J, McGee K. Development of a continuous enzymatic process for the preparation of (R)-3-(4-fluorophenyl)-2-hydroxy propionic acid. Org Process Res Dev 2002;6(4):520–524.

87. DeSantis G, *et al.* An enzyme library approach to biocatalysis: development of nitrilases for enantioselective production of carboxylic acid derivatives. J Am Chem Soc 2002;124(31):9024–9025.

88. Kazlauskas RJ. Engineering a multipurpose catalyst. Nat Chem Biol 2006;2(10): 514–515.

89. Morley KL, Kazlauskas RJ. Improving enzyme properties: when are closer mutations better? Trends Biotechnol 2005;23(5):231–237.

90. Chandran SS, *et al.* Phosphoenolpyruvate availability and the biosynthesis of shikimic acid. Biotechnol Prog 2003;19(3):808–814.

91. Draths K, Knop DR, Frost J. Shikimic acid and quinic acid: replacing isolation from plant sources with recombinant microbial biocatalysis. J Am Chem Soc 1999;121(7):1603–1604.

92. Hoekstra MS, *et al.* Chemical development of CI-1008, an enantiomerically pure anticonvulsant. Org Process Res Dev 1997;1(1):26–38.

93. Burk MJ, *et al.* An enantioselective synthesis of (S)-(+)-3-aminomethyl-5-methylhexanoic acid via asymmetric hydrogenation. J Org Chem 2003;68(14): 5731–5734.

94. Noorden RV. Keep it Green. Chemistry World. Available at http://www.rsc.org/chemistryworld/news/2007/may/25050701.asp. 2007 May 25.

95. Dunn P. 13th Annual Green Chemistry and Engineering Conference, Maryland, 2009.

CHAPTER 4

REACTION EFFICIENCIES AND GREEN CHEMISTRY METRICS OF BIOTRANSFORMATIONS

ROGER A. SHELDON
Department of Biotechnology, Delft University of Technology, The Netherlands

4.1 INTRODUCTION TO REACTION EFFICIENCY AND SELECTIVITY IN ORGANIC SYNTHESIS

The roots of organic synthesis are generally traced back to Wöhler's synthesis of the natural product urea from ammonium isocyanate in 1828. This marked the demise of the *vis vitalis* (vital force) theory, which maintained that a substance produced by a living organism could not be produced synthetically. The discovery had monumental significance because it showed that, in principle, organic compounds are amenable to synthesis in the laboratory. Another landmark in the development of organic synthesis was the preparation of the first synthetic dye, mauveine (aniline purple), by Perkin in 1856, generally regarded as the first industrial organic synthesis [1]. Interestingly, Perkin's goal was to synthesize the antimalarial drug, quinine, by oxidation of *N*-allyl toluidine with potassium dichromate. This serendipitous discovery marked the advent of the synthetic dyestuffs industry based on coal tar, a waste product from steel manufacture. The development of mauveine was followed by the industrial synthesis of the natural dyes alizarin and indigo by Graebe and Liebermann in 1868 and Adolf Baeyer in 1870, respectively. The commercialization of these dyes marked the demise of their agricultural production and the birth of a science-based, predominantly German, chemical industry.

The modern pharmaceutical and allied fine chemical industries evolved as spin-offs of the synthetic dyestuffs industry. At first, these were relatively simple

Biocatalysis for Green Chemistry and Chemical Process Development, First Edition.
Edited by Junhua (Alex) Tao and Romas Kazlauskas.

molecules but in the ensuing decades the synthetic targets became increasingly complicated, as exemplified by the introduction of the semisynthetic beta-lactam antibiotics and the steroid hormones in the 1940s and in the 1950s. Indeed, another landmark in organic synthesis was the regio- and enantiospecific microbial hydroxylation of progesterone to 11α-hydroxyprogesterone by Peterson and Murray [2] at the Upjohn Company, which enabled the commercially viable synthesis of cortisone and paved the way for the subsequent commercial success of steroid hormones. It marked the beginning of the development of drugs of ever increasing molecular complexity and, to meet this challenge, synthetic organic chemists aspired to achieve increasing levels of sophistication. For example, introduction of the anticancer drug, taxol, into medical practice in the 1990s was enabled by the invention [3] of a commercially viable and sustainable semisynthesis from baccatin III, a constituent of the needles of the English yew, *Taxus baccata*. It is reasonable to conclude, therefore, that the success of the modern pharmaceutical industry is largely due to the remarkable achievements of organic synthesis over the last century. However, many of these time-honored and widely applicable synthetic methodologies were developed at a time when the toxic properties of many reagents and solvents were not known and waste minimization and sustainability were not significant issues.

In the last two decades, however, attention has been drawn to the fact that many of the widely used classical synthetic methodologies generate copious amounts of waste and there has been increasing pressure to minimize or, preferably, eliminate this detritus. An illustrative example is provided by the manufacture of phloroglucinol, a reprographic chemical and pharmaceutical intermediate [4]. Up until the mid-1980s, it was produced mainly from 2,4,6-trinitrotoluene (TNT) by the process shown in Figure 4.1, a perfect example of vintage nineteenth-century organic chemistry that would have appealed to Perkin.

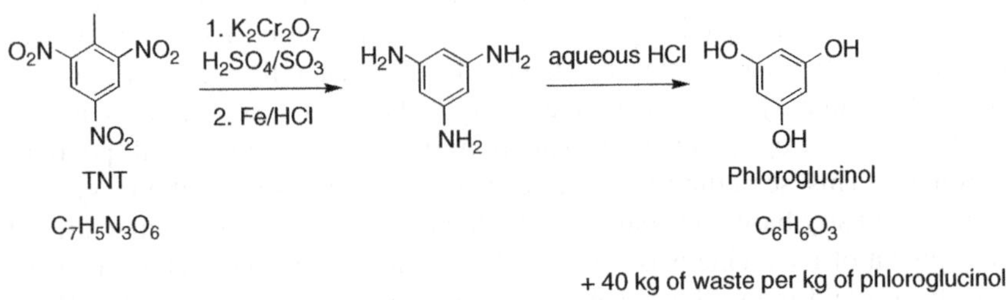

Stoichiometry:

$$C_7H_5N_3O_6 + K_2Cr_2O_7 + 5H_2SO_4 + 9Fe + 21HCl \longrightarrow$$

$$\underset{126}{C_6H_6O_3} + \underset{392}{Cr_2(SO_4)_3} + \underset{272}{2KHSO_4} + \underset{1143}{9FeCl_2} + \underset{161}{3NH_4Cl} + \underset{44}{CO_2} + \underset{144}{8H_2O}$$

Atom economy = 126/2282 = about 5%

Figure 4.1 Manufacture of phloroglucinol in 1980.

This process affords phloroglucinol in an overall yield of >90% over the three reaction steps and, according to the classical concepts of selectivity and reaction efficiency, would generally be considered to be a selective and efficient process. However, for every kilogram of phloroglucinol, this process generates about 40 kg of solid waste, containing $Cr_2(SO_4)_3$, NH_4Cl, $FeCl_2$, and $KHSO_4$. It was eventually discontinued because of the prohibitive costs associated with the disposal of this chromium-containing waste. A cursory examination of the stoichiometric equation (Figure 4.1) of the overall process, something rarely done, reveals the source of this waste: large amounts of inorganic salts derived from the stoichiometric inorganic reagents used. The reaction stoichiometry predicts the formation of about 20 kg of waste per kilogram of phloroglucinol, assuming 100% chemical yield and exact stoichiometric quantities of the various reagents. In practice, an excess of the oxidant and reductant, and a large excess of sulfuric acid, which has to be subsequently neutralized with base, are used and the isolated yield of phloroglucinol is less than 100%. This explains the observed 40 kg of waste per kilogram of desired product.

This poignant example of wasteful use of raw materials led us to develop the E factor concept (see below) for assessing the environmental impact of chemical manufacturing processes. Indeed, in the light of this, and many other examples of processes that generate copious amounts of waste, it became clear that a paradigm shift was needed from traditional concepts of reaction efficiency and selectivity that focus largely on chemical yield, to one that is motivated by maximization of raw materials utilization, elimination of waste, and avoiding the use of toxic and/or hazardous substances. Synthetic organic chemists were well acquainted with terms such as reaction selectivity (yield divided by conversion) and chemo-, regio-, stereo-, and enantioselectivity. In contrast, prior to the mid-1990s, they were not used to considering what we could call the atom selectivity of a reaction, that is, how much of the mass of the reactants actually ends up in the product, the rest being, by definition, waste. This selectivity is of the utmost importance in the context of environmental impact and recognition of this fact led to the concept of atom economy (see below).

4.2 GREEN CHEMISTRY AND SUSTAINABILITY

The World Commission for Environment and Development was founded in 1983 by the United Nations and was given the task of preparing a report on the perspectives of long-term, sustainable, and environmentally friendly development on a global scale by 2000 and after. This culminated in the publication, four years later, of the report, *Our Common Future* [5], also known as the *Brundtland Report* after the Prime Minister of Norway, who was Chairman of the commission at that time. The report defined sustainable development as

> *Meeting the needs of the present generation without compromising the ability of future generations to meet their own needs.*

In the next two decades, the concept of sustainability became the focus of considerable attention both in industry and in society as a whole.

The term "Green Chemistry" was coined in the early 1990s by Anastas *et al.* [6] of the US Environmental Protection Agency (EPA). In 1993, the EPA officially adopted the name "US Green Chemistry Program," which has served as a focal point for activities within the United States, such as the Presidential Green Chemistry Challenge Awards and the annual Green Chemistry and Engineering Conference. This does not mean that research on green chemistry did not exist before the early 1990s; merely, it did not have the name. The guiding principle is *benign by design* of both products and processes [7]. This concept is embodied in the 12 Principles of Green Chemistry [6], which can be paraphrased as

1. Waste prevention instead of remediation
2. Atom efficiency
3. Less hazardous/toxic chemicals
4. Safer products by design
5. Innocuous solvents and auxiliaries
6. Energy efficient by design
7. Preferably renewable raw materials
8. Shorter syntheses (avoid derivatization)
9. Catalytic rather than stoichiometric reagents
10. Design products for degradation
11. Analytical methodologies for pollution prevention
12. Inherently safer processes

Subsequently, Anastas and Zimmerman [8] proposed the 12 principles of green engineering, which embody the same underlying features—conserve energy and resources and avoid waste and hazardous materials—as those of green chemistry, but from an engineering viewpoint. More recently, a mnemonic, PRODUCTIVELY, was proposed by Poliakoff *et al.* [9], which captures the spirit of the 12 principles of green chemistry:

P prevent wastes
R renewable materials
O omit derivatization steps
D degradable chemical products
U use of safe synthetic methods
C catalytic reagents
T temperature, pressure ambient
I in-process monitoring
V very few auxiliary substrates
E E-factor, maximize feed in product

L low toxicity of chemical products
Y yes, it is safe

Alternatively, the essence can be reduced to a working definition in a single sentence [4].

Green chemistry efficiently utilizes (preferably renewable) raw materials, eliminates waste and avoids the use of toxic and/or hazardous reagents and solvents in the manufacture and application of chemical products.

Raw materials include the source of the energy used in the process, as this leads to waste in the form of carbon dioxide emissions. Green chemistry eliminates waste at source, that is, it is primary pollution prevention rather than waste remediation (end-of-pipe solutions), as described by the first principle of green chemistry: prevention is better than cure. In the last 15 years, the concept of green chemistry has been widely embraced in both industrial and academic circles. One could say that sustainability is our ultimate common goal and green chemistry is a means to achieving it.

Having defined what Green Chemistry is, we need to be able to compare processes (and products) on the basis of their greenness. We note, however, that there is no absolute greenness; one process is greener than another and, as with beauty, greenness is in the eye of the beholder. Nonetheless, as Lord Kelvin remarked, "to measure is to know," and appropriate green metrics are a prerequisite for progress toward a meaningful comparison of the greenness of processes.

4.3 THE METRICS OF GREEN CHEMISTRY: ATOM ECONOMY AND THE ENVIRONMENTAL FACTOR

It is now widely accepted that two useful measures of the potential environmental impact of chemical processes are the *E factor* [10–15], defined as the mass ratio of the waste to the desired product and the *atom economy* [16,17], calculated by dividing the molecular weight of the desired product by the sum of the molecular weights of all the substances produced in the stoichiometric equation.

A knowledge of the stoichiometric equation enables prediction of, without performing any experiments, the theoretical amount of waste that can be expected. Our experience with the phloroglucinol process (see above) led us to use what we called *atom utilization* to quickly assess the environmental acceptability of alternative processes to a particular chemical [18]. It was a logical elaboration of the concepts of *syn gas utilization* [19] and *oxygen availability* in different oxidants [20]. However, *atom economy* (*AE*), introduced by Trost in 1991 [16], has become the widely accepted terminology although *atom efficiency* (also abbreviated as AE) is also used. In Figure 4.2, we compare the AE of the classical chlorohydrin route to propylene oxide with that of oxidation with the green oxidant hydrogen peroxide, where the coproduct is water [21].

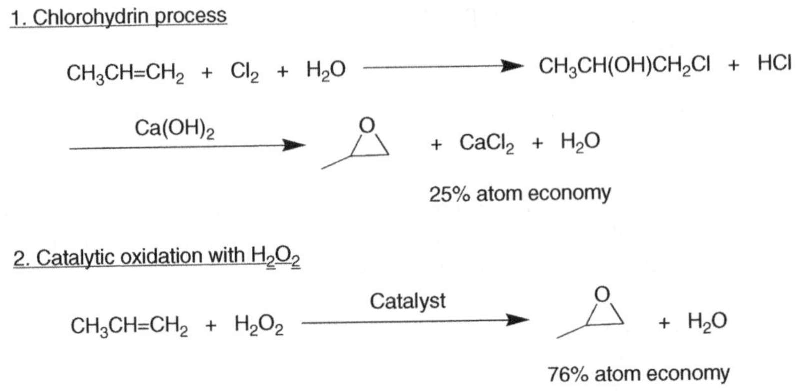

Figure 4.2 Atom economies of processes for the production of propylene oxide.

It should be emphasized that atom economy is a theoretical number based on a chemical yield of 100% of theoretical. It further assumes that reactants are used in exactly stoichiometric amounts and it disregards substances, such as solvent and acids or bases used in workup, which do not appear in the stoichiometric equation.

The E factor, in contrast, is the actual amount of waste produced in the process, defined as everything but the desired product. It takes the chemical yield into account and includes reagents, solvent losses, all process aids, and in principle, even the energy required (although this is often difficult to quantify). We generally excluded water from the calculation of the E factor as we thought that its inclusion would lead to exceptionally high E factors in many cases and make meaningful comparisons of processes difficult. For example, when considering an aqueous waste stream, only the inorganic salts and organic compounds contained in the water are counted, while the water is excluded. However, there is a definite trend in the pharmaceutical industry toward the inclusion of water in the E factor and we note that water usage can be a crucial issue in biomass conversion and in fermentation processes in general (see below).

A higher E factor means more waste, and consequently, greater negative environmental impact. The ideal E factor is zero. Put quite simply, it is kilograms (of raw materials) in, minus kilograms of desired product, divided by kilograms of the product out. It can be easily calculated from a knowledge of the number of tons of raw materials purchased and the number of tons of product sold, for a particular product or a production site or even a whole company. We note that this method of calculation will automatically exclude water used in the process but not water formed. The sheer magnitude of the waste problem in chemical manufacture is readily apparent from a consideration of typical E factors in various segments of the chemical industry shown in Table 4.1, which we published almost two decades ago.

It is also clear that the E factor increases substantially on going downstream from bulk chemicals to fine chemicals and pharmaceuticals. Indeed, when we began an inventory of the E factors of fine chemicals manufacture back in the

TABLE 4.1 The E factor

Industry Segment	Volume (tons/annum)[a]	E factor (kg Waste/kg Product)
Bulk chemicals	10^4-10^6	<1–5
Fine chemical industry	10^2-10^4	5 to >50
Pharmaceutical industry	$10-10^3$	25 to >100

[a]Annual production of the product world wide or at a single site.

late 1980s, it soon became clear that tens of kilograms of waste per kilogram of product were more the norm than the exception and in pharma they were even higher. On the one hand, this is a consequence of the fact that the target compounds are more complicated molecules compared to bulk chemicals and hence, their production involves multistep syntheses, which can be expected to generate more waste. On the other hand, it is a direct consequence of the more widespread use of stoichiometric reagents in these industry segments. In contrast, in bulk manufacture of chemicals, the production volumes are enormous and the use of many stoichiometric reagents is, for purely economic reasons, generally prohibitive.

The E factor has been widely adopted by the chemical industry and, in particular, by the pharmaceutical industry [22] as a useful barometer for gauging the environmental impact of manufacturing processes [23]. To quote from a recent article [24] "Another aspect of process development mentioned by all pharmaceutical process chemists who spoke with C&EN, is the need for determining an E factor." The Green Chemistry Institute Pharmaceutical Round Table has conducted an inventory of the waste generated in processes used by its members (see http://www.epa.gov/greenchemistry/pubs/gcinstitute.html). The latter include several leading pharmaceutical companies (Eli Lilly, Glaxo Smith Kline, Pfizer, Merck, AstraZeneca, Schering Plough, and Johnson & Johnson). The aim was to use this data to drive the greening of the pharmaceutical industry.

Other metrics have also been proposed for measuring the environmental acceptability of processes (Figure 4.3) [25–27]. They can be categorized into two types: (i) metrics that constitute a refinement of the AE concept that is, those based on the stoichiometric equation of the reaction concerned and (ii) metrics that are variations on the theme of the E factor, that is, they address the actual amount of waste formed in the process. As examples of the former, Constable *et al*. [28] at Glaxo Smith Kline (GSK) introduced the terms *reaction mass efficiency* (RME) and *carbon efficiency* (CE). RME is defined as the mass of product obtained divided by total mass of the reactants in the stoichiometric equation expressed as a percentage. It is a refinement of AE that takes into account the chemical yield of the product and the actual quantities of reactants used. A disadvantage of RME compared to AE is that it cannot be used for a quick analysis of different processes before any experimental work is performed. CE is similar to RME but takes only carbon into account, that is, it is the mass of carbon in the product obtained divided by the total mass of carbon present in the reactants.

E factor

$$E = \frac{\text{Total mass of waste}}{\text{Mass of final product}}$$

Atom efficiency (AE)

$$AE\ (\%) = \frac{\text{Molecular weight of product} \times 100}{\Sigma\ \text{Molecular weight of reactants}}$$

Mass intensity (MI)

$$MI = \frac{\text{Total mass used in a process}}{\text{Mass of product}}$$

Reaction mass efficiency (RME)

$$RME(\%) = \frac{\text{Mass of product C} \times 100}{\text{Mass of A} + \text{Mass of B}}$$

Mass productivity (MP)

$$MP = \frac{\text{Mass of product}}{\text{Total mass used in process}}$$

Carbon efficiency (CE)

$$CE(\%) = \frac{\text{Amount of carbon in product} \times 100}{\text{Total carbon present in reactants}}$$

Effective mass yield (EMY)

$$EMY(\%) = \frac{\text{Mass of product} \times 100}{\text{Mass of hazardous reagents}}$$

Figure 4.3 Green metrics.

The GSK group [29] also proposed, as an example of the second category, the use of *mass intensity* (MI), defined as the total mass of materials used in a process divided by the mass of the product obtained expressed as a percentage, that is, MI = E factor + 1 and the ideal MI is 1 as against 0 for the E factor. The same authors also suggested the use of so-called *mass productivity*, which is the reciprocal of the MI. Hudlicky *et al*. [30] proposed an analogous metric: the *effective mass yield* (EMY), which is defined as the mass of the desired product divided by the total mass of the nonbenign reactants used in its preparation. As proposed, it does not include the so-called environmentally benign compounds, such as NaCl and acetic acid. This is questionable as the environmental impact of such substances is very volume dependent. Moreover, defining nonbenign is difficult and arbitrary and it was concluded [28], therefore, that EMY suffers from a lack of definitional clarity.

Summarizing, it is our considered opinion that none of these alternative metrics offers any particular advantage over atom economy and the E factor for assessing how wasteful a process is. The former is a quick tool that can be used before conducting any experiment and the latter is a measure of the total waste that is actually formed in practice. Thus, AE and the E factor are complementary green metrics. This can be illustrated by reference to the phloroglucinol process referred to earlier. From the stochiometry of the overall process (Figure 4.1), we can calculate the AE as about 5%. This would predict an E factor of about 20 but in practice, the E factor is 40. This is because the overall yield is not 100% and a molar excess of the various reactants is used and the sulfuric acid (which is used in large excess) has to be neutralized with base in the workup.

This example is not atypical; the high E factors of processes for the manufacture of fine chemicals and pharmaceuticals, and even some bulk chemicals, consist primarily of inorganic salts such as sodium chloride, sodium sulfate, and ammonium sulfate formed in the reaction or in subsequent neutralization steps.

4.4 THE NATURE OF THE WASTE

All of the metrics discussed above take into account only the mass of the waste generated. However, the environmental impact of this waste is determined not only by its amount but also by its nature. One kilogram of sodium chloride is obviously not equivalent to 1 kg of a chromium salt. Hence, we introduced [13] the term "environmental quotient," EQ, obtained by multiplying the E factor by an arbitrarily assigned unfriendliness quotient, Q. For example, one could arbitrarily assign a Q value of 1 to NaCl and, say, 100–1000 to a heavy metal salt, such as chromium, depending on its toxicity, ease of recycling, and so on. The magnitude of Q is obviously debatable and difficult to quantify but importantly, "quantitative assessment" of the environmental impact of chemical processes is, in principle, possible [31]. It is also worth noting that Q for a particular substance can be influenced by both the production volume and the location of the production facilities. For example, the generation of 100–1000 tons per annum of sodium chloride is unlikely to present a waste problem but 10,000 tons per annum, in contrast, may already present a disposal problem, thus warranting an increase in Q. Ironically, when very large quantities of sodium chloride are generated, the Q value could decrease again as recycling by electrolysis becomes a viable proposition, for example,. in propylene oxide manufacture *via* the chlorohydrin route (see above). Thus, generally speaking, the Q value of a particular waste will be determined by its ease of disposal or recycling. We also mention that, in our experience, organic waste is, generally speaking, easier to dispose of than inorganic waste. This is important when we consider the green metrics of biocatalytic processes (see below).

Another approach to assessing the environmental impact and thereby the sustainability of both products and processes in general, is *Life Cycle Assessment* (*LCA*) [32,33]. This involves the evaluation of products and processes within defined domains, for example, cradle-to-gate, cradle-to-grave, and gate-to-gate, on the basis of quantifiable environmental impact indicators, such as energy usage, global warming, ozone depletion, acidification, eutrophication, smog formation, and ecotoxicity, in addition to the waste generated.

4.5 THE ROLE OF CATALYSIS AND ALTERNATIVE REACTION MEDIA IN WASTE MINIMIZATION

As noted above, the waste generated in the manufacture of fine chemicals and pharmaceuticals is primarily a consequence of the use of stoichiometric inorganic and (organic) reagents that are not (or partially) incorporated into the product. A cursory examination of the phloroglucinol process reveals typical examples: oxidations with chromium(VI) reagents and other inorganic oxidants such as permanganate and manganese dioxide, and stoichiometric reductions with metals (Na, Mg, Zn, and Fe) and metal hydride reagents ($LiAlH_4$ and $NaBH_4$). Similarly, a multitude of reactions, for example, sulfonations, nitrations, halogenations, diazotizations, and Friedel–Crafts acylations, employing stoichiometric amounts of

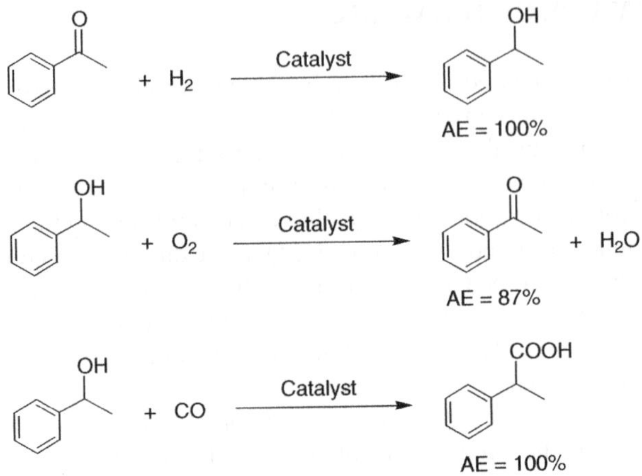

Figure 4.4 Atom-efficient catalytic processes.

mineral acids (H_2SO_4, HF, H_3PO_4) and Lewis acids ($AlCl_3$, $ZnCl_2$, BF_3) are major sources of waste. The solution is evident: substitution of antiquated stoichiometric methodologies with cleaner catalytic alternatives [11–15, 34]. For example, catalytic hydrogenation, oxidation, and carbonylation (Figure 4.4) are highly atom-efficient, low-salt processes.

The generation of copious amounts of inorganic salts can similarly be largely circumvented by replacing stoichiometric mineral acids, such as H_2SO_4, Lewis acids, and stoichiometric bases, such as NaOH and KOH, with recyclable solid acids and bases, preferably in catalytic amounts [35]. Indeed, recyclable solid catalysts have an important role to play in waste minimization of many basic reactions in industrial organic synthesis [36].

Another major source of waste is solvent losses, which generally end up in the atmosphere or in ground water. Indeed, solvent losses are a major contributor to the high E factors of pharmaceutical manufacturing processes [37]. Furthermore, health and/or safety issues associated with many traditional organic solvents, such as chlorinated hydrocarbons, have led to their use being severely curtailed. Consequently, pharmaceutical companies are focusing their effort on minimizing solvent use and in replacing many traditional organic solvents, such as chlorinated and aromatic hydrocarbons, by more environmentally friendly alternatives. In our original inventory of E factors of various processes, we assumed, if details were not known, that solvents would be recycled by distillation and that this would involve a 10% loss. However, the organic chemist's penchant for using different solvents for the various steps in multistep syntheses makes recycling difficult owing to cross contamination.

The best solvent is no solvent but if a solvent is needed, it should be safe to use and there should be provisions for its efficient removal from the product and reuse. Various nonconventional reaction media have been intensely studied in recent years, including *water* [38], *supercritical* CO_2 [39], *fluorous biphasic*

[40], *and ionic liquids* [41], alone or in liquid–liquid biphasic combinations. The use of water and supercritical carbon dioxide as reaction media fits with the current trend toward the use of renewable, biomass-based raw materials, which are ultimately derived from carbon dioxide and water.

As noted above, the relatively large E factors in pharmaceuticals manufacture can also be ascribed to the multistep syntheses that are widely used in this industry segment. This led Wender [42] to advocate the development of more step economic syntheses. This can often be achieved by replacing classical multistep syntheses with shorter catalytic alternatives. The ultimate is the development of catalytic cascade processes, whereby several catalytic steps are integrated in step-economical, one-pot procedures without the need for isolation of intermediates [43]. This is truly emulating the elegant orchestration of enzymatic steps in metabolic pathways in a living cell. Such "telescoping" of multistep syntheses into catalytic cascades has several advantages—fewer unit operations, less solvent, low and reactor volume, shorter cycle times, higher volumetric and space-time yields, and less waste (lower E factor)—that afford substantial economic and environmental benefits. Furthermore, coupling of reactions can be used to drive equilibria toward the product, thereby avoiding the need for excess reagents. On the other hand, there are problems to be overcome: catalysts are often incompatible with one another (e.g., an enzyme and a metal catalyst), rates and optimum conditions can be very different, and catalyst recovery and recycle complicated. Nature solves the problem of compatibility by compartmentalization of enzymes in different parts of the cell, which suggests that the key to developing catalytic cascades is immobilization of the different catalysts.

To summarize, since waste can be attributed mainly to the use of stoichiometric reagents and multistep syntheses, accompanied by substantial solvent losses, the solution to the waste problem is evident: substitution of stoichiometric reagents with atom and step-economic catalytic methodologies in alternative reaction media [44,45]. As we and others have noted elsewhere [10,46], this is true elegance and efficiency in organic synthesis.

4.6 BIOCATALYSIS AND GREEN CHEMISTRY

Biocatalysis has many attractive features in the context of green chemistry. Reactions are performed under mild conditions (physiological pH and ambient temperature and pressure) with a biodegradable catalyst (an enzyme) that is derived from renewable resources and in an environmentally compatible solvent (water). Furthermore, reactions of multifunctional molecules proceed with high activities and chemo-, regio-, and stereoselectivities generally without the need for functional group activation, protection, and deprotection steps required in traditional organic syntheses. This affords processes that are more step economical, generate less waste, and are therefore, both environmentally and economically more attractive than conventional routes. As a direct consequence of the high regio and stereoselectivities, coupled with milder reaction conditions, they often afford products in higher quality than traditional chemical or chemocatalytic processes. For example, they avoid the problem of contamination with traces of (noble) metals, which is often a serious issue in the manufacture of pharmaceuticals. Finally, enzymatic processes (but not fermentations) can be conducted in standard multipurpose batch reactors and hence, do not require any extra investment.

Advances in biotechnology have, in recent decades, provided the basis for the widespread application of biocatalysis in industrial organic synthesis. Recombinant DNA techniques have made it possible, in principle, to produce virtually any enzyme for a commercially acceptable price. Modern protein engineering techniques, such as *in vitro* evolution [47], have enabled the manipulation of enzymes such that they exhibit the desired substrate specificity, activity, stability, pH profile, and so on. This has made it possible to optimize the enzyme to fit the predefined optimum process, that is, genuinely benign by design. Furthermore, the development of effective immobilization techniques has paved the way for optimizing the storage and operational stability and the recovery and recycling of enzymes [48]. Moreover, since most biocatalytic processes are performed under roughly the same conditions of (ambient) temperature and pressure, it is eminently feasible to integrate multiple steps into enzymatic cascade processes [49]. Coimmobilization of two or more enzymes then affords multifunctional solid biocatalysts capable of catalyzing such cascade processes [50].

Indeed, there is a clear trend in the manufacture of pharmaceuticals [51], agrochemicals [52], flavors and fragrances [53], and cosmetic ingredients [54] toward the development of biocatalytic processes as greener, more economical alternatives to less efficient and more costly classical syntheses. A pertinent example of the former is the recently developed chemoenzymatic process (Figure 4.5) for the manufacture of pregabalin [55], the active ingredient of the CNS drug Lyrica. The new route afforded a dramatic improvement in process efficiency by setting the stereocentre early in the synthesis (the golden rule of chirotechnology [56]) and enabling the facile racemization and reuse of the wrong enantiomer. The key enzymatic step was conducted at a 3 M substrate concentration that is at a quite staggering 765 g/L. Moreover, organic solvent usage was dramatically reduced in a largely aqueous process. Compared to the first generation manufacturing process, the new process afforded a higher yield and a fivefold reduction in the E factor from 86 to 17.

4.7 GREEN METRICS OF BIOTRANSFORMATIONS

Biocatalytic processes can be performed with isolated enzymes or as whole cell biotransformations. Isolated enzymes have the advantage of not being contaminated with other enzymes present in the cell. On the other hand, the use of whole cells is less expensive, as it avoids separation and purification of the enzyme. In the case of dead cells, the E factors of the two methods are essentially the same; the waste cell debris is separated before or after the biotransformation. In contrast, when growing microbial cells are used, that is, in fermentation processes, substantial amounts of biomass can be generated. We note, however, that the

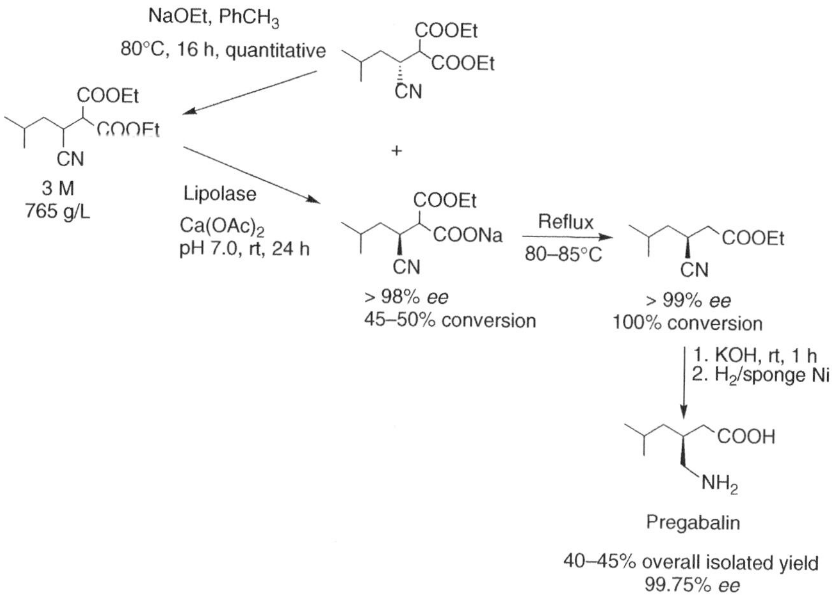

Figure 4.5 Chemoenzymatic process for Pregabalin.

waste biomass is generally easy to dispose of, for example, as animal feed or can, in principle, be used as a source of energy for the process. Many fermentation processes also involve the formation of copious amounts of inorganic salts that may even be the major contributor to waste. To our knowledge, there are no reported E factors for fermentation processes. This would seem to be a hiatus that needs to be filled.

Mass balances of a few fermentation processes have been documented by Petrides [57] from which E factors can be calculated. For example, the E factor for the bulk fermentation product, citric acid, is 1.4, which compares well with the E factor range of <1–5, typical of bulk petrochemicals (Table 4.1). Roughly 75% of the waste is accounted for by calcium sulfate. During the process, calcium hydroxide is added to control the pH, affording calcium citrate, which is reacted with sulfuric acid to produce citric acid and calcium sulfate. Inclusion of water in the calculation afforded an E factor of 17.

Another product of biomass conversion that is currently the subject of much debate is second-generation bioethanol, that is, ethanol derived from lignocellulose (wood) as feedstock. It was recently reported [58] that the E factor of cellulosic ethanol is a whopping 42. However, the major components of this waste are water (36.8 kg/kg ethanol) and carbon dioxide (4.1 kg/kg ethanol). The remainder amounts to 1.1 kg/kg ethanol. The formation of 4 kg of molasses, 0.1 kg of furfural, and 0.2 kg of acetic acid as by-products was not included in the E factor calculation. It was further noted that a cellulosic ethanol plant processing 10,000 tons of lignocellulose feedstock per day to produce 870 tons of ethanol a day would generate 32 milion liters of wastewater daily, that is, water sufficient to supply a town of 300,000 inhabitants. This water contains several organic by-products, the concentrations of which have to be decreased to the ppm level or below in order to enable reuse of the water. This means that a rather sophisticated industrial waste water treatment is mandatory.

Small-volume fermentation processes for biopharmaceuticals can have very high E factors, even when compared with those observed in the production of small-molecule drugs, and water usage tends to be high. The fermentative production of recombinant human insulin [57], for example, involves an E factor of about 6600. The most important contributors to the waste are (in kilograms per kilogram insulin) urea (1692), acetic acid (1346), formic acid (968), phosphoric acid (713), guanidine hydrochloride (445), glucose (432), sodium chloride(430), acetonitrile (424), and sodium hydroxide (140). If water is included, the E factor becomes an astronomical 50,000.

Biotransformations involving the use of isolated enzymes, in contrast, tend to involve significantly higher substrate concentrations and combine a higher productivity with a lower water usage compared to fermentations (see the pregabalin process discussed above and the case studies in the next section).

4.8 CASE STUDIES OF ENZYMATIC VERSUS CLASSICAL CHEMICAL PROCESSES

The following two case studies involve (i) an enzymatic process for the synthesis of an intermediate for the cholesterol-lowering agent, Lipitor®, the first drug with annual worldwide sales exceeding $10 billion and (ii) an enzymatic process for the production of a fatty acid ester used in personal care products.

4.8.1 Enzymatic Synthesis of an Intermediate for Atorvastatin, the Active Ingredient of Lipitor

The development of a green-by-design, three-enzyme process for the synthesis of a key intermediate (Figure 4.6) in the manufacture of atorvastatin, the active ingredient of the cholesterol-lowering drug Lipitor, was recently described [59,60]. The process has been successfully commercialized on a multiton scale by Codexis. The first step involves the biocatalytic reduction of ethyl-4-chloroacetoacetate, using a ketoreductase (KRED) in combination with glucose and an NADP-dependent glucose dehydrogenase (GDH) for cofactor regeneration. The (*S*) ethyl-4-chloro-3-hydroxybutyrate product was obtained in 96% isolated yield and >99.5% *ee*. In the second step, a halohydrin dehalogenase (HHDH) was employed to catalyze a nucleophilic substitution of chloride by cyanide, using HCN at neutral pH and ambient temperature.

All previous manufacturing routes to the hydroxynitrile product (HN) involved, as the final step, a standard S_N2 substitution of halide with cyanide ion in alkaline solution at elevated temperatures. However, both substrate and product are base-sensitive and extensive by-product formation is observed. Moreover, the product is a high-boiling oil, and a troublesome high-vacuum

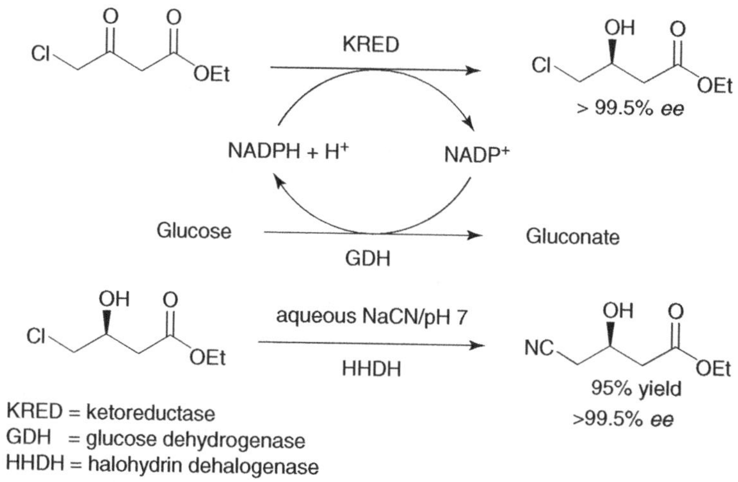

Figure 4.6 A two-step three-Enzyme Process for Atorvastatin Intermediate.

fractional distillation is required to recover a product of acceptable quality, resulting in further yield losses and waste.

The key to designing an economically and environmentally attractive process for HN was to develop a methodology for conducting the cyanation reaction under mild conditions at neutral pH, by employing the enzyme, HHDH. Combination of this with the synthesis of the chlorohydrin substrate by highly enantioselective KRED-catalyzed reduction of the corresponding keto ester, and cofactor regeneration with glucose/GDH, afforded an elegant two-step, three-enzyme process. Unfortunately, the wild-type KRED and GDH exhibited prohibitively low activities (Table 4.2) and large enzyme loadings were required to obtain an economically viable reaction rate. This led to troublesome emulsion formation in downstream processing. Thus, although the analytical yield was >99%, the recovered yield was only 85%. To enable a practical large-scale process, the enzyme

TABLE 4.2 Directed Evolution of a KRED/GDH Biocatalyst

Parameter	Process Design	Initial Process	Final Process
Substrate loading (g/L)	160	80	160
Reaction time (h)	<10	24	8
Biocatalyst loading (g/L)	<1	9	0.9
Isolated yield (%)	>90	85	95
Chemical purity (%)	>98 (GC)	>98	>98
% *ee* of product	>99.5	>99.5	>99.9
Phase separation time (min)	<10	>60	<1
Catalyst productivity (g product/g catalyst)	>160	9	178
Space-time yield (g product/L/day)	>380	80	480

loadings needed to be drastically reduced and this was achieved with *in vitro* evolution using the DNA shuffling technique [61] to improve the activity and stability of KRED and GDH while maintaining the nearly perfect enantioselectivity exhibited by the wild-type KRED.

Through several generations of DNA shuffling, GDH activity was improved by a factor of 13 and KRED activity by a factor of 7 while maintaining the enantioselectivity at >99.5%. With the improved enzymes the reaction was complete in 8 h with substrate loading increased to 160 g/L and substantially reduced enzyme loadings (Table 4.2) and consequently, with no emulsion problems. Phase separation required <1 min and provided the chlorohydrin in >95% isolated yield of >99.9% *ee*.

Similarly, the activity of the wild-type HHDH in the nonnatural cyanation reaction was extremely low and the enzyme exhibited severe product inhibition and poor stability under operating conditions. As a result of the large enzyme loadings, downstream processing was challenging. However, after many iterative rounds of DNA shuffling, the inhibition was largely overcome and the HHDH activity was increased more than 2500-fold compared to the wild-type enzyme (see Table 4.3).

The greenness of the process was assessed according to the 12 principles of green chemistry.

Principle 1—Waste prevention. The highly selective biocatalytic reactions afforded a substantial reduction in waste. In the final process, raw material is converted to a product in >90% isolated yield to a product that is more than 98% chemically pure with an enantiomeric excess of >99.9%. Furthermore, the avoidance of by-products obviates the need for further yield-sacrificing fractional distillation. The butyl acetate and ethyl acetate solvents, used in the extraction of the product from the aqueous layer in the first and second step, respectively, are recycled with an efficiency of 85%. The E factor (kilogram waste per kilogram product) for the overall process is 5.8 if process water is excluded (2.3 for the reduction and 3.5 for the cyanation). If process water is included, the E factor for the

TABLE 4.3 Evolution of the HHDH Biocatalyst

Parameter	Process Design	Initial Process	Final Process
Substrate loading (g/L)	>120	20	140
Reaction time (h)	8	72	5
Biocatalyst loading (g/L)	1.5	30	1.2
Isolated yield (%)	>90	67	92
Chemical purity (%)	>98	>98	>98
% *ee* of HN	>99.5	>99.5	>99.5
Phase separation time (min)	<10	>60	<1
Catalyst Productivity (g product/g catalyst)	80	0.7	117
Space-time yield (g product/L/day)	>360	7	672

TABLE 4.4 E Factor of the Process for Atorvastatin Intermediate

Waste	Quantity (kg/kg HN)	% contribution to E (Excluding Water)	% Contribution to E (Including Water)
Substrate losses (8%)	0.09	<2%	<1%
Triethanolamine	0.04	<1%	<1%
NaCl and Na_2SO_4	1.29	22%	about 7%
Na gluconate	1.43	about 25%	about 9%
BuOAc (85% recycle)	0.46	about 8%	about 0.3%
EtOAc (85% recycle)	2.50	about 43%	about 14%
Enzymes	0.023	<1%	<1%
NADP	0.005	0.1%	<0.1%
Water	12.250	—	67%
E factor	5.8 (18)[a]	—	—

[a]E factor in parentheses includes water.

whole process is 18 (6.6 for the reduction and 11.4 for the cyanation). The main contributors to the E factor, as shown in Table 4.4, are solvent (EtOAc and BuOAc) losses (51%), sodium gluconate (25%), and NaCl and Na_2SO_4 (combined about 22%). The three enzymes and the NADP cofactor account for <1% of the waste. The main waste streams are aqueous and directly biodegradable.

Principle 2—Atom economy. The use of glucose as the reductant for cofactor regeneration is cost effective but not particularly atom efficient. The atom economy is only 45%. However, glucose is a renewable resource and the gluconate coproduct is fully biodegradable.

Principle 3—Less hazardous chemical syntheses. The reduction reaction uses starting materials that pose no toxicity to human health or the environment. It avoids the use of potentially hazardous hydrogen and heavy metal catalysts throughout the process, thus obviating concern for their removal from waste streams and/or contamination of the product. While cyanide must be used in the second step, as in all practical routes to HN, it is used more efficiently (higher yield) and under less harsh conditions compared to previous processes.

Principle 4—Design safer chemicals. This principle is not applicable as the HN product is the commercial starting material for atorvastatin.

Principle 5—Safer solvents and auxiliaries. Safe and environmentally acceptable butyl acetate is used, together with water, as the solvent in the biocatalytic reduction reaction and extraction of the HN product; no auxiliaries are used.

Principles 6 and 9—Design for energy efficiency and catalysis. In contrast to previous processes, which employ elevated temperatures for the cyanation step and high-pressure hydrogenation for the reduction step, both steps in the process in Figure 4.6 are very efficient biocatalytic transformations. The reactions are run at or close to ambient temperature and pressure and pH 7 and the very high energy demands of high-vacuum distillation are dispensed with altogether, resulting in substantial energy savings. The turnover numbers for the different enzymes are $> 10^5$ for KRED and GDH and $> 5 \times 10^4$ for HHDH.

Principles 7 and 10—Use of renewable feedstocks and design for degradation. The enzyme catalysts and the glucose cosubstrate are derived from renewable raw materials and are completely biodegradable. The by-products of the reaction are gluconate, NADP (the cofactor that shuttles reducing equivalents from GDH to KRED), residual glucose, enzyme, and minerals, and the waste water is directly suitable for biotreatment.

Principle 8—Reduce derivatization. The process avoids derivatization steps, that is, it is step economic and involves fewer unit operations than earlier processes, most notably by obviating the trouble-prone product distillation or the bisulfite-mediated separation of dehydrated by-products.

Principles 11 and 12—Real time analysis for pollution prevention and inherently safer chemistry. The reactions are run in pH-stat mode at neutral pH by computer-controlled addition of base. Gluconic acid generated in the first reaction is neutralized with aqueous NaOH and HCl generated in the second step is neutralized with aqueous NaCN, regenerating HCN ($pK_a \sim 9$) *in situ*. The pH and the cumulative volume of the added base are recorded in real time. Feeding NaCN on demand minimizes the overall concentration of HCN, affording an inherently safer process.

In short, this process provides an excellent example of a *benign-by-design* biocatalytic process for the synthesis of an important pharmaceutical intermediate, the successful commercialization of which has been enabled by the employment of modern protein engineering to optimize enzyme performance. Codexis was awarded a Presidential Green Chemistry Challenge Award in 2006 for the development of this process.

4.8.2 Enzymatic Production of an Emollient Ester

Biotechnological processes are being widely applied in the cosmetics ingredients industry, where the label "natural" can add value to a product. The application of biocatalysis can be beneficial even with relatively simple products such as emollients based on fatty acid esters. For example, in a recent report [62], a classical chemocatalytic esterification was compared with an enzymatic equivalent for the industrial scale synthesis of the widely used emollient ester, myristyl myristate (Figure 4.7).The former involved the use of tin(II) oxalate as a catalyst at 240°C for 4 h and the latter employed an immobilized form of *Candida antarctica* lipase B, Novozyme 435, at 60°C for 12 h.

The atom economy of the process is 96% and the E factor is <0.1 in both cases (even when waste water is included). Consequently, the environmental impacts of the two processes were compared in a cradle-to-gate environmental LCA. The assessment involved a comparison on the basis of five impact

$$CH_3(CH_2)_{12}CO_2H + CH_3(CH_2)_{12}CH_2OH \xrightarrow{\text{Catalyst}} CH_3(CH_2)_{12}CO_2(CH_2)_{13}CH_3 + H_2O$$

Atom efficiency = 96%

Chemocatalytic process: Sn oxalate at 240°C
Biocatalytic process: Novozyme 435 at 60°C

Figure 4.7 Chemo- versus biocatalytic production of myristyl myristate.

TABLE 4.5 Biocatalytic Esterification: Impact on Defined Environmental Parameters

Parameter	Units	Chemocatalytic	Biocatalytic	Savings
Energy	GJ	22.5	8.63	62%
Global warming	kg CO_2equivalent	1518	582	62%
Acidification	kg SO_2equivalent	10.58	1.31	88%
Eutrophication	kg PO_4equivalent	0.86	0.24	74%
Smog formation	kg C_2H_4equivalent	0.49	0.12	76%

categories: (i) energy consumption, (ii) global warming, (iii) acidification, (iv) nutrient enrichment, and (v) smog formation.

As shown in Table 4.5, substantial reductions in all categories were achieved. Energy consumption was reduced by more than 60% and the formation of undesirable pollutants by up to 90%. Replacement of an environmentally unattractive tin catalyst by an enzyme and thc considerably milder conditions were the major factors responsible for the significantly more eco-friendly profile of the biocatalytic process.

In addition to the above-mentioned benefits, product quality was much higher in the biocatalytic process, mainly as a result of the much milder reaction conditions. This meant that purification steps could be largely circumvented resulting in simpler downstream processing. Thus, better product quality and process simplification are added economic benefits of the biocatalytic versus the chemocatalytic process.

4.9 CONCLUSIONS AND PROSPECTS

Hopefully, we have shown that the road to green and sustainable chemicals manufacture is a catalytic one and meaningful green metrics for measuring true reaction efficiency constitute a *conditio sine qua non* for monitoring progress. The concepts of atom economy and E factors constitute complementary and widely accepted metrics in this context. Together with LCA, they provide a sound basis for evaluating different processes and products with regard to their greenness and sustainability. All of the various subdisciplines of catalysis—homogeneous, heterogeneous, organocatalysis, and biocatalysis— will have a role to play in this quest for sustainable technologies. Biocatalysis, in particular, has many added benefits, such as mild reaction conditions employing a catalyst that is biocompatible, biodegradable, and derived from renewable resources. Processes are step economic and highly selective resulting in higher product quality and reduced waste generation coupled with superior economics. Extra benefits also ensue from process intensification as a result of integration of biocatalytic steps in cascade processes. Last but not least, catalysis in general, and biocatalysis, in particular, will play a key role in the transition from a global economy that is largely based on nonrenewable fossil feedstocks to a sustainable one that is based on renewable resources. We conclude, therefore, with the words of Albert Einstein, "the significant problems we face today cannot be solved at the same level of thinking we were at when we created them."

REFERENCES

1. Perkin WH. J Chem Soc; 1862:232, Brit. Pat. 1856, 1984.
2. Peterson DH, Murray HC. J Am Chem Soc 1952;71:1871–1872.
3. See in Holton RA, Biediger RJ, Boatman PD. In: Suffness M, editor. Taxol®: science and applications. Boca Raton (FL): CRC Press; 1995. p. 97.
4. Sheldon RA. C R Acad Sci Paris, IIc, Chim/Chem 2000;3:541–551.
5. Brundtland CG. Our common future. Oxford: The World Commission on Environmental Development, Oxford University Press;1987.
6. Anastas P, Warner JC, editors. Green chemistry: theory and practice. Oxford: Oxford University Press;1998.
7. Anastas PT, Farris CA, editors. Benign by Design: alternative synthetic design for pollution prevention, ACS Symposium Series, No. 577. Washington (DC): American Chemical Society; 1994.
8. Anastas PT, Zimmerman JB. In: Abrahams MA, editor. Sustainability science and engineering defining principles. Elsevier, Amsterdam, Netherlands, 2006. pp. 11–32.
9. Tang SLY, Smith RL, Poliakoff M. Green Chem 2005;7:761.
10. (a) Sheldon RA. Chem Ind (London) 1992: 903; (b) see also Sheldon RA. Chem Ind (London) 1997:12.
11. Sheldon RA. In: Sawyer DT, Martell AE, editors. Industrial environmental chemistry. New York: Plenum; 1992. pp. 99–119.
12. Sheldon RA. In: Weijnen MPC, Drinkenburg AAH, editors. Precision process technology. Dordrecht: Kluwer; 1993. pp. 125–138.
13. Sheldon RA. Chemtech March, 1994:38–47.
14. (a) Sheldon RA. J Mol Catal A Chem 1996;107:75–83; (b) Sheldon RA. J Chem Technol Biotechnol 1997;68:381–388.
15. Sheldon RA. Pure Appl Chem 2000;72:1233–1246.
16. Trost BM. Science 1991;254:1471–1477.
17. Trost BM. Angew Chem Int Ed 1995;34:259–281.
18. See in Caruana CM. Chem Eng Prog 1991;87(12):11.
19. Sheldon RA. Chemicals from synthesis gas. Dordrecht: Reidel; 1983. p. 15.
20. Sheldon RA. Bull Soc Chim Belg 1985;94:651.

21. Notari B. Stud Surf Sci Catal 1998;37:413–425.
22. (a) Alfonsi K, Collberg J, Dunn PJ, Fevig T, Jennings S, Johnson TA, Kleine HP, Knight C, Nagy MA, Perry DA, Stefaniak M. Green Chem 2008;10:31; (b) see also Dugger RW, Ragan JA, Brown Ripin DH. Org Proc Res Dev 2005;9:253–258.
23. Ritter SK. *C&EN*. 2008 Aug 18. pp. 59–68.
24. Thayer AN. *C&EN*. 2007 Aug 6. pp. 11–19.
25. For a recent overview see Calvo-Flores FG. ChemSusChem 2009;2:905–919.
26. Auge J. Green Chem 2008;10:225–231.
27. Andraos J. Org Proc Res Dev 2005;9:149–163.
28. Constable DJC, Curzons AD, Cunningham VL. Green Chem 2002;4:521–527.
29. (a) Curzons AD, Constable DJC, Mortimer DN, Cunningham VL. Green Chem 2001;3:1–6; (b) Constable DJC, Curzons AD, Freitas dos Santos LM, Green GR, Hannah RE, Hayler JD, Kitteringham J, McGuire MA, Richardson JE, Smith P, Webb RL, Yu M. Green Chem 2001;3:7–9.
30. Hudlicky T, Frey DA, Koroniak L, Claeboe CD, Brammer LE. Green Chem 1999;1:57–59.
31. For another approach to assess the environmental impact of waste see Eissen M, Metzger JO. Chem Eur J 2002;8:3580–3585.
32. Jimenez-Gonzalez C, Curzons AD, constable DJC, Cunningham VL. Int J LCA 2004;9:114–121.
33. Moretz-Sohn Monteiro JG, de Queiroz Fernandes Araujo O, Luiz de Medeiros J. Clean Technol Environ Policy 2009;11:209–214 and 459–472.
34. Li CJ, Trost BM. Proc Natl Acad Sci USA 2008;105:13197–13202.
35. Sheldon RA, van Bekkum H, editors. Fine chemicals through heterogeneous catalysis. Weinheim: Wiley-VCH; 2001, Chapters 3–7.
36. (a) For recent reviews see, for example. Thomas JM, Hernandez-garrido JC, Bell RG. Top Catal 2009;52:1630–1639; (b) Kaneda K, Mizugaki T. Energy Environ Sci 2009;2:655–673.
37. Jimenez-Gonzales C, Curzons AD, Constable DJC, Cunningham VL. Int J Life Cycle Assess 2004;9:115–121.
38. Lindström UM. Organic reactions in water. Oxford: Blackwell; 2007.
39. (a) Leitner W. Acc Chem Res 2002; 35: 746; (b)Licence P, Ke J, Sokolova M, Ross SK, Poliakoff M. Green Chem 2003;5:99; (c) Cole-Hamilton DJ. Adv Synth Catal 2006;348:1341.
40. (a) Horvath IT, Rabai J. Science 1994;266:72; (b) Horvath IT. Acc Chem Res 1998;31:641.
41. (a) Sheldon RA. Chem Commun 2001:2399; (b) Parvulescu VI, Hardacre C. Chem Rev 2007;107:2615–2665.
42. Wender PA, Handy ST, Wright DL. Chem Ind (London) 1997:765–769.
43. Bruggink A, Schoevaart R, Kieboom T. Org Proc Res Dev 2003;7:622.
44. Sheldon RA. Green Chem 2005;7:267–278.
45. Leitner W, Seddon KR, Wasserscheid P. Special issue on green solvents for catalysis. Green Chem 2003;5:99–284.
46. See also Noyori R. Chem Commun 2005:1807–1811.

47. (a) For recent reviews see Arnold FA. Curr Opin Chem Biol 2009;13:3–9; (b) Turner NJ. Nature Chem Biol 2009;5:567–577; (c) Alexeeva M, Carr R, Turner NJ. Org Biomol Chem 2003;1:4133–4137; (d) Turner NJ. Trends Biotechnol 2003;21:474–478; (e) Luetz S, Giver L, Lalonde J. Biotechnol Bioeng 2008;101:647–653; (f) Reetz MT. J Org Chem 2009;74:5767–5778–; (g) Reetz MT. Adv Catal 2006;49:1–69; (h) Bershtein S, Tawfik DS. Curr Opin Chem Biol 2008;101:647–653.
48. Sheldon RA. Adv Synth Catal 2007;349:1289–1307.
49. Sheldon RA. In: Garcia-Junceda E, editor. Multi-step enzyme catalysis: biotransformations and chemoenzymatic synthesis. Weinheim: Wiley-VCH; 2008. pp. 109–135.
50. (a) For pertinent examples see: Mateo C, Chmura A, Rustler S, van Rantwijk F, Stolz A, Sheldon RA. Tetrahedron: Asymmetry 2006;17:320–323; (b) van Pelt S, van Rantwijk F, Sheldon RA. Adv SynthCatal 2009;351:397–404.
51. (a) Tao J, Xu J-H. Curr Opin Chem Biol 2009;13:43–50; (b) Ran N, Zhao L, Chen Z, Tao J. Green Chem 2007;9:1–14; (c) Tao J, Zhao L, Ran N. Org Proc Res Dev 2007;11:259–267; (d) Yazbeck DR, Martinez CA, Hu S, Tao J. Tetrahedron: Asymmetry 2004;15:2757–2763; (e) Pollard DJ, Woodley JM. Trends Biotechnol 2006;25:66–73; (f) Hailes HC, Dalby PA, Woodley JM. J Chem Technol Biotechnol 2007;82:1063–1066; (g) Patel R, Hanson R, Goswami A, Nanduri V, Banerjee A, Donovan M-J, Goldberg S, Johnston R, Brzozowski D, Tully T, Howell J, Cazzulino D, Ko R. J Ind Microbial Biotechnol 2003;30:252–259.
52. Aleu J, Bustillo AJ, Hernandez-Galan R, Collado IG. Curr Org Chem 2006;10:2037–2054.
53. (a) Franssen MCR, Alessandrini L, Terraneo G. Pure Appl Chem 2005;77:273–279; (b) Vandamme EJ, Soetaert W. J Chem Technol Biotechnol 2002;77:1323–1332; (c) Serra S, Fuganti C, Brenna E. Trends Biotechnol 2005;23:193–198.
54. (a) Veit T. Eng Life Sci 2004;4:508–511; (b) Heinrichs V, Thum O. Lipid Technol 2005;17:82–87.
55. Martinez CA, Hu S, Dumond Y, Tao J, Kellcher P, Tully L. Org Proc Res Dev 2008;12:392–398.
56. Sheldon RA. Chirotechnology: the industrial synthesis of optically active compounds. New York: Marcel Dekker; 1993.
57. Petrides B. In: Harrison RG, Todd PW, Rudge SR, Petrides D, editors. Bioseparations science and engineering. Oxford University Press, Oxford, UK, 2003. Chapter 11.
58. Rinaldi R, Schüth F. Energy Environ Sci 2009;2:610–626.
59. Fox RJ, Davis SC, Mundorff EC, Newman LM, Gavrilovic V, Ma SK, Chung LM, Ching C, Tam S, Muley S, Grate J, Gruber J, Whitman JC, Sheldon RA, Huisman GW. Nat Biotechnol 2007;25: 338.
60. Ma SK, Gruber J, Davis C, Newman L, Gray D, Wang A, Grate J, Huisman GW, Sheldon RA. Green Chem 2010; 12:81–86.
61. Stemmer WP. Nature 1994; 370:389–391.
62. Thum O, Oxenbøll KM. SOFW J 2008; 134:44–47.

PART II

APPLICATION AND CASE STUDIES—PHARMACEUTICALS AND FINE CHEMICALS

CHAPTER 5

BIOCATALYTIC ROUTES TO CHIRAL INTERMEDIATES FOR DEVELOPMENT OF DRUGS

RAMESH N. PATEL
SLRP Associates, Biotechnology Consultation, Bridgewater, New Jersey

5.1 INTRODUCTION

In the development of drugs, the synthesis of single enantiomers of chiral intermediates has become increasingly important in the pharmaceutical industry [1]. Single enantiomers can be produced by either chemical or biocatalytic routes. The advantages of biocatalysis over chemical synthesis are that enzyme-catalyzed reactions are often highly enantioselective and regioselective. They can be carried out under mild conditions at ambient temperature and atmospheric pressure, thereby avoiding the use of more extreme reaction conditions, which could cause problems in isomerization, racemization, epimerization, and rearrangement of the compound. Microbial cells and a wide variety of enzymes derived from them can be used for chiral synthesis. Enzymes can be immobilized and reused for many cycles. In addition, enzymes can be overexpressed to make biocatalytic processes economically efficient, and enzymes with modified activity can be tailor made. Directed evolution of biocatalysts can lead to increased enzyme activity, selectivity, and stability [2–11]. A biocatalytic process generally follows the principles of green chemistry. It prevents waste and is designed to maximize the incorporation of most materials into the final product. It uses safer, less hazardous, and non-toxic solvents and can be carried out in aqueous systems. Many biocatalytic processes use raw materials for production of biocatalysts that are derived from renewable resources and feed stock. In addition, the derivatization, the use of blocking groups, and the protection/deprotection required in chemical processes are avoided in biocatalytic processes.

Biocatalysis for Green Chemistry and Chemical Process Development, First Edition.
Edited by Junhua (Alex) Tao and Romas Kazlauskas.

A number of review articles [12–26] have been published on the use of enzymes in organic synthesis. This chapter provides some examples of the use of enzymes for the synthesis of chiral intermediates that uses, in general, green processes.

5.2 ENZYMATIC PREPARATION OF CHIRAL UNNATURAL AMINO ACIDS

5.2.1 Glucagon-Like Peptide-1 (GLP-1): Preparation of (S)-2-Amino-3-[3-{6-(2-methylphenyl)} pyridyl]-propionic Acid

(*S*)-amino acids are useful intermediates for the synthesis of pharmaceuticals [27]. Many enzymatic approaches have been applied for their preparation, including (*S*)-hydantoinases combined with (*S*)-carbamoylases or HNO_2 [28], (*S*)-acylases [29], (*S*)-amidases [30], (*S*)-transaminases [31], and (*S*)-amino acid dehydrogenases [32,33], and dynamic resolution [34] has been developed. The (*S*)-amino-3-[3-{6-(2-methylphenyl)} pyridyl]-propionic acid 1 (Figure 5.1) is a key intermediate required for the synthesis of glucagon-like peptide-1 (GLP-1) mimics or GLP-1 receptor modulators. Such receptor modulators are potentially useful for the treatment of type II diabetes [35,36].

(*S*)-Amino-3-[3-{6-(2-methylphenyl)}pyridyl]-propionic acid was prepared by the enzymatic deracemization process [37] with a 72% isolated yield with a >99.4% *ee* from racemic amino acid **2** using a combination of the two enzymes, (*R*)-amino acid oxidase from *Trigonopsis variabilis* expressed in *Escherichia coli* and (*S*)-aminotransferase from *Sporosarcina ureae* cloned and expressed in *E. coli*. (*S*)-aspartate was used as the amino donor (Figure 5.1). An (*S*)-aminotransferase was also purified from a soil organism identified as *Burkholderia* sp. and cloned and expressed in *E. coli* and used in this process [37]. This process was scaled up to 100 L at a substrate input of 1.5 kg.

In an alternative process, the enzymatic dynamic resolution of racemic amino acid **2** was also demonstrated. (*R*)-selective oxidation with celite-immobilized

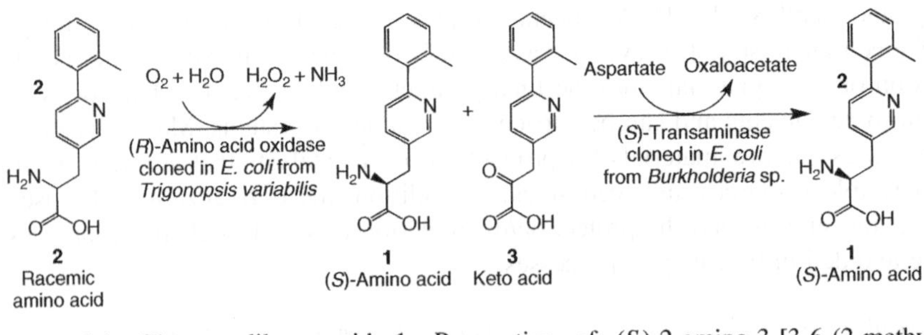

Figure 5.1 Glucagon-like peptide-1: Preparation of (*S*)-2-amino-3-[3-6-(2-methylphenyl)} pyridyl]-propionic acid.

(*R*)-amino acid oxidase from *Trigonopsis variabilis* expressed in *E. coli* in combination with chemical imine reduction with borane-ammonia gave a 75% process yield and a 100% *ee* of (*S*)-amino acid **1** [37].

5.2.2 Calcitonin Gene-Related Peptide Receptors (Antimigraine Drugs): Enzymatic Preparation of (*R*)-2-Amino-3-(7-methyl-1 *H*-indazol-5-yl)propanoic Acid

(*R*)-Amino acids are useful intermediates for the synthesis of β-lactam antibiotics and other pharmaceuticals [38–40]. Many enzymatic approaches have been applied for their preparation, including (*R*)-hydantoinases combined with (*R*)-carbamoylases or HNO_2 [40,41], (*R*)-acylases [42], (*R*)-amidases [43], and (*R*)-transaminases [38], and recently an (*R*)-amino acid dehydrogenase has been developed [44]. Racemic amino acids have also been deracemized to give (*R*)-amino acids using an (*S*)-amino acid oxidase to selectively deplete the (*S*)-amino acid combined with an excess of reducing agent to recycle the imine product to the racemic amino acid [45]. The combination of a (*R*)-amino acid oxidase and an (*S*)-amino acid dehydrogenase to convert a racemic amino acid to an (*S*)-amino acid has been demonstrated [46]. The (*R*)-amino acid (*R*)-2-amino-3-(7-methyl-1 *H*-indazol-5-yl)propanoic acid (*R*)-**4** (Figure 5.2)

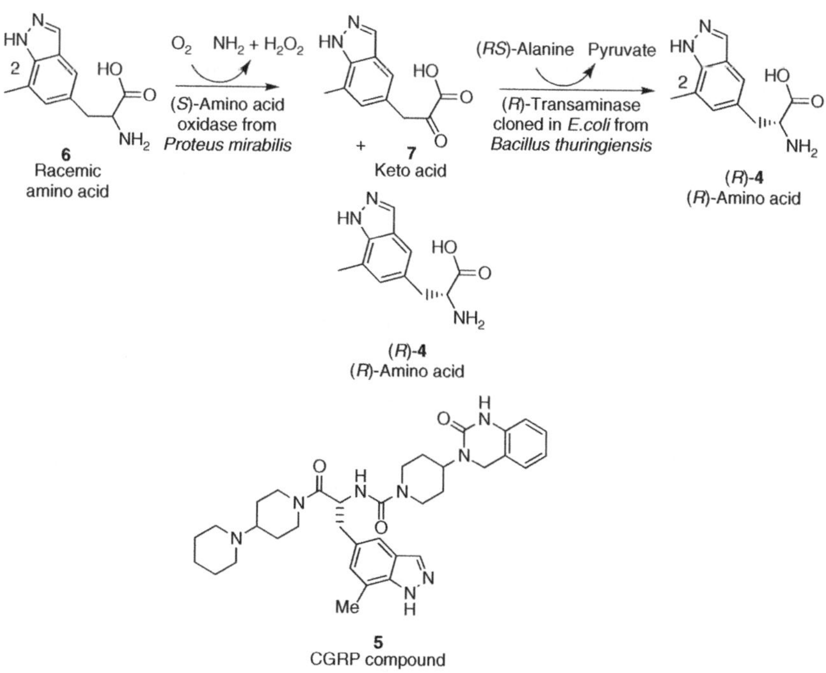

Figure 5.2 Calcitonin gene-related peptide receptors: Enzymatic preparation of (*R*)-2-amino-3-(7-methyl-1 *H*-indazol-5-yl)propanoic acid.

is a key intermediate needed for the synthesis of antagonists of calcitonin gene-related peptide receptors **5** [47]. Such antagonists are potentially useful for the treatment of migraine and other maladies [47,48].

(*R*)-Amino acid **4** was prepared with a 68% isolated yield with a >99% *ee* from racemic amino acid **6** using (*S*)-amino acid oxidase from *Proteus mirabilis* expressed in *E. coli* in combination with a commercially available (*R*)-transaminase using (*R*)-alanine as amino donor [49]. The (*R*)-enantiomer was also prepared with a 79% isolated yield with a >99% *ee* from the corresponding keto acid **7** using the (*R*)-transaminase with racemic alanine as the amino donor. The rate and yield of this reaction could be accelerated by addition of lactate dehydrogenase (with NAD, formate, and formate dehydrogenase (FDH) to regenerate NADH) to remove the inhibitory pyruvate produced during the reaction. An (*R*)-transaminase was identified and purified from a soil organism identified as *Bacillus thuringiensis* and cloned and expressed in *E. coli*. The recombinant (*R*)-transaminase was very effective for the preparation of **4** and gave a nearly complete conversion of **7** to **4** without the need for additional enzymes for pyruvate removal [49].

5.2.3 Vanlev (Antihypertensive Drug): Enzymatic Synthesis of (S)-6-Hydroxynorleucine

Vanlev **8** (Figure 5.3) is an antihypertensive drug that acts by inhibiting angiotensin-converting enzyme (ACE) and neutral endopeptidase (NEP) [50]. (*S*)-6-Hydroxynorleucine **9** is a key intermediate in the synthesis of Vanlev.

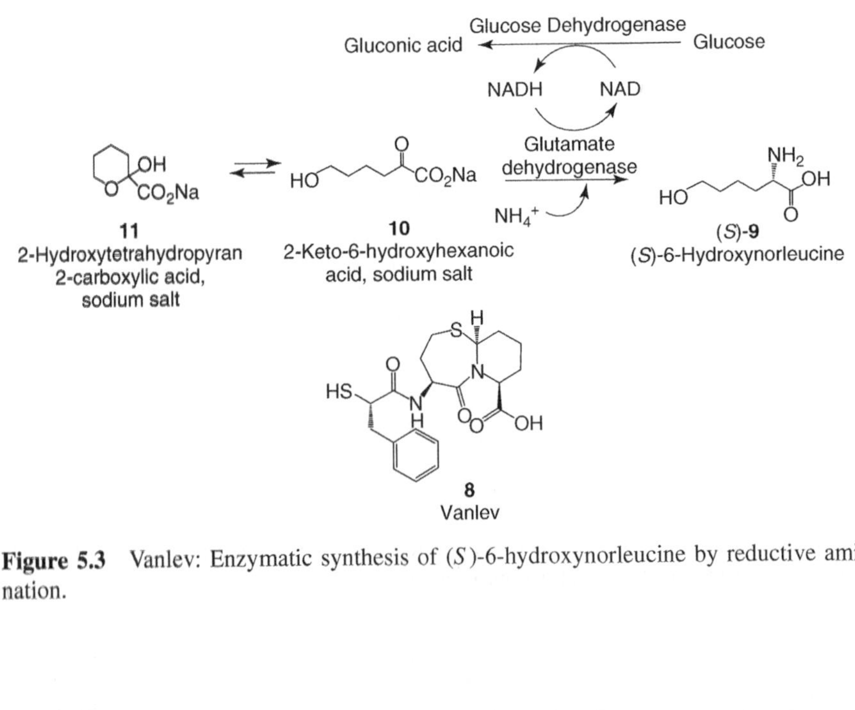

Figure 5.3 Vanlev: Enzymatic synthesis of (*S*)-6-hydroxynorleucine by reductive amination.

The synthesis and complete conversion of 2-keto-6-hydroxyhexanoic acid **10** to (*S*)-6-hydroxynorleucine **9** was demonstrated by reductive amination, using beef liver glutamate dehydrogenase [51]. Compound **10**, in equilibrium with 2-hydroxytetrahydropyran-2-carboxylic acid sodium salt **11**, was converted to **9**. The reaction requires ammonia and NADH. NAD produced during the reaction was recycled to NADH by the oxidation of glucose to gluconic acid using glucose dehydrogenase from *Bacillus megaterium*. The reaction was complete in about 3 h at a substrate input of 100 g/L with a reaction yield of 92% and an *ee* of 99.8% for (*S*)-6-hydroxynorleucine.

The synthesis and isolation of keto acid **10** required several steps. In a second, more convenient process, the ketoacid was prepared by the treatment of racemic 6-hydroxy norleucine **11** [produced by hydrolysis of 5-(4-hydroxybutyl) hydantoin **12**] with (*R*)-amino acid oxidase (Figure 5.4) After the *ee* of the unreacted (*S*)-6-hydroxynorleucine reached 99.8%, the reductive amination procedure was used to convert the mixture containing the 2-keto-6-hydroxyhexanoic acid entirely to (*S*)-6-hydroxynorleucine with a 97% yield with 99.8% *ee* from racemic 6-hydroxynorleucine at 100 g/L substrate input [51]. The (*S*)-6-hydroxynorleucine prepared by the enzymatic process was converted chemically to Vanlev **8** [52].

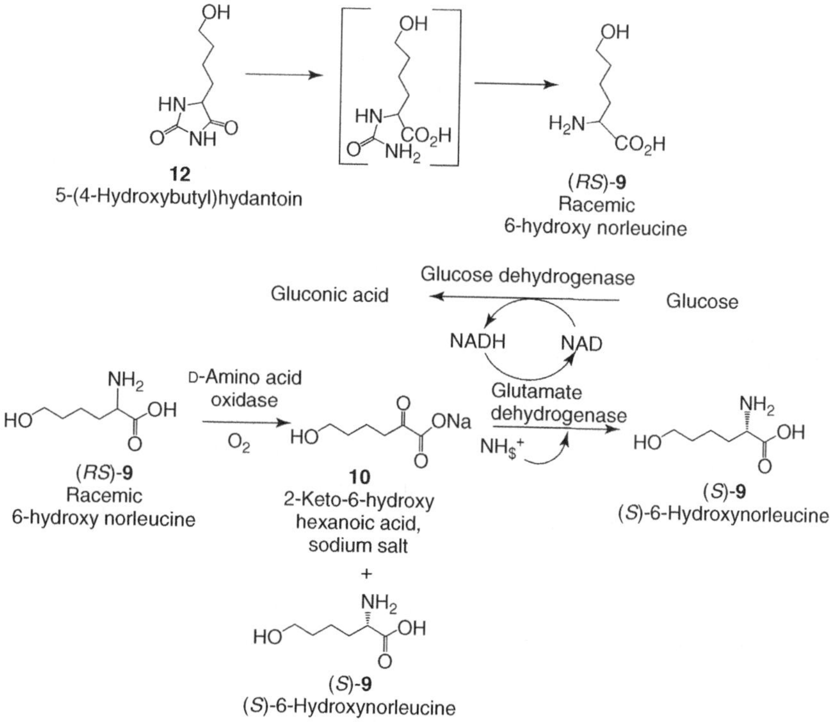

Figure 5.4 Vanlev: Enzymatic synthesis of (*S*)-6-hydroxynorleucine by deracemization.

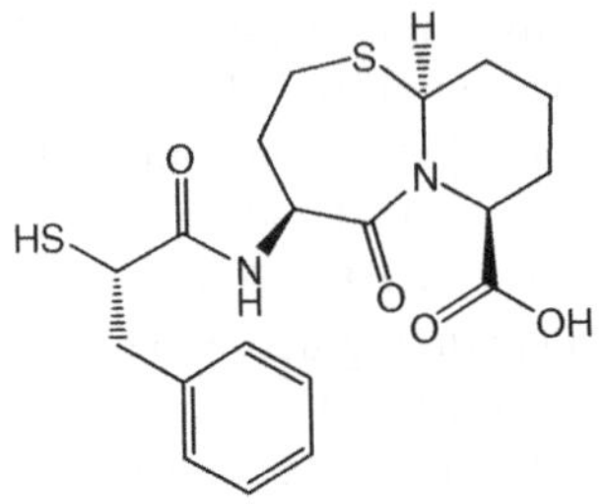

8
Vanlev

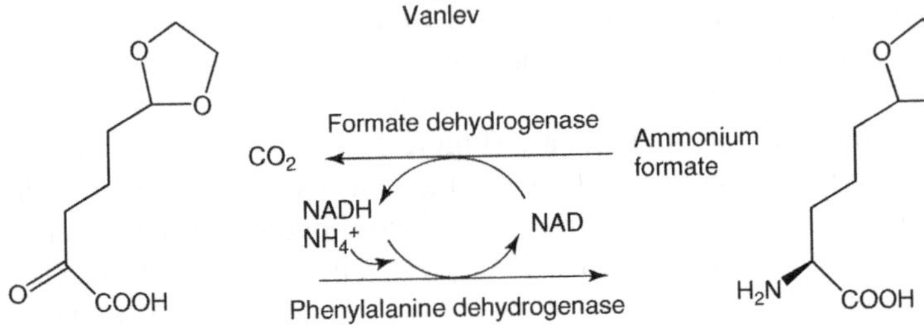

14
Ketoacid acetal

(*S*)**13**
(*S*)-Allysine ethylene acetal

Figure 5.5 Vanlev: Enzymatic synthesis of allysine ethylene acetal.

5.2.4 Vanlev (Antihypertensive Drug): Enzymatic Synthesis of Allysine Ethylene Acetal

(*S*)-2-Amino-5-(1,3-dioxolan-2-yl)-pentanoic acid [(*S*)-allysine ethylene acetal] (**13**, Figure 5.5) is one of the three building blocks used in an alternative synthesis of Vanlev **8**. Synthesis of **13** was demonstrated by reductive amination of ketoacid acetal **14** using phenylalanine dehydrogenase (PDH) from *Thermoactinomyces intermedius* [53]. The reaction required ammonia and NADH; NAD produced during the reaction was recycled to NADH by the oxidation of formate to CO_2 using FDH. *T. intermedius* PDH was cloned and expressed in *E. coli* and recombinant culture was used as a source of PDH. *Pichia pastoris* [27,28] grown on methanol is also a useful source of FDH. Expression of *T. intermedius* PDH in *P. pastoris*, inducible by methanol, allowed generation of both enzymes in a single fermentation. A total of 197 kg of **13** was produced in three 1600-L batches, using a 5% concentration of substrate **14** with an average yield of 91 M% with *ee* >98% [53]. (*S*)-allysine ethylene acetal was converted to Vanlev **8** [52].

5.2.5 Saxagliptin (Antidiabetic Drug): Enzymatic Reductive Amination of 2-(3-hydroxy-1-adamantyl)-2-oxoethanoic Acid

Dipeptidyl peptidase 4 (DPP-4) is a ubiquitous proline-specific serine protease responsible for the rapid inactivation of glucagon-like peptide 1 (GLP-1) and

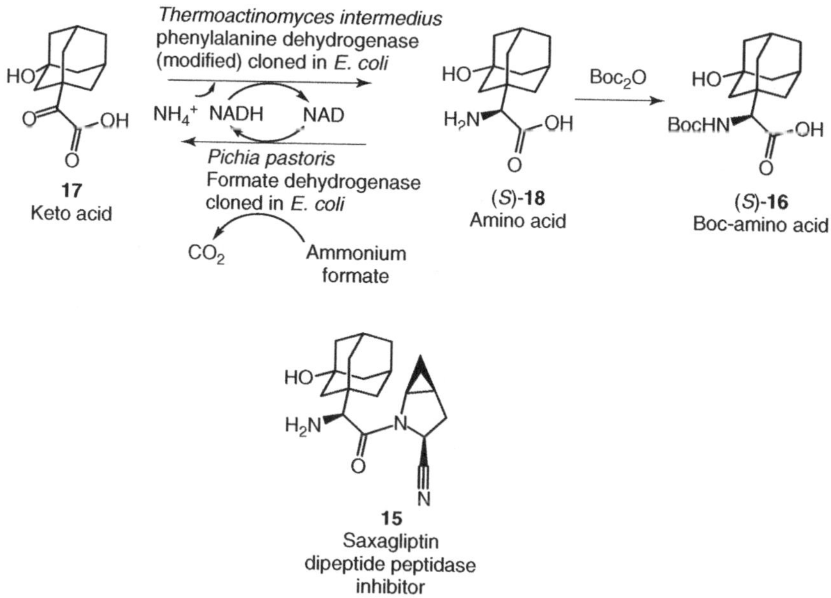

Figure 5.6 Saxagliptin: Enzymatic reductive amination of 2-(3-hydroxy-1-adamantyl)-2-oxoethanoic acid.

glucose-dependent insulinotropic peptide. To alleviate the inactivation of GLP-1, inhibitors of DPP-4 are being evaluated for their ability to provide improved control of blood glucose for diabetic patients [54–56]. Januvia, developed by Merck, is a marketed DPP-4 inhibitor [55].

Saxagliptin **15** [54,56] (Figure 5.6), a DPP-4 inhibitor under development by Bristol-Myers Squibb, requires (*S*)-*N*-boc-3-hydroxyadamantylglycine **16** as an intermediate. A process for conversion of the keto acid **17** to the corresponding amino acid **18** using (*S*)-amino acid dehydrogenases was developed. A modified form of a recombinant PDH cloned from *T. intermedius* and expressed in *P. pastoris* or *E. coli* was used for this process. NAD produced during the reaction was recycled to NADH using FDH. The modified PDH contains two amino acid changes at the C-terminus and a 12 amino acid extension of the C-terminus [57,58].

Production of multikilogram batches was originally carried out with extracts of *P. pastoris* expressing the modified PDH from *T. intermedius* and endogenous FDH. The reductive amination process was further scaled up using a preparation of the two enzymes expressed in single recombinant *E. coli*. The amino acid **18** was directly protected as its boc derivative without isolation to afford intermediates. Yields before isolation were close to 98% with 100% *ee* [57,58]. This process has now been used to prepare several hundred kilograms of boc-protected amino acid **16** to support the development of Saxagliptin.

The use of PDH in the production of (*S*)-phenylalanine and other related (*S*)-amino acids has been demonstrated by Asano [59]. PDH was engineered by directed evolution to use in the organic solvent and to carry out enzymatic reactions in biphasic and homogeneous systems for the preparation of (*S*)-amino acids [60].

5.2.6 Atazanavir (Antiviral Drug): Enzymatic Synthesis of (S)-*tertiary*-Leucine

(*S*)-*tertiary* leucine **19**, owing to its bulky and hydrophobic side-chain, is a key chiral amino acid required for the synthesis of a number of drugs containing peptides [61,62]. An enzymatic reductive amination process for the synthesis of (*S*)-*tertiary* leucine has been developed [61,62]. Atazanavir **20** is an acyclic aza-peptidomimetic, a potent HIV protease inhibitor [63,64]. Synthesis of atazanavir required (*S*)-*tertiary* leucine (Figure 5.7). We have also developed the enzymatic reductive amination of ketoacid **21** to amino acid **19** by recombinant *E. coli* expressing leucine dehydrogenase from *T. intermedius*. The reaction required ammonia and NADH as a cofactor. NAD produced during the reaction was recycled back to NADH using recombinant *E. coli* expressing FDH from *P. pastoris*. A reaction yield of >95% with an *ee* of >99.5% was obtained for **19** at a 100 g/L substrate input (R. Hanson, S. Goldberg, R. Patel, unpublished results). Leucine dehydrogenase from *Bacillus* strain has also been cloned and expressed and used in reductive amination process [65,66].

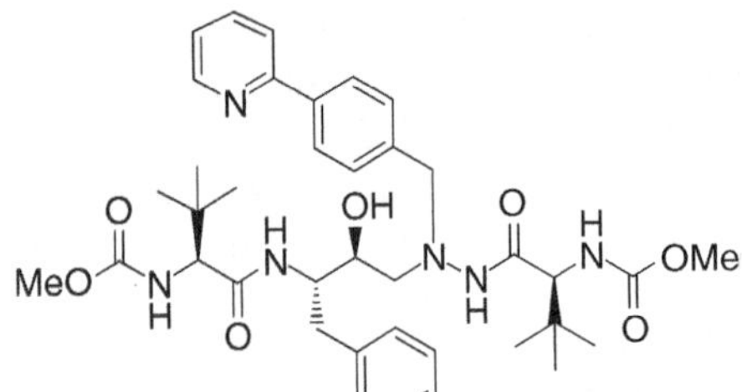

20
Atazanavir

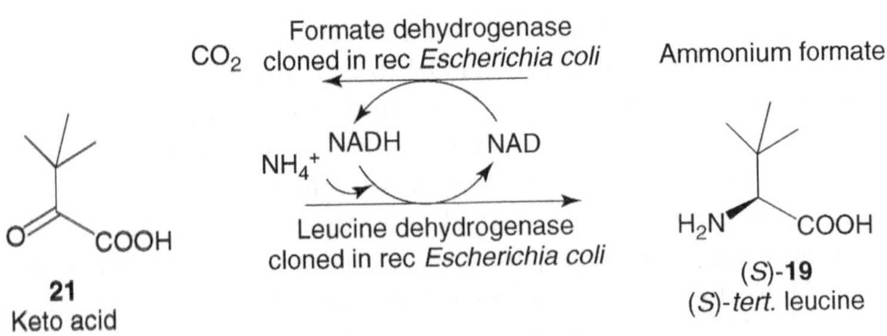

Figure 5.7 Atazanavir: Enzymatic synthesis of (*S*)-tertiary-leucine.

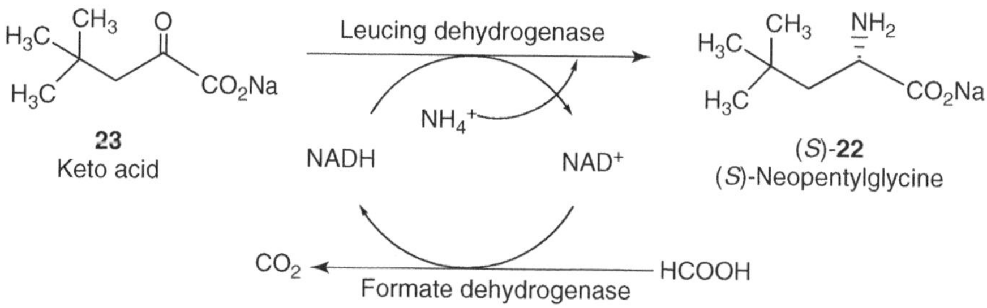

Figure 5.8 Cathepsin S inhibitors: Enzymatic synthesis of (*S*)-neopentylglycine.

5.2.7 Cathepsin S Inhibitors (Treatment of Autoimmune Diseases): Enzymatic Synthesis of (S)-neopentylglycine

A "second-generation process" for the enantioselective synthesis of (*S*)-neopentylglycine **22** (Figure 5.8) was developed by Groeger *et al.* [67]. A recombinant whole cell containing leucine dehydrogenase and FDH was used in the reductive amination of the corresponding α-keto acid **23**. The desired (*S*)-neopentylglycine was obtained with a >95% conversion and a high enantioselectivity of >99% *ee* at substrate concentrations of up to 88 g/L. Spiroheterocyclic compounds (morpholine-4-carboxylic acid amides of heterocyclic cyclohexylalanine and neopentylglycine derivatives and their analogs) are useful as reversible inhibitors of cysteine proteases such as cathepsin S useful in the treatment of a variety of autoimmune diseases [68].

5.2.8 Synthesis of Derivatives of Other (S)-amino Acids

A novel *N*-methyl-(*S*)-amino acid dehydrogenase activity in various bacterial strains has been reported [69]. This enzyme was cloned from *Pseudomonas putida* ATCC12633 into *E. coli* and used in the synthesis of various *N*-alkyl-(*S*)-amino acids from the corresponding α-keto acids (e.g., pyruvate, phenylpyruvate, and hydroxypyruvate) and alkylamines (e.g., methylamine, ethylamine, and propylamine). Ammonia was used as an amino donor and NADPH was more than 300 times more efficient than NADH as a hydrogen donor in the enzymatic reductive amination process.

An opine dehydrogenase (ODH) from *Arthrobacter* sp. was discovered that used pyruvate as an amino acceptor and was active toward short-chain aliphatic (*S*)-amino acids and those substituted with acyloxy, phosphonooxy, and halogen groups [70]. Other substrates for the enzyme were 3-aminobutyric acid and (*S*)-phenylalaninol. Optically pure opine-type secondary amine carboxylic acids were synthesized from amino acids and their analogs such as (*S*)-Met, (*S*)-Ile, (*S*)-Leu, (*S*)-Val, (*S*)-Phe, (*S*)-Ala, (*S*)-Thr, (*S*)-Ser, and (*S*)-phenylalaninol, and α-keto acids such as glyoxylate, pyruvate, and 2-oxobutyrate using the enzyme. The regeneration of NADH was carried out by FDH from *Moraxella* sp. C-1. The

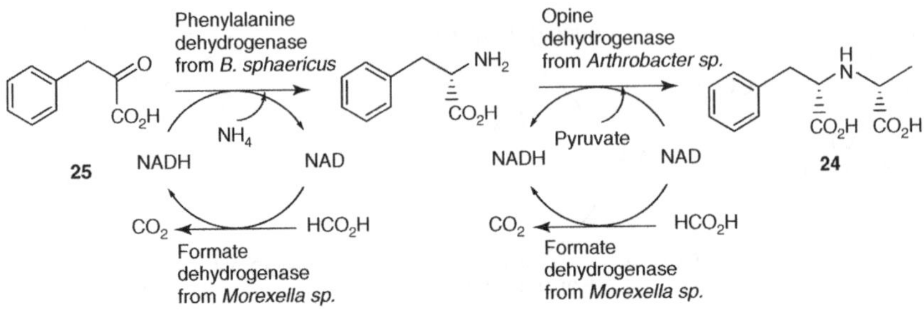

Figure 5.9 Enzymatic synthesis of *N*-[1-(*R*)-(carboxyl)ethyl]-(*S*)-phenylalanine.

absolute configuration of the asymmetric center of the opines was of (*R*) stereochemistry with a >99.9% *ee*. One-pot synthesis of *N*-[1-(*R*)-(carboxyl)ethyl]-(*S*)-phenylalanine **24** from phenylpyruvate **25** and pyruvate (Figure 5.9) using ODH, FDH, and PDH has been described [70].

5.2.9 Inogatran (Thrombin Inhibitor) Preparation of (*R*)-Amino Acid

(*R*)-Amino acids are increasingly becoming important building blocks in the production of pharmaceuticals and fine chemicals, and as chiral-directing auxiliaries and chiral synthons in organic synthesis. Applications of (*R*)-amino acids include their use as key components in β-lactam antibiotics, fertility drugs, and anticoagulants [71,72]

Two important (*R*)-amino acids used in semisynthetic antibiotics, (*R*)-phenylglycine (ampicillin) and *p*-hydroxy-(*R*)-phenylglycine (amoxicillin), are currently produced on a ton scale per year. In addition, there are more than 20 (*R*)-amino acids currently produced at pilot- or full-scale levels. In many of these cases, the (*R*)-enantiomer is not only often more potent than the corresponding (*S*)-enantiomer but also often more stable *in vivo* against enzyme degradation [73,74].

Using both rational and random mutagenesis, Rozzell and Novick [44] have created a broad substrate range, nicotinamide-cofactor-dependent and highly stereoselective (*R*)-amino acid dehydrogenase. This new enzyme is capable of producing (*R*)-amino acids via reductive amination of the corresponding 2-keto acid with ammonia. This biocatalyst was the result of three rounds of mutagenesis and screening performed on the enzyme *meso*-diaminopimelate (*R*)-dehydrogenase from *Corynebacterium glutamicum*. The very high selectivity toward the (*R*)-enantiomer (an *ee* of 95–>99%) was shown to be preserved in the three rounds of mutagenesis and screening [44]. This new enzyme was active against a variety of amino acids and could complement and improve upon current methods for (*R*)-amino acid synthesis. The synthesis of (*R*)-cyclohexylalanine **26** (Figure 5.10) was developed by reductive amination of cyclohexylpyruvate **27** to yield (*R*)-**26** with 98% yield and >99% *ee*. (*R*)-**26** is a potential chiral intermediate for the synthesis of thrombin inhibitor Inogatran **28** [75].

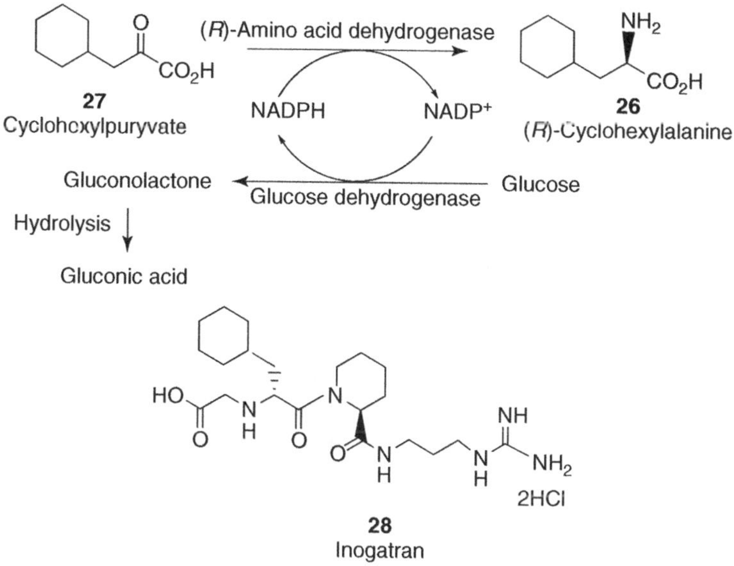

Figure 5.10 Inogatran (thrombin inhibitor): Enzymatic synthesis of (*R*)-cyclohexylalanine.

5.3 ENZYMATIC DESYMMETRIZATION PROCESS

5.3.1 NK1/NK2 Dual Antagonists: Enzymatic Desymmetrization of Diethyl 3-[3′,4′-dichlorophenyl] glutarate

Tachykinins are a group of biologically active neuropeptide hormones that are widely distributed throughout the nervous system. They are implicated in a variety of biological processes such as pain transmission, inflammation, vasodialation, and secretion [76]. The effect of tachykinins is modulated via the specific G-protein coupled receptors such as Neurokinin 1 (NK1) and Neurokinin 2 (NK2). Thus nonpeptide NK-receptor antagonists are potentially useful in the treatment of a variety of chronic diseases including asthma, bronchospasm, arthritis, and migraine [77]. The structure–activity relationship of several nonpeptide NK1/NK2 antagonists has led to the discovery of a new class of oxime-based NK1/NK2 dual antagonists [78,79] such as compound **29** (Figure 5.11). The biological activity of **29** resides mainly in the R,R-diastereomer. An enzymatic process for desymmetrization of the prochiral diethyl 3-(3′,4′-dichlorophenyl) glutarate **30** to the corresponding (*S*)-monoester **31** has been developed using lipase B from *Candida antarctica* with either the free or the immobilized enzyme. At 100 g/L substrate input, a reaction yield of 97% and an *ee* of >99% were obtained for the desired (*S*)-monoester. The process was scaled up to produce 200 kg of product with an overall isolated yield of 80% [80].

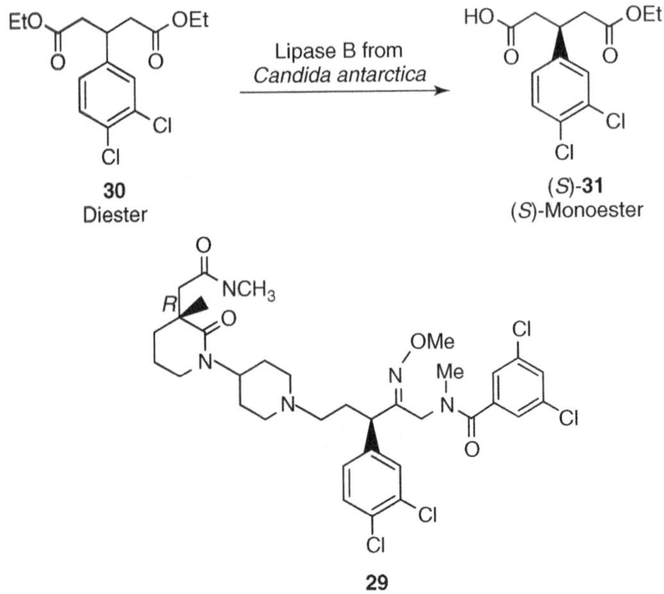

Figure 5.11 NK1/NK2 dual antagonists: Enzymatic desymmetrization of diethyl 3-[3′,4′-dichlorophenyl]glutarate.

DNA family shuffling was used to create a chimeric lipase B protein with improved activity toward diethyl 3-(3′,4′-dichlorophenyl)glutarate. Three homologous lipases from *C. antarctica* ATCC 32657, *Hyphozyma* sp. CBS 648.91, and *Cryptococcus tsukubaensis* ATCC 24555 were cloned and shuffled to generate a diverse gene library, and using a high-throughput screening assay, a chimeric lipase B protein having 20-fold higher activity toward the substrate was identified [81]. The thermostability of the lipase was also improved by directed evolution [82].

5.3.2 Cholesterol-Lowering Drugs: Enzymatic Desymmetrization of Prochiral 3-Hydroxyglutaronitrile

Enzymatic desymmetrization of prochiral 3-hydroxyglutaronitrile **32** using a nitrilase [83,84] has been demonstrated (Figure 5.12). Following esterification of the resulting (*R*)-3-hydroxy-4-cyanobutyric acid **33**, an intermediate useful for the manufacture of the cholesterol-lowering drug Lipitor **34** (atorvastatin calcium) was produced. Nitrilases were identified in genomic libraries created by extraction of DNA directly from environmental samples and were expressed in *E. coli*. The resulting library was screened for highly enantioselective (*R*)-specific nitrilases [85]. (*R*)-3-Hydroxy-4-cyanobutyric acid was produced using a 100-mM initial nitrile concentration with a 98% yield and a 94.5% *ee*. The enantioselectivity of this wild-type nitrilase decreased with increasing nitrile concentration, where an

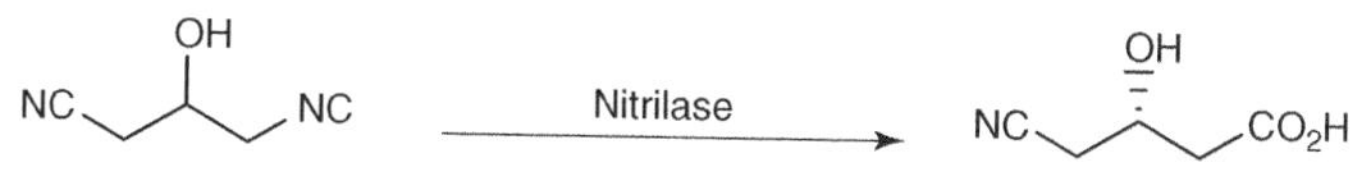

32 (R)-33

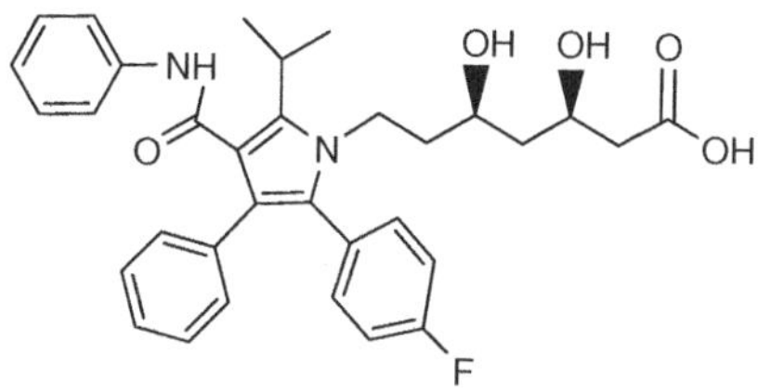

34
Lipitor

Figure 5.12 Cholesterol-lowering drugs: Enzymatic desymmetrization of prochiral 3-hydroxyglutaronitrile.

ee of only 87.8% was obtained at a 2.25 M substrate concentration. Mutagenesis of the nitrilase using a technique that combinatorially saturated each amino acid in the protein to each of the other 19 amino acids resulted in an improved variant (Ala190His) that was expressed in *E. coli*. This variant nitrilase gave an enantiomeric excess of 98.5% at 3 M substrate concentration with a volumetric productivity of 619 g/L/day. Nitrilases from this library have been used to produce a range of (*R*)-mandelic acid derivatives and analogs and (*S*)-phenyllactic acid, with high yields and enantioselectivities [86]

5.3.3 β3 Receptor Agonist: Desymmetrization of Diethyl Methyl-(4-methoxyphenyl) Propanedioate

The (*S*)-monoester **35** (Figure 5.13) is a key intermediate for the synthesis of β3-receptor agonists **36** [87,88]. The enantioselective enzymatic hydrolysis of diester **37** to the desired acid ester **35** by pig liver esterase [89] was demonstrated. The reaction yields and the *ee* of monoester were dependent on the cosolvent used. High *ee*'s (>91%) were obtained with methanol, ethanol, and toluene as cosolvents. Ethanol gave the highest reaction yield (96.7%) and *ee* (96%) for the desired acid ester **35**. It was observed that the *ee* of the (*S*)-monoester was increased by decreasing the temperature from 25 to 10°C; when biotransformation was conducted in a biphasic system using 10% ethanol as a cosolvent, a reaction yield of 96 M% and an *ee* of 96.9% were obtained [89].

5.3.4 Antiviral Drug (Hepatitis B Viral Inhibitor: Enzymatic Asymmetric Hydrolysis and Acylation

Chiral monoacetate esters **38** and **39** are key intermediates for the total chemical synthesis of **40** (Baraclude), a potential drug for hepatitis B viral infection

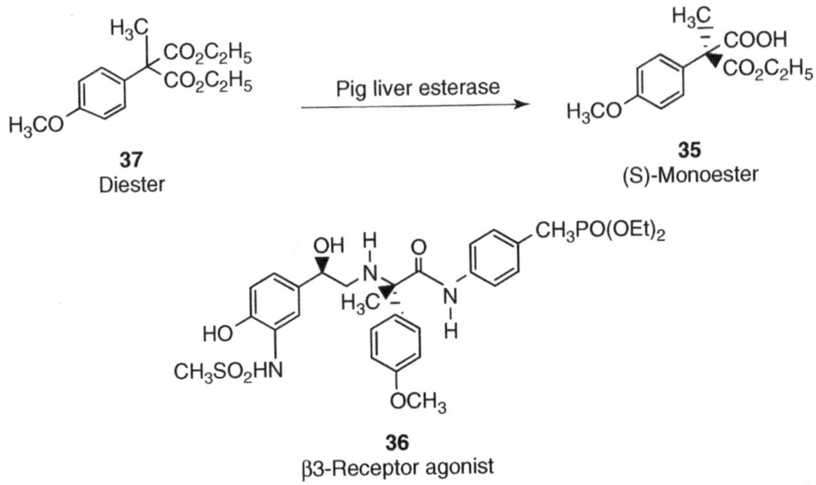

Figure 5.13 β3-receptor agonist: Desymmetrization of diethyl methyl-(4-methoxyphenyl) propanedioate.

[90–92]. Baraclude is a carboxylic analog of 2′-deoxyguanosine in which the furanose oxygen is replaced with an exocyclic double bond; it has been recently approved by the FDA for treatment of HBV infection.

Enzymatic hydrolysis of (1α,2β,3α)]-2-[(benzyloxy)methyl]-4-cyclopenten-1,3-diol diacetate has been demonstrated by Griffith and Danishefsky to obtain the corresponding monoester using acetylcholine esterase from electric eels [93,94]. Recently we have described the enantioselective asymmetric hydrolysis of (1α,2β,3α)-2-[(benzyloxy)methyl]-4-cyclopenten-1,3-diol diacetate **41** (Figure 5.14) to the corresponding (+)-monoacetate **38** by lipase PS-30 from *Pseudomonas cepacia* and pancreatin. A reaction yield of 85 M% and an *ee* of 98% were obtained using lipase PS-30. Using pancreatin, a reaction yield of 75 M% and an *ee* of 98.5% were obtained. We also demonstrated the enzymatic asymmetric acetylation of (1α,2β,3α)-2-[(Benzyloxy) methyl]-4-cyclopenten-1,3-diol **42** to the corresponding (−)-monoacetate **39** using lipase PS-30 [95].

5.4 ENZYMATIC HYDROXYLATION

Hydroxylation reactions are catalyzed by monooxygenases and dioxygenases [18,96–108]. Aromatic ring hydroxylating dioxygenases initiate the degradation of a wide range of aromatic compounds, including polycyclic aromatic hydrocarbons, nitroaromatic and chlorinated aromatic compounds, aromatic acids, and heterocyclic aromatic compounds [18,100–105]. Toluene dioxygenase (TDO) from *P. putida* F1 was the first aromatic hydrocarbon dioxygenase to be characterized [100]. TDO (Figure 5.15) was shown to catalyze the *cis*-dihydroxylation of benzene and toluene [100,101]. A great deal of our understanding of the

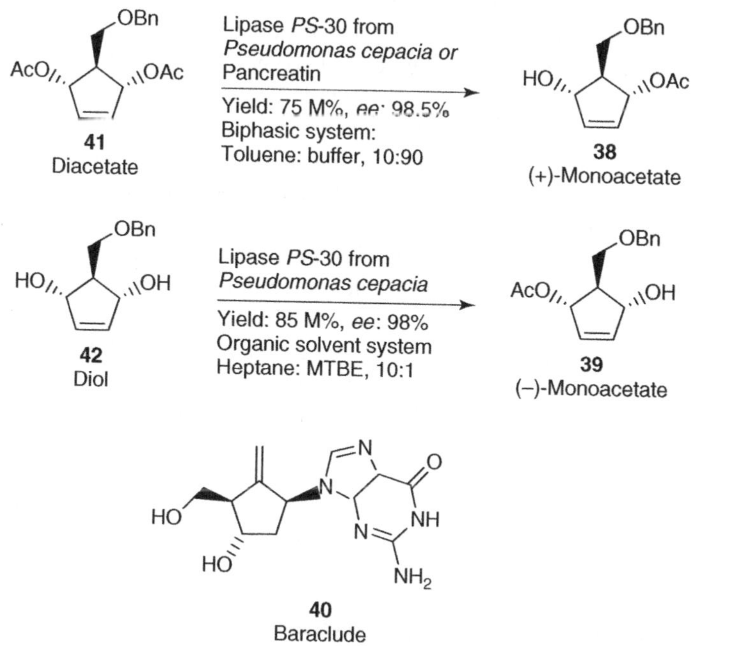

Figure 5.14 Antiviral: Enzymatic asymmetric hydrolysis and acylation.

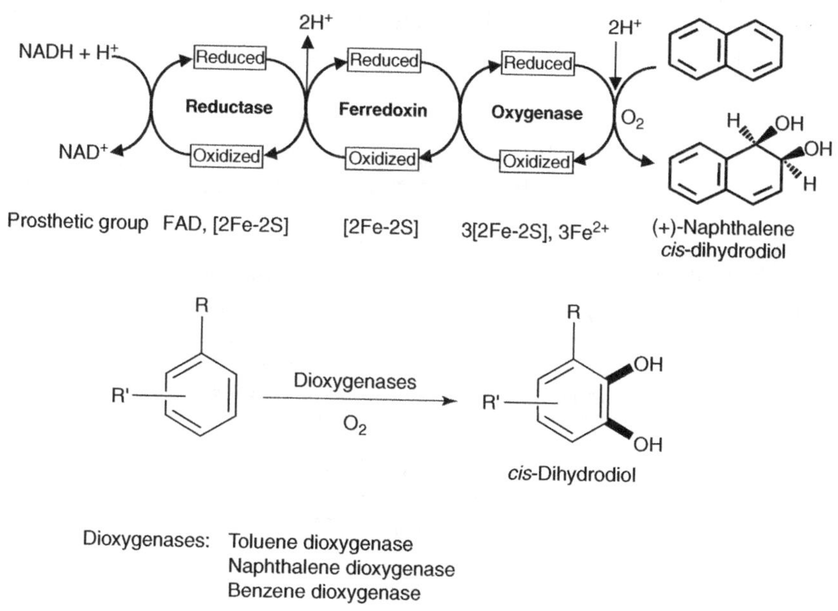

Figure 5.15 Dioxygenase-catalyzed synthesis of *cis*-dihydrodiols.

structure, function, and substrate specificity of aromatic hydrocarbon dioxygenases has come from studies of TDO and naphthalene dioxygenase (NDO). To date, more than 100 Rieske nonheme iron oxygenases have been identified on the basis of biological activity or nucleotide sequence identity. To date, more than 300 arene *cis*-dihydrodiols have been obtained using TDO and these have been extensively used as chiral precursors in organic synthesis [18,100–105]. Benzene dioxygenase, TDO, and NDO have also been reported to catalyze the dihydroxylation of alkenes and recently, biocatalytic asymmetric dihydroxylation of conjugated mono-and poly-alkenes has been demonstrated to yield enantiopure cyclic *cis*-diols. The diol metabolites were obtained from monosubstituted, gem-disubstituted, *cis*-disubstituted, and trisubstituted alkene substrates, using whole cells of *P. putida* strains containing toluene and NDOs [18,100–105]. Dioxygenase selection and alkene type were established as important factors, in preference to dioxygenase-catalyzed 1,2-dihydroxylation of conjugated alkene or arene groups, and monohydroxylation at benzylic or allylic centers. *Cis*-1,2-diol metabolites from arenes, cyclic alkenes, and dienes were generally observed to be enantiopure (>98% *ee*), while 1,2-diols from acyclic alkenes had lower enantiomeric excess values (<88% *ee*) [18,104,105].

5.4.1 Crixivan (Antiviral Drug): Enzymatic Preparation of *cis*-(1*S*,2*R*)-Indandiol and *trans*-(1*R*,2*R*)-indandiol

Cis-(1*S*,2*R*)-indandiol **44** or *trans*-(1*R*,2*R*)-indandiol **45** (Figure 5.16) are both potential precursors to *cis*-(1*S*,2*R*)-1-aminoindan-2-ol **46**, a key chiral synthon for Crixivan (Indinavir) **47**, an HIV protease inhibitor [106]. Enrichment and isolation of microbial cultures yielded two strains, *Rhodococcus* sp. B 264-1

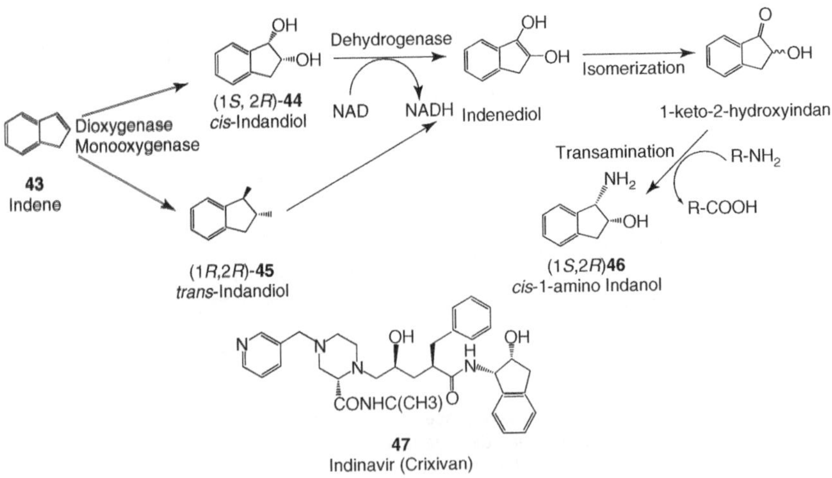

Figure 5.16 Crixivan: Enzymatic preparation of *cis*-(1*S*,2*R*)-indandiol and *trans*-(1*R*,2*R*)-indandiol.

(MB 5655) and I-24 (MA 7205), capable of biotransforming indene **43** to *cis*-(1*S*,2*R*)-indandiol and *trans*-(1*R*,2*R*)-indandiol, respectively [106]. Isolate MB 5655 was found to have a TDO, while isolate MA 7205 was found to harbor both toluene and NDOs as well as a naphthalene monooxygenase, which catalyzes the above biotransformation. When scaled up in a 14-L fermentor, MB5655 produced up to 2.0 g/L of *cis*-(1*S*,2*R*)-indandiol **44** with an *ee* of >99%. *Rhodococcus* sp MA 7205 cultivated under similar conditions produced up to 1.4 g/L of *trans*-(1*R*,2*R*)-indandiol **45** with an *ee* of >98%. Process development studies yielded titers of >4.0 g/L of *trans*-(1*R*,2*R*)-indandiol [107]. A metabolic engineering approach [108] and a directed evolution technique [109] were evaluated to avoid side reactions, block degradative pathways, and to enhance the key reaction to convert indene to *cis*-amino indanol **46** or *cis*-indanediol. The application of multiparameter flow cytometry was employed for the measurement of indene toxicity and it was discovered that a concentration of up to 0.25 g/L of indene (0.037 g indene/g dry cell wt.) in batch bioconversions does not influence cell physiology. Using this information, the implementation of a single-phase indene fed-batch bioconversion was carried out. The results indicated that indene supply at a rate of 0.1 g/L/h was feasible without any deleterious effects. *cis*-(1*S*,2*R*)-Indandiol **44** production rates were enhanced up to 200 mg/L/h by a combination of suitable indene feeding rates in the stationary phase and operating with a high biomass concentration [110].

5.4.2 Carbovir (Antiviral Drug): Enzymatic Hydroxylation 2-Cyclopentylbenzoxazole

By employing protein engineering and substrate engineering, the biohydroxylation of 2-cyclopentylbenzoxazole **48** (Figure 5.17) to compound **49** was demonstrated by Munzer *et al.* to prepare a key intermediate for the synthesis of (−)-carbovir **50** [111]. Removal of the protecting group provided the end product **51**. In the initial screening, regardless of the microorganisms and fermentation conditions used, only a single diastereoisomer, (*S,S*)-**49**, could be obtained with synthetically useful diastereoselectivities. Side products were also formed during this transformation [112]. To circumvent this problem, the cytochrome *P*450 BM-3, a medium chain (C_{12}–C_{18}) fatty acid hydroxylase from *B. megaterium* engineered to accept nonnatural substrates with enhanced regioselectivity, enantioselectivity, catalytic rates, and total turnover, was evaluated [113,114]. Mutations in the active site, for example, enable the enzyme to hydroxylate alicyclic, heterocyclic, aromatic, and even polyaromatic compounds [115]. Biohydroxylation using the wild-type enzyme expressed in *E. coli* gave very low enantioselectivity (1.5%, *S,S*) and high diastereoselectivity (87%). Mutant 139–3 also afforded (*S*,*S*)-**49**, but with a high *ee* (86%) and *de* (96%). In dramatic contrast, mutant 1-12G was found to produce (*R*,*R*)-**49** in high selectivities (89% *ee*, 94% *de*) [111].

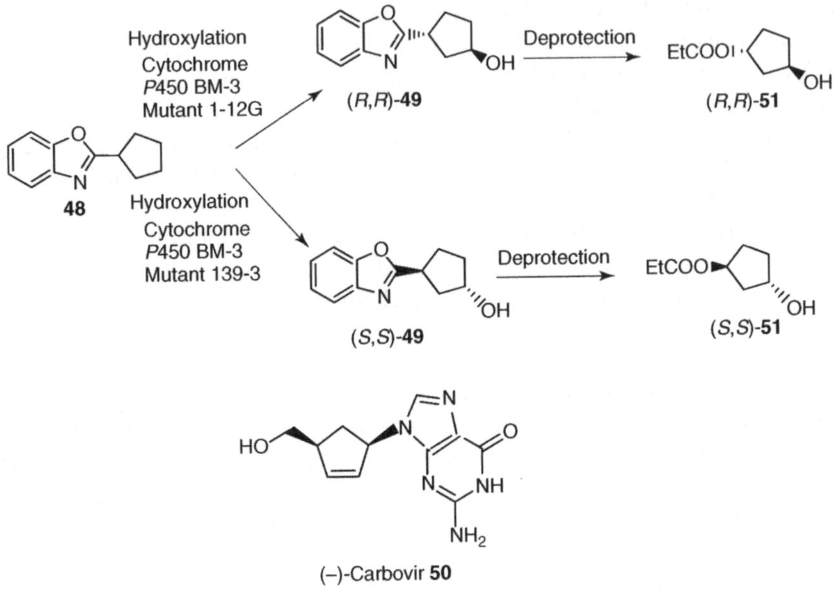

Figure 5.17 Carbovir: Enzymatic hydroxylation 2-cyclopentylbenzoxazole.

5.4.3 Epothilones (Anticancer Drugs): Microbial Hydroxylation of Epothiolone B to Epothilone F

The clinical success of paclitaxel has stimulated research into compounds with similar modes of activity in an effort to emulate its antineoplastic efficacy while minimizing its less desirable aspects, which include non-water solubility, difficult synthesis, and emerging resistance. The epothilones are a novel class of natural product cytotoxic compounds derived from the fermentation of the myxobacterium *Sorangium cellulosum* that are non-taxane microtubule-stabilizing compounds that trigger apoptosis [116,117]. The natural product epothilone B **52** (Figure 5.18) has demonstrated broad spectrum antitumor activity *in vitro* and *in vivo*, including in tumors with paclitaxel resistance [118]. The role of **75** as a potential paclitaxel successor has initiated interest in its synthesis, resulting in several total syntheses of **52** and various derivatives thereof [119,120]. The epothilone analogs were synthesized in an effort to optimize the water solubility, *in vivo* metabolic stability, and antitumor efficacy of this class of antineoplasic agents [121,122].

A fermentation process was developed for the production of epothilone B, and the titer of epothilone B was optimized and increased by a continuous feed of sodium propionate during fermentation. The inclusion of XAD-16 resin during fermentation to adsorb epothilone B and to carry out volume reduction made recovery of the product very simple [123]. A microbial hydroxylation process was developed for conversion of epothilone B **52** to epothilone F **53** by *Amycolatopsis orientalis* SC 15847. Epothilone F is a key intermediate for the synthesis of the

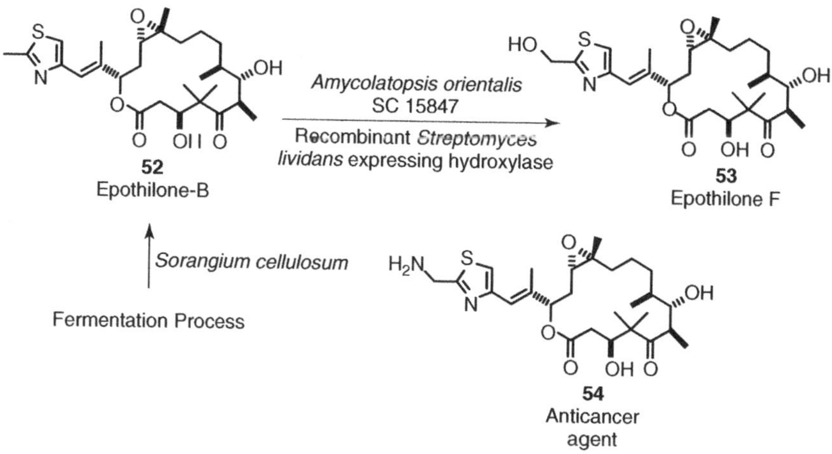

Figure 5.18 Epothilones: Microbial hydroxylation of epothiolone B to epothilone F.

anticancer drug **54**. A bioconversion yield of 37–47% was obtained when the process was scaled up to 100–250 L (Patel *et al.*, unpublished results). Recently, the epothilone B hydroxylase along with the ferredoxin gene has been cloned and expressed in *Streptomyces rimosus* from *A. orientalis* SC 15847 by our colleagues in Bristol-Myers Squibb. Mutants and variants of this cloned enzyme have been used in the hydroxylation of epothilone B to epothilone F to obtain even higher yields of the product [124].

5.4.4 Hydroxumutulins (Anti-Infective Drugs): Microbial Hydroxylation of Pleuromutilin or Mutilin

Pleuromutilin **55** (Figure 5.19) is an antibiotic from *Pleurotus* or *Clitopilus* basidiomycetes strains, which kills mainly gram-positive bacteria and mycoplasms. A more active semisynthetic analog, tiamulin, has been developed for the treatment of animals and poultry infection and has been shown to bind to prokaryotic ribosomes and inhibit protein synthesis [125]. Metabolism of pleuromutilin derivatives results in hydroxylation by microsomal cytochrome *P*-450 at the 2- or 8-position and inactivates the antibiotics [126]. Modification of the 8-position of pleuromutilin and analogs is of interest as a means of preventing the metabolic hydroxylation. Microbial hydroxylation of pleuromutilin **55** or mutilin **56** would provide a functional group at this position to allow further modification. The target analogs would maintain the biological activity of the parent compounds but would not be susceptible to metabolic inactivation.

Biotransformation of mutilin and pleuromutilin by microbial cultures was investigated to provide a source of 8-hydroxymutilin or 8-hydroxypleuromutilin [127]. *Streptomyces griseus* strains SC 1754 and SC 13971 (ATCC 13273) hydroxylated mutilin to (8*S*)-hydroxymutilin **59**, (7*S*)-hydroxymutilin **58**, and (2*S*)-hydroxymutilin **57**. *Cunninghamella echinulata* SC 16162 (NRRL 3655)

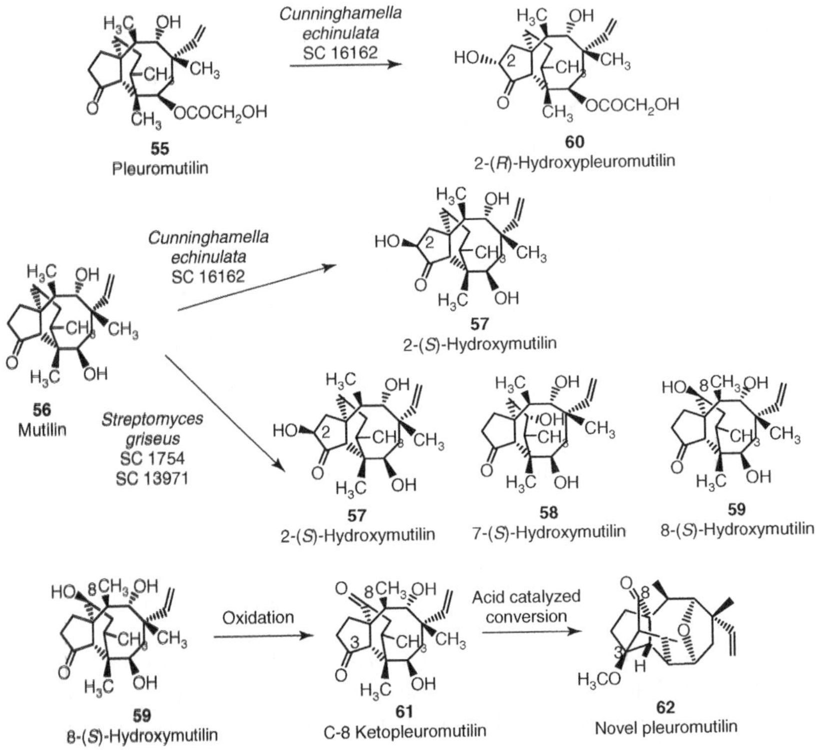

Figure 5.19 Hydroxumutulins: Microbial hydroxylation of pleuromutilin or mutilin.

gave (2*S*)-hydroxymutilin or (2*R*)-hydroxypleuromutilin **60** from biotransformation of mutilin or pleuromutilin, respectively. The biotransformation of mutilin by the *S. griseus* strain SC 1754 was scaled up to produce a total of 49 g of (8*S*)-hydroxymutilin, 17 g of (7*S*)-hydroxymutilin, and 13 g of (2*S*)-hydroxymutilin from 162 g of mutilin [127].

A C-8 ketopleuromutilin **61** derivative has been synthesized from the biotransformation product 8-hydroxymutilin [128]. A key step in the process was the selective oxidation at C-8 of 8-hydroxymutilin using tetrapropylammonium perruthenate. The presence of the C-8 keto group precipitated interesting intramolecular chemistry to afford **62** with a novel pleuromutilin-derived ring system by acid-catalyzed conversion of C-8 ketopleuromutilin.

5.5 ENZYMATIC REDUCTION PROCESS

5.5.1 Thrombin Inhibitor: Enzymatic Preparation of (*R*)-2-Hydroxy-3,3-dimethylbutanoic Acid

Thrombin is a trypsin-like protease enzyme that plays a critical role in intrinsic and extrinsic blood coagulation. As a result of the enzymatic activation of numerous coagulation factors, thrombin is activated to cleave fibrinogen, producing fibrin, which is directly responsible for blood clotting. An imbalance between these factors and their endogenous activators and inhibitors can give rise to a number of disease states such as myocardial infarction, unstable angina, stroke, ischemia, restenosis following angioplasty, pulmonary embolism, deep vein thrombosis, and arterial thrombosis [129,130]. Consequently, the aggressive search for a potent, selective, and bioavailable thrombin inhibitor is widespread [131]. An intensive effort by Merck has led to the identification of thrombin inhibitor **63** [132]. The synthesis of **63** (Figure 5.20) required a key chiral intermediate (*R*)-hydroxy ester **64**. An enzymatic process was developed for the asymmetric reduction of ketoester **65** to (*R*)-**64** using commercially available ketoreductase KRED1001. The cofacator NADPH required for this reaction was regenerated using glucose dehydrogenase. The hydroxy ester (*R*)-**64** was isolated as an oil and then saponified to the corresponding enantiomerically pure hydroxy acid (*R*)-**66** without epimerization [133]. The enantiomerically pure (*R*)-**66** was obtained with an 82% isolated yield (>99.5% *ee*).

5.5.2 Hydroxy Buspirone (Antianxiety Drug): Enzymatic Preparation of 6-Hydroxybuspirone

Buspirone (Buspar®, **67**, Figure 5.21) is a drug used for the treatment of anxiety and depression, which is thought to produce its effects by binding to the serotonin 5HT1A receptor [134–136]. Mainly as a result of hydroxylation reactions, it is extensively converted to various metabolites, and blood concentrations return to low levels a few hours after dosing [137]. A major metabolite, 6-hydroxybuspirone **68**, produced by the action of liver cytochrome P450

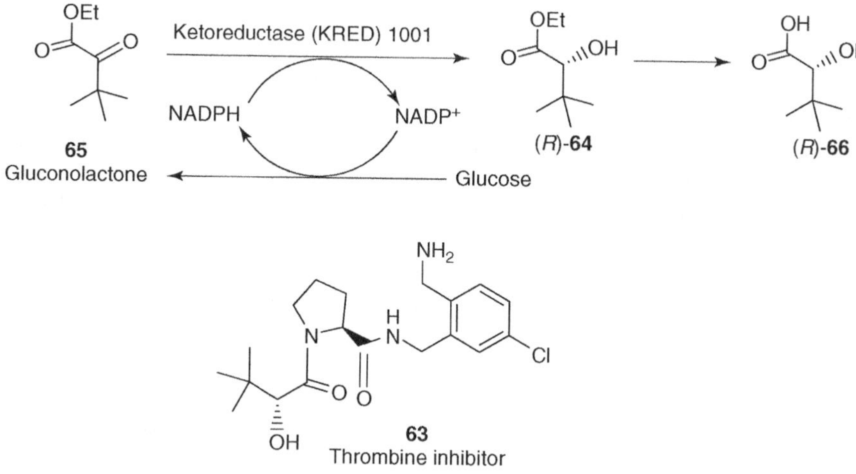

Figure 5.20 Thrombin inhibitor: Enzymatic preparation of (*R*)-2-hydroxy-3,3-dimethylbutanoic acid.

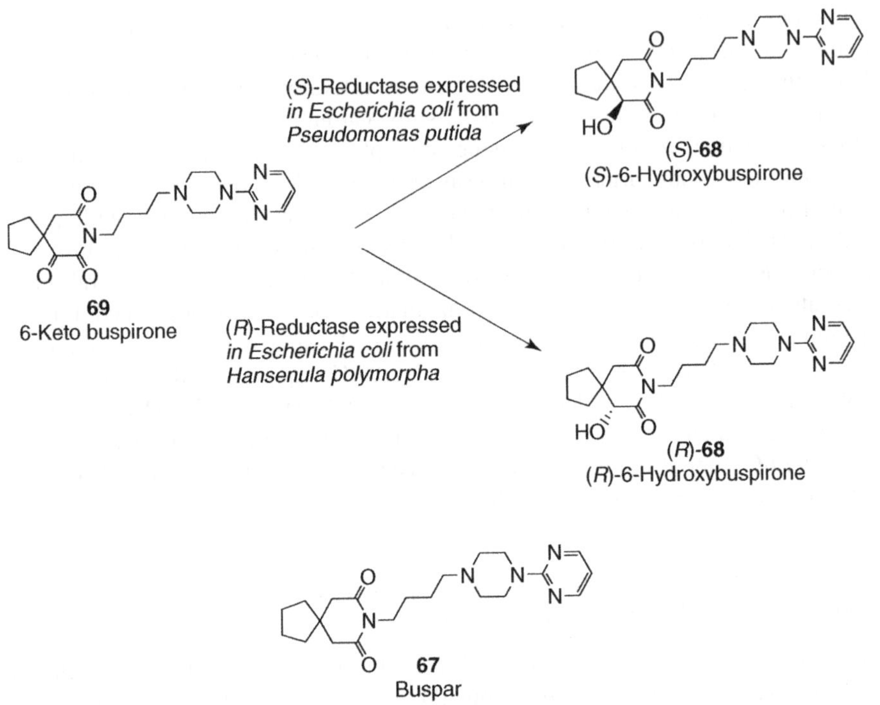

Figure 5.21 Hydroxy buspirone: Enzymatic preparation of 6-hydroxybuspirone.

CYP3A4, is present at much higher concentrations in human blood than buspirone itself. For the development of 6-hydroxybuspirone as a potential antianxiety drug, preparation and testing of the two enantiomers as well as the racemate was of interest. An alternate enantioselective microbial reduction process was developed for the reduction of 6-oxobuspirone **69** to either (*R*)- or (*S*)-6-hydroxybuspirone **68**. About 150 microorganisms were screened for the enantioselective reduction of **69**. *Rhizopus stolonifer* SC 13898, *Neurospora crassa* SC 13816, *Mucor racemosus* SC 16198, and *P. putida* SC 13817 gave >50% reaction yields and >95% *ee*'s of (*S*)-6-hydroxybuspirone. The yeast strains *Hansenula polymorpha* SC 13845 and *Candida maltosa* SC 16112 gave (*R*)-6-hydroxybuspirone **68** with a >60% reaction yield and a >97% *ee* [138]. The NADP-dependent (*R*)-hydroxy buspirone reductase (RHBR) from *Hansenula polymorpha* SC 13845 was cloned and expressed in *E. coli*. To regenerate the cofactor NADPH required for reduction, we have also cloned and expressed the glucose-6-phosphate dehydrogenase gene from *Saccharomyces cerevisiae* in *E. coli*. In addition, the NAD-dependent (*S*)-hydroxy buspirone reductase (SHBR) from *P. putida* SC 16269 was cloned and expressed in *E. coli*. To regenerate the cofactor NADH required for reduction, we have also cloned and expressed the FDH gene from *P. pastoris* in *E. coli*. Recombinant *E. coli* expressing (*S*)-reductase and (*R*)-reductase were used to catalyze the reduction

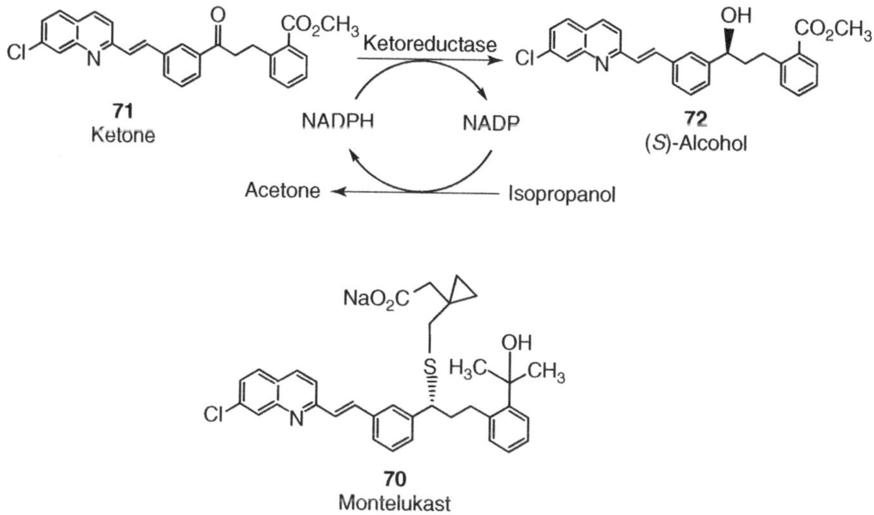

Figure 5.22 Montelukast: Enantioselective enzymatic reduction process.

of 6-ketobuspirone to (*S*)-6-hydroxybuspirone and (*R*)-6-hydroxybuspirone, respectively, with a >98% yield and a >99.9% *ee* [139].

5.5.3 Montelukast (Treatment of Asthma): Enzymatic Reduction Process

Montelukast **70** (Figure 5.22), a selective reversible cysteinyl leukotriene D_4-receptor (LTD_4 receptor) antagonist, is used in the treatment of asthma. Merck has described the synthetic route for the production of montelukast, using a stereoselective chemical reduction of a ketone to the (*S*)-alcohol [140,141]. The reduction of the ketone Montelukast II (MLK-II) **71** to produce the chiral alcohol Montelukast III (MLK III) **72** requires stoichiometric amounts of the chiral reducing agent (−)-chloro diisopinocampheylborane [(−)-DIP-chloride]. In addition to being expensive, (−)-DIP-chloride is both corrosive and moisture sensitive, causing burns if allowed to come in contact with the skin. The reaction must be carried out at −20 to −25°C to achieve the best stereoselectivity and the exothermicity of the reaction. In addition, the quench and extractive workup generate large volumes.

An enzyme-catalyzed reduction of the ketone **71** was developed with *Microbacterium campoquemadoensis* [142].

Recently, a ketoreductase was developed using directed evolution technology, which facilitated the rapid improvement of enzymes in order to achieve the requisite activity, selectivity, and stability using the gene shuffling procedure. The enzyme was improved 2000-fold and the reduction of ketone **71** to chiral alcohol **72** (Figure 5.22) was carried out in a mixture of water, isopropanol, and toluene as a solvent at 45°C. A reaction yield of 99.3% and an *ee* of 99.9% were obtained at a substrate input of 100 g/L [143,144].

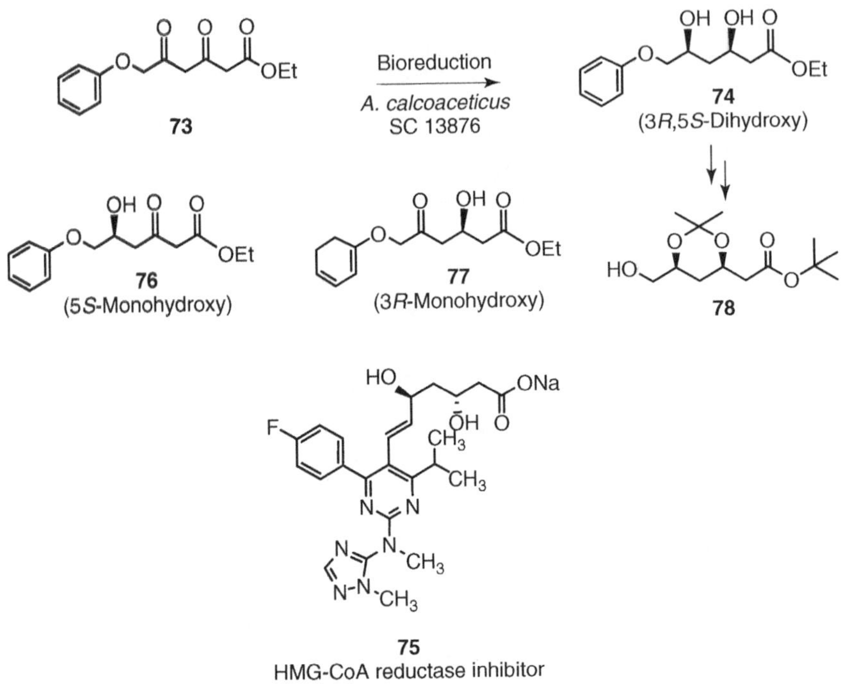

Figure 5.23 Chloesterol-lowering agents: Enzymatic preparation of (3*R*,5*S*)-dihydroxy-6-(benzyloxy) hexanoic acid, ethyl ester.

5.5.4 Cholesterol-Lowering Agents: Enzymatic Preparation of (3S,5R)-dihydroxy-6-(benzyloxy) Hexanoic Acid, Ethyl Ester

The enantioselective reduction of a diketone 3,5-dioxo-6-(benzyloxy) hexanoic acid, ethyl ester **73** to (3*R*,5*S*)-dihydroxy-6-(benzyloxy) hexanoic acid, ethyl ester **74** (Figure 5.23) was demonstrated using cell suspensions of *Acinetobacter calcoaceticus* SC 13876 [145]. Compound **74** is a key chiral intermediate required for the chemical synthesis of **75** and Lipitor, both of which are anticholesterol drugs that act by inhibition of Hydroxymethyl glutaryl CoA (HMG CoA) reductase [146,147]. A reaction yield of 85% and an *ee* of 97% were obtained. Cell extracts of *A. calcoaceticus* SC 13876 in the presence of NAD^+, glucose, and glucose dehydrogenase reduced **73** to the corresponding monohydroxy compounds [3-hydroxy-5-oxo-6-(benzyloxy) hexanoic acid ethyl ester **76** and 5-hydroxy-3-oxo-6-(benzyloxy) hexanoic acid ethyl ester **77**]. Both **76** and **77** were further reduced to the (3*R*,5*S*)-dihydroxy compound **74** with a 92% yield and a 99% *ee* using cell extracts. (3*R*,5*S*)-**74** was converted to **78**, a key chiral intermediate for the synthesis of **75**. Three different ketoreductases were purified to homogeneity from cell extracts, and their biochemical properties were compared. Reductase I catalyzes only the reduction of ethyl diketoester **73** to its monohydroxy products, whereas reductase II catalyzes the formation of dihydroxy products from monohydroxy substrates. A third reductase (III) was identified that catalyzes the reduction of diketoester **73** to *syn*-(3*R*,5*S*)-dihydroxy ester **74** [148], which now has been cloned and expressed in *E. coli* [149] and the reduction of diketoester **73** to *syn*-(3*R*,5*S*)-dihydroxy ester **74** was demonstrated using the recombinant enzyme.

5.6 MICROBIAL BAEYER–VILLIGER OXIDATIONS

5.6.1 Recombinant *E. coli* Expressing Cyclopentanone Monooxygenase

Enzyme-mediated Baeyer–Villiger oxidation offers a "green chemistry" approach for the production of chiral lactones [150–161]. In contrast to the classical protocols of this reaction, which require potentially explosive oxidants (such as hydrogen peroxide or various peracids), display limited functional group tolerance, and lack high stereoselectivity with *de novo* designed chiral catalysts, biocatalysis offers a very mild, environment friendly, and stereoselective entry to this process by utilizing molecular oxygen as the oxidant. An increasing number of flavin-dependent Baeyer–Villiger monooxygenase (BVMOs) with a remarkably broad profile of non-natural substrates has been identified during recent years [150–161]. This has led to the identification of an increasing variety of structurally diverse substrate ketones that are converted efficiently and in high stereoselectivity and that can serve as precursors for the subsequent asymmetric synthesis of natural and bioactive compounds. Enantiodivergent whole cell Baeyer–Villiger oxidation of *rac*-bicyclo[3.2.0]hept-2en-6one **79** (Figure 5.24a) by cyclohexanone (CH) monooxygenase has been demonstrated on a kilogram scale [157]. Regioselective Baeyer–Villiger oxidation of *rac*-3-methylcyclohexanone **80** (Figure 5.24b) by cyclopentanone (CP) monooxygenase yielding only the proximal lactone has also been demonstrated [158].

Microbial Baeyer–Villiger oxidation of different substrates with cyclopentanone monooxygenase (CPMO) from *Comamonas* NCIMB 9872 was upscaled to a benchtop fermenter. Conditions for cell growth and biocatalyst production were optimized, and a convenient, environmentally friendly and applicable methodology for the preparation of chiral building blocks for natural product and bioactive compound synthesis was developed by applying a resin based on the concept of *in situ* "substrate feeding-product removal" (SFPR). Three different ketones (4-methylcyclohexanone, *rac*-3-methylcyclohexanone, and 8-oxabicyclo[3.2.1]oct-6-en-3-one) were converted on a 5–15 g/L scale in a conventional bioreactor, with a volumetric productivity of up to 1 g/L/h with good to excellent yield and enantiomeric purity [159].

Studies of the regiodivergent and enantiocomplementary biooxidations in the CH series, in particular, the biooxidation of pure enantiomers of *trans*- and *cis*-dihydrocarvone **81**, carvomenthone **82**, and menthone **83** (Figure 5.24c), have been reported utilizing a collection of BVMO producing recombinant *E. coli* strains to provide access to all regio- and enantiomeric lactones from

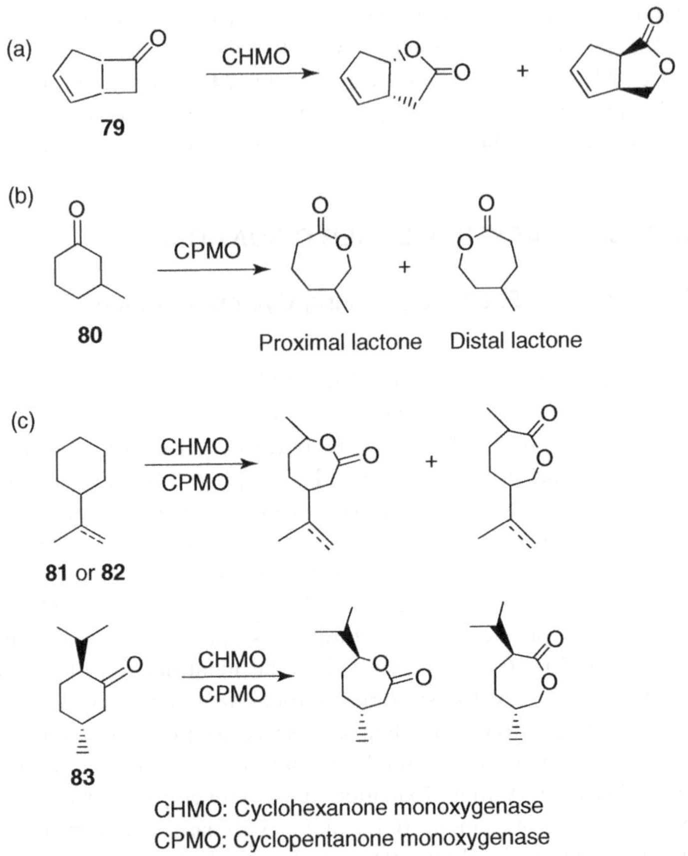

Figure 5.24 Microbial Baeyer–Villiger oxidations catalyzed by cyclohexanone and cyclopentanone monooxygenase.

terpenone precursors [160]. Substrate acceptance profiles and stereopreferences of CH and CP monooxygenases originating from *Acinetobacter, Arthrobacter, Brachymonas, Brevibacterium*, and *Comamonas* species in whole-cell-mediated Baeyer–Villiger oxidations using recombinant *E. coli* were compared to get a variety of products.

By means of a comparison of the phenylacetone monooxygenase (PAMO) structure and a modeled structure of the sequence-related CP monooxygenase, several active-site residues were selected for a mutagenesis study in order to alter the substrate specificity. The M446G PAMO mutant was found to be active with a number of aromatic ketones, amines, and sulfides for which wild-type PAMO shows no activity. In addition to an altered substrate specificity, the enantioselectivity toward several sulfides was dramatically improved. This newly designed Baeyer–Villiger monooxygenase extends the scope of oxidation reactions feasible with these atypical monooxygenases [161].

5.7 ENZYMATIC CONDENSATION REACTIONS

5.7.1 Enzymatic Acyloin Condensation

Stereoselective carbon–carbon bond forming reactions are among the most useful methods in asymmetric synthesis as this allows the simultaneous creation of up to two adjacent stereocenters [162]. Several classes of enzymes are able to catalyze stereoselective C–C bond formation reactions such as aldolases [162], oxynitrilases [163], or thiamin diphosphate-dependent decarboxylases [164–178]. Acyloin formation mediated by bacterial benzoylformate decarboxylase [164,166–168,170–172,177,178], yeast pyruvate decarboxylase [168–175], and phenylpyruvate decarboxylase (PPD) have been reported [173].

The industrial process based on yeast pyruvate decarboxylase for the production of (*R*)-phenylacetyl carbinol **84** (Figure 5.25a), a precursor to (−)-ephedrine, is still used on an industrial scale [166]. Bacterial and yeast decarboxylases have been evaluated as biocatalysts not only for synthesis of acyloin-type compounds, but also for production of hydroxyl ketones, α-arylacetate, *R*-amino acids, dopamine, and organic acids [166–169,175].

Benzoylformate decarboxylase carries out the conversion of benzoylformate to benzaldehyde. Wilcocks *et al.* [176] demonstrated the benzoylformate-decarboxylase-dependent formation of acyloin compound (*S*)-2-hydroxypropiophenone **85** (Figure 5.25b) or (*R*)-benzoin from benzoylformate and acetaldehyde. The synthesis of benzoin and substituted benzoins was demonstrated [171]. Mutant enzymes *P. putida* benzoylformate decarboxylase L476Q and M365L-L461S accepted *ortho*-substituted benzaldehyde derivatives, such as 2-chloro-benzaldehyde or 2-bromo-benzaldehyde, as donor substrates, leading to the production of 2-hydroxyketones and produced enantiopure (*S*)-2-hydroxy-1-(2-methylphenyl)propan-1-one with excellent yield and *ee* [178]. A review article describing various benzylformate-catalyzed acyloin condensation reactions has been published [164–168,175].

Asymmetric acyloin condensation of phenylpyruvate with various aldehydes to produce optically active acyloins $PhCH_2COCH(OH)R$ **86** (Figure 5.25c) has been demonstrated [173]. The acyloin condensation yield decreased with increasing chain length for straight-chain aliphatic aldehydes from 76% for acetaldehyde to 24% for valeraldehyde. The *ee*'s of the acyloin products ranged from 87% to 98%. Indole-3-pyruvate was a substrate for the PPD and provided the acyloin condensation product 3-hydroxy-1-(3-indolyl)-2-butanone **87** with acetaldehyde with a 19% yield.

5.7.2 Enzymatic Aldol Condensation

Aldolases achieve the activation of an aldol donor substrate by two mechanistically different pathways. Class I aldolases bind their substrates covalently at an active-site lysine residue via an imine–enamine intermediate to initiate C–C bond formation or cleavage whereas class II aldolases utilize transition metal ions as a Lewis acid cofactor, which facilitates the stereospecific deprotonation

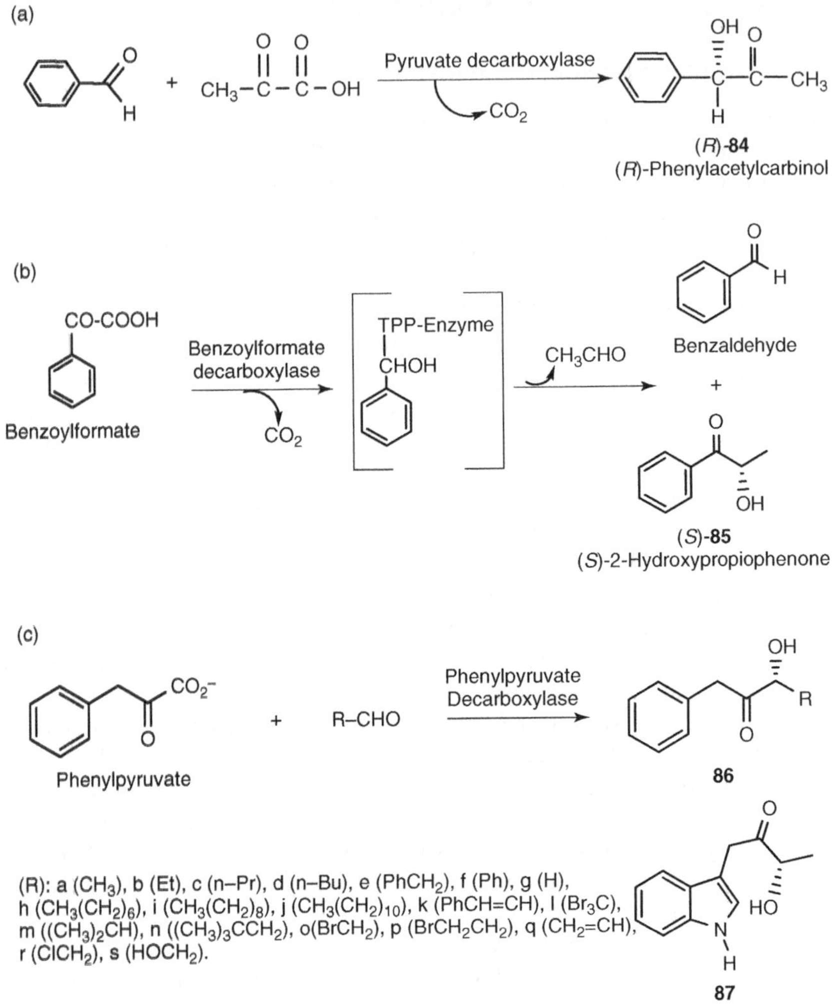

Figure 5.25 Enzymatic acyloin condensations catalyzed by pyruvate-, benzylformate-, and phenyl pyruvate decaroxylases.

of the donor by bidentate coordination to stabilize the enediolate nucleophile [153]. Aldolases have been subdivided into certain families, depending on their nucleophilic substrate: dihydroxyacetone phosphate-dependent aldolases, pyruvate- (and phosphoenolpyruvate-) dependent lyases, aldehyde-dependent aldolases, and glycine-dependent enzymes. Members of the first two families add three-carbon ketone fragments onto the carbonyl group of an aldehyde, yielding 1,3,4-trihydroxylated methyl ketone derivatives or 3-deoxy-2-oxoacids, respectively. The latter two families use the carbon fragments acetaldehyde and glycine, respectively, as the nucleophilic component. Today, many useful synthetic applications have been demonstrated for all classes of aldolases and for related C–C bond-forming enzymes [162,179]. The following select examples demonstrate the synthetic utility of alodolases.

The synthesis of rare and novel "aza sugars," which represent powerful glycosidase inhibitors or show potent antiviral activity, has been developed using D-fructose 1,6-bisphosphate aldolase [180,181]. This flexible synthetic strategy consists of the addition of an aldol to an *N*-functionalized aldehyde, which usually contains an azide group, followed by hydrogenolytic intramolecular reductive amination of the keto functionality (including prior azide to amine reduction) to close an *N*-heterocyclic ring structure stereoselectively [181]. Prominent examples concern the FruA-catalyzed synthesis of deoxynojirimycin **88** and deoxymannojirimycin **89** (Figure 5.26) from 3-azido-2-hydroxypropanal [182]. The stereodivergent preparation of a large variety of diastereomeric aza sugars of the nojirimycin type from any single azido aldehyde can be achieved by choosing from the Dihydroxy acetone phosphate (DHAP) aldolases having different stereospecificity [183].

Starting from (*S*)- and (*R*)-3-azido-2-acetamidopropanal in D-fructose 1,6-bisphosphate aldolase-catalyzed reactions, aza sugar derivatives corresponding to *N*-acetyl-glucosamine and *N*-acetyl-mannosamine have been prepared [182]. Similarly, a D-fructose 1,6-bisphosphate aldolase-mediated enenatioselective dihydroxy acetone phosphate addition to the intermediate served as the key step in the chemoenzymatic synthesis of 3-epiaustraline **90** and australine **91** (Figure 5.27) [184]. Australine and 3-epiaustraline are naturally occurring pyrrolizidine alkaloids having potent glycosidase inhibitory and antiviral activities.

Phospho-2-keto-3-deoxygluconate (KDPG) aldolase has been utilized for the highly enantioselective synthesis of chiral hydroxyl amino acid **92**, which represents the *N*-terminal amino acid part of the nucleoside antifungal nikkomycin K_x and K_z (Figure 5.28) [185,186]. In the one-pot two-step synthesis, the enantioselective aldol reaction is followed by a reductive amination, catalyzed by a PDH, and gives an overall yield of 75.9% and an *ee* of >99.7%.

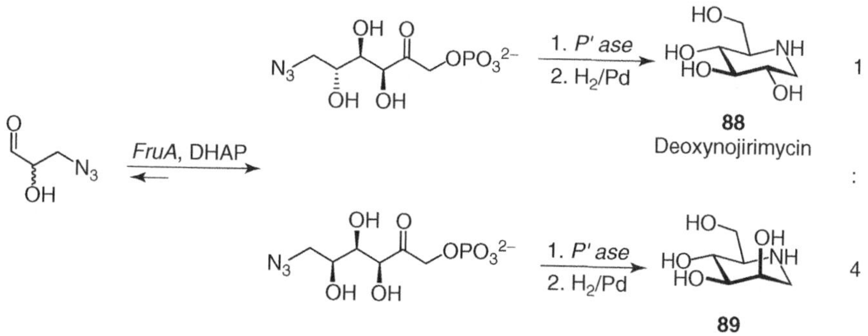

Figure 5.26 Aldolase-catalyzed synthesis of deoxynojirimycin and deoxymannojirimycin.

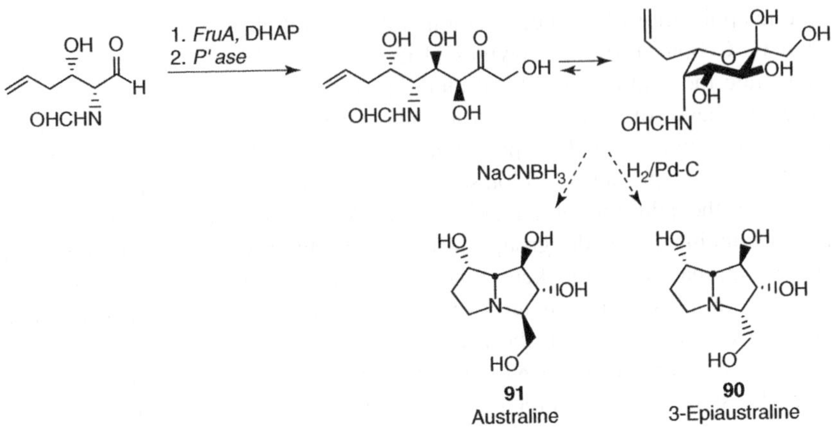

Figure 5.27 Aldolase-catalyzed synthesis of 3-epiaustraline and australine.

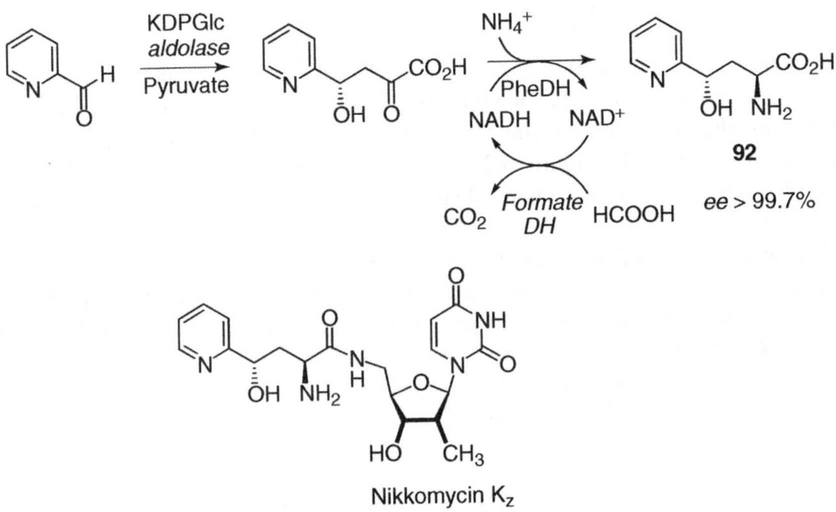

Figure 5.28 KDPG-aldolase-catalyzed synthesis of chiral hydroxyl amino acid.

Zanamivir **93** (Figure 5.29), the first medicinally used sialic acid derivative was developed by Glaxo with an antiviral activity effective against influenza. This required the cost-effective synthesis of the expensive *N*-acetylneuraminic acid, for which NeuA from *E. coli* was used on an industrial scale [187–189]. The expensive *N*-acetylmannosamine was produced by combining an enzymatic *in situ* isomerization of inexpensive *N*-acetylglucosamine catalyzed by *N*-acetyl glucosamine 2-epimerase with the NeuA reaction in an enzyme membrane reactor (Scheme 42) in the presence of excess of pyruvate [180].

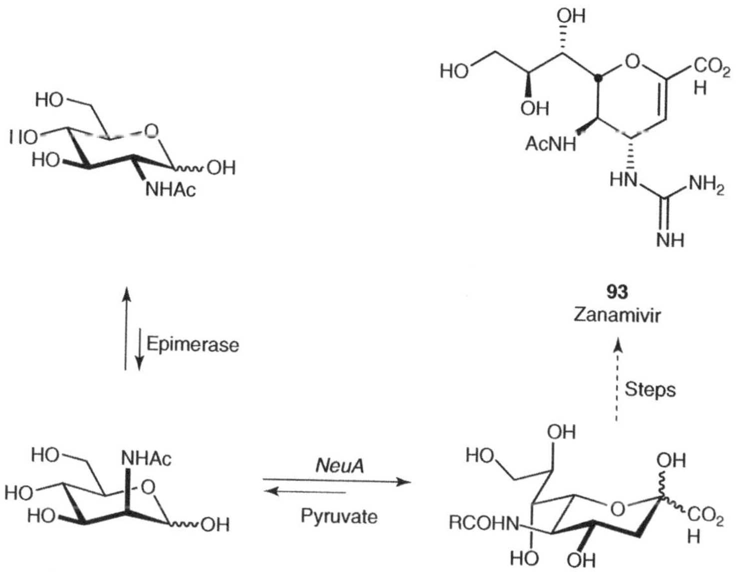

Figure 5.29 Enzymatic synthesis of *N*-acetylmannosamine.

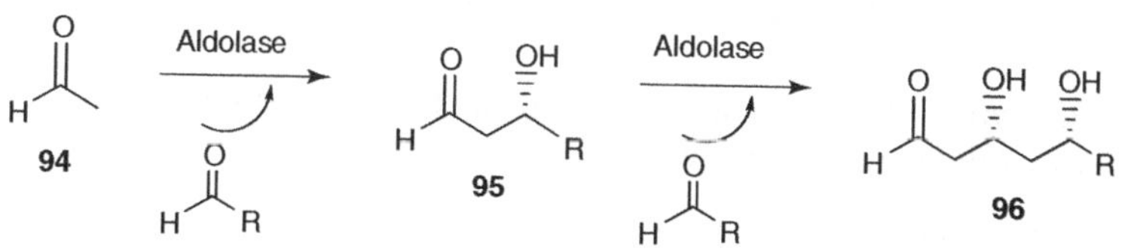

Figure 5.30 Enzymatic synthesis of 2,4-dideoxyhexose derivative.

The chiral 2,4-dideoxyhexose derivative required for the HMG CoA reductase inhibitors has also been prepared using 2-deoxyribose-5-phosphate aldolase (DERA). The reactions start with a stereospecific addition of acetaldehyde **94** (Figure 5.30) to a substituted acetaldehyde to form a 3-hydroxy-4-substituted butyraldehyde **95**, which reacts subsequently with another acetaldehyde to form a 2,4-dideoxyhexose derivative **96**. DERA has been expressed in *E. coli* [190].

The above process has been improved and optimized. An increase of almost 400-fold in volumetric productivity relative to the published enzymatic reaction conditions has been achieved, resulting in an attractive process that has been run on a scale of up to 100 g in a single batch at a rate of 30.6 g/L/h. The catalyst load was improved by 10-fold, from 20 to 2.0 wt% DERA. These improvements were achieved by a combination of the discovery of a DERA with improved activity and reaction optimization to overcome substrate inhibition. The two stereogenic centers are set by DERA with an *ee* of >99.9% and a diastereomeric excess of 96.6% [191].

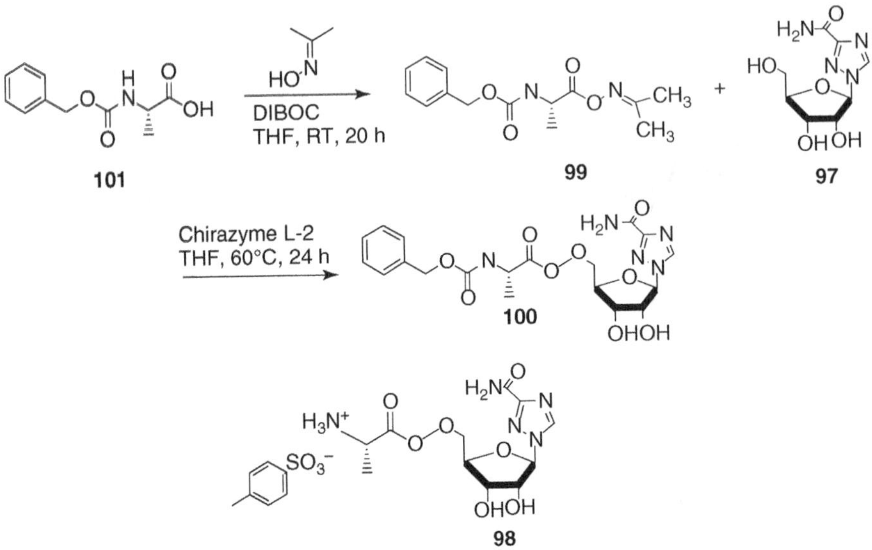

Figure 5.31 Ribavirin prodrug: Regioselective enzymatic acylation of ribavirin.

5.8 ENANTIOSELECTIVE ENZYMATIC ACYLATION

5.8.1 Ribavirin Prodrug (Antiviral Drug): Regioselective Enzymatic Acylation of Ribavirin

Ribavirin **97** (Figure 5.31) is an antiviral agent used in combination with α-2β interferon to treat hepatitis C [192,193]. Although this therapy is effective against the hepatitis C virus, it has several side effects [194]. To improve the pharmacokinetic profile and reduce side effects, a ribavirin prodrug was considered for development. In a series of preclinical evaluations, the alanine ester of ribavirin **98** showed improved bioavalibility and reduced side effects. The synthesis of **98** required the acylation of unprotected ribavirin. The chemical acylation gave a mixture of mono-, di-, and triacylated products. An enzymatic process was developed for the regioselective acylation of ribavirin **97** with the oxime ester of L-carbobenzyloxy-alanine **99** to give the desired **100** using Novozym 435 (*C. antarctica* lipase B or Chirazyme L-2). Chemical deprotection of **100** gave **98**. On a preparative scale, the coupling of **101** with acetone oxime in the presence of di-*tert*-butyl dicarbonate in THF (tetrahyrofuran) was carried out, giving **99** with a >96% yield. At the end of the reaction, the reaction mixture was diluted threefold with THF, ribavirin was added, and the acylation was initiated by the addition of Novozyme 435. After 24 h at 60°C, the product **100** was isolated with an 85% yield [195].

5.8.2 Xemilofiban: Enzymatic Preparation of a β-Aminoester

Xemilofiban **102** and related compounds are peptido-mimetics based on the arginine-glycine-aspartate (RGD) sequence of fibrinogen, which effectively disrupts the platelet–fibrinogen interaction, leading to the inhibition of platelet aggregation and prevention of thrombus formation [196,197].

The optically active β-amino ester (*S*)-**103** (Figure 5.32) has been used in the synthesis of xemilofiban. The resolution of (±)-**104** has been efficiently achieved by enzymatic acylation using immobilized penicillin G amidase and phenylacetic acid as the acyl donor. The reaction was carried out at pH 5.7 and 28°C to yield the phenylacetamide (*R*)-**105** and the unreacted β-amino ester (*S*)-**103**, which were isolated with a >99% *ee* and a 98% *ee*, respectively. This bioconversion was used to resolve 6.1 kg of amine, from which the required (*S*)-**103** was isolated with a >96% *ee* and a >43% yield [198].

5.8.3 Farnesyl Transferase Inhibitor: Enzymatic Resolution of Substituted (6,11-dihydro-5*H*-benzo-[5,6] cyclohepta[1,2-β]pyridin-11-yl)piperidines

An enzymatic process has been developed for the preparation of chiral intermediate (*R*)-**106** (Figure 5.33) for the synthesis of compound **107**, a selective farnesyl transferase inhibitor (involved in binding Ras, which moderates cell proliferation) [199]. Resolution of racemic-substituted (6,11-dihydro-5H-benzo-[5,6]cyclohepta[1,2-β]pyridin-11-yl)piperidines **108** has been demonstrated to yield the desired (*R*)-**106** and unreacted (*S*)-**109** by an enzymatic acylation process using a Toyobo lipase LIP-3000 in *t*-butyl methyl ether with trifluoroethyl isobutyrate [200]. A reaction yield of 46% with an *ee* of 97% was obtained for the *N*-isobutyryl derivative (*R*)-**106**, which was isolated and subsequently hydrolyzed to yield the desired product.

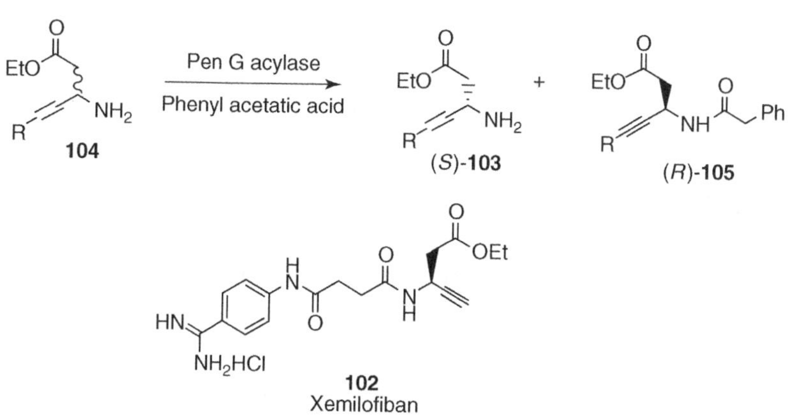

Figure 5.32 Enzymatic synthesis of chiral β-amino ester.

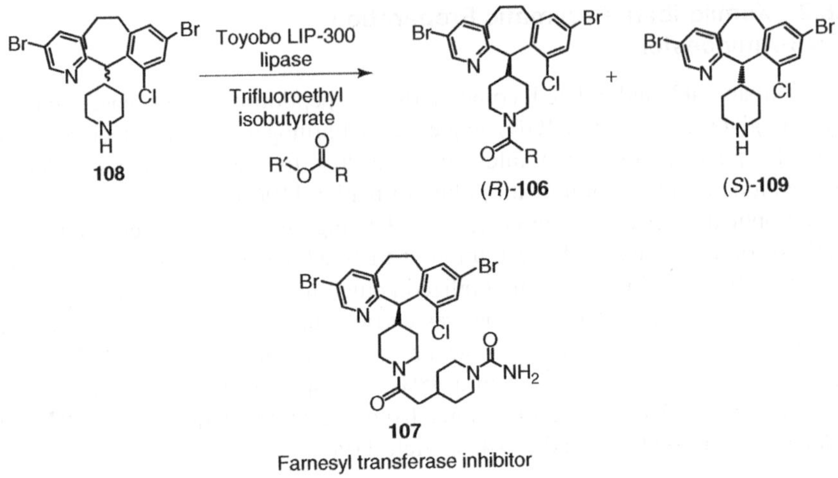

Figure 5.33 Enzymatic resolution of substituted (6,11-dihydro-5*H*-benzo-[5,6]cyclohepta[1,2-β]pyridin-11-yl)piperidines.

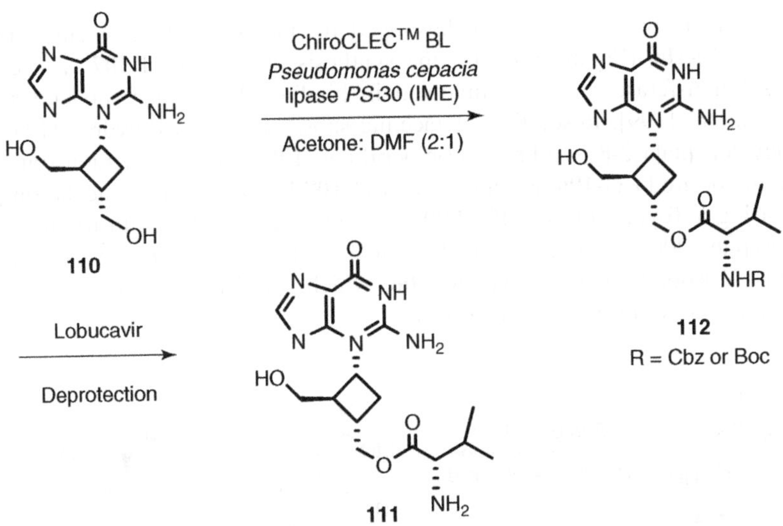

Figure 5.34 Regioselective enzymatic aminoacylation and hydrolysis.

5.8.4 Regioselective Enzymatic Aminoacylation and Hydrolysis

Lobucavir **110** (Figure 5.34) is a cyclobutyl guanine nucleoside analog under development as an antiviral agent for the treatment of herpes virus and hepatitis B [201]. A prodrug **111** in which one of the two hydroxyls is coupled to valine was considered for drug development. Regioselective aminoacylation, difficult to achieve by chemical procedures, appeared to be suitable for an enzymatic approach [95]. Synthesis of the lobucavir L-valine prodrug **111** requires regioselective coupling of one of the two hydroxyl groups of lobucavir **110** with valine. The synthesis of lobucavir prodrug was achieved by enzymatic transesterification of lobucavir with Cbz- or Boc-valine using ChiroCLEC™ BL (61% yield) or more selectively by using lipase from *P. cepacia* (84% yield) to give **112** [202].

5.8.5 Tryptase Inhibitor: Enzymatic Preparation of (S)-*N*-(*tert*-butoxycarbonyl)-3-hydroxymethylpiperidine

(*S*)-*N*-(*tert*-butoxycarbonyl)-3-hydroxymethylpiperidine **113** (Figure 5.35) is a key intermediate in the synthesis of a potent tryptase inhibitor **114** [203]. Lipase PS from *P. cepacia*-catalyzed esterification of *R,S-N*-(*tert*-butoxycarbonyl)-3-hydroxy methylpiperidine **113** with succinic anhydride **115** was demonstrated to provide *R-N*-(*tert*-butoxycarbonyl)-3-hydroxy methylpiperidine **113** and the (*S*)-hemisuccinate ester **116**, which could be easily separated and hydrolyzed by a base to *S*-**113**. The yield and the *ee* could be improved greatly by repetition of the process after isolation of the product. Using the repeated esterification/resolution procedure, *S*-**113** was obtained with a 32% yield (maximum theoretical yield 50%) and a 98.9% *ee* [204].

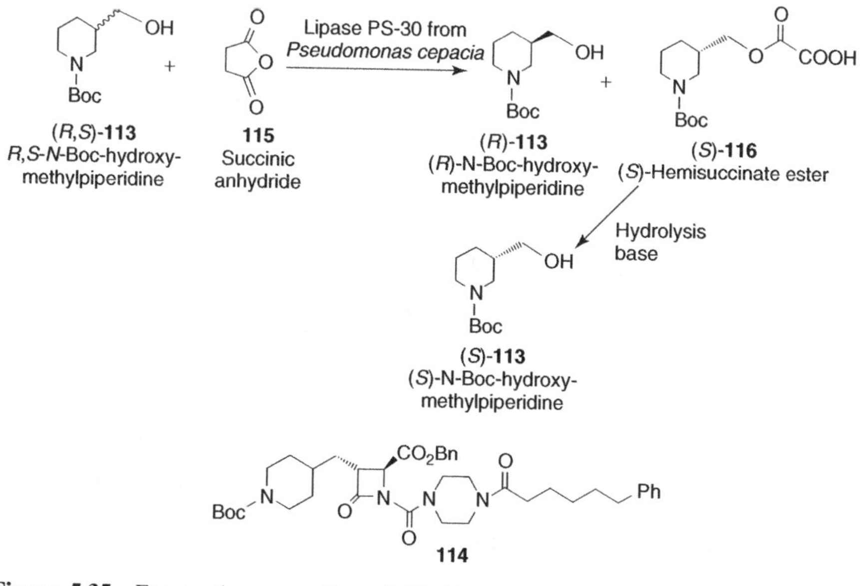

Figure 5.35 Enzymatic preparation of (*S*)-*N*-(tert-butoxycarbonyl)-3-hydroxymethylpiperidine.

5.9 DYNAMIC KINETIC RESOLUTION (DKR)

5.9.1 Enzymatic Synthesis of Intermediates for Propanolol and Denopamine

The deracemization of a number of pharmaceutically valuable building blocks has been carried out by biocatalytic processes. This includes epoxides, alcohols, amines, and acids [11,13,205,206]. DKR (dynamic kinetic resolution) involves a combination of an enantioselective transformation with an *in situ* racemization process such that, in principle, both enantiomers of the starting material can be converted to the product with high yield and *ee*. The racemization step can be catalyzed either enzymatically or nonenzymatically by transition metals.

Propanolol **117** (Figure 5.36) belongs to the group of β-adrenergic blocking agents of the general structure $ArOCH_2CH(OH)CH_2NHR$, where Ar is aryl and R is alkyl. These compounds are potentially useful for the treatment of hypertension. β-Adrenergic receptor blocking activity mainly resides in the (*S*)-enantiomers. Synthesis of (*S*)-**117** was achieved by DKR of (±)-**118** using Novozyme-435 in toluene at 80°C and *p*-chlorophenyl acetate as acyl donor in the presence of the ruthenium complex **119**. The (*R*)-acetate **120** was produced with an *ee* of >99% and an 86% isolated yield [207]. The enzyme was recycled and used again for another cycle without any loss of activity. Using the same procedure, racemic **121** was subjected to DKR to produce the (*R*)-acetate **122** (*ee*>99%, 92% conversion, and 84% isolated yield), a precursor of (*R*)-denopamine **123**, a potent, orally active β_1 receptor agonist for the treatment of hypertension [208].

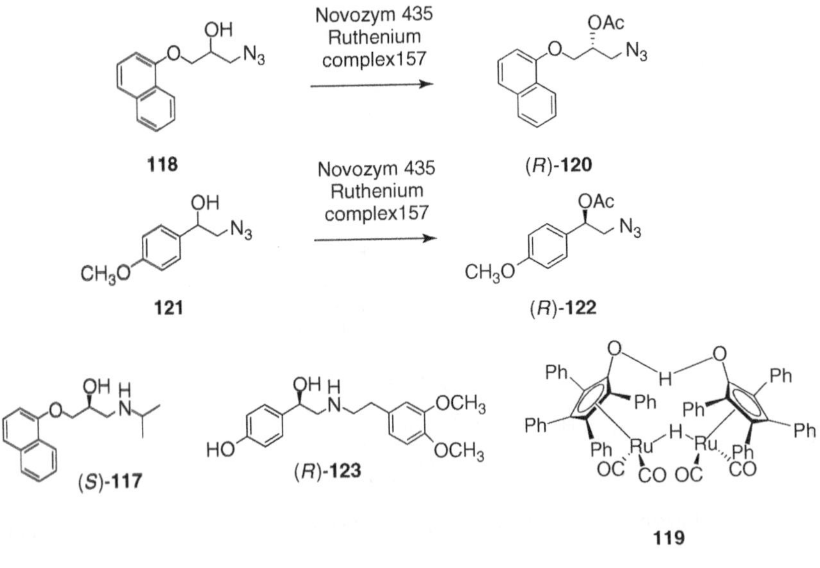

Figure 5.36 Enzymatic synthesis of intermediates for propanolol and denopamine.

5.9.2 Conversion of Racemic α-Methylbenzylamine to (*R*)-α-Methylbenzylamine

A robust and scalable technology has been developed for the preparation of optically active chiral amines by deracemization of racemic mixtures. The approach employs the simultaneous use of a highly selective oxidase biocatalyst and a chemical reducing agent or catalyst, and can be used for the preparation of a wide range of optically pure amines with yields often approaching 100%. An example is the conversion of racemic **α**-methyl benzylamine **124** (Figure 5.37) to (**R**)-α-methyl benzylamine **124** by amine oxidase from *Aspergillus niger* [208–210]. The key advantages of the technology lie in the coordinated action of already proven industrial catalysts and efficient methods of genetic screening to adapt the approach for the preparation of valuable industrial targets [208–210].

5.9.3 Dynamic Kinetic Resolution of Secondary Alcohols

The DKR of secondary alcohols has been demonstrated by combining enantioselective lipases with transition-metal-based racemization catalysts by various groups [211–214]. Kim has recently shown that (*S*)- as well as (*R*)-configured alcohols **125** (Figure 5.38) can be prepared by the use of a commercially available (*S*)-selective subtilisin from *Bacillus licheniformis* as the enantioselective acylating catalyst [215]. The reactions were carried out in THF as solvent, using trifluoroethyl butyrate as the acyl donor and an aminocyclopentadienyl ruthenium complex as the racemizing catalyst. In addition, it was found that pretreatment of the biocatalyst with a nonionic surfactant significantly improved the activity of the biocatalyst (about 4000-fold) under the reaction conditions.

Deracemization of racemic secondary alcohols by two different alcohol dehydrogenases with complementary enantiospecificity has been well described

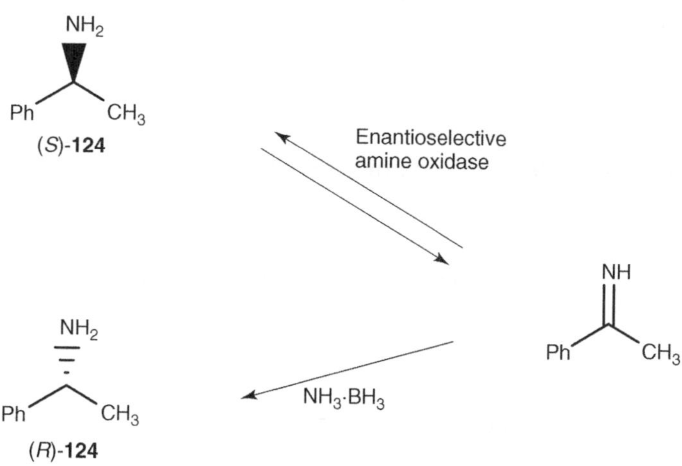

Figure 5.37 Enzymatic deracemization of α-methylbenzylamine.

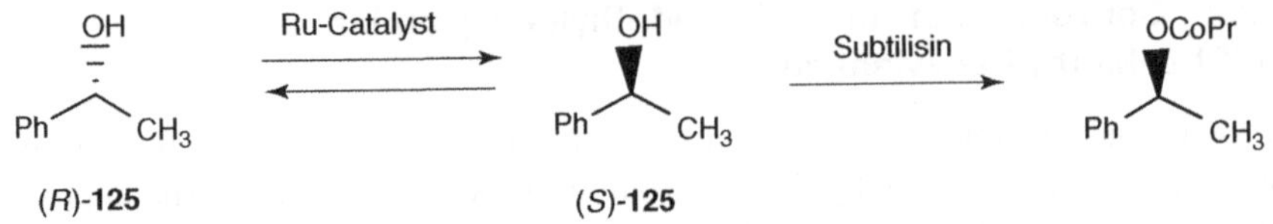

Figure 5.38 Dynamic kinetic resolution of secondary alcohols.

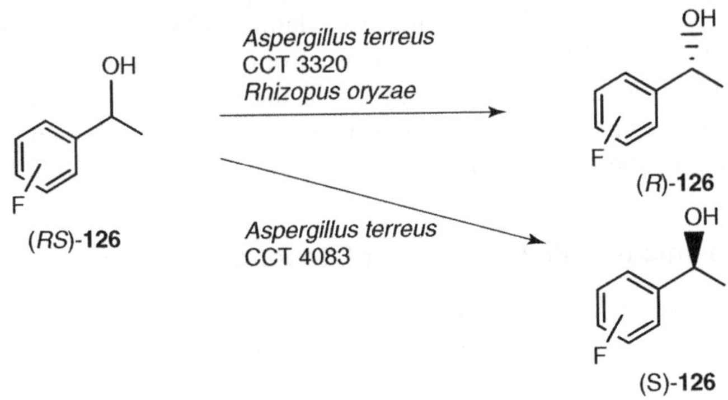

Figure 5.39 Enzymatic deracemise *ortho*- and *meta*-fluorophenyl-1-ethanol.

[216–220]. Porto *et al.* have shown that *Aspergillus terreus* CCT 3320 and *Aspergillus terreus* CCT 4083 are able to deracemize *ortho*- and *meta*-fluorophenyl-1-ethanol **126** (Figure 5.39) with good yields (98%) and high *ee*'s. (99%) [220].

The development of a large-scale process for the DKR of alcohols using various lipases in combination with a range of ruthenium catalysts has been demonstrated. The reactions can be carried out at concentrations up to 1 M with lower catalyst loadings [221]. The process for the preparation of (*R*)-3,5-bis-trifluoromethyl-phenylethan-1-ol using $[RuCl_2(p\text{-cymene})]_2$ as the racemization catalyst in combination with *Candida antarctica* lipase B as the acylating catalyst has also been demonstrated [221].

5.9.4 Deracemization of Racemic Benzoin

The deracemization of racemic benzoin **127** (Figure 5.40) using *Rhizopus oryzae* ATCC 9363 has been demonstrated. The pH of the medium was used to control the absolute configuration of the enantiomer produced. Thus, at pH 7.5–8.5, the (*R*)-enantiomer was obtained with a 73–76% yield and a 97% *ee*, whereas at pH 4–5 the (*S*)-enantiomer was produced with a 71% yield and an 85% *ee* [222].

5.9.5 Chiral Carboxylic Acid by Deracemization

An enzymatic method for the preparation of optically active (*R*)-substituted carboxylic acids **128** (Figure 5.41) using deracemization reaction was reported

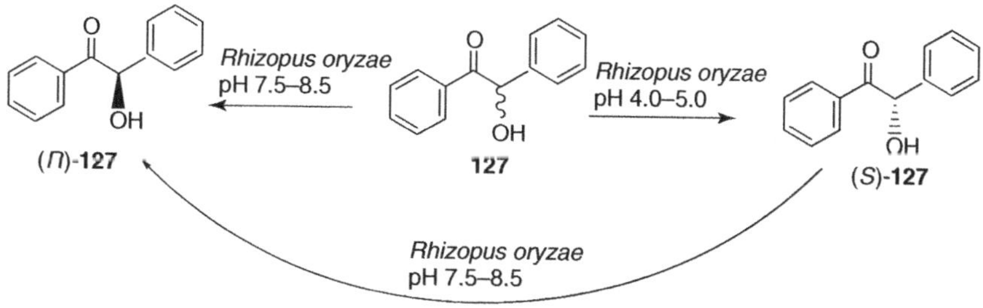

Figure 5.40 Enzymatic deracemization of racemic benzoin.

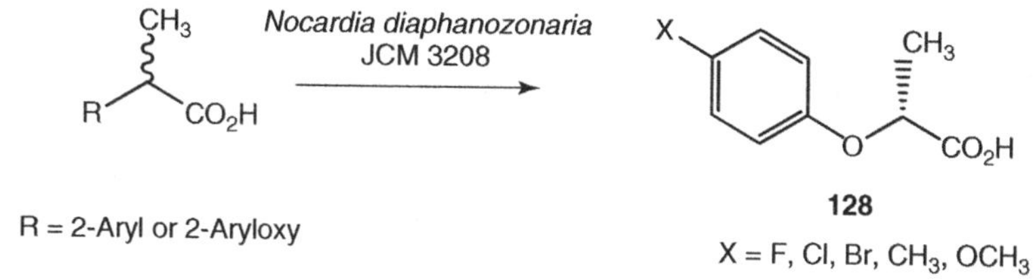

Figure 5.41 Enzymatic preparation of optically active (*R*)-substituted carboxylic acids.

by Kato *et al.* [223,224]. Using the growing cell system of *Nocardia diaphanozonaria* JCM3208, racemates of 2-aryl- and 2-aryloxypropanoic acid are deracemized and (*R*)-form-enriched products recovered with a high chemical yield of >70% and an *ee* of >85%. [223,224].

5.9.6 Enzymatic Preparation of Diols and Epoxides

The kinetic resolution of 2,2-disubstituted epoxides **129** (Figure 5.42) by epoxide hydrolase has been demonstrated with excellent enantio- and regioselectivity, yielding the corresponding (*S*)-diol and (*R*)-epoxide. Subsequently in the second step, the remaining (*R*)-epoxide was transformed by acid catalysis with *inversion*

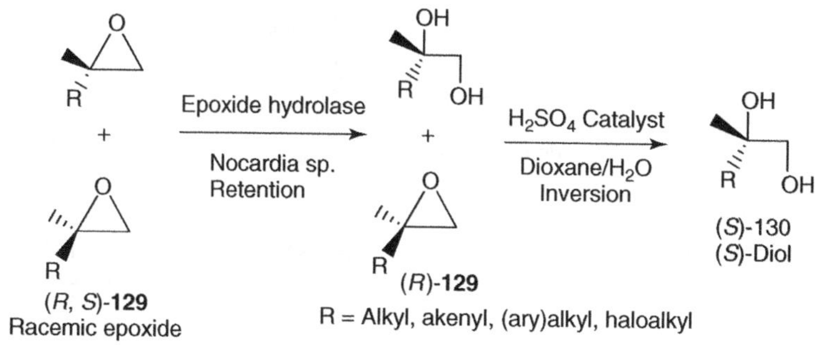

Figure 5.42 Enzymatic resolution of 2,2-disubstituted epoxides.

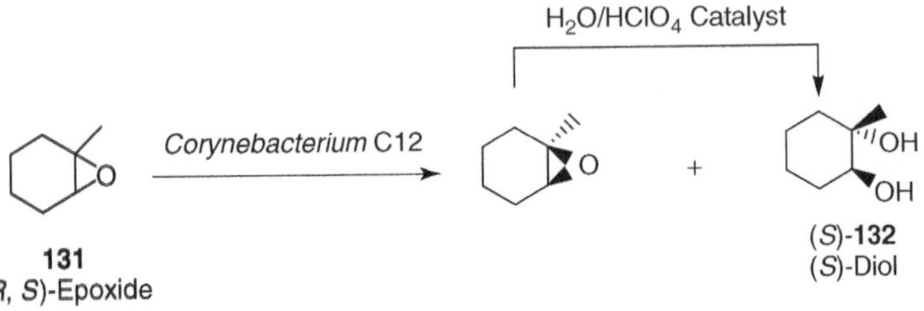

Figure 5.43 Enzymatic conversion of *rac*-1-methyl-1,2-epoxycyclohexane to (1-*S*,2-*S*)-1-methylcyclohexane-1,2-diol.

of configuration under carefully controlled reaction conditions to yield the corresponding (*S*)-diol **130** in a virtually enantiopure form with high chemical yields (>90%). This methodology proved to be highly flexible and was also applicable to styrene oxide substrates [225,226].

The deracemization of *rac*-1-methyl-1,2-epoxycyclohexane **131** (Figure 5.43) has been demonstrated by cells of *Corynebacterium* sp. C12 and *Methylobacterium* sp. [227]. A one-pot combination of *Corynebacterium* C12 epoxide hydrolase and acid-catalyzed ring-opening converted *rac*-1-methyl-1,2-epoxycyclohexane to (1-*S*, 2-*S*)-1-methylcyclohexane-1,2-diol **132**. Alternatively, instead of the chemical conversion of the unreacted oxirane, the diol formed can be converted into the remaining epoxide [228].

A highly productive bioprocess was developed for the preparation of enantiopure azole antifungal chiral synthons [229]. These are key building blocks for the synthesis of new triazole drug derivatives known to display valuable activity against such infections as fluconazole-resistant oro-esophageal candidiasis. Using recombinant *A. niger* epoxide hydrolase under optimized conditions, the hydrolytic kinetic resolution of 1-chloro-2-(2,4-difluorophenyl)-2,3-epoxypropane **133** (Figure 5.44) was performed at a substrate concentration as high as 500 g/L (i.e., 2.5 M) and afforded the (unreacted) (*S*)-epoxide **133** and the corresponding vicinal diol **134**, both in a nearly enantiopure form with quantitative yield.

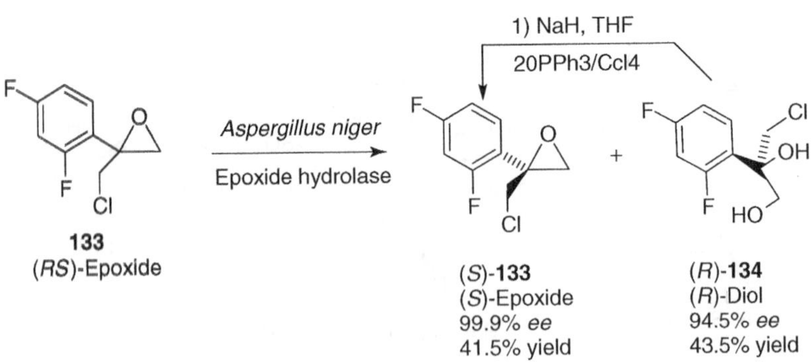

Figure 5.44 Kinetic resolution of 1-chloro-2-(2,4-difluorophenyl)-2,3-epoxypropane.

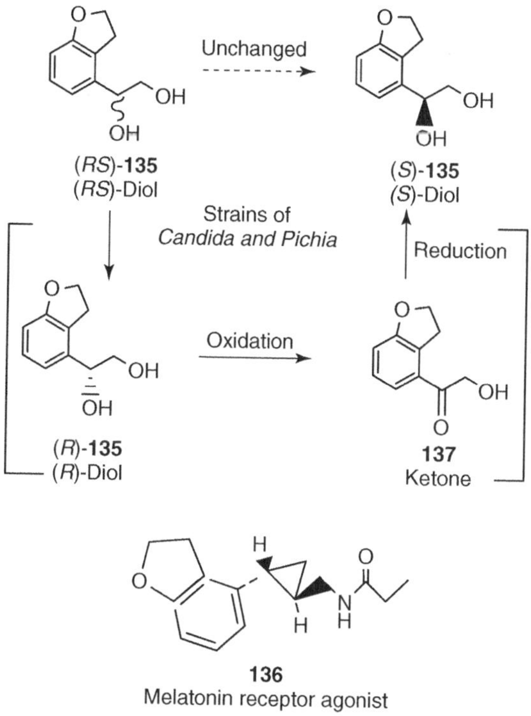

Figure 5.45 Enzymatic deracemization of dihydrobenzo[*b*]furan-4′-yl) ethane-1,2-diol.

The (*S*)-diol *S*-1-2′,3′-dihydrobenzo[*b*]furan-4′-yl)ethane-1,2-diol **135** (Figure 5.45) is a potential precursor in the synthesis of a melatonin receptor agonist **136**, useful for the treatment of sleep disorders [230]. The DKR of the racemic diol **135** to the (*S*)-enantiomer **135** has been demonstrated [231] by *Candida boidinii* strains SC 13822 and *C. boidinii* SC 16115 with a >70% yield and a 90–100% *ee*. A new compound was formed during these biotransformations and was identified as the hydroxy ketone **137** by LC-MS. The area of the HPLC peak for hydroxy ketone **137** first increased with time, reached a maximum, and then decreased, as expected for the proposed DKR.

The enantioselectivity of halohydrin dehalogenases in the conversion of haloalcohols makes them promising biocatalysts for the production of optically active epoxides and β-substituted alcohols [232]. The possibility to use halohydrin dehalogenase for ring-opening reactions with an alternative nucleophile cyanide has been explored [233–238]. The epoxide ring opening of epoxybutyronitrile using an enzyme from a *Corynebacterium* to yield (*R*)-4-chloro-3-hydroxybutyronitrile, an intermediate in the synthesis of L-carnithine, was demonstrated [233].

The epoxide-ring-opening reactions were further explored using halohydrin dehalogenase from *Agrobacterium radiobacter* (HheC). This enzyme catalyzes epoxide-ring-opening reactions with several alternative nucleophiles, such as

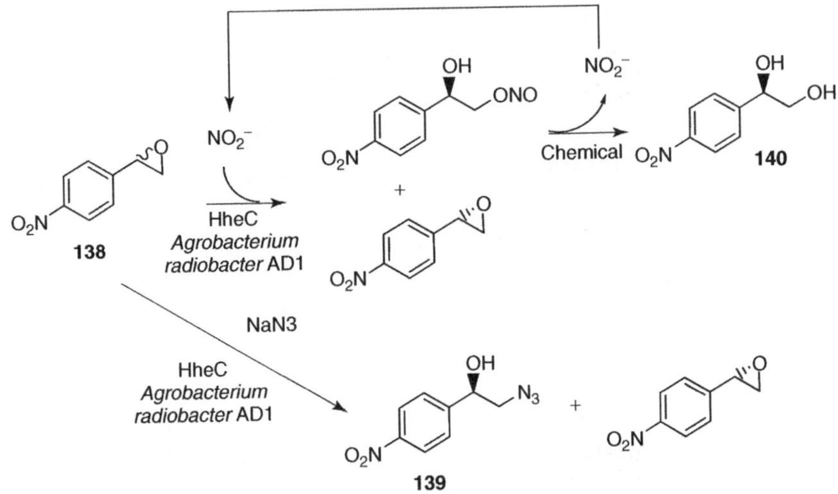

Figure 5.46 Halohydrin-dehalogenase-catalyzed preparation of (*R*)-2-azido-1-phenylethanols.

azide, cyanide, and nitrite, which allow the production of a range of β-substituted derivatives [234–238]. The enantioselective azidolysis of styrene oxide and substituted derivates **138** (Figure 5.46) yields 2-azido-1-phenylethanols **139** of the (*R*) configuration, since the enzyme is enantioselective for (*R*)-epoxides in ring-opening reactions [236]. HheC has been used in a DKR of epihalohydrins using an azide for ring opening to yield (*S*)-1-azidohaloalcohol. With epibromohydrin as the substrate, the racemization rate was higher, which allowed an efficient dynamic resolution and production of (*S*)-1-azido-3-bromo-2-propanol at a >99% *ee* with a yield of 77% [235,236].

Halohydrin dehalogenases (HheC) were used in the opening of epoxide **138** with nitrite as the nucleophile. Nitrite predominantly attacks the epoxide (Figure 5.45) through its nitrite oxygen, yielding an unstable nitrite ester intermediate that spontaneously hydrolyzes to form the corresponding diol **140** with a 99% *ee* [237]. The HheC-mediated conversion of epoxides with cyanide was recently found to yield a variety of β-cyanoalcohols [238].

5.10 CONCLUSIONS

The production of single enantiomers of drug intermediates is increasingly important in the pharmaceutical industry. Biocatalysis provides organic chemists with an alternative opportunity to prepare pharmaceutically important chiral compounds. The examples presented in this review are only from a few select articles. Different types of biocatalytic reactions are capable of generating a wide variety of chiral compounds useful in the development of drugs. Hydrolytic enzymes such as lipases, esterases, proteases, dehalogenases, acylases, amidases, nitrilases, epoxide hydrolases, and decarboxylases are used for the resolution of a variety of racemic compounds and in the asymmetric synthesis of enantiomerically enriched chiral compounds. Dehydrogenases and aminotransferases have been successfully used along with cofactors and cofactor-regenerating enzymes for the synthesis of chiral alcohols, aminoalcohols, amino acids, and amines. Aldolases and decarboxylases have been effectively used in asymmetric synthesis by aldol condensation and acyloin condensation reactions. Monooxygenases have been used in enantioselective and regioselective hydroxylation, epoxidation, and Baeyer–Villiger reactions. Dioxygenases have been used in the chemoenzymatic synthesis of chiral diols. Several approaches have been described such as, deracemization and dynamic resolution, to achieve a >50% yield and high *ee* by a combination of chemo- and/or biocatalysts in sequential reactions or by a single biocatalyst. Stereoinversion *via* an oxidation–reduction sequence has also been demonstrated. In the last decade, progress in biochemistry, protein chemistry, molecular cloning, random and site-directed mutagenesis, directed evolution of biocatalysts under desired process conditions, and fermentation technology has opened up unlimited access to a variety of enzymes and microbial cultures as tools in organic synthesis. The future of biocatalysis for synthesis of chiral compounds looks very promising.

REFERENCES

1. Food & Drug Administration. FDA's statement for the development of new stereoisomeric drugs. Chirality 1992;4:338–340.
2. Oliver M, Voigt CA, Arnold FH. Enzyme engineering by directed evolution. In: Drauz K and Waldman H, editors. Volume 1, Enzyme catalysis in organic synthesis. 2nd ed. Wiley-VCH; 2002. pp. 95–138.
3. Kazlauskas RJ. Enhancing catalytic promiscuity for biocatalysis. Curr Opin Chem Biol 2005;9(2):195–201.
4. Schmidt M, Bauman M, Henke E, Konarzycka-Bessler M, Bornscheuer UT. Directed evolution of lipases and esterases. Methods Enzym (Protein Eng) 2004;388:199–207.
5. Reetz MT, Torre C, Eipper A, Lohmer R, Hermes M, Brunner B, Maichele A, Bocola M, Arand M, Cronin A, Genzel Y, Archelas A, Furstoss R. Enhancing the enantioselectivity of an epoxide hydrolase by directed evolution. Org Lett 2004;6(2):177–180.
6. Otey CR, Bandara G, Lalonde J, Takahashi K, Arnold FH. Preparation of human metabolites of propranolol using laboratory-evolved bacterial cytochromes P450. Biotechnol Bioeng 2006;93(3):494–499.
7. Huisman GW, Lalonde JJ. Enzyme evolution for chemical process applications. In: Patel RN, editor. Biocatalysis in the pharmaceutical and biotechnology industries. Boca Raton (FL): CRC Press; 2007. pp. 717–742.
8. Alphand V, Carrea G, Wohlgemuth R, Furstoss J, Woodley J. Towards large-scale synthetic applications of Baeyer-Villiger monooxygenases. Trends Biotechnol 2003;21(7):318–323.

9. DiCosimo R. Nitrilases and nitrile hydratases. In: Patel RN, editor. Biocatalysis in the pharmaceutical and biotechnology industries. Boca Raton (FL): CRC Press; 2007. pp. 1–26.
10. Patel RN. Biocatalysis: synthesis of chiral intermediates for pharmaceuticals. Curr Organ Chem 2006;10(11):1289–1321.
11. Simeo Y, Kroutil W, Faber K. Biocatalytic deracemization: dynamic resolution, stereoinversion, enantioconvergent processes, and cyclic deracemization. In: Patel RN, editor. Biocatalysis in the pharmaceutical and biotechnology industries. Boca Raton (FL): CRC Press; 2007. pp. 27–51.
12. Simons C, Hanefeld U, Arends I, Maschmeyer T, Sheldon R. Towards catalytic cascade reactions: asymmetric synthesis using combined chemo-enzymatic catalysts. Top Catal 2006;40(1–4):35–44.
13. Turner NJ. Enzyme catalyzed deracemization and dynamic kinetic resolution reactions. Curr Opin Chem Biol 2004;8(2):114–119.
14. Robertson DE, Bornscheuer UT. Biocatalysis and biotransformation new technologies, enzymes and challenges. Curr Opin Chem Biol 2005;9(2):164–165.
15. Ishige T, Honda K, Shimizu S. Whole organism biocatalysis. Curr Opin Chem Biol 2005;9(2):174–180.
16. Stewart JD. Green chemical manufacturing with biocatalysis. In: Grassian VH, editor. Environmental catalysis. Boca Raton (FL): CRC Press; 2004. pp. 649–665.
17. Fessner W-D, Jennewein S. Biotechnological applications of aldolases. In: Patel RN, editor. Biocatalysis in the pharmaceutical and biotechnology industries. Boca Raton (FL): CRC Press; 2007. pp. 363–400.
18. Boyd DR, Bugg TDH. Arene *cis*-dihydrodiol formation:from biology to application. Org Biomol Chem 2006;4(2):181–192.
19. Zhao H, Chockalingom K, Chen Z. Directed evolution of enzymes and pathways for industrial biocatalysis. Curr Opin Biotechnol 2002;13(2):104–110.
20. Schulze B, Wubbolts M. Biocatalysis for industrial production of fine chemicals. Curr Opin Biotechnol 1999;10(6):609–611.
21. Kroutil W, Mang H, Edegger K, Faber K. Recent advances in the biocatalytic reduction of ketones and oxidation of sec-alcohols. Curr Opin Chem Biol 2004;8(2):120–126.
22. Tao J, Lin G-Q, Liese A. In: Tao J, Lin G-Q and Liese A, editors. Biocatalysis for the pharmaceutical industry: discovery, development, and manufacturing. Wiley; 2008.
23. Hall M, Stueckler C, Ehammer H, Pointner E, Oberdorfer G, Gruber K, Hauer B, Stuermer R, Kroutil W, Macheroux P, Faber K. Asymmetric Bioreduction of C=C bonds using enoate reductases OPR1, OPR3 and YqjM: Enzyme-Based Stereocontrol. Adv Synth Catal 2008;350(3):411–418.
24. Leo'n R, Fernandes P, Pinheiro HM, Cabral JMS. Whole cell biocatalysis in organic media. Enzyme Microb Technol 1998;23:483–500.
25. Gotor V. Pharmaceuticals through enzymatic transesterification and enzymatic aminolysis reactions. Biocatal Biotransformation 2000;18:87–103.
26. Zak A. Industrial biocatalysis. Curr Opin Chem Biol 2001;5:130–136.
27. Patel RN. Synthesis of chiral pharmaceutical intermediates by biocatalysis. Coord Chem Rev 2008;252(5–7):659–701.

28. Rusnak M, Mueller M, May O, Hermsen P-J, Straatmann HMGG, Skranc W, Boesten WHJ, Heemskerk T. PCT Int. Appl., WO 2008067981 A2 20080612 CAN 149:52046 AN 2008:702845. 2008. p. 18.

29. Nakamatsu T, Kawasaki H, Watanabe K, Nakazawa M, Izawa K. Eur. Pat. Appl., EP 1624052 A1 20060208 CAN 144:190698 AN 2006:122116. 2006. p. 23.

30. Yeom S-J, Kim H-J, Oh D-K. Enantioselective production of 2,2-dimethyl cyclopropane carboxylic acid from 2,2-dimethylcyclopropane carbonitrile using the nitrile hydratase and amidase of *Rhodococcus erythropolis* ATCC 25544. Enzyme Microb Technol 2007;41(6–7):842–848.

31. Kaulmann U, Smithies K, Smith EB, Hailes HC, Ward JM. Substrate spectrum of ω-transaminase from *Chromobacterium violaceum* DSM30191 and its potential for biocatalysis. Enzyme Microb Technol 2007;41(5):628–637.

32. Thayer AN. Biocatalysis helps reach a resolution: Enzymatic reactions combined with racemization can generate enantiopure materials in high yields. Chem Eng News 2006;84(33):29–31.

33. Yamaguchi S, Komeda SH, Asano HY. New enzymatic method of chiral amino acid synthesis by dynamic kinetic resolution of amino acid amides: Use of stereoselective amino acid amidases in the presence of α-amino-ε-caprolactam racemase. Appl Environ Microbiol 2007;73(16):5370–5373.

34. Oliver M, Verseck S, Bommarius A, Drauz AK. Development of dynamic kinetic resolution processes for biocatalytic production of natural and nonnatural L-amino acids. Org Process Res Dev 2002;6(4):452–457.

35. Haque TS, Ewing WR, Mapelli C, Lee VG, Sulsky RB, Riexinger DJ, Martinez RL, Zhu YZ. PCT Int. Appl., WO 2007082264 A2 20070719 CAN 147:189414 AN 2007:789960. 2007. p. 193.

36. Qian F, Ewing WR, Mapelli C, Riexinger DJ, Lee VG, Sulsky RB, Zhu Y, Haque TS, Martinez RL, Naringrekar V, Ni N, Burton LS. US patent Appl. Publ., Cont.-in-part of U.S. Ser. No. 442,017 US 2007099835 A1 20070503 CAN 146:482245 AN 2007:488639. 2007. p. 154.

37. Patel RN, Chen Y, Goldberg SL, Hanson RL, Goswami A, Tully TP, Parker WL. PCT Int. Appl., WO 2007112299 A2 20071004 CAN 147:425649 AN 2007:1120352. 2007. p. 59.

38. Yagasaki M, Ozaki MA. Industrial biotransformations for the production of **D**-amino acids. J Mol Catal B Enzym 1998;4:1–11.

39. Taylor PP, Pantaleone DP, Senkpeil RF, Fotheringham IG. Novel biosynthetic approaches to the production of unnatural amino acids using transaminases. Trends Biotechnol 1998;16:412–417.

40. Drauz K. Chiral amino acids: a versatile tool in the synthesis of pharmaceuticals and fine chemicals. Chimia 1997;51:310–315.

41. Pietzsch M, Syldatk C. In: Drauz K, editor. Volume 2, Enzyme catalysis in organic synthesis. 2nd ed. Weinheim: Wiley-VCH; 2002. pp. 761–775.

42. Moriguchi M, Ideta K. Production of D-Aminoacylase from *Alcaligenes Denitrificans* subsp. *xylosoxydans* MI-4. Appl Environ Microbiol 1988;54:2767–2770.

43. Komeda H, Asano Y. Gene cloning, nucleotide sequencing, and purification and characterization of the D-stereospecific amino-acid amidase from *Ochrobactrum anthropi* SV3. Eur J Biochem 2000;267:2028–2035.

44. Vedha-Peters K, Gunawardana M, Rozzell JD, Novick SJ. Creation of a broad-range and highly Stereoselective D-amino acid dehydrogenase for the one-step synthesis of D-amino acids. J Am Chem Soc 2006;128:10923–109297.

45. Alexandre F-R, Pantaleone DP, Taylor PP, Fotheringham IG, Ager DJ, Turner NJ. Amine–boranes: effective reducing agents for the deracemisation of dl-amino acids using l-amino acid oxidase from *Proteus myxofaciens*. Tetrahedron Lett 2002;43:707–710.

46. Hanson RL, Schwinden MD, Banerjee A, Brzozowski DB, Chen B-C, Patel BP, McNamee CG, Kodersha GA, Kronenthal DR, Patel RN, Szarka LJ. Enzymatic synthesis of L-6-hydroxynorleucine. Bioorg Med Chem 1999;7:2247–2252.

47. Chaturvedula PV, Chen L, Civiello R, Degnan AP, Dubowchik GM, Han X, Jiang JJ, Macor JE, Poindexter GS, Tora GO, Luo G. Anti-migraine spirocycles. US patent Appl. Publ. U.S.2007/0149503 A1. 2007.

48. Chaturvedula PV, Dubowchik GM, Degan AP, Han X, Conway D, Cook C, Davis R, Denton R, Macci NR, Mathias S, Pin L, Signor G, Thalody R, Schartman KA, Widmann C, Xu C, Macor JE. Abstracts of Papers; 234th National Meeting of the American Chemical Society; 2007 Aug 19–23; Boston, MA. Washington (DC): American Chemical Society; 2007.

49. Hanson RL, Davis BL, Goldberg SL, Johnston RM, Parker WL, Tully TP, Montana MA, Patel RN. Enzymatic preparation of a D-amino acid from a racemic amino acid or keto acid. Org Process Res Dev 2008;12(6):1119–1129.

50. Robl J, Sun C, Stevenson J, Ryono D, Simpkins L, Cimarusti M, Dejneka T, Slusarchyk W, Chao S, Stratton L, Misra R, Bednarz M, Asaad M, Cheung H, Aboa-Offei B, Smit P, Mathers P, Fox M, Schaeffer T, Seymour A, Trippodo N. Dual metalloprotease inhibitors: mercaptoacetyl-based fused heterocyclic dipeptide mimetics as inhibitors of angiotensin-converting enzyme and neutral endopeptidase. J Med Chem 1997;40:1570–1577.

51. Hanson RL, Schwinden MD, Banerje A, Brzozowski DB, Chen B-C, Patel BP, McNamee CG, Kodersha GA, Kronenthal DR, Patel RN, Szarka LJ. Enzymatic synthesis of L-6-hydroxynorleucine. Bioorg Med Chem 1999;7(10):2247–2252.

52. Patel RN. Enzymatic synthesis of chiral intermediates for Omapatrilat, anantihypertensive drug. Biomol Eng 2001;17:167–182.

53. Hanson RL, Howell J, LaPorte T, Donovan M, Cazzulino D, Zannella V, Montana M, Nanduri V, Schwarz S, Eiring R, Durand S, Wasylyk J, Parker L, Liu M, Okuniewicz F, Chen B-C, Harris J, Natalie K, Ramig K, Swaminathan S, Rosso V, Pack S, Lotz B, Bernot P, Rusowicz A, Lust D, Tse K, Venit J, Szarka L, Patel RN. Synthesis of allylsine ethylene acetal using phenylalanine dehydrogenase from *Thermoactinomyces intermedius*. Enzyme Microb Technol 2000;26:348–358.

54. Gallwitz B. Glucagon-like peptide-1-based therapies for the treatment of type 2 diabetes mellitus. Treat Endocrinol 2005;4:361–370.

55. Sinclair EM, Drucker DJ. Glucagon-like peptide 1 receptor agonists and dipeptidyl peptidase IV inhibitors: new therapeutic agents for the treatment of type 2 diabetes. Curr Opin Endocrinol Diabetes 2005;12:146–151.

56. Augeri DJ, Robl JA, Betebenner DA, Magnin DR, Khanna A, Robertson JG, Wang A, Simpkins LM, Taunk P, Huang Q, Han S-P, Abboa-Offei B, Cap M, Xin L, Tao L, Tozzo E, Welzel GE, Egan DM, Marcinkeviciene J, Chang SY, Biller SA, Kirby MS, Parker RA, Hamann LG. Discovery and preclinical profile of Saxagliptin

(BMS-477118): A highly potent, long-acting, orally active dipeptidyl peptidase IV inhibitor for the treatment of type 2 diabetes. J Med Chem 2005;48:5025–5037.

57. Vu TC, Brzozowski DB, Fox R, Godfrey JD Jr, Hanson RL, Kolotuchin SV, Mazzullo JA Jr, Patel RN, Wang J, Wong K, Yu J, Zhu J, Magnin RD, Augeri DJ, Hamann LG. Preparation of cyclopropyl-fused pyrrolidine-based inhibitors of dipeptidyl peptidase IV. PCT Int. Appl., CODEN: PIXXD2 WO 2004052850 A2 20040624 CAN 141:54618 AN 2004:515478. 2004. p. 101.

58. Hanson RL, Goldberg SL, Brzozowski DB, Tully TP, Cazzulino D, Parker WL, Lyngberg OK, Vu TC, Wong MK, Patel RN. Preparation of an amino acid intermediate for the dipeptidyl peptidase IV inhibitor, saxagliptin, using a modified phenylalanine dehydrogenase. Adv Synth Catal 2007;349(8+9):1369–1378.

59. Asano Y. Enzymatic synthesis of (S)-phenylalanine and related (S)-amino acids by phenylalanine dehydrogenase. Methods Biotechnol (Microbial Enzymes and Biotransformations) 2005;17:141–150.

60. Cainelli G, Engel PC, Galletti P, Giacomini D, Gualandi A, Paradisi F. Engineered phenylalanine dehydrogenase in organic solvents: homogeneous and biphasic enzymatic reactions. Org Biomol Chem 2005;3(24):4316–4320.

61. Bommmarius AS, Schwarm M, Strigl K, Kottenhahn M, Huthmacher K, Drauz K. Synthesis and use of enantiomerically pure tertiary leucine. Tetrahedron: Asymmetry 1995;6(12):7851–7858.

62. Kragl U, Vasic-Racki D, Wandrey C. Continuous production of (S)-tert-leucine in series of two enzyme membrane reactor. Bioprocess Eng 1996;14:291–197.

63. Bold G, Faessler A, Capraro H-G, Cozens R, Klimkait T, Lazdins J, Mestan J, Poncioni B, Roesel J, Stover D, Tintelnot-Blomley M, Acemoglu F, Beck W, Boss E, Eschbach M, Huerlimann T, Masso E, Roussel S, Ucci-Stoll K, Wyss D, Lang M. New aza-dipeptide analogs as potent and orally absorbed HIV-1 protease inhibitors: candidates for clinical development. J Med Chem 1998;41(18):3387–3401.

64. Robinson BS, Riccardi KA, Gong YF, Guo Q, Stock DA, Blair WS, Terry BJ, Deminie CA, Djang F, Colonno RJ, Lin PF. BMS-232632, a highly potent human immunodeficiency virus protease inhibitor that can be used in combination with other available antiretroviral agents. Antimicrob Agents Chemother 2000;44(8):2093–2099.

65. Stoyan T, Recktenwald A, Kula MR. Cloning, sequencing and overexpression of the leucine dehydrogenase gene from *Bacillus cereus*. J Biotechnol 1997;54(1):77–80.

66. Menzel A, Werner H, Altenbuchner J, Groeger H. From enzymes to "designer bugs" in reductive amination: a new process for the synthesis of L-tert-leucine using a whole cell-catalyst. Eng Life Sci 2004;4(6):573–576.

67. Groeger H, May O, Werner H, Menzel A, Altenbuchner J. A "second-generation process" for the synthesis of L-neopentylglycine: asymmetric reductive amination using a recombinant whole cell catalyst. Org Process Res Dev 2006;10(3):666–669.

68. Emmanuel MJ, Frye LL, Hickey ER, Liu W, Morwick TM, Spero DM, Sun S, Thomson DS, Ward, YD, Young ERR. Novel spiroheterocyclic compounds [morpholine-4-carboxylic acid amides of heterocyclic cyclohexylalanine and neopentylglycine derivatives and their analogs], useful as reversible inhibitors of cysteine proteases such as cathepsin S. PCT Int. Appl., WO 2001019816 A1 20010322 CAN 134:252348 AN 2001:208265. 2001. p. 361.

69. Mihara H, Muramatsu H, Kakutani R, Yasuda M, Ueda M, Kurihara T, Esaki N. **N**-methyl-L-amino acid dehydrogenase from *Pseudomonas putida*. A novel

member of an unusual NAD(P)-dependent oxidoreductase superfamily. FEBS J 2005;272(5):1117–1123.

70. Kato Y, Yamada H, Asano Y. Stereoselective synthesis of opine-type secondary amine carboxylic acids by a new enzyme opine dehydrogenase, use of recombinant enzymes. J Mol Catal B Enzym 1996;1(3–6):151–160.
71. Merviel P, Najas S, Campy H, Floret S, Brasseur F. Use of GNRH antagonists in reproductive medicine. Minerva Ginecol 2005;57(1):29–43.
72. Hauptmann J. Pharmacokinetics of an emerging new class of anticoagulant antithrombotic drugs A review of small-molecule thrombin inhibitors. Eur J Clin Pharmacol 2002;57(11):751–758.
73. Straathof AJJ, Panke S, Schmid A. The production of fine chemicals by biotransformations. Curr Opin Biotechnol 2002;13(6):548–556.
74. Bommarius AS, Schwarm M, Drauz K. Biocatalysis to amino acid-based chiral pharmaceuticals-examples and perspectives. J Mol Catal B Enzym 1998;5(1–4): 1–11.
75. Gustafsson D, Elg M, Lenfors S, Boerjesson I, Teger-Nilsso A-C. Effects of inogatran, a new low-molecular-weight thrombin inhibitor, in rat models of venous and arterial thrombosis, thrombolysis and bleeding time. Blood Coagul Fibrinolysis 1996;7(1):69–79.
76. Veronesi B, Carter JD, Devlin RB, Simon SA, Oortgiesen M. Neuropeptides and capsaicin stimulate the release of inflammatory cytokines in a human bronchial epithelial cell line. Neuropeptides (Edinburgh) 1999;33(6):447–456.
77. Yuan L, Burcher E, Nail BS. Characterization of tachykinin receptors mediating bronchomotor and vasodepressor responses to neuropeptide γ and substance P in the anesthetized rabbit. Pulm Pharmacol Ther 1998;11(1):31–39.
78. Ting PC, Lee JF, Anthes JC, Shih N-Y, Piwinski JJ. Synthesis and NK1/NK2 receptor activity of substituted 4(Z)-(methoxyimino)pentyl-1-piperazines. Bioorg Med Chem Lett 2000;10(20):2333–2335.
79. Reichard GA, Ball ZT, Aslanian R, Anthes JC, Shih N-Y, Piwinski JJ. The design and synthesis of novel NK1/NK2 dual antagonists. Bioorg Med Chem Lett 2000;10(20):2329–2332.
80. Homann MJ, Vail R, Morgan B, Sabesan V, Levy C, Dodds DR, Zaks A. Enzymatic hydrolysis of a prochiral 3-substituted glutarate ester, an intermediate in the synthesis of an NK1/NK2 dual antagonist. Adv Synth Catal 2001;343(6+7):744–749.
81. Suen W-C, Zhang N, Xiao L, Madison V, Zaks A. Improved activity and thermostability of *Candida antarctica* lipase B by DNA family shuffling. Protein Eng Des Sel 2004;17(2):133–140.
82. Zhang N, Suen W-C, Windsor W, Xiao L, Madison V, Zaks A. Improving tolerance of *Candida antarctica* lipase B towards irreversible thermal inactivation through directed evolution. Protein Eng 2003;16(8):599–605.
83. DeSantis G, Wong K, Farwell B, Chatman K, Zhu Z, Tomlinson G, Huang H, Tan X, Bibbs L, Chen P, Kretz K, Burk MJ. Creation of a productive, highly enantioselective nitrilase through gene site saturation mutagenesis (GSSM). J Am Chem Soc 2003;125(38):11476–11477.
84. DeSantis G, Zhu Z, Greenberg WA, Wong K, Chaplin J, Hanson SR, Farwell B, Nicholson LW, Rand CL, Weiner DP, Robertson DE, Burk MJ. An enzyme library approach to biocatalysis: development of nitrilases for enantioselective production of carboxylic acid derivatives. J Am Chem Soc 2002;124(31):9024–9025.

85. Desantis G, Short JM, Burk M, Wong K, Farwell R, Chatman K. Identification, cloning, sequences and design of nitrilases for synthesis of enantioselective amino acids and hydroxy acids. PCT Int. Appl., WO 2003097810 A2 20031127 CAN 140:2345 AN 2003:931482. 2003. p. 445.
86. Robertson DE, Chaplin JA, DeSantis G, Podar M, Madden M, Chi E, Richardson T, Milan A, Miller M, Weiner DP, Wong K, McQuaid J, Farwell B, Preston LA, Tan X, Snead MA, Keller M, Mathur E, Kretz PL, Burk MJ, Short JM. Exploring nitrilase sequence space for enantioselective catalysis. Appl Environ Microbiol 2004;70(4):2429–2436.
87. Bloom JD, Datta MD, Johnson BD, Wissner A, Bruns MG, Largis EE, Dolan JA, Claus TH. Disodium (R,R)5-[2-(3-chlorophenyl)-2-hydroxyethyl]amino]propyl]-1,3 benzodioxole-2,2-dicarboxylate. A potent β-adrenergic agonist virtually specific for β3 receptors. J Med Chem 1989;35:3081–3084.
88. Fisher LG, Sher PM, Skwish S, Michael IM, Seiler S, Dickinson KEJ. BMS 187257, a potent, selective, and novel heterocyclic β-3 adrenergic receptor agonist. Bioorg Med Chem Lett 1994;6:2253–2258.
89. Patel RN, Banerjee A, Chu L, Brzozowski D, Nanduri V, Szarka LJ. Microbial synthesis of chiral intermediates for β-3-receptor agonists. J Am Chem Soc 1998;75:1473–1482.
90. Innaimo SF, Seifer M, Bisacchi GS, Standring DN, Zahler R, Colonno RJ. Identification of BMS-200475 as a potent and selective inhibitor of hepatitis B virus. J Antimicrob Agents Chemother 1997;41(7):1444–1448.
91. Genovesi EV, Lamb L, Medina I, Taylor D, Seifer M, Innaimo S, Colonno RJ, Standring DN, Clark JM. Efficacy of the carbocyclic 2′-deoxyguanosine nucleoside BMS-200475 in the woodchuck model of hepatitis B virus infection. J Antimicrob Agents Chemother 1998;42(12):3209–3217.
92. Pendri YR, Chen C-P, Patel S, Evans JM, Liang J, Kronenthal DR, Powers GL, Prasad S, Bien JT, Shi Z, Patel RN, Chan YY, Rijhwani SK, Singh AK, Wang S, Stojanovic M, Polniaszek R, Lewis C, Thottathil J, Krishnamurty D, Zhou MX, Vemishetti P. PCT Int. Appl. Process for preparing the antiviral agent entecavir. WO 2004052310 A2 20040624 CAN 141:38813. 2004. p. 143.
93. Griffith DA, Danishefsky SJ. The total synthesis of allosamidin: expansions of the methodology of aza-glycosidation pursuant to the total Synthesis of allosamidin. A surprising enantiotopic sense for a lipase-induced deacetylation. J Am Chem Soc 1996;118(40):9526–9538.
94. Danishefsky SJ, Cabal MP, Chow K. Novel stereospecific silyl group transfer reactions: practical routes to the prostaglandins. J Am Chem Soc 1989;111 (9):3456–3457.
95. Patel RN, Banerjee A, Pendri YR, Liang J, Chen C-P, Mueller RH. Preparation of a chiral synthon for an HBV inhibitor: enzymatic asymmetric hydrolysis of (1α,2β,3α)-(benzyloxymethyl)cyclopent-4-ene-1,3-diol diacetate and enzymatic asymmetric acetylation of (1α,2β,3α)-2 (benzyloxymethyl)cyclopent-4-ene-1,3-diol. Tetrahedron: Asymmetry 2006;17(2):175–178.
96. Cashman JR. The role of flavin-containing monooxygenases in drug metabolism and development. Curr Opin Drug Discov Dev 2003;6(4):486–493.
97. Alphand V, Carrea G, Wohlgemuth R, Furstoss R, Woodley JM. Towards large-scale synthetic applications of Baeyer-Villiger monooxygenases. Trends Biotechnol 2003;21(7):318–323.

98. Strand SE. Biochemistry of the soluble methane monooxygenase. Annu Rev Microbiol 1994;48:371–399.

99. Nolan LC, O'Connor KE. [1]Dioxygenase- and monooxygenase-catalysed synthesis of *cis*-dihydrodiols, catechols, epoxides and other oxygenated products. Biotechnol Lett 2008;30(11):1879–1891.

100. Gibson DT, Parales RE. Aromatic hydrocarbon dioxygenases in environmental biotechnology. Curr Opin Biotechnol 2000;11:236–243.

101. Parales RE, Resnick SM. Application of aromatic hydrocarbon dioxygenases in. In: Patel RN, editor. Biocatalysis in the pharmaceutical and biotechnology industries. Boca Raton (FL): CRC Press; 2007. pp. 299–332.

102. Lee KF, Parales JV, Gibson DT, Ramaswamy S. Purification, crystallization and preliminary x-ray diffraction studies of the three components of the toluene 2,3-dioxygenase enzyme system. Acta Crystallogr F 2005;F61(7):669–672.

103. Karlsson Andreas, Parales JV, Parales RE, Gibson DT, Eklund H, Ramaswamy S. Crystal structure of naphthalene dioxygenase: side-on binding of dioxygen to iron. Science (Washington (DC)) 2003;299(5609):1039–1042.

104. Hudlicky T, Gonzalez D, Gibson DT. Enzymatic dihydroxylation of aromatics in enantioselective synthesis: expanding asymmetric methodology. Aldrichimica Acta 1999;32(2):35–62.

105. Boyd DR, Sharma ND, Bowers NI, Brannigan IN, Groocock MR, Malone JF, McConville G, Allen CCR. Biocatalytic asymmetric dihydroxylation of conjugated mono- and poly-alkenes to yield enantiopure cyclic cis-diols. Adv Synth Catal 2005;347(7+8):1081–1089.

106. Chartrain M, Jackey B, Taylor C, Sandfor V, Gbewonyo K, Lister L, Dimichele L, Hirsch C, Heimbuch B, Maxwell C, Pascoe D, Buckland B, Greasham R. Bioconversion of indene to cis-(1*S*,2*R*)-indandiol and trans-(1*R*,2*R*)-indandiol by *Rhodococcus* sp. J Ferment Technol 1998;86:550–558.

107. Buckland BC, Robinson DK, Chartrain M. Biocatalysis for pharmaceuticals–status and prospects for a key technology. Metab Eng 2000;2:42–48.

108. Chartrain M, Salmon PM, Robinson DK, Buckland BC. Metabolic engineering and directed evolution for the production of pharmaceuticals. Curr Opin Biotechnol 2000;11:209–214.

109. Zhang N, Stewart BG, Moore JC, Greasham RL, Robinson DK, Buckland B, Lee C. Directed evolution of toluene dioxygense from *Pseudomonas putida* for improved selectivity toward cis- (1S,2R)-indanediol during indane. Metab Eng 2000;2(4):339–348.

110. Amanullah A, Hewitt CJ, Nienow AW, Lee C, Chartrain M, Buckland BC, Drew SW, Woodley JM. Measurement of strain-dependent toxicity in the indene bioconversion using multiparameter flow cytometry. Biotechnol Bioeng 2003;81(4):405–420.

111. Muenzer DF, Meinhold P, Peters MW, Feichtenhofer S, Griengl H, Arnold FH, Glieder A, de Raadt A. Stereoselective hydroxylation of an achiral cyclopentanecarboxylic acid derivative using engineered P450s BM-3. Chem Commun (Cambridge) 2005;20:2597–2599.

112. de Raadt A, Griengl H, Petsch M, Plachota P, Schoo N, Weber H, Braunegg G, Kopper I, Kreiner M, Zeiser A. Microbial hydroxylation of 2-cycloalkylbenzoxazoles. Part III. Determination of product enantiomeric excess and cleavage of benzoxazoles. Tetrahedron: Asymmetry 1996;7(2):491–496.

113. Urlacher VB, Lutz-Wahl S, Schmid RD. Microbial P450 enzymes in biotechnology. Appl Microbiol Biotechnol 2004;64(3):317–325.

114. Li Q-S, Schwaneberg U, Fischer M, Schmitt J, Pleiss J, Lutz-Wahl S, Schmid RD. Rational evolution of a medium chain-specific cytochrome P-450 BM-3 variant. Biochim Biophys Acta 2001;1545(1–2):114–121.

115. Arnold FH, Otey CR. Libraries of optimized cytochrome P 450 enzymes and the optimized P 450 enzymes exhibiting higher activities and new substrate specificities. PCT Int. Appl., WO 2005017106 A2 20050224 CAN 142:256750. 2005. p. 122.

116. Gerth K, Pradella S, Perlova O, Beyer S, Muller R. Myxobacteria: proficient producers of novel natural products with various biological activities-past and future biotechnological aspects with the focus on the genus *Sorangium*. J Biotechnol 2003;106(2–3):233–253.

117. Benigni D, Stankavage R, Chiang S-J, Hou H, Eagan B, Gu D, Hou D, Mintzmyer L, Tully TP, Davis BL, Hargro I, Mascari M, Galvin G, Stein G, McConlogue CW, Comezoglu FT. Methods for the preparation, isolation and purification of epothilone B, and X-ray crystal structures of epothilone B. PCT Int. Appl., WO 2004026254 A2 20040401 CAN 140:286248 AN 2004:270015. 2004. p. 85.

118. Goodin S, Kane MP, Rubin EH: Epothilones: mechanism of action and biologic activity. J Clin Oncol 2004;22(10):2015–2025.

119. Nicolaou KC, Roschangar F, Vourloumis D. Chemical biology of epothilones. Angew Chem Int Ed 1998;37(15):2014–2045.

120. Altmann K-H. The merger of natural product synthesis and medicinal chemistry: on the chemistry and chemical biology of epothilones. Org Biomol Chem 2004;2(15):2137–2152.

121. Boddy CN, Hotta K, Tse ML, Watts RE, Khosla C. Precursor-directed biosynthesis of epothilone in *Escherichia coli*. J Am Chem Soc 2004;126(24):7436–7437.

122. Lin N, Brakora K, Seiden M. BMS-247550(Bristol-Myers Squibb/GBF). Curr Opin Invest Drugs (Thomson Current Drugs) 2003;4(6):746–756.

123. Low JA, Wedam SB, Lee JJ, Berman AW, Brufsky A, Yang SX, Poruchynsky MS, Steinberg SM, Mannan N, Fojo T, Swain SM. Phase II clinical trial of ixabepilone (BMS-247550), an epothilone B analog, in metastatic and locally advanced breast cancer. J Clin Oncol 2005;23(12):2726–2734.

124. Basch JD, Chiang S-J, Liu SW, Nayeem A, Sun Y-L. The epothilone B hydroxylase and ferredoxin genes of Amycolatopsis and their use in the development of strains for the manufacture of hydroxylated epothilones. PCT Int. Appl., WO 2004078978 A1 20040916 CAN 141:255532 AN 2004:756881. 2004. 192.

125. Hoegenauer G. Mechanism of action of antibacterial agents. Tiamulin and pleuromutilin. Antibiotics (New York) 1979;5(1):344–360.

126. Berner H, Vyplel H, Schulz G, Schneider H. Chemistry of pleuromutilins. V. Photoisomerization of AB-trans-anellated mutilan-11-one. Tetrahedron 1983; 39(10):1745–1748.

127. Hanson RL, Matson JA, Brzozowski DB, LaPorte TL, Springer DM, Patel RN. Hydroxylation of mutilin by *Streptomyces griseus* and *Cunninghamella echinulata*. Org Process Res Dev 2002;6(4):482–487.

128. Springer DM, Sorenson ME, Huang S, Connolly TP, Bronson JJ, Matson JA, Hanson RL, Brzozowski DB, LaPorte TL, Patel RN. Synthesis and activity of a C-8 keto pleuromutilin derivative. Bioorg Med Chem Lett 2003;13(10):1751–1753.

129. Vacca JP. New advances in the discovery of thrombin and factor Xa inhibitors. Curr Opin Chem Biol 2000;4(4):394–400.

130. Gladwell TD. Bivalirudin: a direct thrombin inhibitor. Clin Ther 2002;24(1):38–58.

131. Fevig JM, Wexler RR. Anticoagulants: thrombin and factor Xa inhibitors. Annu Rep Med Chem 1999;34:81–100.

132. Williams PD, Coburn C, Burgey C, Morrissette MM. Preparation of triazolopyrimidines as thrombin inhibitors. PCT Int. Appl., WO 2002064211 A1 20020822 CAN 137:185499 AN 2002:637565. 2002. p. 184.

133. Nelson TD, LeBlond CR, Frantz DE, Matty L, Mitten JV, Weaver DG, Moore JC, Kim JM, Boyd R, Kim P-Y, Gbewonyo K, Brower M, Sturr M, McLaughlin K, McMasters DR, Kress MH, McNamara JM, Dolling UH. Stereoselective synthesis of a potent thrombin inhibitor by a novel P2-P3 lactone ring opening. J Org Chem 2004;69(11):3620–3627.

134. Jajoo HK, Mayol RF, LaBudde JA, Blair IA. Metabolism of the antianxiety drug buspirone in human subjects. Drug Metab Dispos 1989;17(6):634–640.

135. Mayol RF. Buspirone metabolite for the alleviation of anxiety. US Cont.-in-part of U. S. Ser. No. 484,161 US 6150365 A 20001121 CAN 133:359250 AN 2000:819478. 2000. p. 14.

136. Yevich JP, New JS, Lobeck WG, Dextraze P, Bernstein E, Taylor DP, Yocca FD, Eison MS, Temple DL Jr. Synthesis and biological characterization of α-(4-fluorophenyl)-4-(5-fluoro-2-pyrimidinyl)-1-piperazinebutanol and analogs as potential atypical antipsychotic agents. J Med Chem 1992;35(24):4516–4525.

137. Yevich JP, Mayol RF, Li J, Yocca F. (*S*)-6-Hydroxy-buspirone for treatment of anxiety, depression and related disorders. US patent Appl. Publ., US 2003022899 A1 20030130 CAN 138:131151 AN 2003:77553. 2003. p. 7.

138. Patel RN, Chu L, Nanduri V, Jianqing L, Kotnis A, Parker W, Liu M, Mueller R. Enantioselective microbial reduction of 6-oxo-8-[4-[4-(2-pyrimidinyl)-1-piperazinyl]butyl]-8-azaspiro[4.5]decane-7,9-dione. Tetrahedron: Asymmetry 2005; 16(16):2778–2783.

139. Goldberg SL, Nanduri VB, Chu L, Johnston RM, Patel RN. Enantioselective microbial reduction of 6-oxo-8-[4-[4-(2-pyrimidinyl)-1-piperazinyl]butyl]-8-azaspiro[4.5]decane-7,9-dione: cloning and expression of reductases. Enzyme Microb Technol 2006;39(7):1441–1450.

140. King AO, Corley EG, Anderson RK, Larsen RD, Verhoeven TR, Reider PJ, Xiang YB, Belley M, Leblanc Y. An efficient synthesis of LTD4 antagonist L-699,392. J Org Chem 1993;58(14):3731–3735.

141. Zhao M, King AO, Larsen RD, Verhoeven TR, Reider PJ. A convenient and economical method for the preparation of DIP-chloride and its application in the asymmetric reduction of aralkyl ketones. Tetrahedron Lett 1997;34(15):2614–2644.

142. Shafiee A, Motamedi HA. Purification, characterization and immobilization of an NADPH-dependent enzyme involved in the chiral specific reduction of the keto ester M, an intermediate in the synthesis of an anti-asthma drug, Montelukast, from *Microbacterium campoquemadoensis* (MB5614). Appl Microbiol Biotechnol 1998;49(6):709–717.

143. Rozell D, Liang J. Catalysts, Speciality Chemicals Magazine; 2008 April.

144. Liang J, Borup B, Mitchell V, Mundorff E, Lalonde J, Huisman G. Ketoreducytase polypeptides and uses thereof. Application *No. patents/app/20090155863* AC12P1712FI, USPC Class: 43512. 2009.

145. Patel RN, Banerjee A, McNamee CG, Brzozowski DB, Hanson RL, Szarka LJ. Enantioselective microbial reduction of 3,5-dioxo-6-(benzyloxy) hexanoic acid, ethyl ester. Enzyme Microb Technol 1993;15(12):1014–1021.

146. Sit SY, Parker RA, Motoe I, Balsubramanian HW, Cott CD, Brown PJ, Harte WE, Thompson MD, Wright J. Synthesis, biological profile, and quantitative structure activity relationship of a series of novel 3-hydroxy-3-methylglutaryl coenzyme A reductase inhibitors. J Med Chem 1990;33:2982–2999.

147. Roth BD. The discovery and development of atorvastatin, a potent novel hypolipidemic agent. Prog Med Chem 2002;40:1–22.

148. Guo Z, Chen Y, Goswami A, Hanson RL, Patel RN. Synthesis of ethyl and t-butyl (3*R*,5*S*)-dihydroxy-6-benzyloxyhexanoates via diastereo- and enantioselective microbial reduction. Tetrahedron: Asymmetry 2006;17(1):1589–1602.

149. Goldberg S, Guo Z, Chen S, Goswami A, Patel RN. Synthesis of ethyl-(3*R*,5*S*)-dihydroxy-6-benzyloxyhexanoates via diastereo-and enantioselective microbial reduction: Cloning and expression of ketoreductase IIIfrom *Acinetobacter* sp. SC 13874. Enzyme Microb Technol 2008;43:544–549.

150. Rudroff F, Rydz J, Ogink FH, Fink M, Mihovilovic MD. Comparing the stereoselective biooxidation of cyclobutanones by recombinant strains expressing bacterial baeyer-villiger monooxygenases. Adv Synth Catal 2007;349(8+9):1436–1444.

151. Iwaki H, Wang S, Grosse S, Bergeron H, Nagahashi AA, Lertvorachon J, Yang J, Konishi Y, Hasegawa Y, Lau PCK. Pseudomonad cyclopentadecanone monooxygenase displaying an uncommon spectrum of Baeyer-Villiger oxidations of cyclic ketones. Appl Environ Microbiol 2006;72(4):2707–2720.

152. Mihovilovic MD, Snajdrova R, Groetzl B. Microbial Baeyer-Villiger oxidation of 4,4-disubstituted cyclohexan- and cyclohexenones by recombinant whole-cells expressing monooxygenases of bacterial origin. J Mol Catal B Enzym 2006;39(1–4):135–140.

153. Alphand V, Furstoss R. Microbiological transformations 22. Microbiologically mediated Baeyer-Villiger reactions: a unique route to several bicyclic gamma-lactones in high enantiomeric purity. J Org Chem 1989;57:1306–1309.

154. Kamberbeek NM, Janssen DB, van Berkel WJH, Fraaje MW. Baeyer-Villiger monooxygenases, an emerging family of flavin-dependent biocatalysts. Adv Synth Catal 2993;345:667–678.

155. Mihovilovic MD, Mueller B, Stanett C. Monooxygenase-mediated Baeyer-Villiger oxidation. Eur J Org Chem 2002;22:3711–3730.

156. Stewart JD. Cyclohexanone monooxygenase. A useful reagent for asymmetric Baeyer Villiger reactions. Curr Org Chem 1998;2:195–216.

157. Hiker I, Wohlgemuth R, Alphand V, Furstoss R. Microbial transformations 59: first kilogram scale asymmetric microbial baeyer-villiger oxidation with optimized productivity using a resin-based in situ SFPR strategy. Biotechnol Bioeng 2002;92(6):702–710.

158. Rudroff F, Alphand V, Furstoss R, Mihovilovic MD. Optimizing fermentation conditions of recombinant *Escherichia coli* expressing cyclopentanone monooxygenase. Org Process Res Dev 2006;10(3):599–604.

159. Cernuchova P, Mihovilovic MD. Microbial Baeyer-Villiger oxidation of terpenones by recombinant whole-cell biocatalysts-formation of enantiocomplementary regioisomeric lactones. Org Biomol Chem 2007;5(11):1715–1719.

160. Hilker I, Gutierrez MC, Alphand V, Wohlgemuth R, Furstoss R. Microbiological transformations 57. facile and efficient resin-based in situ SFPR: preparative-scale synthesis of an enantiopure "Unexpected" lactone regioisomer via a Baeyer-Villiger oxidation process. Org Lett 2004;6(12):1955–1958.

161. Torres Pazmino DE, Snajdrova R, Rial DV, Mihovilovic MD, Fraaije MW. Altering the substrate specificity and enantioselectivity of phenylacetone monooxygenase by structure-inspired enzyme redesign. Adv Synth Catal 2007;349(8+9):1361–1368.

162. Fessner W-D. Enzyme-catalyzed aldol additions. Mod Aldol React 2004;1:201–272.

163. Griengl H, Schwab H, Fechter M. The synthesis of chiral cyanohydrins by oxynitrilases. Trends Biotechnol 2000;18(6):252–256.

164. Pohl M, Lingen B, Muller M. Thiamin-diphosphate-dependent enzymes: new aspects of asymmetric C–C bond formation. Chem Eur J 2002;8(23):5288–5295.

165. Fessner W-D, Helaine V. Biocatalytic synthesis of hydroxylated natural products using aldolases and related enzymes. Curr Opin Biotechnol 2001;12(6):574–586.

166. Sprenger GA, Pohl M. Synthetic potential of thiamin diphosphate-dependent enzymes. J Mol Catal B Enzym 1999;6(3):145–159.

167. Ward OP, Singh A. Recent development in enzymatic acyloin condensations. In: Patel RN, editor. Biocatalysis in the pharmaceutical and biotechnology industries. Boca Raton (FL): CRC Press; 2007. pp. 605–621.

168. Ohta H, Sugai T. Enzyme-mediated decarboxylase reactions in organic synthesis. In: Patel RN, editor. Stereoselective biocatalysis. New York: Marcel Dekker; 2000. pp 487–526.

169. Yagasaki M, Ozaki A. Industrial biotransformations for the production of D-amino acids. J Mol Catal B Enzym 1998;4(1–2):1–11.

170. Prosen E, Ward OP, Collins S, Dewdney NJ, Hong Y, Wilcocks R. Enantiospecific production of S-(-)-2-hydroxypropiophenone mediated by benzoylformate decarboxylase from *Acinetobacter calcoaceticus*. Biocatalysis 1993;8(1):21–29.

171. Demir AS, Dunnwald T, Iding H, Pohl M, Muller M. Asymmetric benzoin reaction catalyzed by benzoylformate decarboxylase. Tetrahedron: Asymmetry 1999;10(24):4769–4774.

172. Pohl M, Liese A. Industrial processes using lyases for C–C, C–N, and C–C bond formation. In: Patel RN, editor. Biocatalysis in the pharmaceutical and biotechnology industries. Boca Raton (FL): CRC Press; 2007. pp. 661–676.

173. Guo Z, Goswami A, Mirfakhrae KD, Patel RN. Asymmetric acyloin condensation catalyzed by phe nylpyruvate decarboxylase. Tetrahedron: Asymmetry 1999;10(24):4667–4675.

174. Lee S-G, Hong S-P, Sung M-H. Development of an enzymatic system for the production of dopamine from catechol, pyruvate and ammonia. Enzyme Microb Technol 1999;25:298–302.

175. Abelyan VA, Abelyan LA. Production of L-lactic acid and L-malic acid by intact and immobilized cells of *Lactobacillus casei*. Appl Biochem Microbiol 1998;34:578–580.

176. Wilcocks R, Ward OP, Collins S, Dewdney NJ, Hong Y, Prosen E. Acyloin formation by benzoylformate decarboxylase from *Pseudomonas putida*. Appl Environ Microbiol 1992;58(5):1699–1704.

177. Lingen B, Pohl M, Demir AS, Liese A, Mueller M. Enantioselective syntheses of hydroxy ketones via benzoylformate decarboxylase- and benzaldehyde lyase-catalyzed C–C bond formation. Oxid Stress Dis 2004;11(Thiamine):113–129.

178. Lingen B, Groetzinger J, Kolter D, Kula M-R, Pohl M. Improving the carboligase activity of benzoylformate decarboxylase from *Pseudomonas putida* by a combination of directed evolution and site-directed mutagenesis. Protein Eng 2002;15(7):585–593.

179. Dean SM, Greenberg WA, Wong C-H. Recent advances in aldolase-catalyzed asymmetric synthesis. Adv Synth Catal 2007;349(8+9):1308–1320.

180. Look GC, Fotsch CH, Wong C-H. Enzyme-catalyzed organic synthesis: practical routes to aza sugars and their analogs for use as glycoprocessing inhibitors. Acc Chem Res 1993;26:282–290.

181. Kajimoto T, Chen L, Liu KKC, Wong C-H. Palladium-mediated stereocontrolled reductive amination of azido sugars prepared from enzymic aldol condensation: a general approach to the synthesis of deoxy aza sugars. J Am Chem Soc 1991;113(17):6678–6680.

182. Ziegler T, Straub A, Effenberger F. Enzyme-catalyzed reactions. 3. Enzyme-catalyzed synthesis of deoxymannojirimycin, 1-deoxynojirimycin and 1,4-dideoxy-1,4-imino-D-arabinitol. Angew Chem 1988;100(5):737–738.

183. Kajimoto T, Liu KKC, Pederson RL, Zhong Z, Ichikawa Y, Porco JA, Wong C-H. Enzyme-catalyzed aldol condensation for asymmetric synthesis of azasugars: synthesis, evaluation, and modeling of glycosidase inhibitors. J Am Chem Soc 1991;113(16):6187–6196.

184. Romero A, Wong C-H. Chemo-enzymatic total synthesis of 3-epiaustraline, australine, and 7-epialexine. J Org Chem 2000;65(24):8264–8268.

185. Henderson DP, Shelton MC, Cotterill IC, Toone EJ. Stereospecific preparation of the N-terminal amino acid moiety of Nikkomycins KX and KZ via a multiple enzyme synthesis. J Org Chem 1997;62(23):7910–7911.

186. Sugai T, Lin CH, Shen GJ, Wong C-H. CMP-KDO synthetase: overproduction and application to the synthesis of CMP-KDO and analogs. Bioorg Med Chem 1995;3(3):313–320.

187. Blayer S, Woodley JM, Lilly MD, Dawson MJ. Characterization of the chemoenzymic synthesis of N-acetyl-D-neuraminic acid (Neu5Ac). Biotechnol Prog 1996;12(6):758–763.

188. Mahmoudian M, Noble D, Drake CS, Middleton RF, Montgomery DS, Piercey JE, Ramlakhan D, Todd M, Dawson MJ. An efficient process for production of N-acetylneuraminic acid using N-acetylneuraminic acid aldolase. Enzyme Microb Technol 1997;20(5):393–400.

189. Kragl U, Gygax D, Ghisalba O, Wandrey C. Enzymatic, two-step synthesis of N-acetylneuraminic acid in a enzyme-membrane reactor. Angew Chem 1991;103(7):854–855.

190. Gijsen HJM, Wong C-H. Sequential three- and four-substrate aldol reactions catalyzed by aldolases. J Am Chem Soc 1995;117(29):7585–7591.

191. Greenberg WA, Varvak A, Hanson SR, Wong K, Huang H, Chen P, Burk MJ. Development of an efficient, scalable, aldolase-catalyzed process for enantioselective synthesis of statin intermediates. Proc Natl Acad Sci USA 2004;101(16):5788–5793.

192. Pianko S, McHutchison JG. Treatment of hepatitis C with interferon and ribavirin. J Gastroenterol Hepatol 200;15(6):581–586.

193. Ferenci P, Brunner H, Nachbaur K, Datz C, Gschwantler M, Hofer H, Stauber R, Hackl F, Jessner W, Rosenbeiger M, Petra M-S, Hegenbarth K, Gangl A, Vogel W. Combination of interferon induction therapy and ribavirin in chronic hepatitis C. Hepatology (Philadelphia (PA)) 2001;34(5):1006–1011.

194. Bonkovsky HL, Stefancyk D, McNeal K, Banner BF, Liu Q, Zucker GM, Israel J, Stagias J, Colker J. Comparative effects of different doses of ribavirin plus interferon-α2β for therapy of chronic hepatitis C: Results of a controlled, randomized trial. Dig Dis Sci 2001;46(10):2051–2059.

195. Tamarez M, Morgan B, Wong GSK, Tong W, Bennett F, Lovey R, McCormick JL, Zaks A. Pilot-scale lipase-catalyzed regioselective acylation of ribavirin in anhydrous media in the synthesis of a novel prodrug intermediate. Org Process Res Dev 2003;7(6):951–953.

196. Cossy J, Schmitt A, Cinquin C, Buisson D, Belotti D. A very short, efficient and inexpensive synthesis of the prodrug form of SC-54701A a platelet aggregation inhibitor. Bioorg Med Chem Lett 1997;7(13):1699–1700.

197. Yamanaka T, Ohkubo M, Takahashi F, Kato M. An efficient synthesis of the orally-active GpIIb/IIIa antagonist FR184764. Tetrahedron Lett 2004;45(13):2843–2845.

198. Landis BH, Mullins PB, Mullins KE, Wang PT. Kinetic resolution of β-amino esters by acylation using immobilized penicillin amidohydrolase. Org Process Res Dev 2000;6(4):539–546.

199. Njoroge FG, Taveras AG, Kelly J, Remiszewski S, Mallams AK, Wolin R, Afonso A, Cooper AB, Rane DF, Liu Y-T, Wong J, Vibulbhan B, Pinto P, Deskus J, Alvarez CS, del Rosario J, Connolly M, Wang J, Desai J, Rossman RR, Bishop WR, Patton R, Wang L, Kirschmeier P, Bryant MS, Nomeir AA, Lin C-C, Liu M, McPhail AT, Doll RJ, Girijavallabhan VM, Ganguly AK. (+)-4-[2-[4-(8-Chloro-3,10-dibromo-6,11-dihydro-5H-benzo[5,6]cyclohepta[1,2-?]-pyridin-11(R)-yl)-1-pip eridinyl]-2-oxo-ethyl]-1-piperidinecarboxamide (SCH-66336): a very potent farnesyl protein transferase inhibitor as a novel antitumor agent. J Med Chem 1998;41:4890–4902.

200. Morgan B, Zaks A, Dodds DR, Liu J, Megati S, Njoroge FG, Girijavallabhan VM. Enzymatic kinetic resolution of piperidine atropisomers: Synthesis of a key intermediate of the farnesyl protein transferase inhibitor, SCH66336. J Org Chem 2000;65(18):5451–5459.

201. Ireland C, Leeson PA, Castaner J. Lobucavir, antiviral. Drugs Future 1997; 22(4):359–370.

202. Hanson RL, Shi Z, Brzozowski DB, Banerjee A, Kissick TP, Singh J, Pullockaran AJ, North JT, Fan J, Howell J, Durand SC, Montana MA, Kronenthal DR, Mueller RH, Patel RN. Regioselective enzymatic aminoacylation of lobucavir to give an intermediate for lobucavir prodrug. Bioorg Med Chem 2000;8(12):2681–2687.

203. Bisacchi G, Slusarchyk WA, Treuner U, Sutton JC, Zahler R, Seiler S, Kronenthal DR, Randazzo ME, Xu Z, Shi Z, Schwinden MD. Preparation of amidino and guanidino azetidinone compounds as tryptase inhibitors. PCT Int. Appl., WO 9967215 A1 19991229 CAN 132:64103 AN 1999:819347. 1999. p. 326.

204. Goswami A, Howell JM, Hua EY, Mirfakhrae KD, Soumeillant MC, Swaminathan S, Qian X, Quiroz FA, Vu TC, Wang X, Zheng B, Kronenthal DR, Patel RN. Chemical and enzymatic resolution of (*R,S*)-*N*-(tert-butoxycarbonyl)-3-hydroxymethylpiperidine. Org Process Res Dev 2001;5(4):415–420.

205. Kim M-J, Ahn Y, Park J. Dynamic kinetic resolution and asymmetric transformations by enzyme-metal combinations. In: Patel RN, editor. Biocatalysis in the pharmaceutical and biotechnology industries. Boca Raton (FL): CRC Press; 2007. pp. 249–272.

206. Martin-Matute B, Baeckvall J-E. Dynamic kinetic resolution catalyzed by enzymes and metals. Curr Opin Chem Biol 2007;11(2):226–232.

207. Pamies O, Baeckvall J-E. Dynamic kinetic resolution of β-azido alcohols. An efficient route to chiral aziridines and β-amino alcohols. J Org Chem 2000;66(11):4022–4025.

208. Turner N, Fotheringham I, Speight R. Novel biocatalyst technology for the preparation of chiral amines. Innov Pharm Technol 2004;4(14):114–116,118–122.

209. Carr R, Alexeeva M, Dawson MJ, Gotor-Fernandez V, Humphrey CE, Turner NJ. Directed evolution of oxidase for the preparative deracemization of cyclic secondary amines. ChemBioChem 2005;6:637–639.

210. Roff GJ, Lloyd RC, Turner NJ. A versatile chemo-enzymatic route to enantiomerically pure β-branched α-amino acids. J Am Chem Soc 2004;126(13):4098–4099.

211. Zhu Y, Fow K-L, Chuah G-K, Jaenicke S. Dynamic kinetic resolution of secondary alcohols combining enzyme-catalyzed transesterification and zeolite-catalyzed racemization. Chem Eur J 2007;13(2):541–547.

212. Boren L, Martin-Matute B, Xu Y, Cordova A, Baeckvall J-E. (*S*)-selective kinetic resolution and chemoenzymatic dynamic kinetic resolution of secondary alcohols. Chem Eur J 2005;12(1):225–232.

213. Kim M-J, Kim HM, Kim D, Ahn Y, Park J. Dynamic kinetic resolution of secondary alcohols by enzyme-metal combinations in ionic liquid. Green Chem 2004;6(9):471–474.

214. Lee HK, Kim D, Park J. (*S*)-Selective dynamic kinetic resolution of secondary alcohols by the combination of subtilisin and an aminocyclopentadienyl ruthenium complex as the catalysts. J Am Chem Soc 2003;125:11494–11495.

215. Verzijl GKM, De Vries JG, Broxterman QB. Process for the preparation of enantiomerically enriched esters and alcohols. PCT Int. Appl., WO 2001090396 A1 20011129. 99. 2001.

216. Allan GR, Carnell AJ. Microbial deracemization of 1-aryl and 1-heteroaryl secondary alcohols. J Org Chem 2001;66(19):6495–6497.

217. Azerad R, Buisson D. Dynamic resolution and stereoinversion of secondary alcohols by chemo-enzymatic processes. Curr Opin Biotechnol 2000;11(6):565–571.

218. Kroutil W, Faber K. Deracemization of compounds possessing a sec-alcohol or -amino group through a cyclic oxidation-reduction sequence: a kinetic treatment. Tetrahedron: Asymmetry 1998;99(16):2901–2913.

219. Carnell AJ. Stereoinversions using microbial redox-reactions. Adv Biochem Eng Biotechnol 1999;63(Biotransformations):57–72.

220. Comasseto V, Omori AT, Andrade LH, Porto ALM. Bioreduction of fluoroacetophenones by the fungi *Aspergillus terreus* and *Rhizopus oryzae*. Tetrahedron: Asymmetry 2003;14:711–715.

221. Broxterman QB, Verzijl GKM. Process for the synthesis of (*R*)-1-(3,5-bis(trifluoromethyl) phenyl) ethan-1-ol and esters thereof by dynamic kinetic resolution. PCT Int. Appl. WO 2003043575 A2 20030530. 2003.

222. Demir AS, Hamamci H, Sesenoglu O, Neslihanoglu R, Asikoglu B, Capanoglu D. Fungal deracemization of benzoin. Tetrahedron Lett 2002;43:6447–6449.

223. Kato D, Mitsuda S, Ohta H. Microbial deracemization of α-substituted carboxylic acids. Org Lett 2002;4:371–373.

224. Kato D, Mitsuda S, Ohta H. Microbial deracemization of α-substituted carboxylic acids: substrate specificity and mechanistic investigation. J Org Chem 2003;68:7234–7242.

225. Paedragosa-Moreau S, Morisseau C, Baratti J, Zylber J, Archelas A, Furstoss R. Microbiological transformations. 37. An enantioconvergent synthesis of the β-blocker (*R*)-Nifenalol using a combined chemoenzymic approach. Tetrahedron 1997;53(28):9707–9714.

226. Orru RVA, Archelas A, Furstoss R, Faber K. Epoxide hydrolases and their synthetic applications. Adv Biochem Eng Biotechnol 1999;63(Biotransformations):145–167.

227. Archer IVJ, Leak DJ, Widdowson DA. Chemoenzymic resolution and deracemization of (±)-1-methyl-1,2-epoxycyclohexane: the synthesis of (1-*S*,2-*S*)-1-methylcyclohexane-1,2 diol. Tetrahedron Lett 1996;37(48):8819–8822.

228. Ueberbacher BJ, Osprian I, Mayer SF, Faber K. A chemoenzymatic, enantioconvergent, asymmetric total synthesis of (*R*)-fridamycin E. Eur J Org Chem 2005;7:256–259.

229. Monfort N, Archelas A, Furstoss R. Enzymatic transformations. Part 55: Highly productive epoxide hydrolase catalyzed resolution of an azole antifungal key synthon. Tetrahedron 2004;60(3):601–605.

230. Catt JD, Johnson G, Keavy DJ, Mattson RJ, Parker MF, Takaki KS, Yevich JP. Preparation of benzofuran and dihydrobenzofuran melatonergic agents. US 5856529 A 19990105 CAN 130:110151 AN 1999:34508. 1999. p. 24.

231. Goswami A, Mirfakhrae KD, Patel RN. Deracemization of racemic 1,2-diol by biocatalytic stereoinversion. Tetrahedron: Asymmetry 1999;10(21):4239–4244.

232. Janssen DB. Dehalogenases in biodegradation and biocatalysis. In: Patel RN, editor. Biocatalysis in the pharmaceutical and biotechnology industries. Boca Raton (FL): CRC Press; 2007. pp. 441–462.

233. Nakamura T, Nagasawa T, Yu F, Watanabe I, Yamada H. A new catalytic function of halohydrin hydrogen-halide-lyase, synthesis of β-hydroxynitriles from epoxides and cyanide. Biochem Biophys Res Commun 1991;180(1):124–130.

234. Lutje Spelberg JHL, Tang L, van Gelder M, Kellogg RM, Janssen DB. Exploration of the biocatalytic potential of a halohydrin dehalogenase using chromogenic substrates. Tetrahedron: Asymmetry 2002;13(10):1083–1089.

235. Lutje Spelberg JH, van Hylckama Vlieg JET, Tang L, Janssen DB, Kellogg RM. Highly enantioselective and regioselective biocatalytic azidolysis of aromatic epoxides. Org Lett 2001;3(1):41–43.

236. Lutje Spelberg JH, Tang L, Kellogg RM, Janssen DB. Enzymatic dynamic kinetic resolution of epihalohydrins. Tetrahedron: Asymmetry 2004;15(7):1095–1102.

237. Hasnaoui G, Lutje Spelberg JH, de Vries E, Tang L, Hauer B, Janssen DB. Nitrite-mediated hydrolysis of epoxides catalyzed by halohydrin dehalogenase from *Agrobacterium radiobacter* AD1: a new tool for the kinetic resolution of epoxides. Tetrahedron: Asymmetry 2005;16(9):1685–1692.

238. Elenkov MM, Hauer B, Janssen DB. Enantioselective ring opening of epoxides with cyanide catalyzed by halohydrin dehalogenases: a new approach to nonracemic β-hydroxy nitriles. Adv Synth Catal 2006;348(4+5):579–585.

CHAPTER 6

TRANSGLUTAMINASE FOR PROTEIN DRUG MODIFICATION: PEGYLATION AND BEYOND

SEAN HU
Novozymes, Davis, CA

YIFENG YIN
Trinity Biosystems, Menlo Park, CA

6.1 INTRODUCTION

The protein therapeutics market is approaching the 100 billion dollar mark, representing one of the fastest growing segments of the pharmaceutical industry. While protein drugs offer life-saving and life-enhancing treatments with significantly lower toxicity than most small-molecule therapeutics, their therapeutic efficacy is often negatively impacted by various biological processes. The most prominent issue is the rapid clearance by renal excretion, especially for proteins of small size. In addition, many proteins, particularly nonhuman proteins, would induce immune responses. Both renal clearance and immunogenicity, together with other issues such as proteolysis, contribute to a relative short *in vivo* half-life for most protein drugs. As a result, protein drugs have to be injected frequently. This increases not only the difficulty of patients' compliance, but also the cost of treatment.

To address these issues, two approaches have been taken. The first one involves the application of sustained release systems, such as liposome, microparticle, or nanoparticle. The other is protein protraction by protein fusion, glycosylation, and fatty acid or polyethylene glycol (PEG) conjugation. PEG is a polyether prepared from catalyzed polymerization of ethylene oxide. Depending on the initiator of the reaction (methanol or water), the terminal groups of the linear

Biocatalysis for Green Chemistry and Chemical Process Development, First Edition.
Edited by Junhua (Alex) Tao and Romas Kazlauskas.

polymer could be methoxy/hydroxyl group (mPEG-OH), or two hydroxyl groups (HO-PEG-OH). The hydroxyl group(s) can be readily converted to different functional groups for conjugation. In some cases, two chains of linear polymer are linked to one activated functional group to form a branched PEG. To avoid cross-linking, mono-functional PEGs are mostly used for conjugation, which is commonly referred to as *PEGylation*. The pioneering studies of PEGylation by Abuchowsky and Davies [1,2] dates back to the 1970s, although wide acceptance of PEGylation did not occur until the late 1990s when its long-term safety profile was unequivocally established. PEG became the polymer of choice because of its unique physical characteristics: uncharged, hydrophilic, and nonimmunogenic. In addition, PEG has an unusually large hydrodynamic radius as each of the ethylene oxide monomeric unit can coordinate several water molecules. Covalently conjugated PEG not only increases the size of the protein molecule, but shields potential immunogenic epitopes, reduces protein aggregation, and slows down proteolytic degradation. As a result, PEGylated protein drugs usually possess more favorable biological behavior *in vivo* as compared to the parent molecules. Thus, PEGylation has become the most widely utilized modification approach to prepare long-lasting second generation protein therapeutics [3,4].

In this chapter, we first summarize applications and limitations of classic chemical conjugation. Such chemical modifications not only involve the use of hash chemicals, sometimes even toxical reagents, but also create mixtures that require extensive purification. Hence, we focus on the environmentally friendly enzymatic conjugation using microbial transglutaminase (mTGase) [5]. Enzymatic conjugation, mild and green in nature, can accomplish site-specific modifications not obtainable via chemical methods and greatly simplify downstream processing by reducing expensive, power and water hungry HPLC (high performance liquid chromatography) steps.

6.2 CHEMICAL CONJUGATION OF PEG

During the last 30 years, many types of chemistries have been developed for PEG conjugation. Since functional groups on the protein determine which chemistry to be used, our discussion will thus follow such reactive groups.

6.2.1 Chemical Conjugation Reactions

6.2.1.1 Amino Group (N-Terminal α-NH_2 and Lysine ε-NH_2). Earlier PEGylation was mostly directed toward the free amino groups in proteins, and this is still true. There are multiple amino groups on a protein molecule, such as the α-amino group at N-terminus (pK_a, 6–8) and ε-amino group of lysine (pK_a, 8–10.5). When the conjugation reaction is carried out at a pH between the pK_a values of N-terminal and lysine amino groups, N terminal α-NH_2 group will be in nonprotonated form while lysine ε-NH_2 will still be protonated. This allows the selective modification of N terminal α-NH_2 group by careful control of reaction pH.

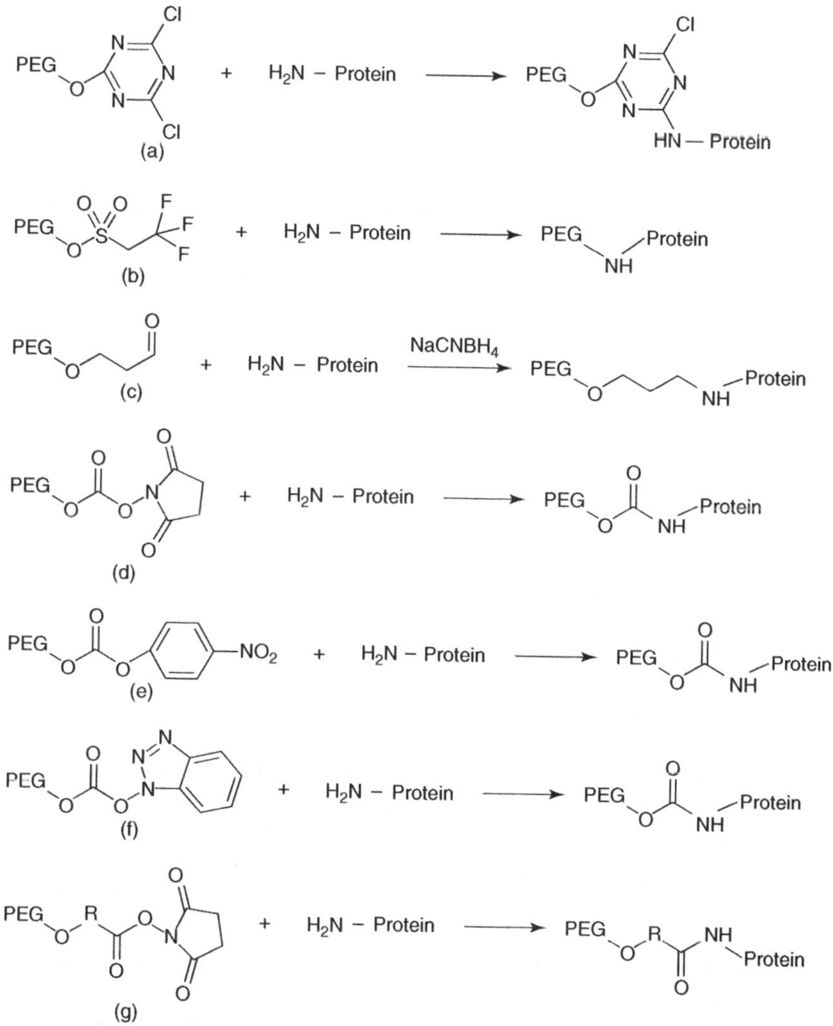

Figure 6.1 PEGylation reactions towards protein amino groups. (a) PEG dicholoro-triazine. (b) PEG PEG tresylate. (c) PEG propionaldehyde. (d) PEG succinimidyl carbonate. (e) PEG p-nitrophenyl carbonate. (f) PEG benzotriazol carbonate. (g) PEG N-hydroxysuccinimide ester.

Two types of reaction can be used to conjugate PEG to a NH_2 group. The first type, alkylation reaction (Figure 6.1a–c), turns the primary amine into a secondary amine while preserving the positive charge on the nitrogen atom. The earliest PEGylation experiments carried out by Abuchowski [2] utilized PEG dichlorotriazine to connect protein to PEG with a triazine ring (Figure 6.1a). This approach is no longer used because of the PEG reagent's toxicity. PEG tresylate (trifluoroethanesulfonate, Figure 6.1b) also alkylates nonprotonated amino groups efficiently as tresylate is a good leaving group. PEG activated with an aldehyde group (Figure 6.1c) can also conjugate with NH_2 group to form a Schiff's base intermediate, which can be readily converted to a stable secondary amine through reduction. This reductive alkylation was successfully used to selectively PEGylate N-terminal α-NH_2 group of granulocyte colony-stimulating factor (GCSF) under acidic condition [6] to create Neulasta®, which has been approved by FDA in 2002 and marketed by Amgen to treat chemotherapy-induced neutropenia.

The other type of widely used conjugation reaction is acylation, resulting in the loss of positive charge (Figure 6.1d–g). Assuming PEGylation sites are mostly away from a protein's active site, loss of charge is not a significant issue. PEG reagents designed for NH_2 acylation are exemplified by carbonate PEGs, such as PEG succinimidyl carbonate (Figure 6.1d), PEG *p*-nitrophenyl carbonate (Figure 6.1e), and PEG benzotriazol carbonate (Figure 6.1f), which form a urethane linkage to the proteins. This type of activated PEG reacts more slowly with amino groups than other PEG compounds, facilitating a better controllable and selective conjugation. PEG *N*-hydroxysuccinimide (NHS) ester (Figure 6.1g) is a more reactive reagent for amino group acylation. It reacts with NH_2 group with high specificity under very mild condition, although its high reactivity makes the reaction difficult to control at large scale.

Some of the above PEGylation reagents, including PEG succinimidyl carbonate, PEG benzotriazol carbonate, and PEG dichlorotriazine, can react with other surface nucleophilic groups, such as hydroxyl groups of serine and tyrosine residues, and imidazole group of histidine residue, and these side reactions can be a concern when carrying out PEGylation reactions.

6.2.1.2 Thiol Group of Cysteine. The low abundance of free cysteine residues in proteins makes the thiol group an ideal target for PEGylation. If applicable, a free cysteine residue can be introduced into the protein by site-directed mutagenesis [7,8]. PEG pyridyldisulfide (Figure 6.2a) reacts specifically with free thiol, leading to a disulfide bond connecting PEG with the protein. This can be used as a releasable PEGylation. Additionally, PEG can be conjugated with a stable thioether linkage. Reagents for this purpose include PEG maleimide (Figure 6.2b), PEG vinylsulfone (Figure 6.2c), and PEG iodoacetamide (Figure 6.2d). However, PEG iodoacetamide reacts slowly, and is not widely used. Just like other thiol cross-linking reagents, both PEG maleimide and PEG vinylsulfone could lead to side reactions with primary amine group at basic pH (>8).

6.2.1.3 Imidazole Group of Histidine. Histidine is frequently indicated in catalytic centers because of its near neutral pK_a (~6). This pK_a allows its PEGylation through acylation to be readily carried out under slightly acidic conditions. For example, interferon α-2b can be conjugated at a single His34 residue by a PEG succinimidyl carbonate (Figure 6.2e) or a PEG benzotriazol carbonate (Figure 6.2f) at pH 4.5–6.8 [9]. The acyl-imidazole linkage is not stable in aqueous environment under physiological pH, and will be slowly hydrolyzed to lose PEG and release unmodified native protein. Such a releasable linker might be responsible for the high potency of Pegintron.

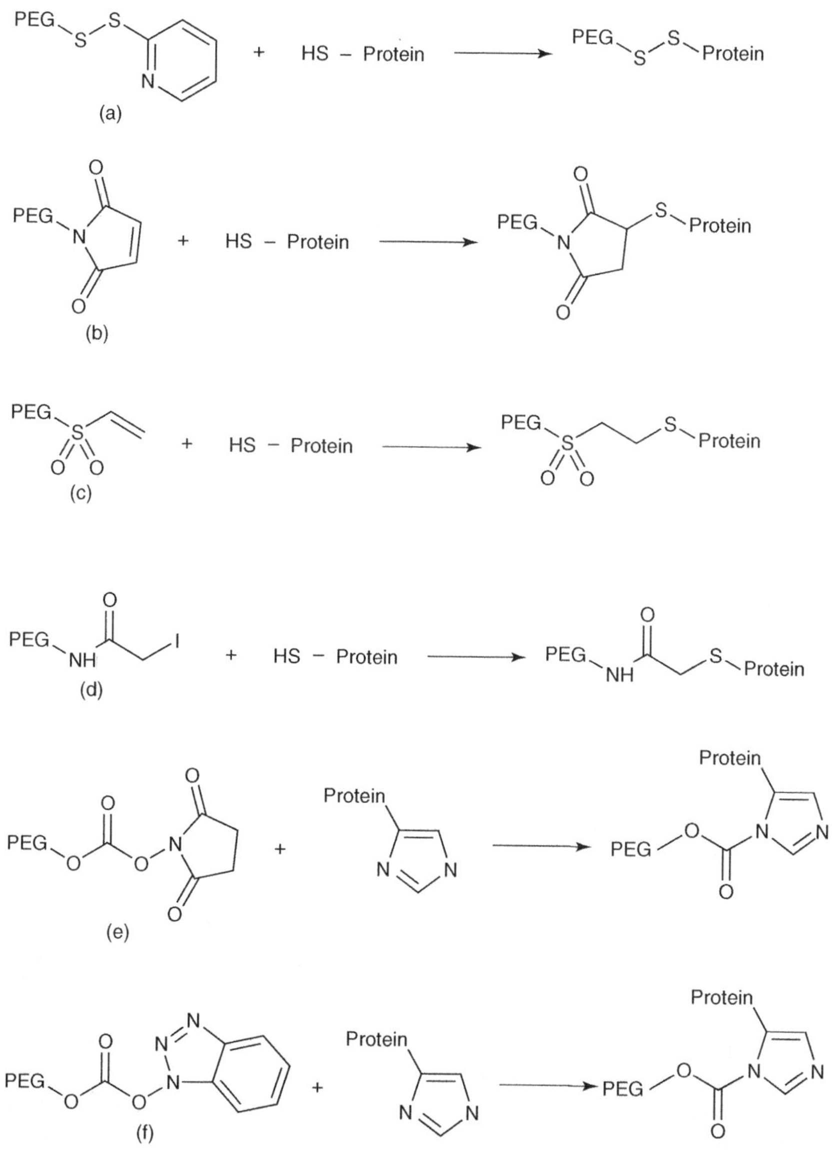

Figure 6.2 Examples of PEGylation reactions towards thiol (a–d) and imidazole (e–f) groups in protein. (a) PEG pyridyldisulfide. (b) PEG maleimide. (c) Vinylsulfone. (d) PEG iodoacetamide. (e) PEG succinimidyl carbonate. (f) PEG benzotriazol carbonate.

6.2.1.4 Carboxyl Group (C-Terminal, Glutamate, and Aspartate). The conjugation reaction happens more readily on the nucleophilic moieties such as NH_2 and SH groups. There is no easy way to activate carboxylic groups without causing intramolecular cross-linking or aggregation. One way to get around this limitation is to use PEG hydrazide to directly react with a COOH group under acidic conditions [10]. Owing to lack of specificity, there is currently no commercial product developed with conjugation to COOH.

6.2.2 PEGylated Protein Drugs On The Market

The first PEGylated protein, Enzon's Adagen® (PEGylated bovine adenosine deaminase), was approved by FDA in March 1990 to treat X-linked severe combined immunodeficiency (SCID). Since then, eight PEGylated protein drugs (Table 6.1) are now on the market [6,11–17]. Among them, Amgen's Neulasta and Roche's Pegasys® have revenues in 2007 of \$3 billion and \$1.6 billion, respectively and rank 10th and 19th of the top biotech drugs, demonstrating the great potential of PEGylated proteins [18].

Adenosine deaminase and asparaginase from Enzon are enzymes of nonhuman origin. PEGylation is needed to prevent rapid neutralization by immunoglobulin against these foreign proteins. Since 2000, PEGylated human proteins have entered the market, and PEGylation improves therapeutic efficacy, mainly by increasing half-life *in vivo*.

Long half-life comes with the sacrifice of biological activity. Increase in half-life and decrease in biological activity are directly related to PEG size. The molecular weight of PEG commonly used in protein modification is between 5 and 40 kDa. For PEG alone, the glomerular filtration rate stays high for 5 kDa PEG, and 10 kDa PEG lowers clearance rate by about half, while 20 kDa PEG only has about 10% of the filtration rate [19]. The effect of PEG size on renal clearance rate is more complicated with PEGylated proteins. For example, conjugation of a single 12 kDa PEG lowers interferon α-2b apparent clearance by 90%, thus increasing half-life by approximately 10-fold [20]. In the case of Mircera®, a 30 kDa PEG is conjugated randomly to an amino group (N-terminus or one of lysines, predominantly Lys45 or Lys52) of erythropoietin (EPO) [21]. Micera has a 50- to 100-fold lower affinity to EPO receptor than EPO. However, PEGylation extends serum half-life to 130 h, compared with 8.5 h for EPO [16], making once a month administration possible (in comparison to once a week injection of Aranesp®).

In addition, the loss of biological activity is greatly affected by PEGylation location on the protein molecule. Roche's Pegasys (PEGylated form of interferon α-2a) was prepared by random linkage of 40 kDa branched PEG-NHS ester to lysine amino groups at pH 9.0. Nine out of the 12 possible mono-PEGylated products (N-terminal and 11 lysine amino groups) were identified with different levels of biological activity. The most active PEGylation product modified at Lys31, has sixfold higher antiviral activity compared to that of the least active PEGylated protein, modified at Lys121 [14]. Therefore, the development of PEGylated drugs is always a balance between improving clearance and immunogenicity profile and avoiding impact on the biological activity of the parent drug.

6.2.3 Random Versus Site-Specific, Challenge with Chemical Conjugation

With more understanding of how PEGylation affects the biological function, immunogenicity and toxicity, the trend favors PEGylation at a defined amino

TABLE 6.1 PEGylated Protein Drugs Currently on US Market

Name	Indication	Protein	PEG Size (kDa) and Functionality	PEGylation Specificity	Linkage Type	Company	Year of Approval
Adagen® (Pegademase)	Severe combined immunodeficiency	Bovine adenosine deaminase	PEG 5-succinimide	Nonspecific	Stable	Enzon	1990
Oncaspar® (Pegaspargase)	Acute lymphoblastic leukemia	*Escherichia coli* asparaginase	PEG 5-succinimide	Nonspecific	Stable	Enzon	1994
Pegintron®	Chronic hepatitis C infections	Interferon α-2b	PEG 12-succinimidyl carbonate	Specific	Releasable	Schering-plough/Merck	2000
Pegasys®	Chronic hepatitis B and C infections	Interferon α-2a	Branched PEG 40-NHS	Nonspecific	Stable	Roche	2001
Neulasta® (Pegfilgrastim)	Neutropenia	Granulocyte colony-stimulating factor	PEG 20-propionaldehyde	Specific	Stable	Amgen	2002
Somavert® (Pegvisomart)	Acromegaly	B2036, an hGH agonist	Several PEG 5-NHS	Nonspecific	Stable	Pfizer	2003
Mircera®	Anemia	Erythropoietin β	PEG 30-butanoic NHS	Nonspecific	Stable	Roche	2007
Cimzia® (Certolizumab pegol)	Crohn's disease	Humanized anti-TNFα Fab′	Branched PEG40-maleimide	Specific	Stable	UCB	2008

acid residue, leading to very closely related molecular entities with clearly defined structure-activity relationship (SARs) and Pharmacokinetics/Pharmacodynamics (PK/PD) profiles. There are other benefits with site-specific PEGylation over a random approach: better quality control, improved therapeutical efficacy, and ease to win the regulatory approval. With the development of many different chemistries and strategies in conjugation reactions, the early random PEGylation on multiple amino acid residues has given way to more specific coupling to a single residue. One method is to take advantage of pK_a difference among amino groups of N-terminus and lysines. This is best exemplified by Neulasta, where reductive alkylation only happens at N-terminal amino with the choice of acidic pH for conjugation. Alternatively, site-specific PEGylation is achieved with a single free cysteine for Cimzia.

However, for many potential PEGylation targets, neither approach is applicable. The difference between pK_a of N-terminal and lysine amino groups is frequently too small to allow selective PEGylation. In addition, N-terminal NH_2 can be completely resistant to conjugation because of steric hindrance, as in the case of interferon α-2a PEGylation [14]. The unpaired cysteine residue could be at the active site and/or is not present in most proteins. Clearly, site-specific PEGylation is still a challenging task.

6.3 TAILOR-MADE mTGase FOR SITE-SPECIFIC MODIFICATION OF PROTEIN DRUGS

6.3.1 Posttranslational Modification of Proteins and Its Application in Protein Conjugation

Chemical modification of a given protein at a specific site is not only very challenging, but sometimes impossible. However, nature does it with precision and efficiency, as in protein posttranslational modification (PTM). PTM enhances, modifies, and extends the functionality of the protein via many enzymatic processes such as glycosylation, lipoylation, acylation, alkylation, and transglutamination. Presumably, all these PTM processes could be adapted to modify therapeutic proteins. However, to be able to compete or to further extend chemical modification, such processes have to meet several criteria: (i) they are easily implemented, (ii) the modification is unique and specific, and (iii) the yield is high. In practice, only a few of enzymatic processes have the potential for industrial application.

Glycolyslation, sometimes necessary for bioactivity, has long been known to increase protein solubility and stability. Protraction by glycosylation also increases serum half-life. This can be well illustrated by a blockbuster drug, Aranesp. EPO is a 35 kDa glycoprotein of 165 amino acid residues which stimulates red blood cell production, differentiation, and maturation. Nonglycosylated EPO may have a higher specific activity *in vitro*, but extremely short biological half-life. Recombinant EPO, approved in 1989 for anemia and its sequel, is expressed in CHO cells with carbohydrate chains at four sites (Asn24,

Asn36, Asn83, and Ser126). By site-directed mutagenesis, Amgen introduced five mutations into EPO: A30N, H32T, P87V, W88N, and P90T. The two extra *N*-glycosyl chains increase the molecular weight of this hyper-glycosylated EPO analog (approved and marketed as Aranesp in 2001) from 34 to 37 kDa, and its half-life from 8.5 to 25.3 h [22].

Genetic engineering may not be a general solution for glycosylation as it involves mutagenesis, new production process, and a mammalian expression host. Hence, the popular and robust production hosts such as *Escherichia coli* and yeast are not suitable for mammalian glycoprotein production. To design an *in vitro* glycosylation process, Ramakrishnan used GlcNAc transferase to selectively transfer UDP-GalNAc to Thr/Ser residues or a C-terminus fusion tag with a single Thr residue [23,24]. Then, a modified sugar (such as 2-keto-Gal) can be transferred to GlcNAc by glycosyltransferases [25,26]. While this approach was utilized to conjugate scFv with small molecules [27], a direct coupling of nonglycoprotein with monodispersed polymers does not seem to be obvious without a fusion tag.

Tyrosinases are oxidative enzymes that convert accessible tyrosine residues of proteins into reactive ortho-quinone moieties. Oxidized tyrosyl residues can undergo nonenzymatic reactions with available nucleophiles such as the nucleophilic amino groups (Figure 6.3).

Even though the cascade of reactions involving *o*-quinones is poorly understood, it does not preclude its application in synthesis of protein—chitosan/gelatin conjugates [28,29]. As the oxidized tyrosyl residue reacts with any nucleophiles such as ε-NH_2 of lysine and thiol of cysteine, tyrosinase can be used in protein labeling, protein cross-linking and protein grafting to polysaccharide (or *vice versa*). It is very unlikely to be useful in site-specific therapeutic protein modification.

Proteases are well known for their proteolytic cleavage of protein. Although it is lesser known, they can also be used in peptide and ester conjugation [30]. A *subtilisin BPN′* variant was engineered by Wells' group to ligate esterified peptides onto the N termini of proteins with little residual protease activity [31]. Its utility in protein PEGylation was illustrated by adding a lysine handle to atrial natriuretic peptide. While this tool finds its way in specific labeling of protein N-termini for proteomics studies [32], adaptation to direct N-terminal PEGylation is possible if the residual protease activity is not a concern.

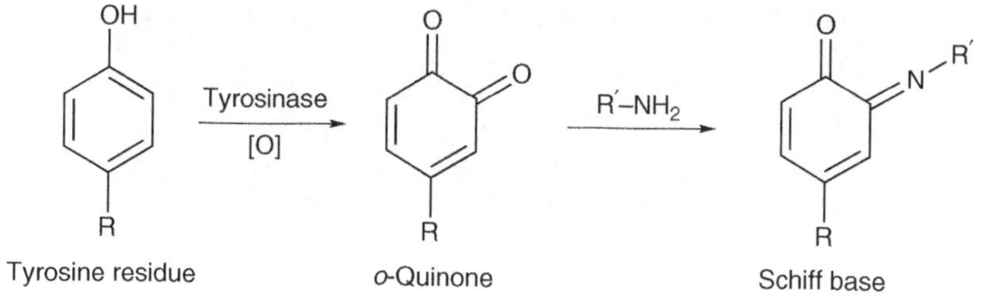

Figure 6.3 Tyrosinase catalyzed oxidation of tyrosine and subsequent cross-linking reaction.

Noncanonical amino acids are potential unique sites for conjugation if they can be incorporated into a protein. Amber, opal, and ochre are three stop codons to signal the termination of protein translation. tRNAs have been identified in some strains that recognize amber codons and insert an amino acid rather than halting translation. An elegant *E. coli* system was designed using an orthogonal phenylalanyl-tRNA synthetase and $tRNA^{phe}$ pair from *Saccharomyces cerevisiae* to incorporate *p*-fluorophenylalanine [33]. On the basis of this principle, *p*-acetyl phenylalanine was incorporated during expression at a specific site of human growth hormone (hGH), and aminooxy PEG was subsequently attached to this handle to give PEGylated hGH [34]. Inherently, the expression level of target protein could be very low.

Alternatively, an *E. coli* leucyl, phenylalanyltransferases was used to append a noncanonical amino acid residue to the N-terminus of protein bearing an unstructured/accessible N-terminus with a basic residue [35]. For example, *p*-ethynylphenylalanine from aminoacyl-tRNA was added to N-terminus of model protein dihydrofolate reductase (DHFR) for PEG conjugation.

Transglutaminase (EC2.3.2.13) catalyzes protein cross-linking reactions between glutamine and lysine residues to form ε(γ-glutamyl) lysyl isopeptide bonds (Figure 6.4) and has been dubbed nature's biological glue [36]. Owing to the formation of an acyl enzyme intermediate, transglutaminase (TGase) would exert substrate specificity toward both acyl donor and acceptor. Mammalian tissue TGases, requiring Ca^{2+} for activity, are involved in broad biological processes such as wound healing, epidermal keratinization, tissue repair, fibrogenesis, apoptosis, inflammation, and cell-cycle control. Guinea pig liver TGase (G-TGase) and factor XIII are two well-studied TGases [37–39]. However, their applications are limited due to their animal origin and stringent substrate specificity [40–44]. For instance, G-TGase could not couple monodansyl cadaverine (MDC) to hGH, EPO, GCSF, and interferon α-2b at enzyme concentration

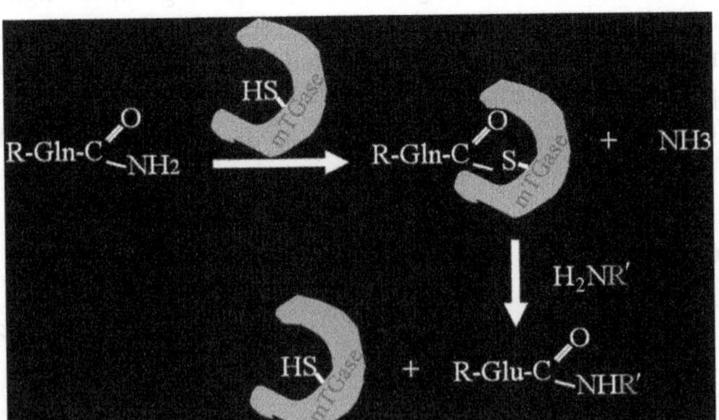

Figure 6.4 Transglutaminase catalyzed protein cross-linking reaction. Gln residue from protein substrate and the sulfhydryl group of the catalytic Cys residue form an acyl-enzyme intermediate. Nucleophilic attack from unprotonated amino group of Lys side chain would result in isopeptide bond formation.

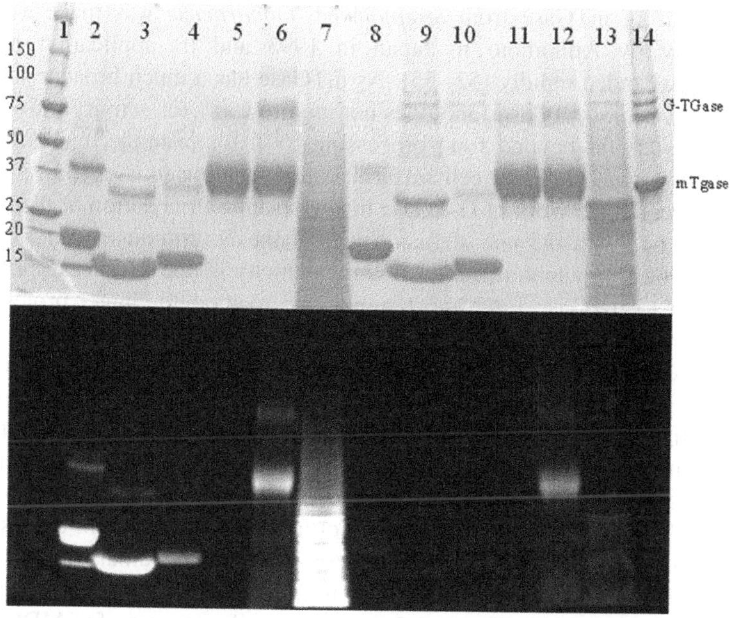

Figure 6.5 mTGase (from *Streptomyces mobaraense*) has a much broader protein substrate specificity than G-TGase does. (a) Gel image of Coomassie Blue stained SDS PAGE of 4–16%. (b) Fluorescent image of the same gel. Lanes 2 to 4, hGH, GCSF, and Interferon a2b could be labeled with fluorescent dye monodansyl cadaverine (MDC) (b) by mTGase and cross-linked dimers were seen (a). However, they were not labeled by G-TGase (lanes 8 to 10) even though enzyme concentration was 40 times higher and incubation time was 7.5 times longer than mTGase reactions. EPO and N-N-dimethylated casein (DMC), which are glycoproteins, were not labeled with MDC by either mTGase or G-TGase unless they were reduced: lanes 5 to 7: EPO, EPO + DTT, and DMC + DTT with mTGase; lanes 11–13: EPO, EPO + DTT, and DMC + DTT with G-GTase. Positions of mTGase (Novozymes) and G-TGase (Sigma) were shown in lane 14. Each protein (1.5 mg/ml) was mixed with acyl acceptor MDC (3 mM) in Tris buffer (50 mM, pH 7.5), CaCl2 (10 mM) was added for G-TGase only. mTGase or G-TGase was added at 0.005 and 0.2 unit/ml, respectively, to initiate the reaction. After 4 h of incubation at 37°C, samples were taken from reactions catalyzed by mTGase. To EPO/mTGase and DMC/mTGase reactions, DTT was added (to 10 mM) and incubation was extended for additional 4 h. After 30 h 37°C, samples were taken from G-TGase catalyzed reactions. To EPO/G-TGase and DMC/G-TGase reactions, DTT was added (to 10 mM) and incubation was extended for additional 4 h. Concentrations stated here were all final concentrations in the reaction mixtures. When Texas Red cadaverine and Z-Gly-Ala-Gln-AMC were used as acyl acceptor or donor, similar gel imagines were obtained.

of 0.2 unit/mL after 30 h incubation at 37°C (Figure 6.5). Even though both G-TGase and factor XIII have been expressed in bacteria [45–47], yeast [48,49], and plants [50], extremely low expression level makes it impossible to produce them recombinantly for industrial purposes. On the other hand, microbial TGase (mTGase) [51] are easily produced by fermentation. Since its discovery in the later 1980s [52], mTGase from *Streptomyces mobaraense* was made available commercially by Ajinomoto in Japan in 1993 and its application in food processing expanded rapidly [53–55]. As mTGase has a much broader substrate specificity than tissue TGases and does not require Ca^{2+} for activity [56–59], its application goes far beyond food processing [60], for example, hapten-protein conjugation *in vitro* [61] and cell surface protein labeling *in vivo* [62].

Sato pioneered the work of G-TGase in site-specific PEGylation of therapeutic proteins [63,64] with the use of fusion tags at the N-terminus of interleukin 2 (IL2) in 1996. The potential risk of immunogenicity of the fusion tag prompted Sato to turn to mTGase for a more applicable method for intact protein conjugation. It was found that IL2 was PEGylated specifically on Q74 (out of six glutamine residues total) by mTGase [65]. Apparently, Johnson & Johnson was persistent with G-TGase. Its 2004 US patent [66] claimed EPO PEGylation using G-TGase. In its first example, EPO was labeled with MDC to give a monomer, and cross-linked multimers, implicating the involvement of multiple glutamine and lysine residues. However, our data indicated that EPO was not a substrate of either G-TGase or mTGase. EPO was not labeled by fluorescent dyes such as MDC, Texas Red cadaverine, and *Z*-Gly-Ala-Gln-Aminocoumarin (AMC) after incubation for 30 h at 37°C with 0.2 unit/mL of G-TGase unless EPO was reduced by dithiothreitol (Figure 6.5, gel images shown were for MDC only). Results were similar when Texas Red cadaverine and *Z*-Gly-Ala-Gln-AMC were used as acyl acceptor and donor, respectively. Under the same conditions, GCSF, hGH, and interferon α-2b were substrates of mTGase, but not of G-TGase. Evidently, bulky glycan chains of EPO block TGase's access to glutamine and lysine residues.

Recently, Bioker moved its pegylated GCSF (BK0026) into phase I trial. BK0026 has a 20 kDa mPEG attached to GCSF via Q134. The reaction was carried out under very mild conditions (4°C at pH 7.4, and PEG-NH_2: GCSF molar ratio was ~10) with mTGase. Q134 was the only residue (out of 17 Qs total) modified by mTGase [67]. It is interesting to note that Q131 is modified alternatively in the double mutant (P132Q/Q134N) by mTGase [68].

Broad substrate specificity of mTGase can be a double-edged sword. On one hand, it increases the chance for a target protein to be a substrate. On the other hand, it would cross link the target protein intermolecularly to give a complex mixture if multiple reactive glutamines and lysines exist. Let us select hGH for our case study. Chemical PEGylation of hGH has been studied extensively [69]. Direct PEGylation of hGH with aminoPEG mediated by mTGase was explored by Wu [70] and the yield was low due to the formation of cross-linked multimers. Among 13 glutamines and 9 lysines on hGH, 2 Gln and 2 Lys residues of hGH are reactive under mTGase [71]. High molecular weight polymer mixture of hGH was evident, even when aminoPEG was 50-fold in excess (our unpublished results). Cross-linked hGH was also observed at low pHs between 5 and 6.5 when protonated $-NH_3^+$ groups of lysine residues were not expected to be nucleophiles. Clearly, low pH does not prevent ε-amino groups from acting as acyl acceptors, and as a matter of fact, they are preferred nucleophiles over the amino groups on PEG. This observation agrees well with a general-base-catalyzed deacylation mechanism [72]. Therefore, cross-linking happens between Gln and Lys residues of hGH, as well as among Gln residues and PEG-amine. A multiple step PEGylation process [71] provided a solution to the reactive lysine residues of hGH. When 1,3-di-amino propanol (DAP) was used as a buffer (molar ratio DAP: hGH > 500), mTGase mediated the incorporation of DAP at Q40 and Q141 only. In other words, saturated DAP protected lysine residues on hGH. The transamidation products were separated on an anion exchange column to yield ~40% of Q141 adduct ($N^{\varepsilon 141}$-2-hydroxy-3-aminopropyl-hGH), and was then chemically oxidized by $NaIO_4$ to an aldehyde ($N^{\varepsilon 141}$-2-oxyethyl-hGH). Upon oximation with PEG hydroxylamine, mono-PEGylated hGH at Q141 was obtained (Figure 6.6).

6.3.2 Semi-rational Design of mTGase for Site-Specific PEGylation

Even though mTGase has been proven to be very useful in site-specific PEGylation, it is really a hit or miss endeavor. As hGH is PEGylated at two sites by wild type (wt) mTGase, we would have a mixture of four products: unmodified hGH, modified Q40, modified Q141, and double modified Q40/141. Although Q141 is preferred over Q40, the reaction could not be allowed to go to completion. Otherwise, the double modified form becomes more prominent. Left-over hGH and Q40 modified product would not only reduce final yield, but also add cost in downstream purification. So, there is a great need for a mono-PEGylation reaction. mTGase's catalytic triad, Cys 64, Asp255, and His 274, sit deep in a groove (Figure 6.7a) [73]. Any Gln or Lys participating in transglutamination has to be accessible to mTGase, presumably on the surface and in a flexible region of hGH. Preliminary protein–protein docking of mTGase/hGH showed that Q40 could not reach mTGase active site, but Q141 could (Figure 6.7b) (Leif Nørskov-Lauritsen, unpublished work). Reaction mixture analysis did indicate that Q141 was preferred to Q40. Hence, the flexibility of mTGase or

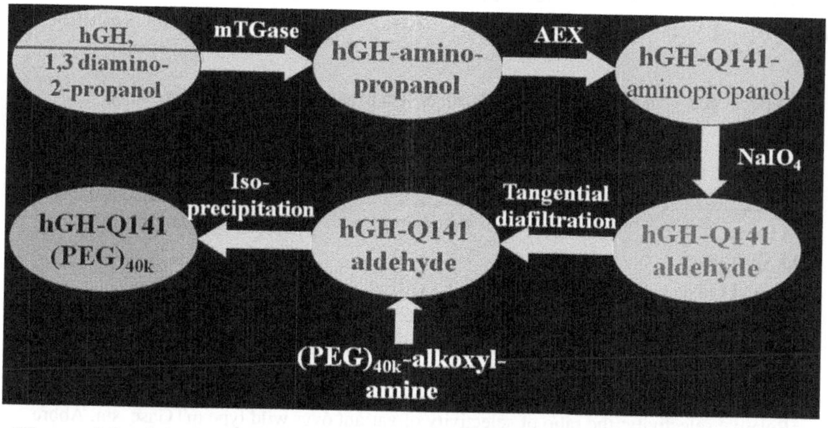

Figure 6.6 Schematic process of mTGase mediated mono-PEGylation of hGH.

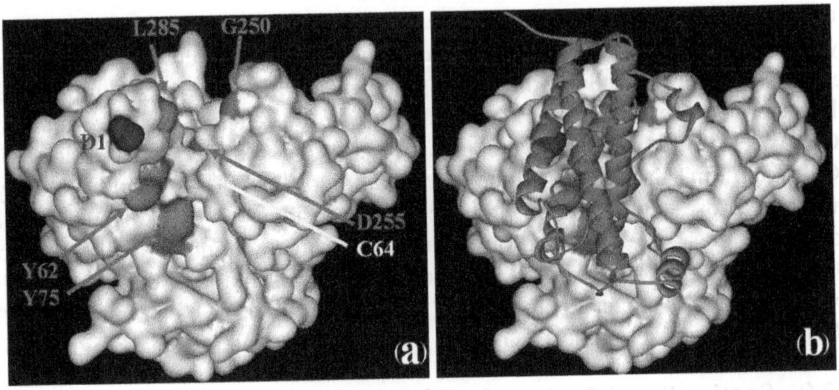

Figure 6.7 mTGase active site (b) and mTGase-hGH Docking (a). Residues marked were expected to interact with hGH closely. Docking was made by fixing Q141 of hGH (ribbon in dark gray) at mTGase (light gray) active site Cys64.

hGH would be responsible for Q40 reactivity. mTGase/GCSF docking from Tramontano's laboratory also confirmed this notion: mTGase had to assume an open configuration in order for Q131 and Q134 of GCSF to reach mTGase's active site [68]. Would the site-specificity of mTGase increase with the decrease in protein chain flexibility by increasing viscosity or decreasing temperature? When the reaction was run at low temperature and/or high concentration of ethylene glycerol, higher selectivity of Q141 over Q40 was confirmed (our unpublished results). Further evidence supporting amenable site-specificity came from mTGase variants. *Streptomyces ladakanum* mTGase (mTGase_sl) is highly homologous (93%) to the widely used *Streptomyces mobaraense* mTGase (mTGase_sm). The most noticeable residue change is G250S on one side of groove to the entrance of the active side (Figure 6.7a).

Surprisingly, mTGase_sl's site-selectivity is almost 2 times higher than that of mTGase_sm. Such high selectivity is retained when G250S mutation is transferred to mTGase_sm (Table 6.2). A similar mutation (G250T) also increased selectivity. Even more interestingly, extension of N-terminus (adding Met,

TABLE 6.2 mTGase Variant and N-Terminus Effect on Selectivity

	Selectivity (Relative Selectivity[a]	
	Streptomyces mobaraensis	*Streptomyces ladakanum*
MTGase	5.6 (1.0[a])	9.5 (1.7)
M_mTGase	7.8 (1.4)	10.0 (1.8)
AP_mTGase	5.7 (1.0)	11.8 (2.1)
GP_mTGase	6.7 (1.2)	15.0 (2.7)
AP-mTGase-G250S	10.8 (1.9)	NA
AP-mTGase-G250T	8.0 (1.4)	NA

[a]Relative selectivity: the ratio of selectivity of variant over wild type mTGase_sm. Abbreviation: NA, not available.

Ala_Pro, or Gly_Pro) also improved selectivity [74] (Table 6.2), as predicted by Shimba [75]. Encouraged by such findings, we made a small focused library of three residues (L285, Y62, and Y75), which were expected to be in close contact with substrate hGH based on a mTGase/hGH-Q141 docking model (Figure 6.7b).

To facilitate genetic manipulation, it would be better to make libraries in *E. coli*. mTGase can be expressed in *E. coli* as inclusion bodies or as inactive but soluble pro-mTGase. Neither form is suitable for screening. The native cleavage site between pro- and mTGase is very similar to an enterokinase restriction site (DDDDK ↓ XX), but EK cuts seven amino acid residues away from the expected site. To express active mTGase in *E. coli*, Zhao and her colleagues [76] inserted a protease 3C cleavage site between Pro and mature portion of mTGase (Figure 6.8). Activation of proenzyme was achieved by either EK or 3C digestion. Alternatively, coexpression of Pro-mTGase with protease 3C or EK also yielded active mTGase.

To define the quantitative measurement of site specificity and design an assay for the screening, two hGH mutants, Q40N and Q141N, were used. The selectivity is measured as the ratio of initial velocities of two reactions carried out in parallel (Figure 6.9). That is,

$$\text{Selectivity} = \frac{\frac{d(\text{Product Q141})}{dt}}{\frac{d(\text{Product Q40})}{dt}}$$

A fluorescence label, dansyl cadaverine is used as glutaminyl acceptor. The two reactions could be easily followed in 96 well plates by monitoring the increase of fluorescence at 510 nm (Figure 6.10).

To confirm the selectivity of mTGase variants obtained with hGH mutants, wild type (wt) hGH was used as substrate and the reactions were carried out under similar conditions. After ~30% of wt-hGH was consumed, the reaction mixture was separated by capillary electrophoresis (C.E.). Selectivity was estimated as the

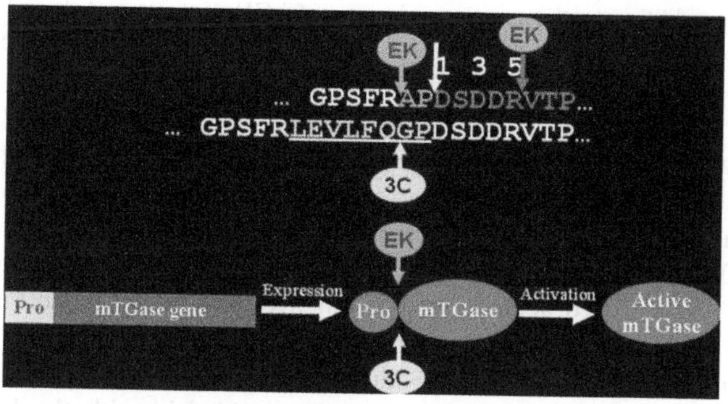

Figure 6.8 Pro-mTGase processing by EK or Protease 3C. Primary sequence at the junction site was shown with native sequence on the top, and 3C site insertion at the bottom. The arrow at position 1 indicates native processing site. The expected EK cleavage site is at position 5 while the actual cleavage site is seven amino acid residues away to the left of the expected site.

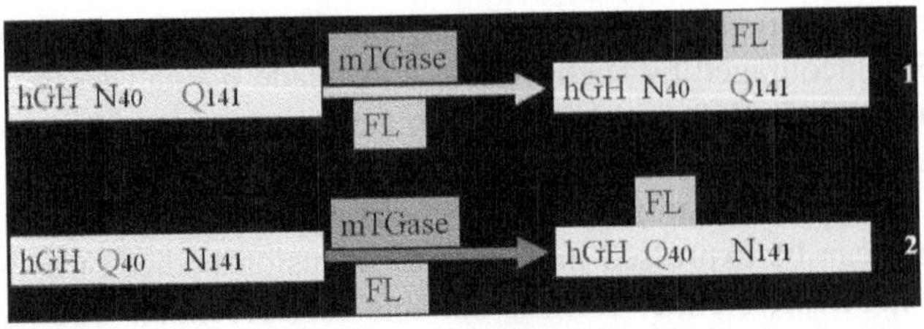

Figure 6.9 Reactions of kinetic assay using two hGH mutants. hGH_Q40N and hGH_Q141N were used as protein substrates, and a fluorescence label (FL) with free amino group was used as an acyl acceptor.

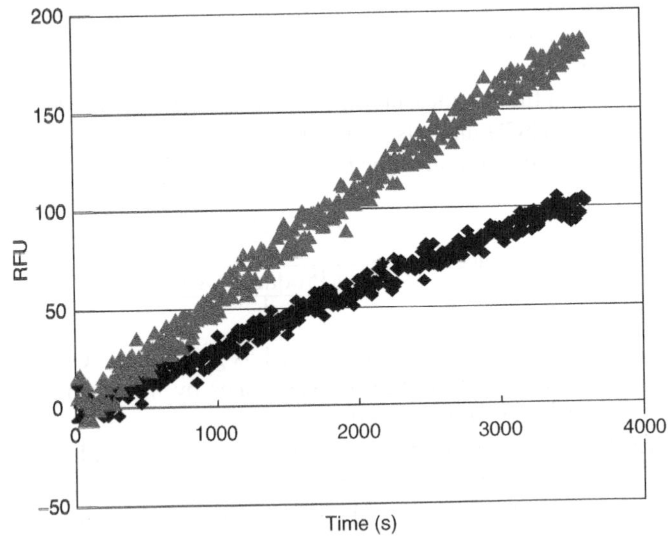

Figure 6.10 Fluorescence of kinetic assay reactions. Fluorescence at 510 nm increases upon mTGase addition to the reaction mixtures of hGH mutants and dansyl cadaverine. Reaction progress curves maintain linearity for >30 min. Selectivity is calculated from the ratio of the slopes of two lines.

ratio of area of product Q141 to that of product Q40 on the chromatogram. Some representative mutants with improved selectivity are listed in Table 6.3 [74].

The data also pointed to a large error from kinetic assay when selectivity was very high. With highly selective mTGase mutants, the reaction rate of Q40 was too low to be measured accurately. Nonetheless, the fluorescence kinetic assay provided a very convenient and reliable primary screening method.

With mutations limited to four residues based on our docking mode, we were able to increase the site selectivity of mTGase 15-fold with the best mutant, named HS_mTGase, and essentially eliminated the Q40 site by-product. By using this mutant, hGH mono-PEGylation process becomes highly efficient. Long acting hGH, pegylated by HS_mTGase, has finished phase II trial, and is heading into phase III.

TABLE 6.3 Mutants with Improved Q141/Q40 Selectivity

	Relative Selectivity	
	C.E.	Kinetics
mTGase_sm	1.0	1.0
Y75N-mTGase_sl	12.8	3.3
Y75A-mTGase_sl	3.0	2.4
Y75F-mTGase_sl	3.8	3.9
Y62H_Y75N-mTGase_sl	4.6	3.3
Y62H_Y75F-mTGase_sl	15	6.1

6.3.3 Prospective and Beyond

Among the few enzymatic and genetic approaches available, mTGase stands out as the preferred choice for several reasons: (i) no need to mutate parent proteins so as no modification to current production processes; (ii) no unwanted protein degradation as in the case of tyrosinase and protease; (iii) surface Gln residues are widely available and are very unlikely involved in biological interactions; and (iv) mTGase, being a small monomeric 38 kDa protein without any cofactor requirement, is easily expressed in *E. coli* and well suited for genetic engineering to modify its specificity. The stable isopeptide bond formed between protein and PEG could give flexibility to further protein drug formulation (spray-dry or solution). With the simple requirement of amino group on PEG, freedom of operation is ensured than other special activated PEGs.

Many approved biotherapeutics, with more under development, have very short half-lives and are highly immunogenic. PEGylation or conjugation could be the best option for these drugs or for investigational drugs. With the advance of mTGase PEGylated hGH into phase III trial by Novo Nordisk, mTGase will gain more acceptance in pharmaceutical protein conjugation. As Pfizer also advanced its N-terminal pegylated hGH into phase II trial (withdrew at the end of 2007) [77], a direct comparison of Novo Nordisk's enzymatic process and Pfizer's chemical process would shed light on how greener the mTGase conjugation process could be. Unfortunately, data from these industrial processes were never disclosed. On the basis of laboratory data from patent filings, the HS-mTGase process could double the yield of the chemical process [78].

Besides PEG, other biological polymers, such as hydroxyethyl starch [79] and dextran [80], could be also conjugated using mTGase. In addition to protein protraction, mTGase could also find its application in targeted drug delivery, where active drug molecules are loaded to a targeting partner such as an antibody. Polymer-anticancer drug conjugates [81] explore tumor cells' enhanced permeability and retention toward polymer. mTGase may be well suited for this purpose to couple multiple (small) molecules to a large polymer backbone. Starting as a food processing enzyme, mTGase may become a major technical enzyme for biotech-pharmaceutical industry.

6.4 CONCLUSION

Unlike small molecule drugs which are relatively easy to make and taken orally, protein drugs are expensive, inconvenient to deliver, and immunogenic. Nonetheless, the market for these large molecule drugs is growing faster than the small counterparts. To address the shortcomings of protein therapeutics, PEGylation emerged as the most successful method to extend half-life and to minimize immunogenicity. Chemical PEGylation successfully brought a few drugs to market since 1990. However, regulatory scrutiny over the drugs' safety and efficacy makes the random chemical PEGylation obsolete. On the other hand, the greener and highly site-specific enzymatic process produces a monoconjugated product with higher bioactivity and better safety profile. Among several promising enzymes for protein modification, mTGase is the preferred choice as it is cheap, stable, robust, and easily expressed in *E. coli*. Its catalysis is well studied and without any adverse side reactions, its structure is known, and it is amenable to protein engineering for optimal substrate specificity. Even when multiple Gln and Lys residues are involved in mTGase-mediated conjugation, we have shown that rational design could readily deliver mTGase variants with the needed selectivity. With two multibillion dollar protein drugs (hGH and GCSF) pegylated by mTGase in clinical trials, mTGase is showing its great potential as a versatile and powerful protein conjugation tool.

REFERENCES

1. Abuchowski A, McCoy JR, Palczuk NC, van Es T, Davis FF. Effect of covalent attachment of polyethylene glycol on immunogenicity and circulating life of bovine liver catalase. J Biol Chem 1977;252:3582–3586.
2. Abuchowski A, van Es T, Palczuk NC, Davis FF. Alteration of immunological properties of bovine serum albumin by covalent attachment of polyethylene glycol. J Biol Chem 1977;252:3578–3581.
3. Bailon P, Won CY. PEG-modified biopharmaceuticals. Expert Opin Drug Deliv 2009;6:1–16.
4. Veronese FM, Mero A. The impact of PEGylation on biological therapies. BioDrugs 2008;22:315–329.
5. Sergi M, Caboi F, Maullu C, Orsini G, Tonon G. Enzymatic techniques for PEGylation of biopharmaceuticals. In: Veronese FM, editor. PEGylated protein drugs: basic science and clinical applications. Springer, Basel, 2009; pp 75–88.
6. Kinstler OB, Brems DN, Lauren SL, Paige AG, Hamburger JB, Treuheit MJ. Characterization and stability of N-terminally PEGylated rhG-CSF. Pharm Res 1996;13:996–1002.
7. Rosendahl MS, Doherty DH, Smith DJ, Carlson SJ, Chlipala EA, Cox GN. A long-acting, highly potent interferon alpha-2 conjugate created using site-specific PEGylation. Bioconjug Chem 2005;16:200–207.
8. Cox G. Derivatives of growth hormone and related proteins. WO 99/03887. 1999.
9. Lee S, McNemar C. Substantially pure histidine-linked protein polymer conjugates. US 5,985,263. 1999.

10. Zalipsky S, Meno-Rudolph S. Hydrazide derivatives of polyethylene glycol and their bioconjugates. In: Harris JM, Zalipsky S, editors. Polyethylene glycol chemistry and biological applications (ACS Symposium Series). 1st ed. Washington (DC): American Chemical Society; 1997. pp. 318–341.

11. Levy Y, Hershfield MS, Fernandez-Mejia C, Polmar SH, Scudiery D, Berger M, Sorensen RU. Adenosine deaminase deficiency with late onset of recurrent infections: response to treatment with polyethylene glycol-modified adenosine deaminase. J Pediatr 1988;113:312–317.

12. Graham ML. Pegaspargase: a review of clinical studies. Adv Drug Deliv Rev 2003;55:1293–1302.

13. Youngster S, Wang YS, Grace M, Bausch J, Bordens R, Wyss DF. Structure, biology, and therapeutic implications of pegylated interferon alpha-2b. Curr Pharm Des 2002;8:2139–2157.

14. Foser S, Schacher A, Weyer KA, Brugger D, Dietel E, Marti S, Schreitmuller T. Isolation, structural characterization, and antiviral activity of positional isomers of monopegylated interferon alpha-2a (PEGASYS). Protein Expr Purif 2003;30:78–87.

15. Pradhananga S, Wilkinson I, Ross RJ. Pegvisomant: structure and function. J Mol Endocrinol 2002;29:11–14.

16. Curran MP, McCormack PL. Methoxy polyethylene glycol-epoetin beta: a review of its use in the management of anaemia associated with chronic kidney disease. Drugs 2008;68:1139–1156.

17. Rivkin A. Certolizumab pegol for the management of Crohn's disease in adults. Clin Ther 2009;31:1158–1176.

18. Harris G. The future of biotech: the 2010 guide to emerging markets and technology. Atlanta, GA: AHC Media LLC; 2009.

19. Jorgensen KE, Moller JV. Use of flexible polymers as probes of glomerular pore size. Am J Physiol 1979;236:F103–F111.

20. Glue P, Fang JW, Rouzier-Panis R, Raffanel C, Sabo R, Gupta SK, Salfi M, Jacobs S, Hepatitis C Intervention Therapy Group. Pegylated interferon-alpha2b: pharmacokinetics, pharmacodynamics, safety, and preliminary efficacy data. Clin Pharmacol Ther 2000;68:556–567.

21. Macdougall IC, Eckardt KU. Novel strategies for stimulating erythropoiesis and potential new treatments for anaemia. Lancet 2006;368:947–953.

22. Egrie JC, Browne JK. Development and characterization of novel erythropoiesis stimulating protein (NESP). Nephrol Dial Transplant 2001;16 Suppl 3:3–13.

23. Ramakrishnan B, Boeggeman E, Qasba PK. Novel method for *in vitro* O-glycosylation of proteins: application for bioconjugation. Bioconjug Chem 2007;18:1912–1918.

24. Ramakrishnan B, Boeggeman E, Qasba PK. Applications of glycosyltransferases in the site-specific conjugation of biomolecules and the development of a targeted drug delivery system and contrast agents for MRI. Expert Opin Drug Deliv 2008;5:149–153.

25. Pasek M, Ramakrishnan B, Boeggeman E, Manzoni M, Waybright TJ, Qasba PK. Bioconjugation and detection of lactosamine moiety using alpha1,3-galactosyltransferase mutants that transfer C2-modified galactose with a chemical handle. Bioconjug Chem 2009;20:608–618.

26. Qasba PK, Boeggeman E, Ramakrishnan B. Site-specific linking of biomolecules via glycan residues using glycosyltransferases. Biotechnol Prog 2008;24:520–526.

27. Boeggeman E, Ramakrishnan B, Pasek M, Manzoni M, Puri A, Loomis KH, Waybright TJ, Qasba PK. Site specific conjugation of fluoroprobes to the remodeled Fc N-glycans of monoclonal antibodies using mutant glycosyltransferases: application for cell surface antigen detection. Bioconjug Chem 2009;20:1228–1236.
28. Chen T, Embree HD, Wu LQ, Payne GF. *In vitro* protein-polysaccharide conjugation: tyrosinase-catalyzed conjugation of gelatin and chitosan. Biopolymers 2002;64:292–302.
29. Demolliens A, Boucher C, Durocher Y, Jolicoeur M, Buschmann MD, De Crescenzo G. Tyrosinase-catalyzed synthesis of a universal coil-chitosan bioconjugate for protein immobilization. Bioconjug Chem 2008;19:1849–1854.
30. Quaedflieg PJLM, Cusan C, Ritzen B, Dias A. Enzymatic conjugation of bioactive moieties. WO2009/101178. 2009.
31. Chang TK, Jackson DY, Burnier JP, Wells JA. Subtiligase: a tool for semisynthesis of proteins. Proc Natl Acad Sci USA 1994;91:12544–12548.
32. Mahrus S, Trinidad JC, Barkan DT, Sali A, Burlingame AL, Wells JA. Global sequencing of proteolytic cleavage sites in apoptosis by specific labeling of protein N termini. Cell 2008;134:866–876.
33. Furter R. Expansion of the genetic code: site-directed p-fluoro-phenylalanine incorporation in Escherichia coli. Protein Sci 1998;7:419–426.
34. Cho HS, Daniel TO, Dimarchi RD, Hays A-M, Wilson TE, Sim B-C, Litzinger DC. Modified human growth hormone. US20060189529. 2006.
35. Connor RE, Piatkov K, Varshavsky A, Tirrell DA. Enzymatic N-terminal addition of noncanonical amino acids to peptides and proteins. Chembiochem 2008;9:366–369.
36. Griffin M, Casadio R, Bergamini CM. Transglutaminases: nature's biological glues. Biochem J 2002;368:377–396.
37. Folk JE, Chung SI. Molecular and catalytic properties of transglutaminases. Adv Enzymol Relat Areas Mol Biol 1973;38:109–191.
38. Folk JE. Transglutaminases. Annu Rev Biochem 1980;49:517–531.
39. Muszbek L, Yee VC, Hevessy Z. Blood coagulation factor XIII: structure and function. Thromb Res 1999;94:271–305.
40. Groenen PJ, Smulders RH, Peters RF, Grootjans JJ, van den Ijssel PR, Bloemendal H, de Jong WW. The amine-donor substrate specificity of tissue-type transglutaminase. Influence of amino acid residues flanking the amine-donor lysine residue. Eur J Biochem FEBS 1994;220:795–799.
41. Hettasch JM, Peoples KA, Greenberg CS. Analysis of factor XIII substrate specificity using recombinant human factor XIII and tissue transglutaminase chimeras. J Biol Chem 1997;272:25149–25156.
42. Folk JE, Cole PW. Structural requirements of specific substrates for guinea pig liver transglutaminase. J Biol Chem 1965;240:2951–2960.
43. Gorman JJ, Folk JE. Structural features of glutamine substrates for transglutaminases. Specificities of human plasma factor XIIIa and the guinea pig liver enzyme toward synthetic peptides. J Biol Chem 1981;256:2712–2715.
44. Gorman JJ, Folk JE. Structural features of glutamine substrates for transglutaminases. Role of extended interactions in the specificity of human plasma factor XIIIa and of the guinea pig liver enzyme. J Biol Chem 1984;259:9007–9010.
45. Board PG, Pierce K, Coggan M. Expression of functional coagulation factor XIII in Escherichia coli. Thromb Haemost 1990;63:235–240.

46. Gillet SM, Chica RA, Keillor JW, Pelletier JN. Expression and rapid purification of highly active hexahistidine-tagged guinea pig liver transglutaminase. Protein Expr Purif 2004;33:256–264.
47. Ikura K, Tsuchiya Y, Sasaki R, Chiba H. Expression of guinea-pig liver transglutaminase cDNA in Escherichia coli. Amino-terminal N alpha-acetyl group is not essential for catalytic function of transglutaminase. Eur J Biochem FEBS 1990;187:705–711.
48. Bishop PD, Teller DC, Smith RA, Lasser GW, Gilbert T, Seale RL. Expression, purification, and characterization of human factor XIII in Saccharomyces cerevisiae. Biochemistry 1990;29:1861–1869.
49. Coggan M, Baker R, Miloszewski K, Woodfield G, Board P. Mutations causing coagulation factor XIII subunit A deficiency: characterization of the mutant proteins after expression in yeast. Blood 1995;85:2455–2460.
50. Gao J, Hooker BS, Anderson DB. Expression of functional human coagulation factor XIII A-domain in plant cell suspensions and whole plants. Protein Expr Purif 2004;37:89–96.
51. Rao RU, Chandrashekar R, Mehta K. Transglutaminases of lower organisms. Prog Exp Tumor Res 2005;38:209–222.
52. Ando H, Adachi M, Umeda K, Matsuura A, Nonaka M, Uchio R, Tanaka H, Motoki M. Purification and characterization of a novel transglutaminase derived from microorganisms. Agric Biol Chem 1989;53:2613–2617.
53. Motoki M, Seguro K. Transglutaminase and its use for food processing. Trends Food Sci Technol 1998;9:204–210.
54. Zhu Y, Rinzema A, Tramper J, Bol J. Microbial transglutaminase—a review of its production and application in food processing. Appl Microbiol Biotechnol 1995;44:277–282.
55. Yokoyama K, Nio N, Kikuchi Y. Properties and applications of microbial transglutaminase. Appl Microbiol Biotechnol 2004;64:447–454.
56. Ohtsuka T, Ota M, Nio N, Motoki M. Comparison of substrate specificities of transglutaminases using synthetic peptides as acyl donors. Biosci Biotechnol Biochem 2000;64:2608–2613.
57. Ohtsuka T, Sawa A, Kawabata R, Nio N, Motoki M. Substrate specificities of microbial transglutaminase for primary amines. J Agric Food Chem 2000;48:6230–6233.
58. Tagami U, Shimba N, Nakamura M, Yokoyama K, Suzuki E, Hirokawa T. Substrate specificity of microbial transglutaminase as revealed by three-dimensional docking simulation and mutagenesis. Protein Eng Des Sel 2009;22:747–752.
59. Taguchi S, Nishihama KI, Igi K, Ito K, Taira H, Motoki M, Momose H. Substrate specificity analysis of microbial transglutaminase using proteinaceous protease inhibitors as natural model substrates. J Biochem 2000;128:415–425.
60. Zhu Y, Tramper J. Novel applications for microbial transglutaminase beyond food processing. Trends Biotechnol 2008;26:559–565.
61. Meusel M. Synthesis of hapten-protein conjugates using microbial transglutaminase. Methods Mol Biol 2004;283:109–123.
62. Lin CW, Ting AY. Transglutaminase-catalyzed site-specific conjugation of small-molecule probes to proteins *in vitro* and on the surface of living cells. J Am Chem Soc 2006;128:4542–4543.
63. Sato H, Ikeda M, Suzuki K, Hirayama K. Site-specific modification of interleukin-2 by the combined use of genetic engineering techniques and transglutaminase. Biochemistry 1996;35:13072–13080.

64. Sato H. Enzymatic procedure for site-specific pegylation of proteins. Adv Drug Deliv Rev 2002;54:487–504.
65. Sato H, Hayashi E, Yamada N, Yatagai M, Takahara Y. Further studies on the site-specific protein modification by microbial transglutaminase. Bioconjug Chem 2001;12:701–710.
66. Pool C. Formation of novel erythropoietin conjugates using transglutaminase. US6995245. 2004.
67. Tonon G, Orsini G, Schrepfer R, Taylor G, Sergi M. G-CSF Site-Specific Mono-Conjugates. WO2008/017603. 2008.
68. Maullu C, Raimondo D, Caboi F, Giorgetti A, Sergi M, Valentini M, Tonon G, Tramontano A. Site-directed enzymatic PEGylation of the human granulocyte colony-stimulating factor. FEBS J 2009;276:6471–6450.
69. Finn RF. PEGylation of human growth hormone: strategies and properties. In: Veronese FM, editor. PEGylated protein drugs: basic science and clinical applications. Springer, Basel, 2009; pp. 187–203.
70. Wu Z, Huang J, Liu C. Biological modifying and recombinant human growth hormone compound and preparing method thereof. CN1528787.
71. Dörwald F, Johansen N, Iversen L. Transglutaminase mediated conjugation of growth hormone. WO/2006/134148. 2006.
72. Leblanc A, Gravel C, Labelle J, Keillor JW. Kinetic studies of guinea pig liver transglutaminase reveal a general-base-catalyzed deacylation mechanism. Biochemistry 2001;40:8335–8342.
73. Kashiwagi T, Yokoyama K, Ishikawa K, Ono K, Ejima D, Matsui H, Suzuki E. Crystal structure of microbial transglutaminase from Streptoverticillium mobaraense. J Biol Chem 2002;277:44252–44260.
74. Hu S, Zhao X, Jianhua W, Chang C, Nørskov-Lauritsen L, Su J. Transglutaminase variants with improved specificity. WO/2008/102007. 2008.
75. Shimba N, Shinohara M, Yokoyama K, Kashiwagi T, Ishikawa K, Ejima D, Suzuki E. Enhancement of transglutaminase activity by NMR identification of its flexible residues affecting the active site. FEBS Lett 2002;517:175–179.
76. Zhao X, Shaw AC, Wang J, Chang CC, Deng J, Su J. A novel high-throughput screening method for microbial transglutaminases with high specificity toward Gln141 of human growth hormone. J Biomol Screen 2010;15:206–212.
77. Touraine P, D'Souza GA, Kourides I, Abs R, Barclay P, Xie R, Pico A, Torres-Vela E, Ekman B. Lipoatrophy in GH deficient patients treated with a long-acting pegylated GH. Eur J Endocrinol 2009;161:533–540.
78. Finn RF, Liao W, Siegel NR. Chemically-modified human growth hormone conjugates. US2004/0038892A1. 2004.
79. Besheer A, Hertel TC, Kressler J, Mader K, Pietzsch M. Enzymatically catalyzed HES conjugation using microbial transglutaminase: proof of feasibility. J Pharm Sci 2009;98:4420–4428.
80. Valdivia A, Villalonga R, Di Pierro P, Perez Y, Mariniello L, Gomez L, Porta R. Transglutaminase-catalyzed site-specific glycosidation of catalase with aminated dextran. J Biotechnol 2006;122:326–333.
81. Greco F, Vicent MJ. Combination therapy: opportunities and challenges for polymer-drug conjugates as anticancer nanomedicines. Adv Drug Deliv Rev 2009;61:1203–1213.

CHAPTER 7

MICROBIAL PRODUCTION OF PLANT-DERIVED PHARMACEUTICAL NATURAL PRODUCTS THROUGH METABOLIC ENGINEERING: ARTEMISININ AND BEYOND

JEFFREY A. DIETRICH
UCB-UCSF Joint Graduate Group in Bioengineering, UC-Berkeley, UC-San Francisco, Berkeley, California

J. L. FORTMAN
Fuels Synthesis Division, Joint BioEnergy Institute, Lawrence Berkeley National Lab, Berkeley, California

DARMAWI JUMINAGA
California Institute of Quantitative Biomedical Research (QB3), University of California-Berkeley, Berkeley, California

JAY D. KEASLING
Fuels Synthesis Division, Joint BioEnergy Institute, Lawrence Berkeley National Lab, Berkeley, California

7.1 THE CASE FOR MICROBIAL PRODUCTION OF PLANT-DERIVED PHARMACEUTICALS

The use of plant-derived compounds for human medicinal purposes has an extremely long and rich history. The discovery of multiple medicinal plant pollens in a Neanderthal grave suggests that knowledge and use of herbal medicines may have been an integral part of human cultural development [1,2]. Detailed written records from both Eastern and Western cultures cast native flora as a

Biocatalysis for Green Chemistry and Chemical Process Development, First Edition.
Edited by Junhua (Alex) Tao and Romas Kazlauskas.

fundamental pillar of modern medicine. For example, *Sushruta Samhita*, written in what is present-day India, and *De Materia Medica* from Greece provide descriptions of hundreds of plant-based remedies. However, it was not until the advent of modern chemistry and the chemical revolution that an accurate description of plant-derived bioactive molecules could be provided. With this information, modern researchers are unearthing plant-derived pharmaceutical compounds described in ancient records while also discovering novel small molecules produced in trace amounts in their native hosts.

The commercial and societal importance of this work should not be understated. A detailed analysis of plant-derived pharmaceuticals that are commercially available and under clinical trials has been conducted, which illustrates the invaluable role these compounds play in stocking the pharmacopia. In the period 1981–2002, of the over 877 small-molecule pharmaceutical compounds introduced, 61% have their origin in a natural product [3]. Plant-derived small molecules continue to play an important role in pharmaceutical development, particularly in the areas of anticancer and anti-infectious agents, where 91 plant-derived compounds were in clinical trials between 2000 and 2006 [4]. The market value of these compounds is immense and has spurred increased innovation in the field. By 2011, the aggregate market value for all plant-derived compounds is expected to exceed $26 billion [5]. Considering that only a sliver of the total plant secondary metabolite pool has been investigated for pharmaceutical activity to date, this field will continue to be an area of active research.

Despite the immense market potential, inefficiencies in current small-molecule production schemes are drivers for establishing microbial production routes, an increasingly important area of study in metabolic engineering. The majority of natural products used today are extracted from native sources or are chemically synthesized; both production regimes have significant long-term disadvantages. First, a paucity of supply can limit economical extraction of the active agent from natural sources. In the case of the highly potent anticancer drug taxol, initial commercial production routes required harvesting the bark from one 100-year-old Pacific yew (*Taxus brevifolia*) to extract a single dose; the establishment of a semisynthetic commercial production route using 10-deacetylbaccatin III extracted from *Taxus baccata* or *Taxus yunnanensis* pine needles provides a more environmentally friendly alternative [6,7]. Similar arguments can be made for the clinically proven anticancer compounds vincristine and vinblastine. These molecules are found only in minute amounts (2×10^{-4}% dry weight) in the native source, the tropical plant *Catharanthus roseus* [8]. Traditional chemical synthesis is also not a viable option currently, as both vincristine and vinblastine are prohibitively complex to be produced economically. However, the arguments for microbial production routes are not purely monetary. Synthetic chemistry and natural product extractions commonly utilize environmentally detrimental catalysts and solvents in their workflows, a feature minimized in microbial production routes.

Against this backdrop has been the rapid development of small-molecule production routes in microbial hosts. Construction of microbial production routes can mitigate issues in paucity of supply by establishing a stable, high-yielding source of the compound of interest. Further, the same complex synthetic reactions can be performed microbially while also benefiting from improved regioselectivities, specificities, and substrate conversion efficiencies that are hallmarks of enzyme catalysts [9]. Lastly, biosynthetic routes can prove highly versatile, serving as production platforms for entire families of small molecules. Exemplifying this argument are efforts to engineer a biosynthetic route for production of precursors to the antimalarial pharmaceutical artemisinin. Artemisinin is a member of the isoprenoid family of natural products, of which there are over 50,000 known members [10]. The same strains used to produce the precursors to artemisinin can be readily modified to target any number of other isoprenoid compounds. Here we focus on the various metabolic engineering strategies used in our laboratory to construct both *Escherichia coli* and *Saccharomyces cerevisiae* artemisinin production hosts. Techniques for engineering both native and heterologous pathways to reroute carbon flux, achieving functional plant-derived enzyme expression, managing product-induced toxicity, and using physical chemistry to develop efficient purification schemes are discussed. We then follow by extrapolating the lessons learned during artemisinin biosynthesis toward engineering more complex plant secondary metabolites, including the terpene indole and benzylisoquinoline alkaloid (BIA) families of natural products.

7.2 USING ARTEMISININ TO ESTABLISH AN ISOPRENOID BIOSYNTHESIS PLATFORM

Between 1999 and 2008, a period in which the vast majority of our laboratory's work on artemisinin was conducted, there existed few cost-effective antimalarial medications for use in the developing world. Increased resistance of the *Plasmodium* parasite to standard pharmaceutical treatments, including chloroquine, was (and remains) widespread. Malaria is a disease endemic to the developing world, with over three billion people living under the threat of malaria, and an estimated 350–500 million new infections are witnessed each year; the majority of cases occur in young children [11]. In sub-Saharan Africa, malaria is the leading cause of mortality, accounting for 18% of all childhood deaths [12]. The economic burden malaria places on developing nations is immense. Cross-country comparisons show that malaria is responsible for a 1.3% decrease in the annual economic growth in countries to which malaria remains endemic [13]. Perhaps of even greater concern are predictions of the future spread of drug-resistant malaria due to human-caused climate change. Predictive models of global warming and malaria show increased incidence in Eurasia and Africa, particularly focused on already poor and vulnerable regions [14].

In 2005, the World Health Organization recommended artemisinin-based combination therapies (ACTs) as the first-line malaria treatment [15]. Even today, numerous obstacles impede large-scale adoption of ACTs, many of which fetter most plant-derived pharmaceuticals. First and foremost is the high cost of ACTs. In the developing world, annual expenditure in healthcare can be as low as \$5 per capita [16], a figure particularly sobering, considering that a typical adult ACT course costs approximately \$2.20 [17]. A number of total chemical syntheses have been reported, but all suffer from yields <10% and are unlikely to provide artemisinin at costs below \$100/kg [18]. As a further complication, naturally sourced artemisinin from *Artemisia annua* is subject to dramatic swings (>\$1000/kg) as farmers respond to changes in supply and demand [19]. In this light, a primary success criterion for the artemisinin project was to provide a significant and reliable cost–benefit over existing production regimes. As analyzed in detail elsewhere [20], a goal of 30–60% decrease in ex-factory ACT costs was established. Not only should microbially produced artemisinin be cost effective, but it should also address a number of ancillary issues. Plant-sourced small molecules are subject to unpredictable growth conditions and supply chain insecurities, features mitigated in the controlled environment of a fermenter. Further, microbial production routes can decrease the redistribution of land away from food crops and toward pharmaceutical-yielding plant species such as *A. annua*. Finally, high-efficiency product extraction is accomplished with decreased use of environmentally detrimental organic solvents. While the recent establishment of an *A. annua* genetic map complete with identification of genetic modifications that impart improved artemisinin yields [21] goes a long way toward decreasing the total acreage necessary for a given harvest, the occupational and environmental risks associated with product extraction remain. The most widely used extraction method today is with petroleum ether, a highly flammable, carcinogenic mixture of compounds that provides only low extraction efficiencies (62–70%) [22]. In contrast, the microbial purification routes described herein can reach efficiencies >90% and are environmentally benign.

7.2.1 Establishing the Roots of Isoprenoid Biosynthesis

In theory, all plant-derived metabolites can be derived from the microbial host's existing metabolic pathways; thus, heterologous production of any small molecule requires a thorough understanding of host metabolism. *E. coli* and *S. cerevisiae* stand as the most well-characterized prokaryotic and eukaryotic hosts for metabolic engineering and as such were the focus of our efforts. All terpene biosynthesis is founded upon production of the universal isoprenoid building blocks isopentyl diphosphate (IPP) and dimethylallyl diphosphate (DMAPP). The metabolic pathways for these compounds differ between eukaryotic and prokaryotic hosts, however. The mevalonate pathway, common to eukaryotic hosts, branches from central metabolism at acetyl-CoA. In comparison, the deoxyxylulose phosphate pathway (DXP), a pathway found in most prokaryotes, is based on glyceraldehyde-3-phosphate and pyruvate (Figure 7.1). Isoprenoid biosynthesis using either pathway is tenable; however, the mevalonate pathway was selected for a number of practical reasons. Most importantly, the DXP pathway was only more recently elucidated [23] and remains poorly characterized with respect to native regulation. In contrast, refactoring—changing promoters, ribosome binding sites, and codon usage, among other approaches—a heterologous mevalonate pathway provides a means

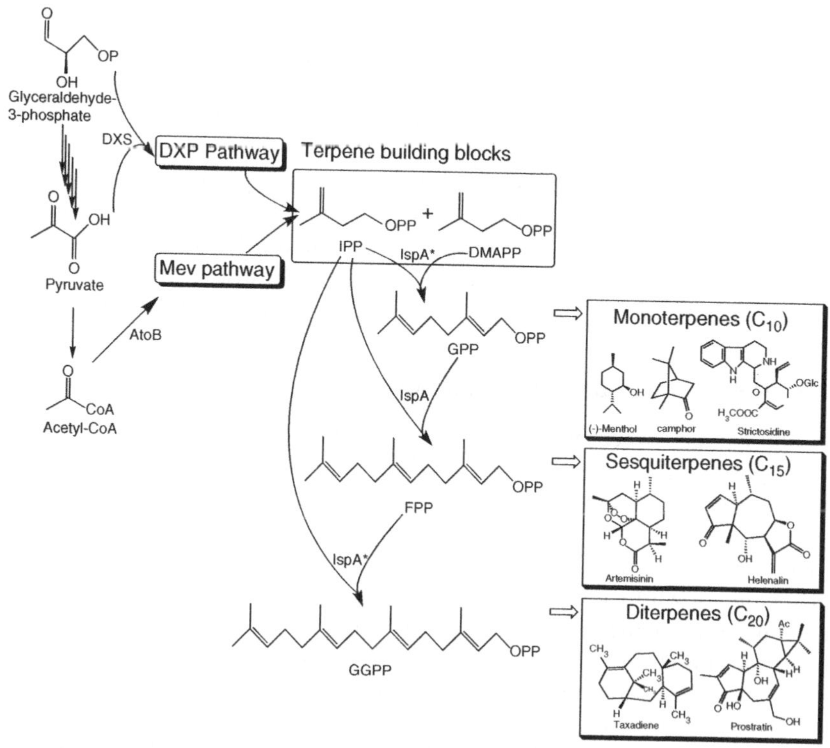

Figure 7.1 Biosynthetic routes from central metabolism to isoprenoids. Engineering either the mevalonate (Mev) or deoxyxylulose phosphate (DXP) pathways necessitates diverting carbon from central metabolism. The Mev pathway, native to most eukaryotes, is based upon acetyl-CoA, while the native DXP pathway common to most prokaryotes is based on glyceraldehyde-3-phosphate and pyruvate. Both pathways yield the universal isoprenoid building blocks, isopentenyl pyrophosphate (IPP) and dimethylallyl pyrophosphate (DMAPP). Condensation of these basic building blocks yields geranyl pyrophosphate (GPP), farnesyl pyrophosphate (FPP), and geranylgeranyl pyrophosphate (GGPP). Native *E. coli* farnesyl diphosphate synthase (IspA) can produce FPP and mutant versions (IspA*) can provide GPP and FPP [25–27]. The cyclic monoterpenes strictosidine, menthol, and camphor; the cyclic sesquiterpenes, artemisinin and helenalin; and prostratin and the diterpene precursor to taxol, taxadiene, are presented to illustrate modularity in isoprenoid-derived pharmaceutical biosynthesis.

of bypassing native regulation and increasing pathway flux. For example, an early laboratory *E. coli* strain engineered for amorphadiene biosynthesis required overexpression of heterologous enzymes derived from both *S. cerevisiae* and *Haematococcus pluvialis* as well as the use of native enzymes [24].

IPP and DMAPP serve as the C_5 building blocks for the entire 50,000+ member spectrum of the isoprenoid family of natural products [10]. The native *E. coli* gene ispA encodes for a farnesyl diphosphate (FPP) synthase used for production of these sesquiterpene (C_{15}) building blocks. Following incorporation of a single point mutation, ispA can be converted into a geranyl diphosphate (GPP) synthase suitable for monoterpene (C_{10}) biosynthesis [25,28]. Further chain elongation toward the diterpene (C_{20}) building block geranylgeranyl diphosphate (GGPP) requires either the heterologous expression of a GGPP synthase or the inclusion of additional mutations in *E. coli* ispA [25–27]. These linear isoprenoid building blocks can then be cyclized following heterologous expression of an appropriate plant-derived synthase (Figure 7.1). Cyclization represents a key committed enzymatic reaction that effectively locks carbon into the downstream biosynthetic pathway. The unmodified terpene skeleton can then be acted upon by a series of tailoring enzymes to add the chemical moieties essential for bioactivity.

7.2.2 Pushing Versus Pulling Carbon

A useful exercise when planning metabolic engineering experiments is dividing the complete biosynthetic pathway—feedstock to product—into two or more congruous segments. Connecting the segments are key enzymatic reactions, termed here *pivot nodes*. Using this framework, metabolic engineering strategies can be divided into two complementary approaches: *pushing* upstream carbon toward, and *pulling* downstream carbon away from, pivot nodes. Fundamental to this paradigm is an intelligent choice of pivot nodes for the pathway and experiment in question (Figure 7.2). In general, irreversible enzymatic reactions are excellent pivot nodes since carbon can be locked into the downstream biosynthetic pathways at these points. The irreversible cyclization of FPP to amorphadiene by an *A. annua* amorphadiene synthase (ADS) served this role in the artemisinin pathway. From amorphadiene, a three-step enzymatic oxidation yields artemisinic acid, and two additional chemical steps produce artemisinin [29]. However,

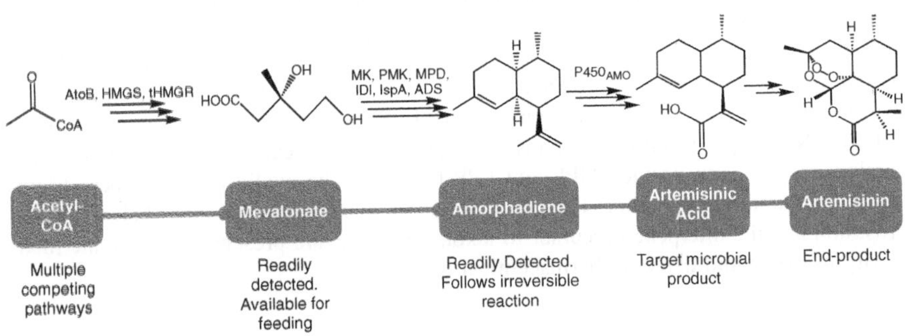

Figure 7.2 Establishment of pivot nodes in artemisinic acid biosynthesis. Pivot nodes are established at key metabolic intermediates to discretize long pathways into manageable segments. Irreversible enzymatic reactions, commercial availability of intermediates for use as feedstock molecules, and availability of measurement techniques and metabolites subject to a high degree of competition make strong pivot points. Acetoacetyl-CoA thiolase (AtoB), HMG-CoA synthase (HMGS), truncated HMG-CoA reductase (tHMGR), mevalonate kinase (MK), phosphomevalonate kinase (PMK), mevalonate pyrophosphate decarboxylase (MPD), IPP isomerase (IDI), FPP synthase (IspA), amorphadiene synthase (ADS), and cytochrome *P*450 monooxygenase CYP71AV1 (*P*450_{AMO}).

irreversibility is not the only criterion for a pivot node—branch points in central metabolism are also excellent candidates. Acetyl-CoA, the mevalonate pathway's root into central metabolism, was a point of extensive engineering. Lastly, some pathway intermediates are selected because they diffuse across the cell membrane or can be easily detected. Mevalonate proved to be readily quantified by Gas Chromatography-Mass Spectrometry (GC-MS) (following lactonization) and could be obtained commercially for feeding experiments and downstream pathway optimization. Establishment of pivot nodes proved highly effective in artemisinin biosynthesis, and this level of abstraction enabled a complex, multistep pathway to be broken down into metabolite-specific goals to be solved in parallel.

7.2.3 Increasing Carbon Flux by Pathway Overexpression and Refactoring

The approach common to most small-molecule production is pulling carbon away from native metabolism by overexpressing a heterologous biosynthetic pathway. These efforts alone can often achieve fairly substantial titers or final product concentrations. Wild-type *S. cerevisiae* overexpressing ADS produced 4.4 mg/L amorphadiene (a key pivot node in this work) and established a baseline production level [30]. Subsequent pathway optimization in *S. cerevisiae* was primarily focused on solubilizing HMGR (HMG-CoA reductase), natively a membrane-associated protein. Overexpression of a truncated, soluble version of the protein (tHMGR, truncated HMG-CoA reductase) increased amorphadiene production titers fivefold; following additional genomic modifications, the gene encoding for tHMGR would be integrated onto the chromosome to provide an additional 1.5-fold improvement in amorphadiene titers. Lastly, to increase substrate availability, FPP synthase was also overexpressed. While amorphadiene titers remained unchanged, specific production was improved by approximately 10% because of decreased cell growth.

In comparison with the *S. cerevisiae* host, overexpression of a heterologous mevalonate pathway and ADS in *E. coli* achieved amorphadiene production levels of 112 mg/L in defined medium [24]. Mevalonate pathway overexpression and refactoring in *E. coli* likely explain much of the initial discrepancy in amorphadiene titers between the two hosts (112 mg/L in *E. coli* vs 4.4 mg/L in *S. cerevisiae*). Refactoring is often done concomitantly with overexpression; replacing native promoters with well-characterized genetic parts removes native transcriptional regulation while providing user-defined transcriptional outputs that enable accurate enzyme titration. Since the completion of the artemisinin project, a number of research tools have been made available that greatly facilitate pathway refactoring. For example, the Registry of Standard Biological Parts provides a range of well-characterized promoter systems in a catalog format [31], and a ribosome-binding site calculator enables increased control over *E. coli* translation initiation rates [32]. As the community's reservoir of biological parts increases, and as we begin to understand how to encode for greater user-defined functionalities, the process of pathway construction and optimization is likely to be greatly accelerated.

Even in light of these significant engineering advances, substantial a priori knowledge of individual enzyme kinetic parameters and expression levels is always desirable. This information, unfortunately, is often unavailable for all of a pathway's constituent enzymes. One approach that reduced the need for enzyme characterization in the mevalonate pathway was the development of a modular *E. coli* vector set [33]. By altering plasmid copy number, operon design, and promoter choice, pathway enzymes were combinatorially titrated to identify bottleneck enzymes. Promoter swapping, in conjunction with codon optimization up through mevalonate, improved production fivefold over the base strain (221 mg/L/OD mevalonate). Gene copy numbers of the titrating pathway enzymes identified mevalonate kinase (MK) as the rate-limiting pathway enzyme for amorphadiene biosynthesis. Following up on this finding, further MK overexpression improved production titers sevenfold over the base strain (293 mg/L/OD amorphadiene). Continued iteration of this process could be used to both balance and increase carbon flux of a target pathway.

While pathway overexpression and refactoring formed the crux of our efforts to engineer the isoprenoid pathway, the pathway also served as a testbed for several fruitful optimization approaches. For example, use of intergenic regions—sequences shown to control RNA stability, translation initiation, and transcription termination—to engineer the top portion of the isoprenoid pathway (acetyl-CoA to mevalonate) was explored [34]. A library of over 600 variants was examined using a mevalonate auxotroph reporter strain, leading to isolation of a variant displaying over a sevenfold increase in mevalonate production. A second method explored the balancing pathway flux using protein scaffolds based on eukaryotic-derived protein–protein interaction domains [35]. By tethering mevalonate pathway enzymes to a synthetic protein scaffold, enzyme stoichiometry could be user defined; in addition, the close proximity of the enzymes putatively increases the local substrate concentration and channels metabolic intermediates. An astounding 77-fold improvement in production levels was achieved while concomitantly decreasing total protein expression levels and metabolic load [35].

7.2.4 Achieving Functional Expression of Plant *P*450s

One particularly challenging problem that can derail any efforts to pull carbon through a biosynthetic pathway is achieving functional expression of heterologous enzymes, a problem often exacerbated when working with plant-derived *P*450s expressed in prokaryotic hosts. Plant *P*450s are often membrane associated, a feature that often results in poor expression and loss of function when they are heterologously expressed in bacterial hosts. Prior to the artemisinin project, only extremely low-level production (<10 mg/L) of oxidized cyclic terpenes was reported in *E. coli* hosts harboring plant-derived *P*450s [36].

Expression of the native *A. annua* cytochrome *P*450 and redox partner (CYP71AV1, $P450_{AMO}$) in *S. cerevisiae* successfully catalyzed the three-step

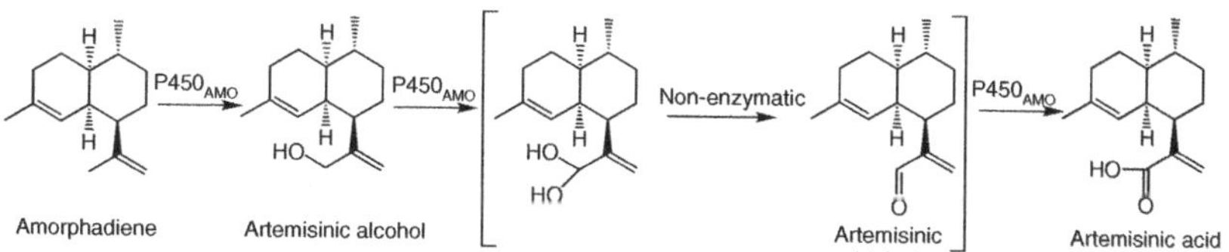

Figure 7.3 Three-step oxidation of amorphadiene to artemisinic acid. Establishing microbial artemisinic acid biosynthesis was facilitated by a substrate promiscuity of $P450_{AMO}$. The cytochrome $P450$ recognized amorphadiene, artemisinic alcohol, and artemisinic aldehyde as substrates and enabled selective oxidation of the amorphadiene skeleton's terminal alkane. Dehydration of the geminal diol and production of artemisinic aldehyde occurred spontaneously *in vivo* [30].

oxidation of amorphadiene to artemisinic acid (Figure 7.3) [30]. This strain produced over 30 mg/L of the acid without detectable production of the alcohol or aldehyde intermediates. Achieving functional expression of the same set of enzymes in an *E. coli* host proved considerably more challenging. A series of protein engineering approaches were explored on both cadinine hydroxylase and $P450_{AMO}$ in an attempt to develop a more general framework for functional expression of plant-derived $P450$s in *E. coli* [37]. Oxidation of cadinene by 8-cadinene hydroxylase was used to establish a set of guidelines. A combination of downstream pathway overexpression (25-fold titer increase), codon optimization for expression in *E. coli* (2.5-fold titer increase), N-terminal membrane anchor engineering (fivefold titer increase), and swapping cognate redox partners (1.8-fold titer increase) collectively resulted in a strain producing 60 mg/L oxidized cadinene from acetyl-CoA; as an additional metric, when using exogenously added mevalonate to the culture broth to measure productivity, >50% substrate conversion to 8-hydroxy cadinene was achieved. While being substantial, this finding demonstrates an imbalance in the biosynthetic pathway and room for further improvement. On the basis of these results, a similar approach was applied to the artemisinic acid-producing $P450_{AMO}$. When fully engineered, the *E. coli* system produced artemisinic acid at titers of over 100 mg/L; yet again, full substrate conversion remained unrealized.

As an alternative to engineering plant-derived $P450$s for functional expression in microbial hosts, one option is to focus on microbial $P450$s with demonstrated activity in heterologous expression hosts. The nontrivial task then becomes reengineering substrate specificity while also maintaining appropriate regio- and stereoselectivities required for end-product biosynthesis. Toward this end, the well-characterized $P450_{BM3}$ from *Bacillus megaterium* was explored as a tool for amorphadiene oxidation [38]. It was discovered that opening the ligand-binding pocket enabled oxidation of amorphadiene to artemisinic epoxide at titers >250 mg/L; a three-step synthetic chemistry transformation of artemisinic epoxide to the acid was demonstrated with a >75% conversion efficiency for each step [38]. In light of this work, target small-molecule overproduction should be explored from multiple directions simultaneously, using both native and user-devised metabolic pathways. This strategy can be particularly useful when encountering difficult-to-express plant enzymes that frustrate reconstruction of the native pathway.

7.2.5 Decreasing Flux through Competing Pathways

The vast majority of the work described thus far focused on pulling carbon flux by using an overexpressed, heterologous pathway. As discussed previously, the complimentary strategy to pulling carbon is to push it toward key pivot nodes. As applied to artemisinin, this approach was best demonstrated by removing competing pathways for FPP and acetyl-CoA in the *S. cerevisiae* production host. Amorphadiene production was improved twofold by down-regulating squalene synthase (*ERG9*) and removing competition for FPP; an additional twofold production increase (up to 105 mg/L) was achieved upon overexpression of the transcription factor *upc*2-1 to further down-regulate squalene biosynthesis [30].

A problem significantly more difficult than decreasing competition for FPP is increasing the available pool acetyl-CoA in the cytosol. The tricarboxylic acid cycle (TCA) in eukaryotes occurs within the mitochondria, effectively partitioning acetyl-CoA away from the cytosol where heterologous protein expression occurs. Maximum carbon flux that can be directed through an acetyl-CoA rooted pathway is effectively capped by the available pool of cytosolic acetyl-CoA. In an attempt to circumvent this issue, use of a cytosolically localized route from pyruvate to acetyl-CoA, the pyruvate dehydrogenase bypass, was investigated (Figure 7.4) [39]. Here pyruvate is decarboxylated to acetaldehyde, oxidized to acetate, and finally ligated with coenzyme A to produce acetyl-CoA. Overexpression of the acetyl-CoA synthetase ACS1 demonstrated functional conversion of acetate to acetyl-CoA as measured by an increase in amorphadiene titer. However, when coexpressed with acetaldehyde dehydrogenase, ACS6, amorphadiene production levels improved by only 10% over control strains while the acetate accumulated in the media. A comparison of enzyme activity levels between the engineered and control strains displayed 40-fold and 1.1-fold increases in acetaldehyde dehydrogenase and acetyl-CoA synthetase activities, respectively: results that again highlight the importance in balancing the expression levels to match intrinsic enzyme kinetics. Only after incorporating an acetyl-CoA synthetase variant from *Salmonella enterica* engineered to relieve allosteric product repression was an improvement in amorphadiene productivity witnessed over the control strains (1.9-fold). While these results are significant, a great deal more work will need to be directed toward this area in order to achieve high-level production of any acetyl-CoA-derived natural product in *S. cerevisiae*.

7.2.6 Identifying Product-Induced Toxicity

Substantial effort was placed on determining the scope of the stress response in both hosts to isoprenoid and artemisinic acid overproduction [24,40]. It became clear early on that IPP accumulation was toxic to *E. coli*: a condition, fortunately, that was readily countered by overexpression of a downstream terpene cyclase drawing from the IPP pool [21]. This strategy can be broadly generalized, and

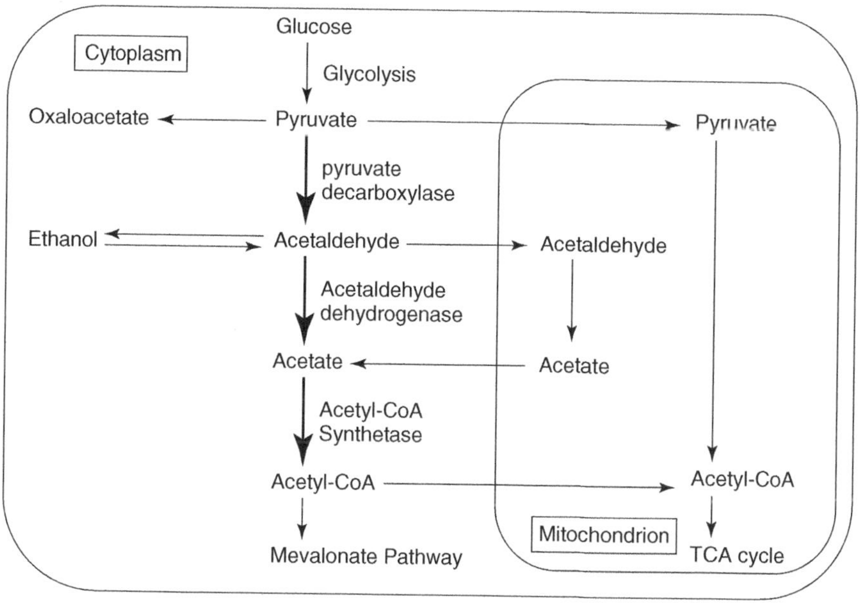

Figure 7.4 Producing cytosolic acetyl-CoA in yeast strains. Overproduction of cytosolic acetyl-CoA is essential for achieving high-level production artemisinin precursors. Overexpressed enzymes are localized in the cytosol; however, the bulk of acetyl-CoA is found in the mitochondrion for use in the tricarboxylic acid (TCA) cycle. The pyruvate dehydrogenase bypass increases the available acetyl-CoA pool in the cytosol. Figure from Ref. [39].

enzymatic conversion of toxic intermediates to innocuous products can always be a first course of action.

A more difficult task is addressing end-product toxicities. In *S. cerevisiae*, genomic transcript data were collected to search for product-induced stress responses and to identify native genes that might be up- or down-regulated to improve resistance [40]. This study was motivated, in part, by the observation that yeast harboring plasmids encoding for artemisinic acid biosynthesis were unstable (72% plasmid loss after 120 h): a condition relieved by inactivating $P450_{\mathrm{AMO}}$ with a single point mutation. This finding indicated that $P450_{\mathrm{AMO}}$ activity, and not overexpression, imposed the stress and plasmid loss. Transcript data confirmed a 2- to 15-fold up-regulation in the expression of genes encoding for several members of the ATP-binding cassette family of proteins associated with pleiotropic drug resistance. Additionally, a greater than twofold increase in expression for various oxidative and osmotic stress response genes was observed. While the initial symptom of toxicity, plasmid instability, was effectively addressed by using different plasmid-born auxotrophic markers, tackling product-induced toxicity poses a more difficult long-term challenge. While this study did not explore application of the transcript data, a number of follow-up experiments addressing product toxicity can be outlined. Genes up-regulated during artemisinic acid overproduction could be refactored for continuous, high-level expression. This strategy may preempt artemisinic acid-induced stress and provide a fitness advantage; however, this approach hinges on the assumption that gene up-regulation counters small-molecule toxicity and is not a more general stress response. A complementary strategy is to screen libraries of transport proteins for more rapid export of artemisinic acid. This work could prove to be broadly applicable: a series of base strains exhibiting increased responsiveness to a broad spectrum of chemical moieties (hydroxyl and carboxylic acid groups, for example) could be used for overproduction of any number of toxic small molecules.

7.2.7 Purifying Amorphadiene and Artemisinic Acid

While the key focus in bench scale metabolic engineering is efficient, high-level production of target small molecules, product purification becomes a key factor during commercial scale-up. Various purification schemes were developed during the course of the artemisinin project. While most were founded on achieving accurate quantification of product titers, many would also prove industrially scalable. It was recognized early on that amorphadiene evaporated from the growth medium (and was stripped out of bioreactors with the sparged air), exhibiting a half-life of 50 min in a bioreactor setting [41]. By adding the linear hydrocarbon dodecane to the culture medium (between 10% and 20% v/v), the hydrophobic amorphadiene was sequestered in the organic (dodecane) phase. The advantages to using a two-phase culture rapidly became clear. First, initial small-molecule production titers are often in the microgram-per-liter range and a hydrophobic overlay mitigates evaporative losses and concentrates the sample for easier detection. The partition coefficient, or log P value, is used to describe a molecule's concentration distribution between the two immiscible phases and can be predicted or experimentally measured when the quantities present are insufficient for empirical testing.

Structurally, artemisinic acid differs from amorphadiene by only a single carboxylic acid moiety, but the chemical properties this functional group imparts are vital from a purification standpoint. Initial experiments demonstrated that 95% of artemisinic acid was associated with the cell pellet and could not be detected in the spent culture broth [30]. Adjusting the pH with an alkaline buffer resulted in release of artemisinate from the cell pellet with high efficiency. Artemisinic acid has a predicted pK_a of 4.34, corroborating the analysis that a pH shift and corresponding deprotonation accounts for release of artemisinate. This experiment demonstrated that the acid was excreted from the yeast host but remained protonated and associated with the cell surface. This was a near-ideal scenario with regard to purification. First, if artemisinic acid was intracellularly localized, a lysis step would be required in the purification workflow; further, accumulation of intracellular acid would almost surely elicit increased stress responses. Second, the association of artemisinic acid with the cell surface eliminated an expensive chromatography-based separation of the acid from the culture broth. Instead, the cell material can be concentrated by centrifugation and washed with an alkaline buffer to collect a concentrated, relatively pure product.

7.2.8 The Artemisinin Project in Retrospect

Only in hindsight is vision 20/20, and only after successful completion of the artemisinin project can a number of key questions regarding research direction be addressed. Perhaps most important is the question of microbial host. At the conclusion of the project, artemisinic acid was being produced at titers greater than 100 mg/L in both *E. coli* and *S. cerevisiae* hosts [30,37]. Both are industrially tractable hosts (yeast being, in general, more widely employed) with an abundance of well-developed genetic tools. Each host, however, has a unique set of advantages and disadvantages. *E. coli* proved to be more readily engineered for production of amorphadiene, but achieving high-level amorphadiene oxidation proved significantly more difficult. In contrast, *S. cerevisiae* displayed numerous advantages: functional expression of the native *A. annua* $P450_{AMO}$ and its cognate redox partner, a readily scalable artemisinic acid purification route, and a more sophisticated genetic network responsive to toxic small-molecule biosynthesis. This said, achieving high-level carbon flux of any acetyl-CoA-derived product remains a significant metabolic engineering hurdle.

While *E. coli* and *S. cerevisiae* will continue to be front-runners during discussions on host selection for plant-derived natural product biosynthesis, the discussion should be expanded. Industrial microbial production routes have long since been built in various strains of *Pseudomonas, Bacillus, Corynebacterium, Streptomyces*, and numerous fungal strains, among others. These hosts exhibit solvent tolerant capabilities superior to those of *E. coli* [42], and are continually being explored as research scientists target more difficult-to-produce and toxic small molecules [43]. Unfortunately, none of these hosts have genetic systems as developed as the one for *E. coli*.

To produce low-cost artemisinin, cheap carbon sources (glucose, glycerol, and galactose) were utilized in production experiments. Mevalonate feeding experiments were used to optimize downstream reactions, an approach that bypasses difficulties in rerouting carbon flux from central metabolism while also simultaneously providing more reproducible assay results. However, a back of the envelope calculation of relative feedstock costs shows the advantages in rooting a commercial production process in central metabolism. When placed in the context of a \$100/kg artemisinin production cost target, the cost of using mevalonate (>\$1000/kg) versus glucose (<\$1/kg) clearly indicates the commercial viability of each process. However, for more valuable pharmaceutical precursors, glucose should not, by default, be considered the primary feedstock option. Feeding high-value intermediates could be used as a simplified, rapid mode of validating biosynthetic production routes for high-value products. Recently, hydroxyvalerate production from levulinic acid, which can be readily derived from wheat straw at reasonable yields [44], was reported using a *Pseudomonas putida* biocatalyst [45].

7.3 TERPENE INDOLE ALKALOIDS: EXPANDING UPON THE MEVALONATE PATHWAY

While all isoprenoids have their roots in central metabolism for IPP and DMAPP biosynthesis, other branches of native metabolism provide additional chemical building blocks. In this sense, heterologous enzymes can be used to link disparate host pathways to create elaborate chemical structures with particularly exquisite pharmacokinetic and bioactive properties. With respect to the isoprenoids, this is perhaps best exemplified by terpene indole alkaloid (TIAs) [46]. TIAs are a large group of chemically related compounds with a broad range of bioactivities. Among the approximately 3000 molecules thus far described from this class are the clinically proven anticancer compounds vincristine and vinblastine, as well as the antihypertensive ajmalicine. In accordance with the general theme of this chapter, TIAs are generally too complex to be produced commercially by traditional chemical synthesis and exist only in minute amounts in their native hosts. Attempts to establish plant cell and tissue culture have also proved difficult and *C. roseus* cell cultures typically produce little to no pharmaceutically relevant small molecules. In fact, <50% of the full metabolite complement of approximately 130 indole alkaloids described from *C. roseus* has been detected in cultured cells [47]. Supply scarcity also hampers continued investigation into the therapeutic potential of many members of this chemical class. At present, only the most potent of the rare alkaloids have been commercially developed. Owing to the rigors of isolation and purification, therapies requiring high or chronic dosing cannot be developed from these molecules.

All TIAs are biosynthesized by a process that uses strictosidine as a key intermediate, establishing a clear pivot node in our abstracted pathway architecture (Figure 7.5). Further, strictosidine is a commercially available intermediate that can be used to conduct feeding experiments for downstream pathway optimization. Strictosidine is produced by the condensation of the tryptophan-derived tryptamine and the monoterpene secologanin by the enzyme strictosidine synthase [48]. In this light, microbial biosynthesis of TIAs draws upon the extensively engineered isoprenoid and amino acid biosynthetic pathways. Compared to the artemisinin, TIAs are structurally much more complex, a feature that diminishes the impetus to establish a microbial production platform. Establishing a 40-step biosynthetic pathway—of which half of the constituent plant enzymes are undiscovered—is a much riskier endeavor than continuing to plant, harvest, and extract natural products from *C. roseus*. Nonetheless, biosynthesis of strictosidine as an advanced intermediate fits well into our pivot node model of metabolic engineering. For example, the asymmetric dimerization of downstream strictosidine derivatives yields vincristine and vinblastine; further, the pathways are rooted in regions of central metabolism well studied for small-molecule overproduction.

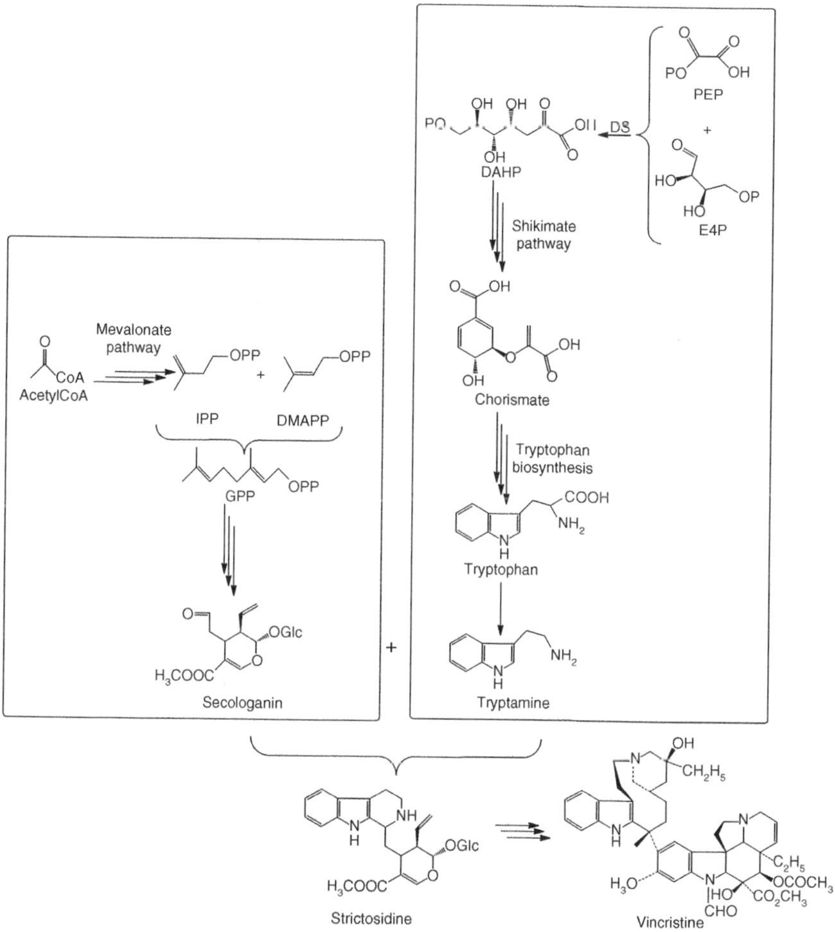

Figure 7.5 Terpene indole alkaloid (TIA) biosynthesis. Biosynthesis of strictosidine, a key intermediate in the production of TIAs, is rooted in carbon derived from the isoprenoid and tryptophan pathways. Tailoring enzymes act upon strictosidine to yield potent pharmaceuticals, for example, vincristine. Metabolic engineering of TIAs requires tying together multiple native microbial pathways with plant-derived enzymes to produce highly complex pharmaceutical natural products.

High-level production of strictosidine from a glucose feedstock will ultimately be dictated by the intracellular abundance of the essential precursors secologanin (derived from GPP) and tryptamine (derived from tryptophan). Secologanin biosynthesis finds its roots in the isoprenoid pathway and can be built on top of the previously discussed, high-yielding mevalonate pathway. Tryptamine is generated by the decarboxylation of the amino acid tryptophan. The gene encoding the enzyme responsible for this conversion has been cloned from a number of organisms, including *C. roseus*, and expressed in an active form in *E. coli* [49].

High-level tryptamine production, however, is dependent on the available tryptophan pool. Conveniently, tryptophan biosynthesis is an area of active research and is extensively reviewed [50,51]. Using our metabolic engineering framework, we highlight important pathway nodes instrumental in achieving high-level tryptophan biosynthesis.

In *E. coli*, most other microorganisms, and plants, the shikimate pathway is used for aromatic amino acid biosynthesis [52,53]. The industrial importance of shikimate in tamiflu, tyrosine-derived degradable biopoplymers, and the amino acids themselves has made it a target of extensive engineering [51,54,55]. The shikimate pathway begins by condensation of phosphoenolpyruvate (PEP), derived from glycolysis, and erythrose 4-phosphate (E4P) from the pentose phosphate pathway. Carbon can be locked into the shikimate pathway through the irreversible condensation of PEP and E4P to 3-deoxy-D-arabino-heptulosonate-7-phosphate (DAHP), a key pivot node, through the action of DAHP synthase. In *E. coli*, the three isozymes of DAHP encoded by *aroF/G/H* are feedback inhibited by the aromatic amino acids and transcriptionally regulated by tryptophan- and tyrosine-responsive TrpR and TyrR, respectively [56,57]. Removal of transcriptional and allosteric regulation can strengthen the shikimate pathway's pull on carbon precursors through to chorismate. Approaching the problem from the other side of the equation, increasing the supply of both PEP and E4P pushes carbon into the shikimate pathway. Central to this issue is the consumption of a molecule of PEP per glucose imported when using the phosphotransferase transport system. This route effectively caps aromatic amino acid production at 50% of maximum stoichiometric theoretical yield. An exemplary study of phenylalanine (also shikimate-derived) overproduction addressed all the above issues by rerouting glucose transport through the PEP-independent galactose permease while concomitantly overexpressing feedback-resistant DAHP synthase, transketolase, and chorismate mutase [58]. Incorporation of all the four alterations enabled the yield cap imposed by use of the phosphotransferase system to be breached, and over 60% of the maximum theoretical phenylalanine yield from glucose was reported.

Next, we consider approaches applied toward the second pivot node, chorismate, from which ubiquinone, folic acid, isochorismate, tryptophan, phenylalanine, and tyrosine are synthesized [53]. As with the shikimate pathway, the tryptophan pathway in *E. coli* is feedback inhibited at both the transcriptional and posttranslational levels. Refactoring the trp operon along with use of a feedback-resistant mutant of anthranilate synthase enabled yields of 0.227 g-Trp/g-Glc, achieving theoretical maximum tryptophan production [59].

By expressing the mevalonate pathway in a high-yielding tryptophan production strain, the onus for high-level production is taken away from engineering the central metabolism and placed on achieving high-flux heterologous pathways. To date, the rate-limiting step in high-level strictosidine production has been identification of the requisite genes from *C. roseus*. Finding these genes will be greatly facilitated by recent publication of three separate expressed sequence tag (EST) analyses of *C. roseus* [60–62]. While there are pathway pieces still missing, research can still be based on a series of key pivot nodes. For example, feeding neptalactol, strictosidine, or other advanced intermediates that can be isolated or obtained commercially enables a thorough exploration of down-stream biosynthetic pathways. Clearly, identification of plant enzymes and pathways encoding for secondary metabolite production remains a significant hurdle in the field.

7.4 BENZYLISOQUINOLINE ALKALOIDS (BIAs): REVISITING PLANT-DERIVED ENZYME OVEREXPRESSION

While the focus of our analysis of TIAs was on their roots in central metabolism, we now shift our attention to the outer branches with the BIAs. Here, we explore how feeding key pivot node compounds is used to develop and optimize the final steps in the microbial expression of plant-derived enzymes. Similar to the TIAs, BIAs are a large family of compounds ($\approx$2500 known to date) with substantial pharmaceutical applications and commercial potential [63]. In aggregate, the opioid drugs commanded a nearly $9.6B market presence in 2008, a number expected to increase to nearly $12 billion by 2018 [64]. As with isoprenoid biosynthesis, targeting a small number of commercially valuable end-products has been a worthwhile strategy toward establishment of platform strains producing the various BIA branch-point intermediates and building blocks.

Metabolic engineering of BIA biosynthetic pathways remains nascent, in part, because of difficulties in isolating and characterizing enzymes catalyzing conversion of tyrosine to dopamine. Bypassing this hurdle, two recent studies used precursor feeding to demonstrate production of the branch-point intermediate reticuline as well as downstream BIAs in *E. coli* and *S. cerevisiae* (Figure 7.6) [65,66]. Both studies sourced their heterologous enzymes from an array of prokaryotic, plant, and mammalian sources. Minami *et al.* [66] used culture-fed dopamine to demonstrate production of (*R,S*)-reticuline and subsequently magnoflorine and scoulerine production. Of particular interest was the use of a *Micrococcus luteus*-derived monoamine oxidase to oxidize dopamine to 3,4-dihydroxyphenylacetaldehyde (DHPAA). DHPAA contains an additional hydroxyl compared to the native substrate, 4-hydroxyphenylacetate. The downstream enzymes recognized the DHPAA substrate, but with decreased specificity. Use of DHPAA does have advantages, however. The native route has a difficult-to-achieve downstream hydroxylation by the poorly expressed *P. somniferum* *P*450 CYP80B1, a step bypassed using the DHPAA substrate. Using this system, microbially produced (*R,S*)-reticuline titers reached 11 mg/L, but with an overall yield from dopamine of only 2.9%. Further transformation of (*S*)-reticuline to more complex end-products was demonstrated by coculturing engineered *E. coli* with a *S. cerevisiae* strain harboring appropriate downstream pathways. Expression of *Coptis japonica*-derived berberine bridge enzyme (BBE) yielded (*S*)-scoulerine; similarly, expression of *C. japonica*-derived cytochrome *P*450 (CYP80G2) and corytuberine-*N*-methyltransferase (CNMT) yielded magnoflorine. Conversion efficiencies from reticuline to (*S*)-scoulerine and magnoflorine were over 75% and 65%, respectively.

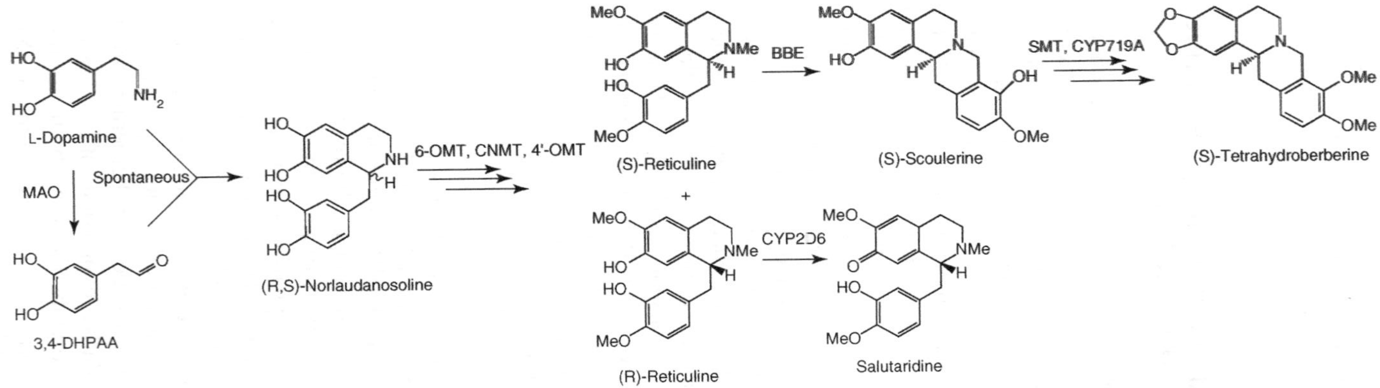

Figure 7.6 Benzylisoquinoline alkaloid (BIA) biosynthesis. Racemic (*R,S*)-reticuline (stereochemistry highlighted in red) was produced from L-dopamine- or (*R,S*)-norlaudanosoline-fed cultures of *E. coli* or *S. cerevisiae*, respectively [65,66]. Reticulin serves as a key branch-point intermediate, and two stereoisomers can be channeled toward precursors of commercially relevant small molecules. Salutaridine, the precursor to morphine, can be derived from (*R*)-reticuline. Similarly, (*S*)-reticuline can be transformed to (*S*)-scoulerine and onward to (*S*)-tetrahydroberberine at a high efficiency. (*S*)-scoulerine is the branch-point for sanguinarine alkaloids and (*S*)-tetrahydroberberine is the precursor to the berberine alkaloids. Monoamine oxidase (MAO), norcoclaurine 6-*O*-methyltransferase (6OMT), coclaurine-*N*-methyltransferase (CNMT), 3′-hydroxy-*N*-methylcoclaurine-4′-*O*-methyltransferase (4′-OMT), berberine bridge enzyme (BBE), (*S*)-scoulerine-9-*O*-methyltransferase (SMT), canadine synthase (CYP719A1), cytochrome *P*450 CYP2D6 (CYP2D6).

Utilizing the same general strategy, Hawkins *et al.* [65] employed an *S. cerevisiae* host expressing *Thalictrum flavum* or *Papaver somniferum*-sourced enzymes and fed norlaudanosoline to produce racemic reticuline and a selection of downstream products (Figure 7.6). The yeast host produced an order-of-magnitude higher reticuline titers (150 mg/L) compared to previously discussed engineered *E. coli*; however, only single-digit conversion efficiencies from norlaudanosoline to reticuline were observed. Racemic reticuline proved to be a versatile intermediate, and (*R*)-reticuline-dependent salutaridine or (*S*)-reticuline-dependent (*S*)-tetrahydroberberine were both demonstrated in the same reticuline-producing host. Incorporation of the requisite plant-derived enzymes enabled production of over 30 mg/L (*S*)-tetrahydroberberine and 20 mg/L salutaridine from (*S*)- and (*R*)-reticuline stereoisomers, respectively. Conversion efficiencies were high (>50%) from reticuline to (*S*)-tetrahydroberberine, but were <10% for reticuline to salutaridine.

As was also found during artemisinin biosynthesis, the BIA biosynthesis conversion efficiencies proved to be highly strain and enzyme dependent. Clearly, *E. coli* and *S. cerevisiae* can both be engineered for reticuline production, but downstream transformations were achieved using only engineered *S. cerevisiae* hosts. Although not explicitly stated, the results of these studies corroborate the analysis of the artemisinin pathway and suggest that *S. cerevisiae* is more readily engineered for functional plant-derived enzyme (and particularly *P*450) expression. This finding is not surprising, considering the much more dramatic change in host physiology between eukaryotic enzyme sources and prokaryotic expression systems. Low conversion efficiencies in all cases, however, indicate that significant pathway imbalances continue to exist in strains of engineered yeast. Perhaps the biggest hurdle is decreased enzyme specificity toward the nonnative substrate norlaudanosoline in the first two steps of reticuline biosynthesis [65], a problem that permanently starves downstream enzymes of their respective substrates. Further, 100% conversion was not observed from reticuline to its downstream intermediates and end-products, indicating that additional points of pathway imbalance also exist.

7.5 CONCLUSIONS

As of this writing, the main thrust of the academic research laboratory work on artemisinin has concluded, achieving artemisinic acid production titers >100 mg/L in both *E. coli* and *S. cerevisiae* hosts [30,37]. Subsequently, the engineered strains were passed on to Amyris Biotechnologies for further optimization and process development. Through collaborative work between the Institute for OneWorld Health and Sanofi-aventis, semisynthetic artemisinin was incorporated into the artemisinin-based combination therapy supply chain in 2010. The success of the artemisinin project is based on, in part, a metabolic engineering paradigm enabling a complex, multistep metabolic pathway to be divided into more readily manageable segments. Division of the pathway into key research thrusts centered on what are termed here *pivot nodes* (acetyl-CoA, mevalonate, amorphadiene, and artemisinic acid) addresses multiple, significant engineering hurdles in parallel. Pushing and pulling carbon through enzyme overexpression, refactoring, genetic knockouts, and protein engineering were all complimentary strategies redirecting carbon flux through the mevalonate and artemisinic acid pathways.

For artemisinin, high-titers were most successfully realized following significant engineering of both native and heterologous biosynthetic pathways. Expanding on this central theme are efforts to produce terpene indole and benzylisoquinoline alkaloids. Efforts to engineer *E. coli* for production of strictosidine—a key branch-point intermediate and pivot node—illustrate how heterologous production of plant secondary metabolites can be rooted in native metabolism. Relieving transcriptional and posttranslational regulation of the shikimate and tryptophan biosynthetic pathways can provide gram per liter titers of tryptophan and approach the theoretical maximum conversion from glucose. Coupling the modifications necessary for tryptophan and isoprenoid overproduction could provide a base strain ideal for microbial production of strictosidine. In contrast, BIAs are a more in-depth case study in the branch points of plant secondary metabolism and heterologous expression of plant-derived enzymes. Achieving functional expression of heterologous plant enzymes remains a difficult task, and poor expression or activity at any point fetters production titers. Continuing anecdotal evidence indicates that an *S. cerevisiae* production host may be more amenable to plant-enzyme expression than *E. coli*.

In summary, efforts to engineer microbial production titers for plant secondary metabolites have been quite successful and titers commonly approach the milligram per liter scale. As titers continue to increase, substantial protein and small molecule-induced host stresses will continue to be observed. Protein scaffolding and pathway balancing have been devised as methods to reduce metabolic load and toxic intermediate accumulation [35], but little has been explored with regard to relieving end-product toxicity. The vinca alkaloids are highly toxic compounds for microbial hosts, and will likely define the limits of microbial production. This represents an area of substantial opportunity to develop methods to enable further order-of-magnitude increases in plant-derived pharmaceutical production titers.

In line with the general theme of this book, a final thought should be placed on the sustainability of microbially versus plant-sourced small-molecule production routes. Even after highly efficient selective plant breeding, achieving high small-molecule concentrations from microbial fermentations appears to be more readily achievable for most molecules. Our work on microbial artemisinin biosynthesis highlights a number of advantages. First, under the controlled conditions of fermentation processes—and one that is also devoid of extraneous plant biomass—extraction and purification efficiencies can readily exceed 90%. Second, in most cases microbial processes promise to be much greener, in large part because of the use of centralized, industrial fermentation units. Unlike distributed plant sourcing, in which diesel-like solvents are commonly employed when purification and monitoring may be difficult, fermentation processes can be much more strictly monitored.

REFERENCES

1. Solecki R. Shanidar IV, a Neanderthal flower burial in northern Iraq. Science 1975;190:880–881.
2. Leroi-Gourhan A. The flowers found with Shanidar IV, a Neanderthal burial in Iraq. Science 1975;190:562–564.
3. Newman J, Cragg G, Snader K. Natural products as sources of new drugs over the period 1981–2002. J Nat Prod 2003;66:1022–1037.
4. Saklani A, Kutty S. Plant-derived compounds in clinical trials. Drug Discov Today 2008;13:161–171.
5. Andrew McWilliams. Plant-derived drugs: products, technology, applications. BCC Research; 2006.
6. Horwitz SB. How to make taxol from scratch. Nature 1994;367:593–594.
7. Jennewein S, Croteau R. Taxol: biosynthesis, molecular genetics, and biotechnological applications. Appl Microbiol Biotechnol 2001;57:13–19.
8. Noble R. The discovery of the vinca alkaloids-chemotherapeutic agents against cancer. Biochem Cell Biol 1990;68:1344–1351.
9. Schoemaker HE, Mink D, Wubbolts MG. Dispelling the myths—biocatalysis in industrial synthesis. Science 2003;299:1694–1697.
10. Conolly J, Hill R. Dictionary of terpenoids. London: Chapman & Hall; 1991.
11. Korenromp E, Nahlen B, Wardlaw T, Young M. World Malaria Report: World Health Organization (WHO), Roll Back Malaria (RBM) and United Nations Children's Fund (UNICEF). World Health Organization; Geneva, Switzerland: 2005.
12. Rowe AK, Rowe SY, Snow RW, Korenromp EL, Schellenberg JRA, Stein C, Nahlen BL, Bryce J, Black RE, Steketee RW. The burden of malaria mortality among African children in the year 2000. Int J Epidemiol 2006;35:691–704.
13. Gallup J, Sachs J. The economic burden of malaria. Am J Trop Med Hyg 2001;64:85–96.
14. van Lieshout M, Kovats RS, Livermore MTJ, Martens P. Climate change and malaria: analysis of the SRES climate and socio-economic scenarios. Glob Environ Change 2004;14:87–99.
15. Olumese P. Guidelines for the treatment of Malaria. World Health Organization; Geneva, Switzerland: 2006.
16. Benatar S. Distributive justice and clinical trials in the third world. Theor Med 2001;22:169–176.
17. Novartis. Novartis accelerates production of life-saving malaria treatment Coartem. Basel, Switzerland: Press Release; 2006.
18. Covello PS. Making artemisinin. Phytochemistry 2008;69:2881–2885.
19. Cohen JM, Singh I, O'Brien ME. Predicting Global Fund grant disbursements for procurement of artemisinin-based combination therapies. Malar J 2008;7: 200.
20. Hale V, Keasling JD, Renninger N, Diagana TT. Microbially derived artemisinin: a biotechnology solution to the global problem of access to affordable antimalarial drugs. Am J Trop Med Hyg 2007;77:198–202.

21. Graham IA, Besser K, Blumer S, Branigan CA, Czechowski T, Elias L, Guterman I, Harvey D, Isaac PG, Khan AM, Larson TR, Pawson YLT, Penfield T, Rae AM, Rathbone DA, Reid S, Ross J, Smallwood MF, Vyas VSTTD, Winzer T, Bowles D. The genetic map of *Artemisia annua L.* identifies loci affecting yield of the antimalarial drug artemisinin. Science 2010;327:328–331.
22. Lapkin AA, Plucinski PK, Cutler M. Comparative assessment of technologies for extraction of artemisinin. J Nat Prod 2006;69:1653–1664.
23. Rohmer M, Knani M, Simonin P, Sutter B, Sahm H. Isoprenoid biosynthesis in bacteria: a novel pathway for the early steps leading to isopentenyl diphosphate. Biochem J 1993: 295.
24. Martin VJJ, Pitera DJ, Withers ST, Newman JD, Keasling JD. Engineering a mevalonate pathway in *Escherichia coli* for production of terpenoids. Nat Biotechnol 2003;21:796–802.
25. Ohnuma S, Narita K, Nakazawa T, Ishida C, Takeuchi Y, Ohto C, Nishino T. A role of the amino acid residue located on the fifth position before the first aspartate-rich motif of farnesyl diphosphate synthase on determination of the final product. J Biol Chem 1996;271:30748–30754.
26. Lee PC, Petri R, Mijts BN, Watts KT, Schmidt-Dannert C. Directed evolution of *Escherichia coli* farnesyl diphosphate synthase (IspA) reveals novel structural determinants of chain length specificity. Metab Eng 2005;7:18–26.
27. Ohnuma S, Nakazawa T, Hemmi H, Hallberg A, Koyama T, Ogura K, Nishino T. Conversion from farnesyl diphosphate synthase to geranylgeranyl diphosphate synthase by random chemical mutagenesis. J Biol Chem 1996;271:10087–10095.
28. Reiling KK, Yoshikuni Y, Martin VJJ, Newman J, Bohlmann J, Keasling JD. Mono and diterpene production in *Escherichia coli*. Biotechnol Bioeng 2004;87:200–212.
29. Roth R, Acton N. A simple conversion of artemisinic acid into artemisinin. J Nat Prod 1989;52:1183–1185.
30. Ro D-K, Paradise EM, Ouellet M, Fisher KJ, Newman KL, Ndungu JM, Ho KA, Eachus RA, Ham TS, Kirby J, Chang MCY, Withers ST, Shiba Y, Sarpong R, Keasling JD. Production of the antimalarial drug precursor artemisinic acid in engineered yeast. Nature 2006;440:940–943.
31. Wentzel A, Ellingsen TE, Kotlar H-K, Zotchev SB, Throne-Holst M. Bacterial metabolism of long-chain n-alkane. Appl Microbiol Biotechnol 2007;76:1209–1221.
32. Salis HM, Mirsky EA, Voigt CA. Automated design of synthetic ribosome binding sites to control protein expression. Nat Biotechnol 2009;27:946–950.
33. Anthony JR, Anthony LC, Nowroozi F, Kwon G, Newman JD, Keasling JD. Optimization of the mevalonate-based isoprenoid biosynthetic pathway in *Escherichia coli* for production of the anti-malarial drug precursor amorpha-4,11-diene. Metab Eng 2009;11:13–19.
34. Pfleger BF, Pitera DJ, Smolke CD, Keasling JD. Combinatorial engineering of intergenic regions in operons tunes expression of multiple genes. Nat Biotechnol 2006;24:1027–1032.
35. Dueber JE, Wu GC, Malmirchegini GR, Moon TS, Petzold CJ, Ullal AV, Prather KLJ, Keasling JD. Synthetic protein scaffolds provide modular control over metabolic flux. Nat Biotechnol 2009;27:753–759.
36. Carter OA, Peters RJ, Croteau R. Monoterpene biosynthesis pathway construction in *Escherichia coli*. Phytochemistry 2003;64:425–433.

37. Chang MCY, Eachus RA, Trieu W, Ro D-K, Keasling JD. Engineering *Escherichia coli* for production of functionalized terpenoids using plant P450s. Nat Chem Biol 2007;3:274–277.

38. Dietrich JA, Yoshikuni Y, Fisher KJ, Woolard FX, Ockey D, McPhee DJ, Renninger NS, Chang MCY, Baker D, Keasling JD. A bovel semi-biosynthetic route for artemisinin production using engineered substrate-promiscuous $P450_{BM3}$. ACS Chem Biol 2009;4:261–267.

39. Shiba Y, Paradise EM, Kirby J, Ro D, Keasling JD. Engineering of the pyruvate dehydrogenase bypass in *Saccharomyces cerevisiae* for high-level production of isoprenoids. Metab Eng 2007;9:160–168.

40. Ro D, Ouellet M, Paradise EM, Burd H, Eng D, Paddon CJ, Newman JD, Keasling JD. Induction of multiple pleiotropic drug resistance genes in yeast engineered to produce an increased level of anti-malarial drug precursor, artemisinic acid. BMC Biotechnol 2008;8:83.

41. Newman JD, Marshall J, Chang M, Nowroozi F, Paradise E, Pitera D, Newman KL, Keasling JD. High-level production of amorpha-4,11-diene in a two-phase partitioning bioreactor of metabolically engineered *Escherichia coli*. Biotechnol Bioeng 2006;95:684–691.

42. de Bont J. Solvent-tolerant bacteria in biocatalysis. Trends Biotechnol 1998;16:493–499.

43. Nielsen DR, Leonard E, Yoon S, Tseng H, Yuan C, Prather KLJ. Engineering alternative butanol production platforms in heterologous bacteria. Metab Eng 2009;11:262–273.

44. Chang C, Cen P, Ma X. Levulinic acid production from wheat straw. Bioresour Technol 2007;98:1448–1453.

45. Martin CH, Prather KLJ. High-titer production of monomeric hydroxyvalerates from levulinic acid in *Pseudomonas putida*. J Biotechnol 2009;139:61–67.

46. O'Connor SE, Maresh JJ. Chemistry and biology of monoterpene indole alkaloid biosynthesis. Nat Prod Rep 2006;23:532–547.

47. van Der Heijden R, Jacobs DI, Snoeijer W, Hallard D, Verpoorte R. The *Catharanthus* alkaloids: pharmacognosy and biotechnology. Curr Med Chem 2004;11:607–628.

48. de Waal A, Meijer AH, Verpoorte R. Strictosidine synthase from *Catharanthus roseus*: purification and characterization of multiple forms. Biochem J 1995;306:571–580.

49. de Luca V, Marineau C, Brisson N. Molecular cloning and analysis of cDNA encoding a plant tryptophan decarboxylase: comparison with animal dopa decarboxylases. Proc Natl Acad Sci USA 1989;86:2582–2586.

50. Sprenger GA. From scratch to value: engineering *Escherichia coli* wild type cells to the production of L-phenylalanine and other fine chemicals derived from chorismate. Appl Microbiol Biotechnol 2007;75:739–749.

51. Ikeda M. Towards bacterial strains overproducing L-tryptophan and other aromatics by metabolic engineering. Appl Microbiol Biotechnol 2006;69:615–626.

52. Herrmann K, Weaver L. The shikimate pathway. Annu Rev Plant Physiol Plant Mol Biol 1999;50:473–503.

53. Dosselaere F, Vanderleyden J. A metabolic node in action: chorismate-utilizing enzymes in microorganisms. Crit Rev Microbiol 2001;27:75–131.

54. Krämer M, Bongaerts J, Bovenberg R, Kremer S, Müller U, Orf S, Wubbolts M, Raeven L. Metabolic engineering for microbial production of shikimic acid. Metab Eng 2003;5:277–283.

55. Sariaslani FS. Development of a combined biological and chemical process for production of industrial aromatics from renewable resources. Annu Rev Microbiol 2007;61:51–69.

56. Gunsalus R, Yanofsky C. Nucleotide sequence and expression of *Escherichia coli* trpR, the structural gene for the trp aporepressor. Proc Natl Acad Sci USA 1980;77:7117–7121.

57. Pittard J, Camakaris H, Yang J. The TyrR regulon. Mol Microbiol 2005;55:16–26.

58. Báez-Viveros JL, Osuna J, Hernández-Chávez G, Soberón X, Bolívar F, Gosset G. Metabolic engineering and protein directed evolution increase the yield of L-phenylalanine synthesized from glucose in *Escherichia coli*. Biotechnol Bioeng 2004;87:516–524.

59. Berry A. Improving production of aromatic compounds in *Escherichia coli* by metabolic engineering. Trends Biotechnol 1996;14:250–256.

60. Shukla AK, Shasany AK, Gupta MM, Khanuja SPS. Transcriptome analysis in *Catharanthus roseus* leaves and roots for comparative terpenoid indole alkaloid profiles. J Exp Bot 2006;57:3921–3932.

61. Murata J, Bienzle D, Brandle JE, Sensen CW, Luca VD. Expressed sequence tags from Madagascar periwinkle (*Catharanthus roseus*). FEBS Lett 2006;580: 4501–4507.

62. Rischer H, Orešič M, Seppänen T, Katajamaa M, Lammertyn F, Ardiles W, Montagu MCEV, Inzé D, Oksman-Caldentey K-M, Goossens A. Gene-to-metabolite networks for terpenoid indole alkaloid biosynthesis in *Catharanthus roseus* cells. Proc Natl Acad Sci USA 2006;103:5614–5619.

63. Sato F, Inui T, Takemura T. Metabolic engineering in isoquinoline alkaloid biosynthesis. Curr Pharm Biotechnol 2007;8:211–218.

64. Datamonitor. Forecast insight: opioids—saturation limits the commercial potential of individual brands; 2009.

65. Hawkins KM, Smolke CD. Production of benzylisoquinoline alkaloids in *Saccharomyces cerevisiae*. Nat Chem Biol 2008;4:564–573.

66. Minami H, Kim J-S, Ikezawa N, Takemura T, Katayama T, Kumagai H, Sato F. Microbial production of plant benzylisoquinoline alkaloids. Proc Natl Acad Sci USA 2008;105:7393–7398.

CHAPTER 8

TOWARD GREENER THERAPEUTIC PROTEINS

SA V. HO and JOSEPH M. McLAUGHLIN
Pfizer Biotherapeutics Pharmaceutical Sciences, Chesterfield, Missouri

JAMES POLLOCK and SUZANNE S. FARID
Department of Biochemical Engineering, University College London, London, UK

8.1 INTRODUCTION

With the advent of molecular biology and supported by innovative development in large-scale bioprocessing technology, biotherapeutics—biological compounds used for treating diseases—have emerged in the last two decades as an important class of drugs and are now an integral part of product portfolios in most, if not all, major pharmaceutical firms. Biotherapeutics complement small-molecule drugs by expanding accessible targets and, for many indications, provide uniquely effective therapies. However, they span a very broad range of compounds, including peptides, proteins, fusion proteins, antibodies and their fragments, nucleotides, and many forms of vaccines.

According to a review by Walsh [1], out of 165 biopharmaceutical products approved in the United States and Europe by 2006 only two are nucleic-acid–based drugs, whereas nine of the 31 therapeutic proteins approved since 2003 are produced in *Escherichia coli*, and 17 are produced by mammalian cell lines. In the 2004 market distribution and manufacturing of therapeutic proteins, nonglycosylated (nonantibody) proteins constituted 40% of total market with a 12% annual growth rate and were produced in *E. coli* or the yeast *Saccharomyces cerevisiae*; glycoproteins (primarily monoclonal antibodies or mAbs) constituted 60% of total market with a 26% annual growth rate and were produced by mammalian cell culture (mostly with cells from Chinese Hamster ovary, or CHO).

Biocatalysis for Green Chemistry and Chemical Process Development, First Edition.
Edited by Junhua (Alex) Tao and Romas Kazlauskas.

Clinically, therapeutic proteins have contributed essential therapies to critical diseases, many life-threatening, including diabetes (insulin), end-stage renal disease (erythropoietin), viral hepatitis (interferon or IFN), cancer (trastuzumab for metastatic breast cancer, bevacizumab for metastatic colorectal cancer, I-131 ch-TNT for advanced lung cancer), growth anomaly (human growth hormone and its antagonist), clotting disorders (Factor VII, VIII, IX), rheumatoid arthritis (anakinra), multiple sclerosis (IFN-β1a and 1b), and inborn errors of metabolism (lysosomal enzymes) [1–3]. Therapeutic vaccines represent an emerging area in which biologics are used to treat infectious diseases, autoimmune diseases, and cancer; for example, Gardasil® is a recently approved cervical cancer vaccine [4]. mAbs have been extensively employed and hold great promise as therapeutic agents for their highly specific binding to cellular receptors as well as for their integral roles in the body's immune system. Examples of important therapeutic mAbs on the market include Herceptin® for treatment of breast cancer, Rituxan® for treatment of B-cell lymphoma, and Remicade® for treatment of rheumatoid arthritis [5].

Not only do therapeutic proteins dominate the current market of pharmaceutical biologics, but they are also produced by biological processes as opposed to chemical synthesis, which is the method of choice for production of peptides and oligonucleotides. While biological processes are generally considered natural and, therefore, inherently green, an overall assessment of the area is warranted, given its growing importance, which is the focus of this chapter for therapeutic proteins as a group.

8.2 CHARACTERISTICS OF THERAPEUTIC PROTEINS

Highly diverse in properties and manufacturing processes, therapeutic proteins have molecular weights typically ranging from around 10 to 200 kDa or even much higher, and are produced by fermentation using primarily microbes or mammalian cells. Therapeutic proteins can be loosely categorized into two main groups: mAbs and nonantibody proteins.

mAbs designate a general class of compounds with defined structure and molecular weight typically around 150 kDa, and are produced by cell culture using mammalian cells. Figure 8.1 shows the general structure for IgG (immunoglobulin G), a common class of therapeutic mAbs. Each IgG molecule comprises a disulfide-linked pair of identical heavy chains, each intertwined with and also disulfide-linked to a light chain. The chemical (amino acid sequence) change from one mAb to another is primarily in the variable regions where impurities, toxins, or foreign substances (called antigens) are bound with high specificity. mAbs may have sugar groups attached at the position shown in the Fc region by a process called glycosylation, which can provide important biologic functions and increase the heterogeneity of the product.

Unlike mAbs, nonantibody proteins represent a very large and diverse group with highly variable properties depending upon their original sources and biological functions. Recombinant proteins produced in microbes, mostly *E. coli* and

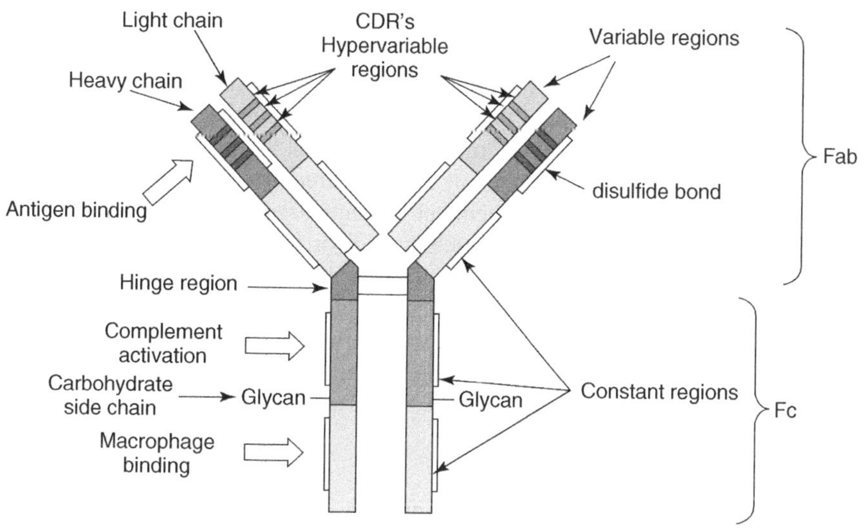

Figure 8.1 General structure of IgG monoclonal antibodies. *Source:* Reproduced with permission from Wiley-VCH.

some in yeasts, can vary widely in size, charge, hydrophobicity, and conformation. Additionally, bioactivity may require the target protein to be in its multimeric forms, that is, dimer or larger. The variation in size, for instance, ranges from 5.6 kDa for insulin and 22 kDa for human growth hormone to several millions for KLH (keyhole limpet hemocyanin, a large, multisubunit, oxygen-carrying protein in the giant keyhole limpet) and virus-like particles (VLP). Each VLP is actually a great assembly of >100 monomeric protein units. Other emerging therapeutics that could involve proteins as part of the drugs include conjugated proteins, such as those attached to a polyethylene glycol (PEG) molecule, and therapeutic vaccines in which protein antigens are used to generate immune responses from the body against certain diseases.

Not included in the above description is an important group of therapeutic compounds called peptides. While made up from the same pool of 20 amino acids as proteins, peptides are smaller than typical proteins, composed of approximately 20–40 amino acids, with overall molecular weight typically below 5000 Da, and produced by chemical synthesis, primarily via solid-phase synthesis. The manufacture of peptides is thus closer to that of small molecules and, from a manufacturing standpoint and environmental assessment, should be treated as such.

8.3 GENERAL FEATURES OF THERAPEUTIC PROTEIN MANUFACTURE

Manufacturing processes for therapeutic proteins can vary greatly, especially for nonantibody proteins. They, however, share some common features that differ significantly from those of small molecules. The general scheme for protein manufacturing is shown in Table 8.1. It typically involves a product synthesis step (bacterial fermentation or mammalian cell culture) followed with a series of processing steps to recover and purify the protein of interest, commonly called downstream processing (DSP). A major requirement is the need to purify the target protein from a large number of different impurities present in the post-fermentation solution, including chemical reagents used in the process, host cell components (proteins, deoxyribose nucleic acids or DNA), and even the various altered forms of the protein product itself. This diverse mixture of impurities is the main reason for the complex DSP, which typically constitutes the bulk of the overall manufacturing process for therapeutic proteins.

Typical raw materials and processing reagents used in the manufacture of therapeutic proteins are shown in Table 8.2. Owing to the complex machinery

TABLE 8.1 General Process Scheme for Therapeutic Protein Manufacturing

Product Synthesis

- Fermentation with bacteria (e.g., *E. coli*), yeasts or fungi for nonantibody proteins
- Mammalian cell culture for production of antibodies

Downstream Processing

- Isolation/recovery
  - Product in fermentation broth: cells and solid removal, volume reduction
  - Product inside cells:
    - Soluble form: cell disruption, solids removal, volume reduction
    - Insoluble form (inclusion bodies): homogenization, differential centrifugation, wash, dissolution
- Purification/reaction
  - *Bulk and intermediate purification*: primarily for removal of process-related impurities, for example, reagents, host cell proteins, DNA, endotoxins; some product-related impurities; common methods:
    - Precipitation, adsorption, extraction
    - Chromatography (bind/elution, flowthrough)
  - *Ultrafiltration/diafiltration (UF/DF)*: used as needed for product concentration (volume reduction) and buffer exchange (prepared for next step or for storage)
  - *Reaction:* used at an appropriate point in purification train for conversion to bioactive forms (e.g., refold/oxidation, dimer formation, PEGylation)
  - *Polishing*: final purification step (invariably using chromatography) to remove close product-related impurities, residual host cell proteins (HCPs) and endotoxins.
  - *Final UF/DF and sterile filtration*: concentration and buffer exchange for long-term product storage or preparation for drug product formulation

Source: Reproduced with permission from Wiley-VCH.

TABLE 8.2 Typical Raw and Processing Materials Used in Manufacture of Therapeutic Proteins

Fermentation: Product synthesis

- Water, inorganic salts, caustic, and acids for pH adjustment and cleaning
- Carbon source (glucose), nitrogen source, complex protein source (yeast extract, serum, etc.), small amounts of organics, antifoam

Downstream Processing: Purification/Reaction

- Processing materials:
  - Water
  - Inorganic salts, bases, and acids: pH adjustment, chromatography column operations, cleaning, and buffer solutions for storage
  - Urea, detergents: enhance solubilization or minimize aggregation of certain proteins
  - C2–C5 alcohols and/or glycols for certain chromatography modalities (hydrophobic interactions, reversed phase)
  - Special organic solvents (e.g., CH_3CN): for postfermentation modification or conjugation reactions such as PEGylation
- Consumables:
  - Dead-end filters; disposable bags, tubing and connectors
  - Ultrafiltration/microfiltration membranes
  - Chromatographic resins

Source: Reproduced with permission from Wiley-VCH.

of biological cells, raw materials required for protein synthesis comprise mainly water, sugar, salts, a few trace minerals, and some supplements. Similarly, DSP uses mostly water and salts in buffer solutions, and consumables such as filters and chromatography resins. Very little organic solvents such as alcohols are used, if at all, and they tend to be nonhazardous.

One major characteristic of the manufacture of biologics is the extensive use of water. Concentrations of the protein product formed in the fermenter or bioreactor, called titers, are typically from 1 to 5 g/L. Titers above 10 g/L are considered highly productive and yet to be routinely achieved; 10 g/L is, however, equal to only 1 wt% of the solution, which means that roughly 100 kg of water is already required per kilogram of unprocessed protein in the fermentation broth. Two commonly used unit operations in bioprocessing—column chromatography and ultrafiltration/diafiltration—also happen to consume large amounts of water. Shown in Table 8.3, typical water usage for these two units can range from 100 to 1000 kg water/kg product for each step. Actual water usage on the basis of the purified protein weight would be even higher because of product loss occurring in these steps as well as in the rest of the process.

TABLE 8.3 Water Usage for Two Common Purification Unit Operations

	Typical Operating Range			
Chromatography Column[a]				
Resin loading, g protein/L resin	10	20	50	100
kg water/kg product	1000	500	200	100
Ultrafiltration/diafiltration[b]				
Protein concentration, g/L	10	20	50	100
kg water/kg product	1000	500	200	100

[a]Chromatography column: number of column volumes (CV) of buffer solution used is assumed to be 10; typical range is 10–20 CVs.

[b]Ultrafiltration/diafiltration: number of turn-over volume (TOVs) of buffer solution used is assumed to be 10; typical range is 5–20 TOVs.

Source: Reproduced with permission from Wiley-VCH.

8.4 ENVIRONMENTAL ASSESSMENT CONSIDERATIONS

A widely adopted concept called the E factor originally developed by Sheldon for analyzing the overall greenness of production in the chemical industry [6–8] could provide a useful index for manufacture of therapeutic proteins. E factor is defined as the total amount in kilograms of organic solvents, reagents, and consumables used per kilogram of the product produced. At the simplest level, the E factor for process water usage could serve as an excellent environmental index for production of therapeutic proteins simply because every manufacturing step involves aqueous solutions that both contain chemical reagents and are processed with tubings, filters, membranes, and resins. However, there are at least four types of water used in a bioprocessing plant: potable, purified, highly purified, and water for injection or WFI, WFI representing the highest in terms of material and energy consumption. For instance, it takes 1 kg of potable water to generate 0.8 kg of WFI. The E factor for water should therefore be weighted with respect to the type of water used. Also, a great deal of water is required for nonprocess operations (e.g., cleaning, sterilization, steam) at a bioprocessing plant; an E factor for nonprocess water usage is thus warranted to help monitor the greenness of this part of the operation.

Useful environmental indices for the manufacture of biologics should also include waste generation and energy consumption, and, for waste generation, would include aqueous wastes and solid wastes. Aqueous wastes are generated from processing, cleaning, and sterilization operations; these wastes are considered innocuous and discharged as municipal wastes after being subjected to biological kill and pH neutralization. Solid wastes include filters, membranes, resins, tubings, and disposable equipment; they are treated as biohazards and are typically autoclaved and then land filled or incinerated. Last but not least, energy consumption would be an important environmental index and should differentiate between energy used in the manufacturing process along with its supporting operations and that for facility maintenance.

From an environmental sustainability standpoint, two main variables have been proposed in other industries to provide a balanced view of the environmental impact of inputs (resource usage) and outputs (emissions, effluents and wastes, and the products and services produced) [9]. They are termed natural resources sustainability (NRS) and environmental burden sustainability (EBS). NRS is based on evaluation of four main variables: energy, materials (excluding fuel and water), water, and land. These variables are related to economical considerations depending upon the different markets. EBS addresses pollutants as they are released to the atmosphere (emissions), water (effluents), and soil (wastes). Each substance is tied to a weighting factor known as the "potency factor" that draws on developments in environmental science to estimate the potential environmental impact rather than being based solely on the quantities of the materials discharged. The following equation has been proposed [9]:

$$\mathrm{EB}_i = \sum W_N \mathrm{PF}_{i,N}$$

where EB_i is the ith environmental burden, W_N the weight of substance N emitted, including accidental and unintentional emissions, and $\mathrm{PF}_{i,N}$ the potency factor of substance N for the ith environmental burden. Normalization procedure was also carried out in which the threshold value for release of each type of substance is used to normalize its total release for comparison with other types.

8.5 THERAPEUTIC PROTEIN MANUFACTURING PROCESS ASSESSMENT

Many factors affect the process design for manufacturing therapeutic proteins; they include protein type and size, production scale, and the type of host cells used. Consideration of the environmental impact is focused on two major groups: nonantibody proteins produced by microbial cells and mAbs produced by mammalian cells.

8.5.1 Microbially Produced Proteins

Recombinant proteins are produced by fermentation using primarily bacterial cells, mostly *E. coli* and some by using yeasts. The manufacturing processes are highly variable in complexity, primarily because DSP has to adapt to the particular host expression system and the properties of the target protein itself [10–14]. They range from the very complex production process for insulin to a typical process for medium-sized proteins and a well-optimized process for a mature product in large-scale commercial production.

Human insulin is a small protein consisting of 51 amino acids with a molecule weight of 5734 and isoelectric point (pI) of 5.4. Insulin is made up of two peptide chains connected by two disulfide bonds: chain A with 21 amino acids and chain B with 30 amino acids. This small recombinant protein has been produced on very large scales for over two decades. The various commercial processes in production have been improved over the years but are still very complex [15–18]. The insulin manufacturing process discussed here is similar to the Eli Lilly's commercial process and is taken from a textbook on bioprocessing [18]. The process consists of a fermentation step to produce proinsulin and a highly complicated DSP train to recover proinsulin from the *E. coli* cells in the fermenter, convert it to insulin, and then to purify it to meet the required product quality for use in humans. For each kilogram of insulin produced, an enormous amount of process water is consumed (>30,000 kg) along with >4000 kg of organic solvents, some of which are hazardous. Also about 15 kg of consumables (solid processing aids) are used per kilogram of insulin produced. These values, especially for water and solvents usage, are very high and represent a unique, extreme case for the manufacture of biologics because of the length of the purification process and the required complex reaction steps that are unusual in biologics processing.

A typical manufacturing process for a medium-sized protein produced by microbial cells would have a fermentation titer of 1–5 g/L and would consist of three to four chromatography with two to three UF/DF (ultrafiltration/diafiltration) steps and up to one reaction step. The overall yield of this process from fermentation to Active Pharmaceutical Ingredients (API) would range from 15% to 30% with no recycle or recovery of used materials. The water usage for this typical process is estimated to range from 10,000 to 20,000 kg for every kilogram of protein produced. The amounts of organic solvents could be substantial if certain purification steps such as reversed phase chromatography are used. Some hazardous solvents may be used to carry out chemical reactions such as conjugating another molecule, such as PEG, to the protein to enhance its specificity and/or stability. For consumables, the amount ranges form 10 to 30 kg/kg protein product. Glucose is the main raw material for cell growth and urea is commonly used for solubilizing inclusion bodies and in assisting the conversion of a protein to its active conformation.

Greener approach in manufacturing would strive toward a minimum number of DSP steps with an associated high overall yield and recycle of water and key processing materials. The commercial production of bovine somatotropin (BST), a growth hormone for increased milk production in dairy cows, represents a real-life case in which these benefits were realized. In addition to high titers (5–10 g/L broth), the protein is produced as inclusion bodies in the cells, which facilitates their recovery with relatively high purity from the fermenter with simple homogenization and differential centrifugation. After dissolution of inclusion bodies followed by a simple refold step to form the bioactive BST, the purification consists primarily of a precipitation step to remove the bulk of the impurities and then a single chromatography column as a polishing step. As shown in Table 8.4, the amount of water used is reduced to <500 kg/kg of BST produced, and with very little urea consumption (Storrs B, Gibb G. Monsanto company, personal communication). Note that these are the values obtained with water/urea recycling. While there is no cost driver for recycling process water, the imposed environmental constraints on discharge of urea necessitate its recycle as aqueous

TABLE 8.4 Water/Materials Usage for the Commercial Bovine Somatotropin (BST) Manufacturing Process [19]

	kg/kg BST
Materials	
Glucose	96
Salts	8
(Reverse Osmosis) Water[a]	454
Urea[a]	26
Consumables	
Fermentation filters	3
Aseptic filters	1
Chromatographic resin	~0.1
UF membrane	~0.1
Total consumables	~4

[a] With urea and water recycle.

Source: Reproduced with permission from Wiley-VCH.

solutions, resulting in water itself being recycled as well. The consumables used are also quite low, about 4 kg/kg of BST. This real-life example demonstrates that a highly optimized biologics production process with incorporated recycle of materials can drastically reduce both usage of water and generation of chemical and solid wastes.

8.5.2 Monoclonal Antibodies and Mammalian Cell Culture Processes

From the manufacturing standpoint, three main characteristics differentiate the production of mAbs by mammalian cells from production of nonantibody proteins by microbial cells. First, the fermentation process during which the protein product is formed is much longer for mammalian cells than for microbial cells, for instance, about 14 days versus 2 days for *E. coli*. Second, the protein produced by mammalian cells is generally secreted into the culture medium, negating the need for cell lysis to recover the product and thus avoiding release of host cell components into the culture solution. Finally, IgG antibodies bind selectively to protein A, enabling the use of a protein A affinity chromatography step to capture mAbs with high yield and purity from the clarified fermentation broth.

Many excellent reviews on mAb manufacturing have been published [19–22]. Given the common properties of mAbs noted above, their manufacturing processes have become more and more standardized over the years with enhanced efficiency. Typically, after the fermentation step (called cell culture for mammalian cells), the cells are removed by centrifugation and depth filtration to obtain the clarified broth containing the product protein. The traditional purification process consists of three bind/elution chromatography columns: a Protein A affinity column where the mAb product is concentrated and host cell protein (HCPs) and genetic components (DNA) along with cell culture media are removed, an ion-exchange column as an intermediate purification to further remove host cell impurities and aggregate forms of mAb, and finally another ion-exchange column (or hydrophobic interactions chromatography) as a polishing step to remove residual impurities. Viral clearance is a major issue with mammalian cell culture processes and is carried out with two different, orthogonal steps, as required by the Food and Drug Administration (FDA), typically a chemical virus inactivation step at low pH and a viral filtration step for physical removal of the viruses.

Manufacturing processes for mAbs span older ones still in production to newer and more optimized processes; the latter group typically consists of one to multiple bioreactors operating in parallel to deliver cell culture broth to a single purification train. The product formation step typically involves one to six bioreactors of up to 20,000 L each, with mAb titers from 2 to 5 g/L broth. DSP includes a 3-column purification train with the last column being either bind/release or flowthrough. Annual throughput ranges from 400 to 5000 kg mAb, with an overall yield from cell culture to purified mAb of around 65%–80%, without recycle or recovery of process chemicals. Typical mAb production processes consume water from over 3000 to almost 7000 kg/kg of mAb produced and consumables from 2 to 8 kg/kg of mAb. Water usage in the cell culture step makes up 20% to 25% of the total, whereas the three chromatography columns use >50% of the total. Furthermore, the large buffer amounts required as scale increases can result in WFI costs representing a surprisingly large proportion of the costs, which can, in some cases, be greater than the cost of other raw materials [21]. Hence, it is expected that strategies that lower buffer, and hence WFI, consumption will potentially lead to significant environmental and financial benefits.

Two current approaches are being pursued for significant process enhancement. In one, strong advocacy has been made for process intensification to handle highly productive cell culture titers of 10 g/L in a large-scale mAb manufacturing plant (10 ton of purified mAb per year) utilizing conventional unit operations [23]. Only two chromatography columns are used in the process and very high resin/membrane loadings are assumed. With this process, the amount of water used drops to 1500 kg/kg of mAb, which is about half of that for a typical mAb process. The total amount of consumables used (chromatography resins, prefilters, viral filters, and membranes), estimated from the data provided in the paper, is around 2 kg/kg mAb, with prefilters constituting 70% of the total weight because of their prevalent use in bioprocessing.

Another approach involves development of a continuous process, most notable of which is the perfusion bioreactor versus the traditional fed-batch bioreactor. Comparison of these two types from an environmental standpoint can yield different rankings depending on the cell densities, titers, pooling strategy, and the scale assumed. Although perfusion bioreactors are typically three- to fivefold smaller than fed-batch ones, the perfusion-based processes can consume more fermentation media because of the daily perfusion exchanges and potentially more cleaning buffers with the more frequent purification lots required to process the pooled daily harvests. As illustrated in Figure 8.2, the perfusion-based process can consume >35% more water and 50% more

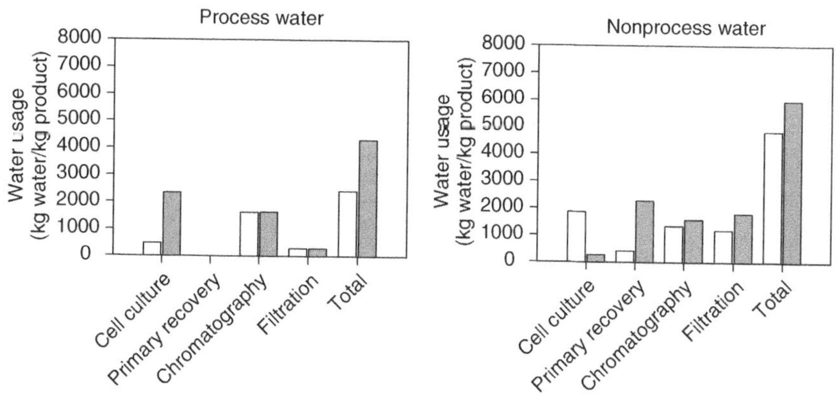

Figure 8.2 E factor for water usage for process water and nonprocess water on an activity basis for a 20,000 L fed-batch (□) and a 4500 L spin-filter perfusion (■) process producing the same annual output.

consumables per kilogram of mAb than fed-batch processes when a 4500 L perfusion-based process is considered equivalent to a 20,000 L fed-batch process. Interestingly, the two approaches can have similar COG (cost of goods) per gram values. However, with the recent introduction of new perfusion reactor technologies capable of achieving higher steady-state cell densities and enhanced perfusate clarification, perfusion-based processes offer the potential to become more environmentally friendly than fed-batch-based processes. For the perfusion option to have a more favorable E factor for water usage in the example in Figure 8.2, the new reactor technologies would have to be capable of producing at least a threefold increase in steady-state cell densities.

8.5.3 Overall Assessment

As noted, within each group (microbial and mammalian), less water is consumed as the process becomes more efficient, indicating that the E factor based on water usage would be a strong indicator of the degree of greenness of the manufacture of biologics. This reflects the fact that every processing step in the manufacture of biologics uses aqueous solutions, which in turn require chemicals (salts and buffers) and consumables (filters, resins, membranes, and disposable bags) for processing. Process water and materials used in the manufacture of both mAbs and nonantibody proteins for the cases discussed above are summarized in Table 8.5.

Total water usage at manufacturing plants, however, includes a great deal more operations than just direct process water to make products, such as in

- equipment cleaning: cleaning in place (CIP), sterilization in place (SIP)
- generation of WFI
- waste disposal (biowastes)

TABLE 8.5 Order-of-Magnitude Estimate of Process Water and Materials Used in Manufacture of Therapeutic Proteins

	Microbially Derived Proteins		mAbs from Cell Culture	
	A highly optimized, large-scale	Typical "composite" process	Optimized large-scale	Highly intensified, large-scale
	kg/kg API			
Water usage	<1000	15,000	4500	1500
Salts + buffers	1	400	300	100
Consumables (solid wastes)	1	20	4	2
Organic solvents	~0	100 (alcohols, may involve some hazardous solvents)	8 (alcohols)	8 (alcohols)

Source: Reproduced with permission from Wiley-VCH.

- treatment of biowaste streams (~1–2 kg steam needed/kg waste solution)
- facility maintenance (cleaning, cooling/heating, etc.)
- evaporative loss

Genentech published on its website [24] the average amount of water usage for all its manufacturing sites for the period 2004–2006. The numbers reported are on the order of several hundred thousand kilograms of water used per kilogram of protein produced. Approximate estimates for a Pfizer pilot plant and a small manufacturing facility appear to be in the same order of magnitude. These large numbers for water usage at a biotherapeutics manufacturing plant highlight the tremendous opportunity for reducing water consumption, and associated chemicals, through better plant design, more streamlined operations, and particularly, water recycle.

Some of the main observations gleaned from the analysis of the environmental characteristics of the manufacture of biotherapeutics are

1. very large amount of water is used in the manufacturing process;
2. significantly more water is used in supporting operations, such as generation of WFI, equipment cleaning and sterilization, wastes processing, and facility maintenance;
3. large amounts of common salts such as NaCl and acids/bases are used in processing, which all end up as salts in the aqueous waste discharge;
4. although the volume of liquid wastes generated is very high, the wastes are mostly aqueous and innocuous;

5. amounts of consumables (resins, membranes, filters, disposable bags, and tubings/connectors) that end up as solid wastes could be large.

In addition to water and consumables, energy is used in the production of biotherapeutics not only to operate the process but also to maintain a controlled environment, to produce clean utilities, and to clean and sanitize equipment. Often the major energy cost considered in economic analysis is the HVAC (heating ventilation and air conditioning) cost. HVAC is significant because process areas such as ISO 7 Grade B require as much as 50 air changes/h; further these areas must be surrounded by similar clean areas requiring 25 air changes/h, whereas a general purpose area requires only 4 air changes/year [25]. These controlled environments must be validated and maintained in continuous operation to assure that the product is not contaminated by the environment. Thus, different levels of environmental control and their associated HVAC requirement strongly affect the overall energy usage at a biotherapeutic manufacturing plant, and hence, costs [25]. A benchmarking study of cleanroom energy use in high-tech and biotech industries for three representative facilities shows that HVAC energy use ranges for almost 40% to >60% of the total energy, whereas energy for process and process utilities ranges from about 50% to <20% of the total [26]. More efficient design and operation of the controlled environment represents an important opportunity in moving toward a "greener" facility.

Clean Utilities such as WFI can be designed for very energy-efficient operation by using approaches such as vapor recompression for water distillation. Equipment cleaning and sanitization consume a significant amount of energy because they typically require high flow rates and large temperature changes. A typical biotherapeutic process uses some energy in the upstream operations for heat and mass transfer (agitation and sparging) and only modest amount of energy for fluid transport and mixing in downstream operations. Overall, an improved understanding of the use of energy in the manufacture of biotherapeutics as well as innovative development to minimize the requirement of cleanroom space itself could significantly reduce the environmental impact of facilities, especially at the earliest stages of architectural design.

Several emerging manufacturing technology areas with potential environmental impacts are listed in Table 8.6. They span from improvement of the existing systems (cell line and bioreactor optimization, process intensification, and better purification technologies) to single-use (disposable) manufacturing approach and alternative production platforms such as cell-free synthesis and transgenic plants or animals. Of these, single-use manufacturing will likely have a large impact in protein production in the near term, and some alternative production platforms, if successful, could potentially alter the landscape of protein manufacture in the long term. While a comprehensive environmental assessment of these two approaches is beyond the scope of this chapter, a brief review of their significance is warranted, which is addressed next.

TABLE 8.6 New Process Developments with Potential Environmental Impact

Process and analytical technologies

- Cell line and bioreactor optimization: increased titers and higher purity
- Host cell proteins: characterization and selective removal via genetic modification
- Continuous processing: for example, perfusion reactor; simulated moving bed
- Nonchromatographic separations: for example, membrane-based purification, selective extraction/precipitation, magnetically enhanced separations, self-processing proteins (such as intein) for enhanced separation
- Process Analytical Technology (PAT)

Single-use manufacturing

Alternative production platforms

- Cell-free synthesis
- Transgenic plants and animals

8.6 SINGLE-USE MANUFACTURING

The adoption of single-use equipment, first with some specific units such as filters and membranes for operational convenience, has now spread to practically every operation used in biotherapeutic manufacturing, including solid and liquid transfers, mixing, solution preparation and storage, bioreaction, chromatography, tangential and normal flow filtration, and freezing. In addition, single-use equipment has been introduced to facilitate tasks such as aseptic and sterile connections, sampling, and process monitoring. The major limitation of single-use equipment is related to the material of construction—plastic—which results in reduced capabilities for operating pressure, heat transfer, mass transfer and scale-up compared with fixed stainless steel equipment. With up to 2000-L scale single-use equipment demonstrated to be capable of meeting biotherapeutic process requirements, production is now possible using all disposable equipment, called single-use manufacturing.

The primary drivers toward single-use manufacturing are speedy reach to market to deliver therapeutic-quality proteins to patients [27] and economics with the potential of lower upfront investment costs and COG/g [28–30]. Studies comparing a single-use plant and a traditional one for mAb manufacture in terms of economics at scales ranging from 200 L to 2 × 5000 L and titers of 0.5–2 g/L indicate that single-use plants have the potential to reduce capital investment requirements by 33%–40% and cost of goods by almost 20%–30% [28,29]. The environmental impact has also been explored and, as expected, the total water usage and chemicals usage are found to reduce by half, primarily through less cleaning [30]. However, other analyses show significant variation in the savings depending on the specific application [31]. The economics varies significantly depending on where single-use components are used in the process and were found to be highly scale and titer dependent, single-use advantages being greatest at low titers and volumes [32].

Many advantages of single-use equipment for the manufacture of biotherapeutics are related to the elimination and simplification of ancillary manufacturing operations such as cleaning, sterilization, and product change-over. Fixed stainless steel equipment requires these additional operations to comply with current good manufacturing practices (cGMP) and assure product quality by controlling environmental, lot-to-lot, and product-to-product contamination. In addition to the cost of executing the manufacturing process and ancillary operations, fixed equipment requires significant regulatory support in the form of installation, operation, and performance qualifications required to provide evidence in support of facility validation, inspection, and licensing. The disposable aspect of single-use equipment eliminates the concern over lot-to-lot and product-to-product contamination.

Environmental assessment of single-use equipment takes a longer term and broader view than economic analysis. In the economic analysis on a 2 × 5000 L scale noted above [30], the waste generated from disposable plastic bags increases by almost 170 kg/kg of protein for the single-use process over the traditional stainless steel one. Additionally, the overall environmental assessment needs to take into account additional water and chemicals used by the equipment suppliers themselves in generating single-use equipment, including environmental contamination resulting from assembly and sterilization operations. Leveen & Cox carried this analysis further for the same case taking into account energy consumption that included manufacture and transport of plastic bags [33]. Expressing the overall energy consumption as carbon footprint, interestingly, they found that the single-use plant would use 35% less than the traditional steel plant.

Assessment of single-use environmental impact is further complicated by the multiple options for disposal: recycling, pyrolysis, incineration, incineration plus cogeneration, and landfill. Recycling has the least environmental impact but is complicated because many single-use components are constructed of multiple polymers. Plastic has a heat value 1.5 times that of coal so incineration plus cogeneration can reduce the environmental impact. However, many single-use components are simply incinerated or sent to a landfill because other options are not available or when the waste stream is not of sufficient magnitude [34,35]. Single-use environmental assessment should also include details such as sterilization techniques for single-use equipment that can include small waste streams containing hazardous materials such as Cobalt 60 for γ radiation generation. A comprehensive evaluation of single-use technology thus requires understanding of the equipment, manufacturing and regulatory operations, economic analysis, and environmental assessment.

8.7 ALTERNATIVE PRODUCTION PLATFORMS

A more drastic shift in the ways therapeutic proteins are produced involves neither microbial nor mammalian cells directly; these alternative production platforms include cell-free synthesis [36–38], transgenic plants [39–43], and transgenic animals [44–46]. In cell-free synthesis, cells are grown primarily to harvest their metabolic and protein production machinery (e.g., ribosomes, RNAs, enzymes, reducing and oxidizing factors) for "chemically" synthesizing the protein of interest from simple raw materials. Significant progress has been made with this production method, moving from simple proteins initially to complex entities such as antibody fragment bioconjugates and VLP. It is possible that cell-free synthesis, if properly designed, may reduce water and chemicals usage and achieve a higher production yield compared to the corresponding fermentation-based process. However, owing to many similarities between the two production means, especially in DSP, cell-free approach is unlikely to result in major beneficial environmental impacts.

Plants genetically modified to express foreign proteins—called transgenic plants—have been developed to produce various protein products in many plant species, such as maize [39,47–49], rice [50], barley [51], alfalfa [52], and tobacco [53,54], to name a few. The protein products have also been expressed in various parts of the plants, such as leaves, stems, seeds, and even roots. The production process in this case involves harvesting the plants and extracting the product from the collected biomass; the typical protein purification operation that follows is not unlike that for proteins produced by fermentation [55–61], perhaps somewhat simpler. The actual synthesis part is obviously different, typically done in the field or in a greenhouse requiring less complex facilities and operations compared to those associated with the traditional manufacture of therapeutic proteins. Thus, protein production using transgenic plants can be less resource intensive and is claimed to have much better economics, with COG at least an order of magnitude lower than traditional fermentation-based processes [59]. However, environmental containment of plant production systems, such as possible genetic exchange with unintended plants and other potential ecological effects, is a critical issue that needs to be properly addressed. Growing aquatic plants such as duckweed in a bioreactor-like environment is closer to the traditional microbial or mammalian cell production methods. In this case, because of the requirement of light for growth and production, aquatic plants tend to grow near the surface, where they are exposed to the light source and consequently do not utilize the full liquid volume in the reactor as microbial and mammalian cells do. So unless they offer great advantages in DSP, aquatic plants will probably retain a small niche in the production of hard-to-make therapeutic proteins [62].

An intriguing emerging application of transgenic plants is in the production of vaccines by plants [63], even in edible forms such as seeds and fruits creating "edible" or "oral" vaccines [64–70]. Plants in which expression of protein antigens with high food value has been found, many of which can be eaten raw, include apple, banana, tomato, and guava (fruits); peanut, corn, soybean, and chickpea (seeds); and cabbage, lettuce, potato, and spinach (vegetables) [71–73]. Plant seeds, in particular, represent a very promising target for producing therapeutic proteins because of the potential long-term stability of the expressed proteins in addition to the possibility of direct consumption as edible vaccines. Single-chain antibodies expressed in seeds of rice and wheat were found to maintain biological activities for years [71]. Compared to traditional vaccines, edible vaccines could offer much lower production cost, convenient storage and transportation (ambient temperatures, no refrigeration), simplicity of use, economic administration, and mucosal immune response. Important technical and logistic problems remain, however, to be addressed in order to make edible vaccines a reality. These challenges include enhanced expression level, yield and quality of the target protein; standardization of dosage; conformation to Good Laboratory Practice (GLP) and Good Manufacturing Practice (GMP) protocols; and the need to establish safety, efficacy, and functional equivalence of the vaccine antigens [63].

Analogous to transgenic plants, transgenic animals are those genetically modified to produce foreign proteins, typically in their milk. This has been successfully demonstrated for over 20 different proteins in cows, goats, sheep, pigs, rabbits, and mice [74–76], and even in insects such as silk worms [77]. Transgenic animals appear particularly promising for the production of large amounts of therapeutic proteins including mAbs [74,78]. Specific examples include antithrombin III in goat milk for preventing blood clotting [79], lysostaphin in mouse milk for preventing *Staphylococcus aureus* infection [80], human growth hormone in the milk of guinea pigs [81]. Goat herds were reported to be developed by Genzyme Transgenics for large-scale production of the tumor necrosis inhibitory mAb Remicade, marketed by Centocor for treating inflammatory conditions such as Crohn's disease and rheumatoid arthritis [82]. Similar to transgenic plant production systems, while the upstream process (product synthesis) with transgenic animals is quite different from that of fermentation-based processes, DSP is similar but somewhat simpler [60]. Key technical issues facing the commercial production of therapeutic proteins using transgenic animals include efficiency and speed of producing a commercial product (genetic modification and breeding to establish the herd), effective handling of infectious diseases, and potential differences in posttranslational characteristics compared to those produced by mammalian cell culture [82–84].

8.8 SUMMARY

Adopting the E factor concept widely used in the chemical industry to measure the environmental efficiency (greenness) of a manufacturing process, we estimated that production of therapeutic proteins using fermentation processes requires approximately 1000 to >10,000 kg of water/kg of product made, but does not use much solvents, especially hazardous ones. Thus, while this index factor for water usage is quite high for therapeutic proteins—about 10–100 times higher than that for small-molecule drugs, for instance—the aqueous wastes generated are mostly innocuous. The E factor for process water represents an appropriate environmental index for production of therapeutic proteins simply because every processing step is carried out in aqueous solutions. Significant reduction in water usage along with chemicals and consumables could be achieved with process enhancement or simplification as well as recycling of materials; the latter is rarely implemented in protein manufacturing because of lack of an economic driver but could play an important role if waste disposal becomes an issue. Significantly much more water is, however, consumed for nonprocess operations at bioprocessing plants, which suggests consideration of an E factor for nonprocess water to help monitor the environmental efficiency of this part of plant operation. Useful environmental indices for the manufacture of biologics should also include waste generation and energy consumption, the latter of which is reportedly dominated by facility operations, especially cleanroom or controlled space because of the required HVAC for its operation.

Important emerging technologies for therapeutic protein production span from improvement of the existing systems and single-use (disposable) manufacturing to alternative production platforms such as cell-free synthesis and transgenic plants or animals. The single-use practice is gaining popularity with manufacturers of biologics because of advantages that include reduced capital infrastructure, increased operational flexibility, and significant reduction in water and chemicals usage. However, more plastic wastes will be generated, which, if not recycled, have to be disposed of via either landfill or incineration. A comprehensive evaluation of single-use technology requires understanding of the equipment, manufacturing and regulatory operations, economic analysis, and environmental assessment.

As alternative production platforms for therapeutic proteins, transgenic crops and transgenic animals could be potentially game-changing, especially for plant-derived oral or edible vaccines, which could greatly simplify production, transportation, storage, and administration of the drugs. However, environmental assessment for these modes of production represents a new and highly complex area because of their potential multidimensional impact that could involve chemical, biochemical, and genetic effects.

ACKNOWLEDGMENTS

The authors would like to thank the following colleagues in Pfizer Global Biologics for their helpful inputs and discussions: Andrew C. Espenschied, Robert E. Kottmeier, James F. Bouressa, and Ferhana Zaman. We also acknowledge the inputs from Brad Storrs and Greg Gibb of Monsanto, Brian Kelley of Wyeth (now with Genentech), and Lindsay Leveen of Genentech. Particular acknowledgement is made to Peter Dunn, Pfizer green chemistry lead, and Berkeley W. Cue, Jr., former cochair ACS GCIPR, for their strong encouragement and support in conducting this work. The support of the UK Engineering and Physical Sciences Research Council (EPSRC) and The Advanced Centre for Biochemical Engineering at University College London (UCL) is also gratefully acknowledged.

REFERENCES

1. Walsh G. Biopharmaceutical benchmarks 2006. Nat Biotechnol 2006;24(7):769–778.
2. Dingermann T. Recombinant therapeutic proteins: production platforms and challenges. Biotechnol J 2008;3(1):90–97.

3. Reichert JM, Valge-Archer VE. Development trends for monoclonal antibody cancer therapeutics. Nat Rev Drug Discov 2007;6:349–356.
4. Rothengass BE. Advocating for the quadravalent HPV vaccination, Gardasil. Int J Pediatr Otorhinolaryngol 2007;71(4):671–672.
5. Reichert JM, Rosensweig CJ, Faden LB, Dewitz MC. Monoclonal antibody successes in the clinic. Nat Biotechnol 2005;23:1073–1078.
6. Sheldon RA. Consider the environmental quotient. Chemtech 1994;24:38.
7. Sheldon RA. Catalysis: the key to waste minimization. J Chem Technol Biotechnol 1997;68:381.
8. Sheldon RA. The E factor: fifteen years on. Green Chem 2007;9:1263–1272.
9. Irabien A, Aldaco R, Dominguez-Ramosa A. Environmental sustainability normalization of industrial processes. In: Jezowski J, Thullie J, editors. 19th European Symposium on Computer Aided Process Engineering—ESCAPE 19. Amsterdam: Elsevier Science B.V; 2009. p. 26.
10. Swartz JR. Advances in *Escherichia coli* production of therapeutic proteins. Curr Opin Biotechnol 2001;12(2):195–201.
11. Andersen DC, Krummen L. Recombinant protein expression for therapeutic applications. Curr Opin Biotechnol 2002;13:117–123.
12. Jana S, Deb JK. Strategies for efficient production of heterologous proteins in *Escherichia coli*. Appl Microbiol Biotechnol 2005;67(3):289–298.
13. Graumann K, Premstaller A. Manufacturing of recombinant therapeutic proteins in microbial systems. Biotechnol J 2006;1(2):164–186.
14. Chou CP. Engineering cell physiology to enhance recombinant protein production in *Escherichia coli*. Appl Microbiol Biotechnol 2007;76(3):521–532.
15. Chance RE, Glazer NB, Wishner KL. Insulin Lispro (Humalog). In: Walsh G, Murphy B, editors. Biopharmaceuticals, an industrial perspective. The Netherlands: Kluwer Academic Publishers; 1999. pp. 149–171.
16. Ladisch MR, Kohlmann KL. Recombinant human insulin. Biotechnol Prog 1992;8(6):469–478.
17. Walsh G. Therapeutic insulins and their large-scale manufacture. Appl Microbiol Biotechnol 2005;67(2):151–159.
18. Harrison RG, Todd P, Rudge SR, Petrides DP. Volume 11, Bioseparations science and engineering. Oxford: Oxford University Press; 2003. pp. 349–362.
19. Shukla AA. Downstream processing of monoclonal antibodies - application of platform approaches. J Chromatogr B 2007;848(1):28–39.
20. Low D, O'Leary R, Pujar NS. Future of antibody purification. J Chromatogr B 2007;848(1):48–63.
21. Farid SS. Process economics of industrial monoclonal antibody manufacture. J Chromatogr B 2007;848(1):8–18.
22. Farid SS. Established bioprocesses for producing antibodies as a basis for future planning. Adv Biochem Eng Biotechnol 2006;101:1–42.
23. Kelley B. Very large scale monoclonal antibody purification: the case for conventional unit operations. Biotechnol Prog 2007;23(5):995–1008.
24. http://www.gene.com/gene/about/environmental/commitment/water.jsp
25. Monk J. Exploring new paradigms for bioprocessing. CEP; July 2009: 19–22.

26. Barnoon BI, Bader B. Lifecycle cost analysis for single-use systems. BioPharm Int Guide 2008:30–43.
27. Tschudi W, Benschine K, Rumsey P. Cleanroom energy benchmarking in high-tech and biotech industries. Presented at the 2001 ACEEE Summer Study on Energy Efficiency in Industry; 2001 July 24–27; Tarrytown, New York. 2001, and published in the Proceedings.
28. Novais JL, Titchener-Hooker NJ, Hoare M. Economic comparison between conventional and disposables-based technology for the production of biopharmaceuticals. Biotechnol Bioeng 2001;75(2):143–153.
29. Farid SS, Washbrook J, Titchener-Hooker NJ. Decision-support tool for assessing biomanufacturing strategies under uncertainty: stainless steel versus disposable equipment for clinical trial material preparation. Biotechnol Prog 2005;21(2):486–497.
30. Sinclair A. A universal cost model for single-use systems in biomanufacturing. Presented at Bioproduction; 2007 Oct 29; Berlin, Germany.
31. Sandstrom C. Disposable vs. traditional equipment - a facility-wide view. CEP; 2009. pp. 30–35.
32. Sinclair A, Monge M. Disposables cost contributions: a sensitivity analysis. Biopharm Int 2009; 22(4).
33. Leveen L, Cox S. Bigfoot or Cinderella—Single use or traditional facilities. Presented at IBC Conference; 2008 Jun 2; Santa Clara (CA).
34. Rawlings B, Pora H. A prescriptive approach to management of solid waste from single-use systems. Bioprocess Int 2009;7(3):40.
35. Rawlings B, Pora H. Environmental impact of single-use and reusable bioprocess systems. Bioprocess Int 2009;7(3):40.
36. Spirin AS, Swartz JR. Cell-free protein synthesis systems: historical landmarks, classification, and general methods. In: Spirin AS, Swartz JR, editors. Cell-free protein synthesis. Weinheim, Germany: Wiley-VCH; 2008. pp. 1–34.
37. Voloshin AM, Swartz JR. Large-scale batch reactions for cell-free protein synthesis. In: Spirin AS, Swartz JR, editors. Cell-free protein synthesis. Weinheim, Germany: Wiley-VCH; 2008. pp. 207–235.
38. Kigawa T, Matsuda T, Yabuki T, Yokoyama S. Bacterial cell-free system for highly efficient protein synthesis. In: Spirin AS, Swartz JR, editors. Cell-free protein synthesis. Weinheim, Germany: Wiley-VCH; 2008. pp. 83–97.
39. Baez J. Biopharmaceuticals derived from transgenic plants and animals. In: Knäblein J, editor. Modern biopharmaceuticals. Weinheim, Germany: Wiley-VCH; 2005. pp. 833–892.
40. Schillberga S, Twymanb RM, Fischera R. Opportunities for recombinant antigen and antibody expression in transgenic plants—technology assessment. Vaccine 2005;23:1764–1769.
41. Howard JA, Hood E. Bioindustrial and biopharmaceutical products produced in plants. Adv Agron 2005;85:91–124.
42. Streatfield SJ. Approaches to achieve high-level heterologous protein production in plants. Plant Biotechnol J 2007;5(1):2–15.
43. Ramessar K, Sabalza M, Capell T, Christou P. Maize plants: an ideal production platform for effective and safe molecular pharming. Plant Sci 2008;174:409–419.
44. Janne J, Hyttinen JM, Peura T, Tolvanen M, Alhonen L, Halmekyto M. Transgenic animals as bioproducers of. therapeutic proteins. Ann Med 1992;24(4):273–280.

45. Das RC. Production of therapeutic proteins from transgenic animals. Am Biotechnol Lab 2001;19(2):60, 62, 64.
46. Lee EG, Leez SH, Park KM, Baek JE, Park JK, Chang WK, Jung JK, Chung BH. Efficient recovery of recombinant human erythropoietin from milk of transgenic pigs by two-step pretreatment. Biotechnol Bioprocess Eng 2008;13(2):189–196.
47. Bailey MR, Woodard SL, Callaway E, Beifuss K, Magallanes-Lundback M, Lane JR, Horn ME, Mallubhotla H, Delaney DD, Ward M, Van Gastel F, Howard JA, Hood EE. Improved recovery of active recombinant laccase from maize seed. Appl Microbiol Biotechnol 2004;63:390–397.
48. Hood EE, Kusnadi A, Nikolov Z, Howard JA. Molecular farming of industrial proteins from transgenic maize. Adv Exp Med Biol 1999;464:127–147.
49. Kusnadi AR, Hood EE, Witcher DR, Howard JA, Nikolov ZL. Production and purification of two recombinant proteins from transgenic corn. Biotechnol Prog 1998;14:149–155.
50. Dodge C, Baez J, Sudarshana M, McDonald K. Production of recombinant human gelatin in rice cell cultures. Presented at AAPS Annual Meeting; 2006; San Antonio (TX).
51. Holkeri H, Wahlstrom E, Ritala A, Baez J, Makinen K, Nuutila AM. Production of non-hydroxylated recombinant collagen in barley. J Biotechnol 2005;118:S84–S84.
52. Khoudi H, Laberge S, Ferullo JM, Bazin R, Darveau A, Castonguay Y, Allard G, Lemieux R, Vezina LP. Production of a diagnostic monoclonal antibody in perennial alfalfa plants. Biotechnol Bioeng 1999;20:135–143.
53. Merle C, Perret S, Lacour T, Jonval V, Hudaverdian S, Garrone R, Ruggiero F, Theisen M. Hydroxylated human homotrimeric collagen I in Agrobacterium tumefaciens-mediated transient expression and in transgenic tobacco plant. FEBS Lett 2002;515:114–118.
54. Ruggiero F, Exposito JY, Bournat P, Gruber V, Perret S, Comte J, Olagnier B, Garrone R, Theisen M. Triple helix assembly and processing of human collagen produced in transgenic tobacco plants. FEBS Lett 2000;469:132–136.
55. Zhang C, Baez J, Glatz CE. Purification and characterization of a recombinant gelatin expressed in transgenic maize. Presented at AIChE Annual Meeting; 2006; San Francisco (CA). P. 425i.
56. Zhong QX, Xu L, Zhang C, Glatz CE. Purification of recombinant aprotinin from transgenic corn germ fraction using ion exchange and hydrophobic interaction chromatography. Appl Microbiol Biotechnol 2007;76:607–613.
57. Olsen D, Jiang J, Chang R, Duffy R, Sakaguchi M, Leigh S, Lundgard R, Ju J, Buschman F, Truong-Le V, Pham B, Polarek JW. Expression and characterization of a low molecular weight recombinant human gelatin: development of a substitute for animal-derived gelatin with superior features. Protein Expr Purif 2005;40:346–357.
58. Perret S, Merle C, Bernocco S, Berland P, Garrone R, Hulmes DJS, Theisen M, Ruggiero F. Unhydroxylated triple helical collagen I produced in transgenic plants provides new clues on the role of hydroxyproline in collagen folding and fibril formation. J Biol Chem 2001;276:43693–43698.
59. Gu ZR, Glatz CE. A method for three-dimensional protein characterization and its application to a complex plant (Corn) extract. Biotechnol Bioeng 2007;97: 1158–1169.

60. Nikolov ZL, Woodard SL. Downstream processing of recombinant proteins from transgenic feedstock. Curr Opin Biotechnol 2004;15:479–486.
61. Hood EE, Woodard SL, Horn ME. Monoclonal antibody manufacturing in transgenic plants—myths and realities. Curr Opin Biotechnol 2002;13:630–635.
62. Gasdaska JR, Spencer D, Dickey L. Advantages of therapeutic protein production in the aquatic plant Lemna. Bioprocessing J 2003;2:49–56. Also the company website: Biolex Therapeutics, http://www.biolex.com.
63. Tiwari S, Verma PC, Singh PK, Tuli R. Plants as bioreactors for the production of vaccine antigens. Biotechnol Adv 2009;27(4):449–467.
64. Ma JKC, Drake PMW, Christou P. The production of recombinant pharmaceutical proteins in plants. Nat Rev Genet 2003;4:794–805.
65. Ma JKC, Barros E, Bock R, Christou P, Dale PJ, Dix PJ. Molecular farming for new drugs and vaccines. Current perspectives on the production of pharmaceuticals in transgenic plants. EMBO Rep 2005;6:593–599.
66. Ma JKC, Chikwamba R, Sparrow P, Fischer R, Mahoney R, Twyman RM. Plant-derived pharmaceuticals—the road forward. Trends Plant Sci 2005;10:580–585.
67. Koprowski H. Vaccines and sera through plant biotechnology. Vaccine 2005;23:1757–1763.
68. Lal P, Ramachandran VG, Goyal R, Sharma R. Edible vaccines: current status and future. Ind J Med Microbiol 2007;25:93–102.
69. Mishra N, Gupta PN, Khatri K, Goyal AK, Vyas SP. Edible vaccines: a new approach to oral immunization. Ind J Biotechnol 2008;7:283–294.
70. Houdebine LM. Production of pharmaceutical proteins by transgenic animals. Comp Immunol Microbiol Infect Dis 2009;32:107–121.
71. Stoger E, Ma JK-C, Fischer R, Christou P. Sowing the seeds of success: pharmaceutical proteins from plants. Curr Opin Biotechnol 2005;16:167–173.
72. Kumar GBS, Ganapathi TR, Bapat VA. Production of hepatitis B surface antigen in recombinant plant systems: an update. Biotechnol Prog 2007;23:532–539.
73. Floss DM, Falkenburg D, Conrad U. Production of vaccines and therapeutic antibodies for veterinary applications in transgenic plants: an overview. Transgenic Res 2007;16:315–332.
74. Pollock DP, Kutzko JP, Birck-Wilson E, Williams JL, Echelard Y, Meade HM. Transgenic milk as a method for the production of recombinant antibodies. J Immunol Methods 1999;231:147–157.
75. John DCA, Watson R, Kind AJ, Scott AR, Kadler KE, Bulleid NJ. Expression of an engineered form of recombinant procollagen in mouse milk. Nat Biotechnol 1999;17:385–389.
76. Toman PD, Pieper F, Sakai N, Karatzas C, Platenburg E, de Wit I, Samuel C, Dekker A, Daniels GA, Berg RA, Platenburg GJ. Production of recombinant human type I procollagen homotrimer in the mammary gland of transgenic mice. Transgenic Res 1999;8:415–427.
77. Tomita M, Munetsuna H, Sato T, Adachi T, Hino R, Hayashi M, Shimizu K, Nakamura N, Tamura T, Yoshizato K. Transgenic silkworms produce recombinant human type III procollagen in cocoons. Nat Biotechnol 2003;21:52–56.
78. Das RC. Production of therapeutic proteins from transgenic animals. Biobusiness 2001;19:60–64.

79. Yeung PK. Technology evaluation: transgenic antithrombin III (rhAT-III), Genzyme Transgenics. Curr Opin Mol Ther 2000;2:336–339.

80. Kerr DE, Plaut K, Bramley AJ, Williamson CM, Lax AJ, Moore K, Wells KD, Wall RJ. Lysostaphin expression in mammary glands confers protection against staphylococcal infection in transgenic mice. Nat Biotechnol 2001;19:66–70.

81. Hens JR, Amstutz MD, Schanbacher FL, Mather IH. Introduction of the human growth hormone gene into the guinea pig mammary gland by in vivo transfection promotes sustained expression of human growth hormone in the milk throughout lactation. Biochim Biophys Acta 2000;1523:161–171.

82. Larrick JW, Thomas DW. Producing proteins in transgenic plants and animals. Curr Opin Biotechnol 2001;12:411–418.

83. Houdebine LM. Transgenic animal bioreactors. Transgenic Res 2000;9:305–320.

84. Lubon H, Palmer C. Transgenic animal bioreactors—where we are. Transgenic Res 2000;9:301–304.

PART III

APPLICATION AND CASE STUDIES—FLAVOR & FRAGRANCE, AGROCHEMICALS AND FINE CHEMICALS

CHAPTER 9

OPPORTUNITIES FOR BIOCATALYSIS IN THE FLAVOR, FRAGRANCE, AND COSMETIC INDUSTRY

STEFANO SERRA
C.N.R. Istituto di Chimica del Riconoscimento Molecolare, Milano, Italy

9.1 INTRODUCTION

The preparation of flavors and fragrances and their use in foods and perfumes, respectively, began in ancient times. Our ancestors found these compounds in nature and discovered how they could be obtained either from vegetal or from animal resources. Concurrently, the production of fermented foods (beer, wine, and others) allowed the generation of new aromas and formed the roots of modern biotechnology. For many centuries, these were the only methods for obtaining this type of compound albeit in complex mixtures. Rapid progress began with the development of synthetic organic chemistry. More than a century ago, the preparation of coumarin (1868) and vanillin (1874) provided the first fragrance and flavor compounds available by synthesis [1]. From the very beginnings of chemistry, the improvement in analytical and synthetic knowledge allowed the isolation and preparation of an impressive number of aromas on an industrial scale [2]. Moreover, the nonnatural fragrances were created by emulation of natural structures, systematic studies on the relationship between odor and chemical structure, and by serendipity [3]. Until recently, all these compounds found widespread application in food, beverages, cosmetic, detergent, and pharmaceutical products with a worldwide industrial size estimated at US$20.3 billion in 2008 [4]. If we consider the latter data, the economic relevance of flavors and fragrances is unquestionable. Moreover, modern production of both taste and odor active compounds is oriented to the use of high-quality ingredients.

Biocatalysis for Green Chemistry and Chemical Process Development, First Edition.
Edited by Junhua (Alex) Tao and Romas Kazlauskas.

Consumers are demanding "natural-like" formulations for food flavoring, while new and unnatural fragrances are required under the influence of the perfumery trends. The first goal is achievable by accurate reconstruction of the required flavor and the chemical study of the so-called "character-impact compounds" has become a fundamental part of flavor research. Otherwise, the discovery of new fragrances prompted the studies of the chemical synthesis of new odorous molecules. Moreover, since both flavors and fragrances are often a complex mixture of isomers, the identification and the sensory evaluation of each component are necessary. In this context, chiral chemistry plays a primary role. It is known that the enantiomeric composition of an odorous compound greatly affects its olfactory properties in terms of either the quality of its perceived odor or its odor threshold [5,6]. Carvone and limonene are classical examples of flavors widely used in industry whose enantiomeric forms show different odor properties. The (+)-enantiomers of the latter compounds smell of caraway and of orange, respectively, while the (−)-enantiomers smell of spearmint and of turpentine, respectively. Consequently, only one enantiomer, or a balanced combination of the enantiomers, might be used in aroma, fragrances, or cosmetic formulations. Otherwise, preparation of enantioenriched compounds needs asymmetric synthesis or the use of optically active starting materials, greatly increasing the cost of production. Although the majority of these products were prepared by chemical synthesis or by extraction from plants, the employment of new biotechnological processes has increased considerably in the past decades [7–11]. Biocatalysis represents a useful tool in this field catalyzing a large number of stereo and regioselective chemical manipulations that are not suitably obtained by the less selective classical synthetic procedures. Moreover, contemporary microbiological techniques, including genetic engineering, have now enhanced the efficiency of biocatalysis. Furthermore, the increasing ecological sensitivity supports the choice of environmentally friendly processes and consumers have developed a preference for "natural or organic" products, thus developing a market for flavors of biotechnological origin [12]. A great number of key flavor or fragrance compounds were found in nature as trace components. Consequently, for many compounds, extraction is not a suitable or realistic approach and only the synthetic or microbiological pathways allow their production.

Therefore, a question arises: when should flavors and fragrances be prepared by biocatalytic pathways? Economic, chemical, and marketing considerations affected the answer to the latter question. The cost and availability of the starting materials and of the microorganisms/enzymes employed in the biotransformations as well as the technological feasibility of the chemical processes should be economically supported by the market preference. Biocatalysis helps to reduce the use or the generation of hazardous compounds and well suits both the use of renewable starting materials and the employment of safer chemical transformations. Thus, biocatalytic processes are much more "green" than the corresponding classical chemical syntheses though, for many instances, they are still more expensive. The laws that regulate the protection of the environment, the availability of the raw materials, and the needs of consumers differ markedly from state to state, accounting for the cost of a given process and the preference for a biocatalytic or chemical synthesis. Since the legislation, the production mode and the market of natural flavor are very different from those of artificial flavors and fragrances, the more relevant biocatalytic methods for the production of the latter substances are described in two separate paragraphs.

9.2 NATURAL FLAVOR

9.2.1 Legislation

It is important to realize that the origin of flavors also affects their commercial values. The legislations of different states [13] have officially stated that "natural" flavor substances can be prepared either by physical processes (extraction from natural sources) or by enzymatic or microbial processes, which involve precursors isolated from nature. This classification has created a dichotomy in the market because compounds labeled as "natural" become profitable products, whereas other flavors that occur in nature but are obtained through chemical methods must be indicated as "nature-identical" and are less appreciated by the consumers. From a chemical point of view, there is no difference between a compound found in nature and the identical molecule synthesized in the laboratory, but still the price of a flavor sold as natural is often significantly higher than those prepared by chemical synthesis. Although United States [14] and the new European Directive [15] do not recognize the term "nature identical," the classification described above is still widely used and has stimulated much research aimed at developing new biotechnological processes for flavor synthesis.

9.2.2 Natural Routes

The "natural" routes for flavor production are the bioconversions of natural precursors using biocatalysis, *de novo* synthesis (fermentation), and isolation from plants and animals. All these methods are used in industrial production and often the most relevant flavors are prepared both by them and by classical chemical synthesis. These are the cases of the flavors of phenylpropanoid origin and of those deriving by degradation of fatty acids.

The most relevant components of the former class of compounds are vanillin, phenylethanol, raspberry ketone, benzaldehyde, and dihydrocoumarin (Figure 9.1), whereas the last kind of biosynthetic pathway are responsible of the natural production of γ and δ-lactones and of the so-called "green notes."

9.2.2.1 Flavors from Phenylpropanoids. Vanillin [16] is the most important flavor in terms of consumption. The compound occurs in the pods of tropical *Vanilla* orchid (principally *Vanilla planifolia*) at a level of 2% by weight but the extracted compound covers <1% of the global market. The value of vanillin extracted from pods is variously calculated as being between $1200/kg and $4000/kg, whereas the price of synthetic vanillin, that is, vanillin prepared

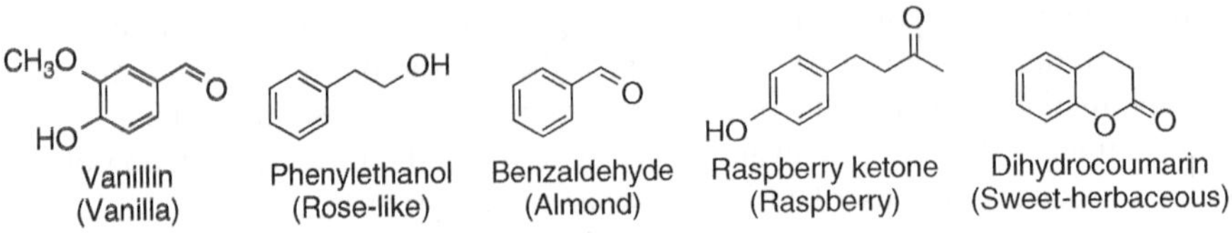

Figure 9.1 Examples of some relevant natural flavors of phenylpropanoid origin prepared by biotransformation.

mainly from guaiacol, is less than $15/kg. Therefore, a number of biotechnological processes for natural vanillin production have recently been investigated although none of them still allows satisfying the whole market demand. The most relevant methods of preparation of natural vanillin are based on the bioconversion of natural compounds of phenylpropanoid origin [16–18] (lignin, ferulic acid, eugenol, isoeugenol, and phenolic stilbenes isorhapontin; Figure 9.2), since its production by *de novo* biosynthesis affords disappointing results.

Three significant routes have been developed. The first is based on the degradation of ferulic acid, frequently occurring in cell walls of wood, grass, corn hulls, and mainly extracted from rice bran. Different white rot fungi and bacteria are able to convert ferulic acid into vanillic acid and/or vanillin, whereas other microorganisms reduce vanillic acid to vanillin and vanillic alcohol. Since vanillin is very reactive, it is toxic to most microorganisms, thus justifying its rapid degradation during the biotransformations. In order to overcome the overoxidation/reduction steps, the fermentation processes were optimized using various worthwhile (bio) technological tricks. For instance, some processes are based on a two-step transformation. First, one microorganism metabolizes ferulic acid into vanillic acid that is then reduced to vanillin by means of a second microorganism.

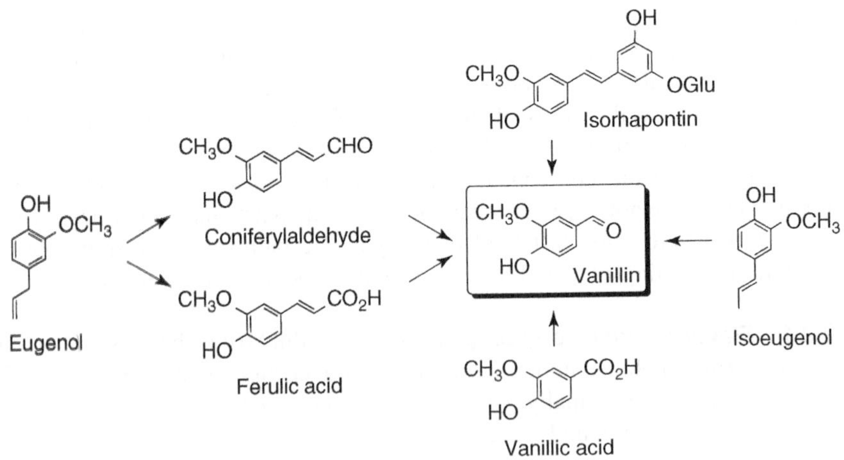

Figure 9.2 Examples of some relevant routes for the bioproduction of vanillin.

In addition, the *in situ* adsorption on the resins of the obtained vanillin reduces its toxicity, thus increasing the yield of the biotransformation.

A second possible route is the bioconversion of eugenol or isoeugenol [18], which are the main constituents of the essential oil of the clove tree. These propenylbenzene derivatives are cheap and commercially available raw materials for biotransformation processes, although with a major limitation due to their toxicity toward a great number of microorganisms. In spite of this fact, recent studies have elucidated the metabolic pathways responsible for the degradation of these propenylbenzene isomers by a restricted number of microorganisms. Eugenol is converted into coniferylaldehyde and/or ferulic acid, which are then degraded to vanillin and vanillic acid. Otherwise, isoeugenol is converted directly to vanillin and vanillic acid through the intermediacy of the corresponding isoeugenol-epoxide. The microbial degradations of both eugenol and isoeugenol suffer from the overoxidation reaction. In order to overcome this problem, some new processes in which the produced vanillin is not further transformed have been developed. The use of biotransformations that employ isolated enzymes or the microbial degradation by recombinant strains in which the vanillin dehydrogenase activity is suppressed were the most successful approaches. A third route is founded on the oxidation of the natural stilbene isorhapontin that is commonly found in spruce bark. Certain Pseudomonas strains cleave the double bond of the latter stilbene by means of a stilbene dioxygenase, thus affording vanillin and the glycoside of the resorcyl aldehyde.

As stated for vanillin, phenylethanol, raspberry ketone, benzaldehyde, and dihydrocoumarin are also relevant phenylpropanoids used in flavor industries but differently; their extraction is unsuitable and their main mode of production is the bioconversion of some natural precursors. Phenylethanol shows a rose-like odor and occurs in fermented foodstuffs and in many essential oils, whereas natural benzaldehyde is usually liberated from the cyanogenic glycosides present in fruit kernels, and is used as a key ingredient in cherry and almond flavors.

The biotechnological production of the latter compound (Figure 9.3) requires the isolation of the glycoside amygdalin from apricot kernels followed by its

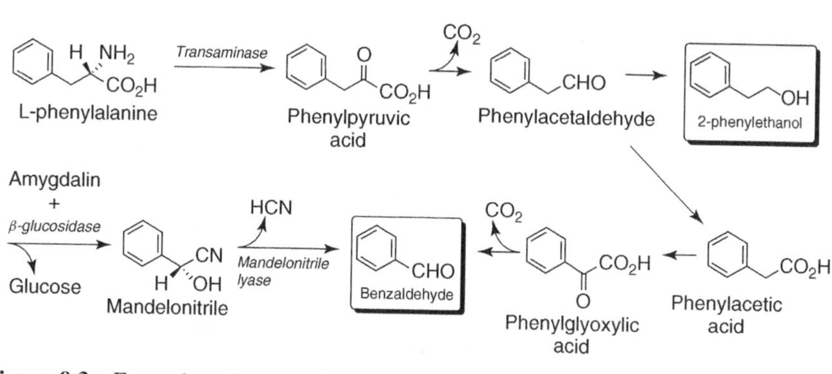

Figure 9.3 Examples of some relevant routes for the bioproduction of 2-phenylethanol and benzaldehyde.

hydrolysis by means of β-glucosidase catalysis. The mandelonitrile obtained is then converted to benzaldehyde with a concurrent generation of an equimolar amount of the toxic hydrocyanic acid. Both the safety problems related to the latter compound and the rather high price of the natural raw material prompted new studies on the degradation of different natural substrates. The microbial transformation of L-phenylalanine is the best alternative, offering a direct access to both phenylethanol [19] and benzaldehyde. Indeed, natural L-phenylalanine is cheap and easily available since its microbial production is a key step in the industrial synthesis of the sweetener aspartame. Different microorganisms, particularly yeast, are able to degrade phenylalanine following the Ehrlich pathway. The first step is catalyzed by a transaminase and affords phenylpyruvate that by means of decarboxylase eliminates carbon dioxide to give phenylacetaldehyde. The reduction of the latter compound by yeast dehydrogenase affords the suitable 2-phenylethanol, whereas a restricted number of microorganisms, comprising bacterium and white rot fungi, convert the intermediate phenylacetaldehyde into phenylacetic acid. The following oxidation to phenylglyoxylic acid and decarboxylation reaction provides the accumulation of benzaldehyde in the culture medium and the *in situ* adsorption on resins minimizing its subsequent reduction to benzylic alcohol by the action of alcohol dehydrogenase, thus increasing the yield of the biotransformation.

Raspberry ketone is the key flavor molecule of raspberry, in which it occurs in trace amounts (<4 mg of ketone from kilograms of berries). Biotransformation processes offer the sole alternative for the production of this compound in its natural form (Figure 9.4).

Both 4-(4-hydroxyphenyl)butan-2-ol (betuligenol) and its *O*-glucoside (betuloside) are quite widespread in nature and are usually extracted from birch and maple bark or from rhododendron and taxus spp. The β-glucosidase-mediated hydrolysis of betuloside gives betuligenol that can be further converted into the suitable ketone by means of the catalysis of alcohol dehydrogenase [20], using either isolated enzymes or microbial systems. 4-Hydroxybenzalacetone, prepared

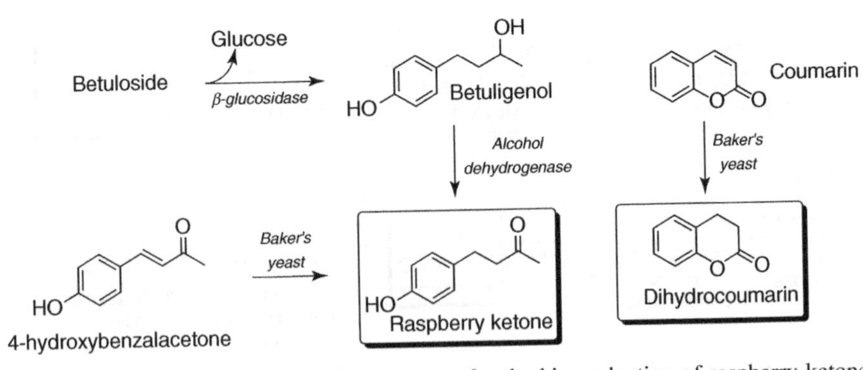

Figure 9.4 Examples of some relevant routes for the bioproduction of raspberry ketone and dihydrocoumarin.

upon condensation of 4-hydroxybenzaldehyde of extractive botanical origin with acetone produced by sugar fermentation, is a further possible precursor in the biotechnological production of raspberry ketone. Indeed, baker's yeast and other microorganisms [21] reduce the double bond of 4-hydroxybenzalacetone although with partial over-reduction of the carbonyl functional group. Similarly, a yeast-mediated biohydrogenation is the key reaction in the preparation of dihydrocoumarin from coumarin. This compound is the impact flavor of tonka bean absolute (*Dipteryx odorata* Willd), a widely used ingredient in the flavor industry. Although the unique odor features of coumarin make it irreplaceable, the use of this compound is regulated with limit of concentration by law for toxicological reasons. Therefore, dihydrocoumarin is used to implement or replace coumarin in many formulations for foods and fragrances. Since the latter compound is present in nature only in sweet clover (*Melilotus officinalis*) and tonka beans in trace amount, its extraction is very expensive. Recent studies have shown that different microorganisms [22,23] are able to biohydrogenate coumarin although the best result has been obtained with the use of baker's yeast. Hence, the combined use of extractive coumarin with the yeast-mediated reduction gives a valid access to the natural form of dihydrocoumarin.

9.2.2.2 *Flavors Derived by Degradation of Fatty Acids.* Fatty acids are the biochemical precursors of a large number of flavors occurring in products of vegetal and animal origin. These acids are cheap and easily available since they are obtainable by hydrolysis (usually catalyzed by lipases) of oils and fats. Consequently, both their biosynthetic pathways of degradation and their use as raw starting materials in the biotechnological preparation of natural flavors have received growing attention.

In particular, γ- and δ-lactones (Figure 9.5), with up to 12 carbon atoms are widespread in fermented food, milk products, and in a variety of fruits in minute amounts.

All lactones originate [24] from their corresponding 4- or 5-hydroxy carboxylic acids respectively, those are in turn deriving from longer chain hydroxy-fatty acids by β-oxidation degradation. The latter intermediates are formed by either

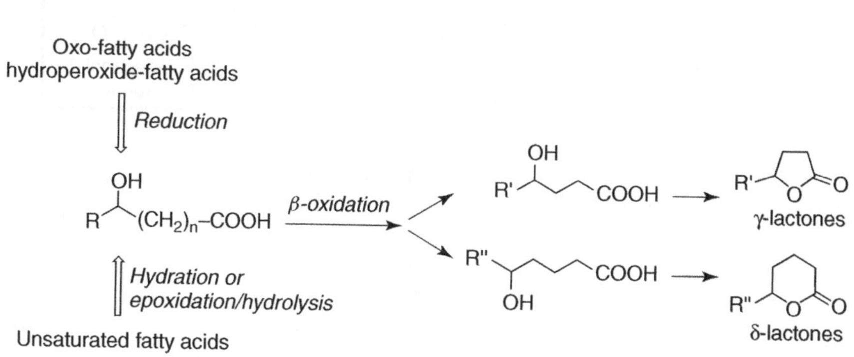

Figure 9.5 Biosynthetic pathways of γ and δ-lactones.

reduction of oxo-acids (or hydroperoxide fatty acids) or hydratation (or epoxidation followed by hydrolysis) of unsaturated fatty acids. According to the latter biochemical pathways, some relevant lactones are manufactured by degradation, *via* β-oxidation, of natural hydroxy-fatty acids [25,26] (Figure 9.6).

γ-Decalactone is the most important lactone for flavor application showing an oily- peachy aroma with a very powerful creamy-fruity, peach-like taste. The compound is obtained by chain shortening of C18 ricinoleic acid (from castor oil) by different microorganisms. Similarly, coriolic acid is the precursor of δ-decanolide, whereas some precious γ-lactones containing an odd number of carbon atoms are accessible by degradation of natural hydroxy acids [27]. Interestingly, δ-decalactone can be obtained in the natural modification using baker's yeast either by reduction of its corresponding α,β-unsaturated compound (massoia lactone), which is the main component of massoi bark oil [28], or by oxidation of 11-hydroxy palmitic acid.

Similar to the above-described lactone production, biocatalyzed degradation of unsaturated and polyunsaturated fatty acids is the process responsible for the formation in nature of the "green" organoleptic notes [24]. From a chemical point of view, these notes are due to C6 aldehydes and their corresponding alcohols (Figure 9.7).

The most relevant compound belonging to this class is *cis*-3-hexen-1-ol (leaf alcohol), which has an odor of freshly cut grass and is essential to obtain flavor and fragrance material used for green top notes. Since the "green notes" obtainable by distilling plant oil are expensive and do not meet the increasing market demand for the natural product, different biotransformations have been developed. Following the biosynthetic pathway of degradation, the C13-lipoxygenase-mediated oxidation of linolenic and linoleic acid affords (*S*)-13-hydroperoxy-octadeca-9(*Z*),11(*E*),15(*Z*)-trienoic acid and (*S*)-13-hydroperoxy-octadeca-9(*Z*), 11(*E*)-dienoic acid, respectively. The hydroperoxyde lyase-catalyzed cleavage of the C13 hydroperoxide functional group converts the compounds into *cis*-3-hexen-1-al and hexan-1-al respectively that can be reduced by yeast to the

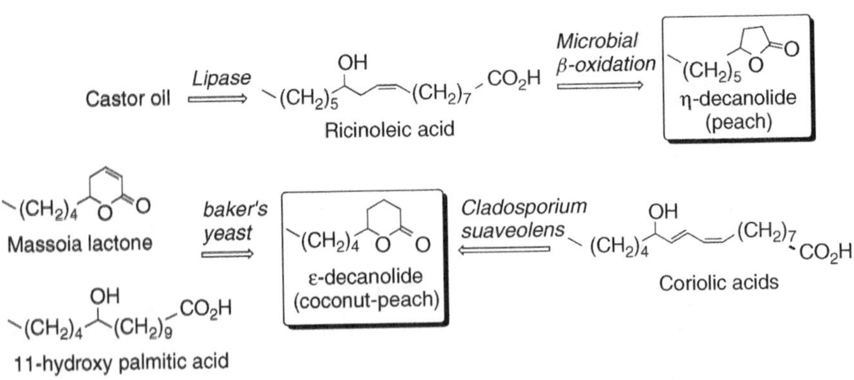

Figure 9.6 Examples of some relevant routes for the bioproduction of γ and δ-decanolides.

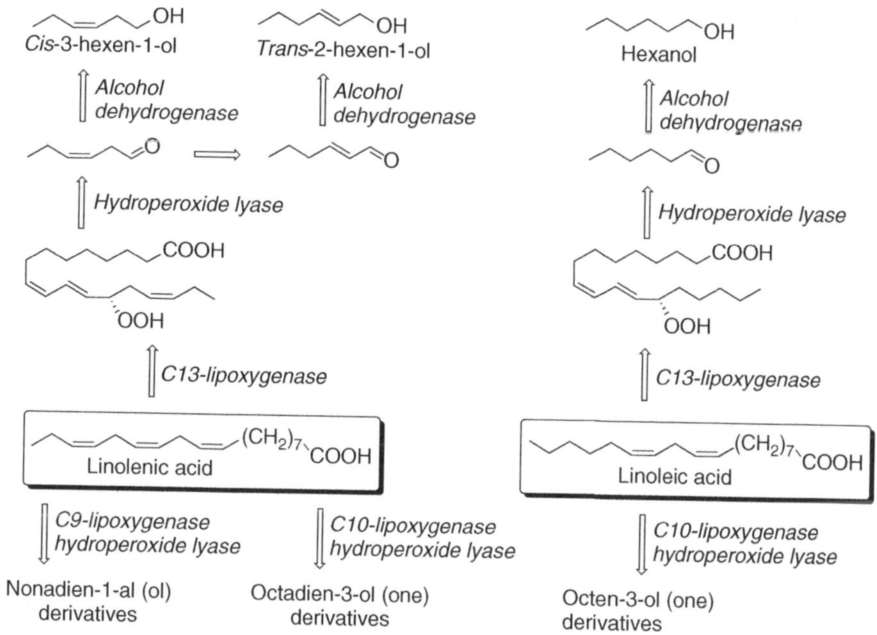

Figure 9.7 Bioproduction of relevant flavors from linolenic and linoleic acids.

corresponding alcohols. Eventually, the double bond isomerization of *cis*-3-hexen-1-al gives *trans*-2-hexen-1-al, which after reduction affords *trans*-2-hexen-1-ol. As the isolated enzymes are expensive, the action of lipoxygenases contained in soya flour was combined with the hydroperoxyde lyase activity of guava fruit homogenate followed by baker's-yeast-mediated reduction [29]. It is noteworthy that the action of C9 and C10 lipoxygenases on linolenic and linoleic acid could afford a number of C9 and C8 aldehyde, ketone, and alcohol derivatives of relevance in the flavor industry. For instance, the principal mushroom flavor components, oct-1-en-3-ol and oct-1-en-3-one, are accessible from linoleic acid using the lipoxygenase and lyase action of mushroom homogenates. However, since isolated enzymes are ideal for the synthesis of single odorants and their availability is still restricted, studies on the larger-scale production of lipoxygenase and lyase could represent a new chance for the natural flavor preparation.

9.2.2.3 Authentication of Natural Flavors. To date, a great number of biocatalytic processes to further attractive flavors have been described. In spite of this fact, the number of industrial applications is limited and the cases illustrated above comprise both the more established and the more promising ones. Moreover, an additional problem in this area is the occurrence of adulterations with readily available "natural-identical" products. The achievement of new analytical methods [30] capable of discriminating between natural and natural-identical flavors has become necessary. Different studies based on stable isotope characterization of aromas have showed encouraging results and are now applied by specialized laboratories to prove authenticity [31–36].

9.3 NATURE-IDENTICAL FLAVORS, ARTIFICIAL FLAVORS, AND FRAGRANCES

As mentioned above, the compounds found in nature and prepared by chemical synthesis are defined nature-identical, whereas those not found in nature are artificial. Although different state legislations do not recognize the term "nature identical," the classification described above is still used and could justify the market preference for nature-identical flavors over artificial flavors. The methods of production of both nature-identical and artificial fragrant compounds are affected by stringent economic limitations. Biocatalytic approaches to the latter compounds are often expensive, although environmental-friendly conditions and high chemical selectivity are the relevant features that make the latter methods attractive. Moreover, since chiral flavors and fragrance play a primary role, biocatalytic transformations showing high enantioselectivity could meet the industrial requirement. Furthermore, increasing awareness of the preservation of the environment and human health has favored the trend of employing a single active enantiomer in flavors, fragrances, and cosmetic ingredients. In this context, biocatalysts show a number of advantages over chemical catalysts. They allow transformations with higher regio- and enatioselectivity, thus affording only the wanted isomers. This reduces the overall amount of the chemicals wasted in the environment that arise both from reagents not completely converted and from unwanted products. Moreover, biocatalytic syntheses themselves are less hazardous and often are based on the use of renewable feedstocks.

In conclusion, the use of biocatalysis could be advantageous also in the production of nature-identical or artificial odorants, since its "green" aspects could support both the industrial process requirements and the consumer's preferences.

9.3.1 Use of Lipases

In this context, lipases [37,38] are the favorite catalysts, since they show remarkable chemoselectivity, regioselectivity, and enantioselectivity toward a broad range of substrates. They are readily available in large quantities, because many of them can be produced in high yields by gene expression in an appropriate microorganism, do not require cofactors nor catalyze side reactions, and remain enzymatically active in organic solvents with great advantages as for the dissolution of the substrates and the recovery of the final products. Lipases are green catalysts usually employed to catalyze hydrolysis or acylation reactions under mild conditions (atmospheric pressure and room temperature) and using common organic solvents. No specific reaction apparatus is needed and in most cases, the enzyme can be recovered and employed again, without loss of activity.

Here, we describe some relevant applications of lipase-catalyzed reactions in the synthesis of chiral flavors and fragrances.

9.3.1.1 Menthol. (−)-Menthol is one of the most important flavors and it is used extensively as food additive, in pharmaceuticals, cosmetics, toothpastes, and chewing gum. The desired organoleptic properties of this monoterpene are related to its absolute configuration, and of the eight possible isomers, only the natural (1*R*,3*R*,4*S*) isomer is suitable as aroma.

The majority of (−)-menthol is still obtained by freezing the oil of *Mentha arvensis* to crystallize the menthol present (Figure 9.8). Although many efforts have been devoted to the production of (−)-menthol from other readily available raw materials, only Haarmann & Reimer (H&R) and Takasago processes are the commercial synthetic sources of this flavor [39].

The first one is based on a classical resolution procedure and starts from inexpensive *m*-cresol and propylene to produce thymol. The catalytic hydrogenation of the latter compound gives the mixture of the eight isomers of menthol. The following fractional distillation affords racemic menthol (two isomers), which is converted into the corresponding benzoate. The large-scale industrial process utilizes selective crystallization of (−)-menthyl benzoate by seeding a saturated solution of supercooled melt of racemic benzoate with crystals of (−)-menthyl benzoate. Saponification of the obtained crystalline material affords (−)-menthol, whereas the mother liquor gives (+)-menthol. The latter compound and the other six isomers are recycled in a separated racemization step. Otherwise, Takasago uses asymmetric synthesis [40] in the key step of its process. Myrcene is converted into diethygeranylamine, which is isomerized to (+)-citronellal in the presence of a chiral rhodium phosphine catalyst (RhI-(*S*)-BINAP). The transformation of citronellal into (−)-menthol is performed through cyclization to (−)-isopulegol followed by hydrogenation.

Close to the route described first, two new processes based on lipases resolution have been recently proposed. Haarmann & Reimer accomplished the resolution of racemic menthol benzoate by lipase-mediated (e.g., *Candida rugosa*) enantioselective hydrolysis to provide (−)-menthol with essentially complete enantioselectivity [41,42]. Furthermore, AECI LIMITED process starts directly from the mixture of the eight isomers of menthol [43]. The enantio- and diastereoselective acylation of this mixture using lipases yields menthyl acetate in a 96% enantiomeric excess. The ester is separated from the unreacted isomers by distillation and then hydrolyzed to yield (−)-menthol. In both processes, the undesired isomers are recycled by isomerization. Although neither the H&R nor the AECI biocatalytic process has been commercialized, these means are supported on a well-established route and will certainly be the subject of further developments.

9.3.1.2 *p*-Menthane Monoterpenes. The monoterpene of the *p*-menthane family are widespread in nature and are well known as flavoring ingredients and as valuable synthetic intermediates. Many industrial processes are correlated with this class of compounds because of the high commercial requirement of (−)-menthol. Otherwise, some new findings on the peculiar organoleptic properties of different *p*-menthane alcohols, lactones, and ethers have prompted the studies of their synthesis.

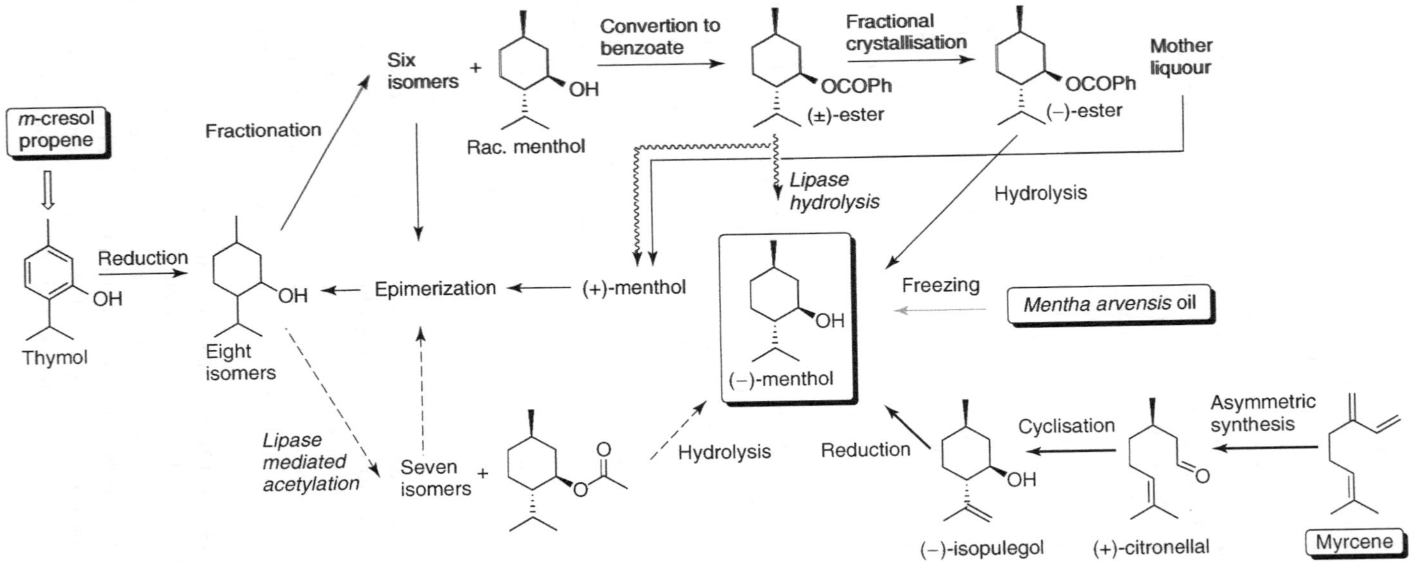

Figure 9.8 Industrial production of (−)-menthol. Black, gray, and bold arrows indicate Haarmann & Reimer, extractive, and Takasago processes, respectively. Wavy and dashed arrows give the new biocatalytic processes of Haarmann & Reimer and AECI, respectively.

For instances, the mixture of the eight isomers of isopulegol is used conventionally as a perfume ingredient, whereas recent investigations have shown that its main component (−)-isopulegol is odorless and can be used as a cooling agent [44] (Figure 9.9). Furthermore, the lactone **1** [45], (−)-mint lactone **2**, and (+)-isomint lactone **3** [46] are found as minor components in the essential oil of peppermint and are used in commercial flavors for their appreciated coumarin-like and mint-like olfactive properties. The (−)-wine lactone **4** was recognized as a key flavor compound of different white wines and synthetic studies revealed [47] that natural **4** is the most powerful isomer with a odor threshold <0.04 pg/L. A similar case is that of dill ether **5**, where the evaluation of its isomeric forms allows establishing that (+)-**5** shows the higher odor-activity value and is the character-impact compound of dill flavor [48]. This research demonstrates that both the quality and intensity of the latter odorants are related to relative and absolute configuration. Taking advantage of the known selectivity of biocatalysis, many specific preparation of the latter monoterpenes have been developed. (−)-Isopulegol was prepared in an optically pure form by lipase PS (lipase from *Pseudomonas* sp.) mediated enantio- and diastereoselective acetylation of the commercial mixture of its eight isomers [49]. *Trans* and *cis* piperitol and isopiperitenol are valuable intermediates in the synthesis of menthol, whereas diols **6**–**8** are useful precursors of the *p*-menthane lactones. The latter compounds were prepared in a single diastereoisomeric form and were resolved by lipase-mediated acetylation of the corresponding racemic materials [49,50].

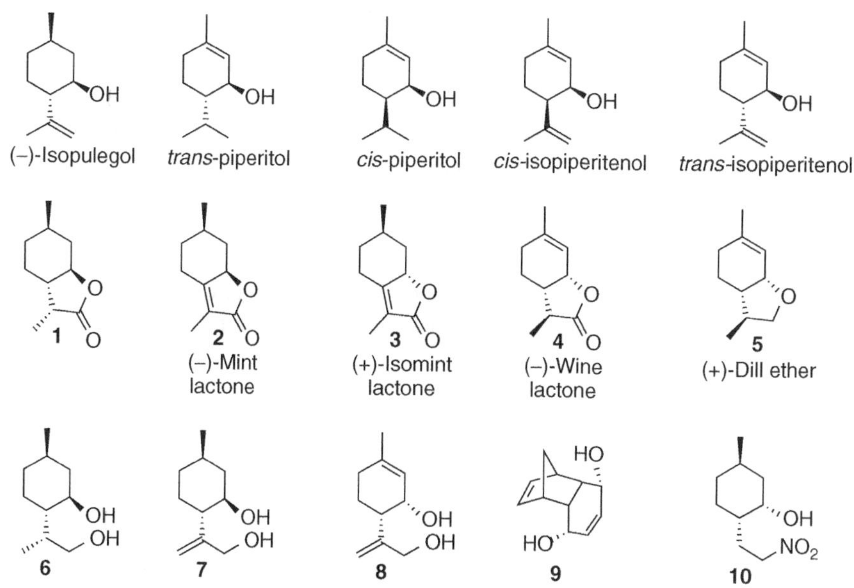

Figure 9.9 Natural *p*-menthane monoterpenes (alcohols, lactones, and ether) prepared in enantiopure form by the use of the biocatalysis. Compounds **6**–**10** are key intermediates in their synthesis.

The enantiopure diols **6** and **7** obtained were converted by chemical manipulation to lactone **1** and mint lactone **2**, respectively [50] while the more versatile diol **8** was transformed either in the wine lactone **4** or in the dill ether **5** [51].

Lipase PS also catalyzed the resolution of the *meso* tricyclic diol **9** [52], whereas the alcohol **10** was obtained by stereoselective baker's-yeast-mediated reduction of the corresponding ketone [53]. These two enantiopure materials were converted in the mint and isomint lactone, respectively.

9.3.1.3 Carotenoids Deriving Flavors and Fragrances. Carotenoids are C40 terpenic compounds widespread in nature whose degradation is the source of a great number of flavors and fragrances. The primary odor constituents derived from carotenoids are the cyclohexenic derivatives formed either via enzymatic or via chemical oxidation of their polyenic chain (Figure 9.10). Among the latter group of compounds the C13-, C11-, C10-, and C9-derivatives are the most relevant ones.

The possibility of three different positions of the cyclohexenic double bond, the presence of a stereogenic center in position 6 (carotenoids numbering), and the eventual structural rearrangements afford a large number of isomers. Concerning the synthesis of these compounds, the main problem is the enantioselective formation of the stereocentre in position 6. Since the corresponding racemic materials are often easily available, the lipase-mediated resolution process has been shown to be a valid method for their preparation.

In this context, the cases of irone and ionone isomers are of outstanding relevance. Iris essential oil is considered among the historically most important flavor and fragrance materials. Although this natural product is very expensive, it is still used in fine formulations since it gives better performances as compared with the corresponding synthetic materials. This superiority is due to the complexity of the natural isomeric mixture where each component might show different olfactory features. For instance, until the end of nineteenth century, the only source of violet fragrance was the violet flower oil and the iris essential oil. The principles of odor of these raw materials are the norterpenoids ionone and irone, respectively (Figure 9.11).

Ionones have been found in a number of plants, whereas irones are formed during aging and manufacturing of the iris rhizomes. Comprehensively, 5-ionone

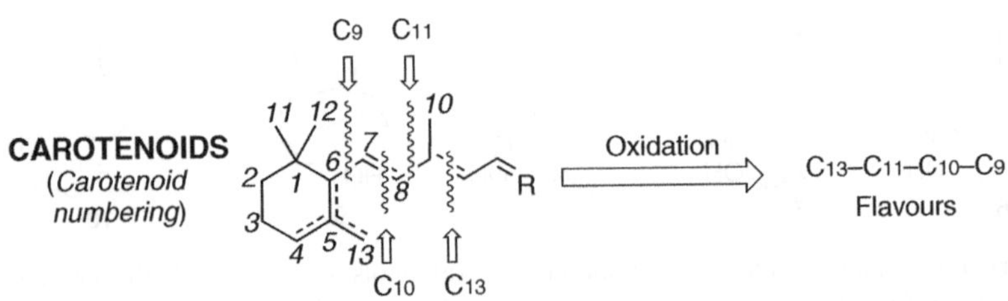

Figure 9.10 Generation of relevant flavors and fragrances by degradation of carotenoids.

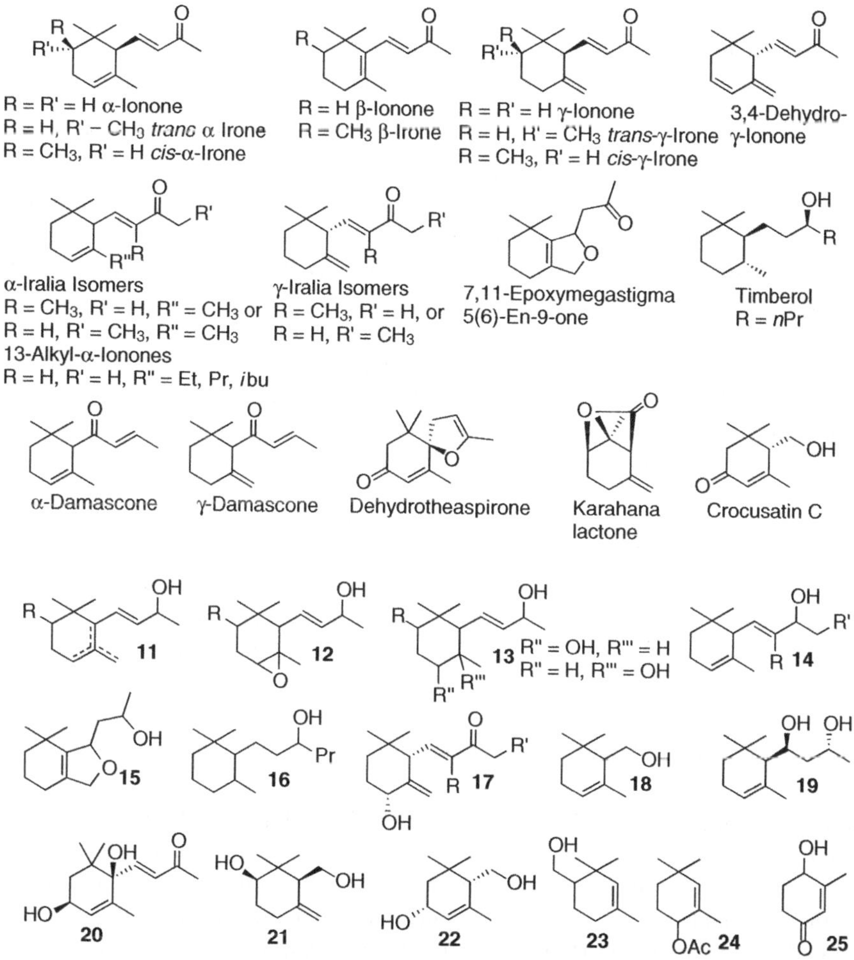

Figure 9.11 Natural and artificial odorants with carotenoids deriving frameworks prepared in enantiopure form by the use of lipase-mediated resolution. Compounds **11–25** are the racemic starting materials.

and 10-irone stereoisomers are possible. Thanks to the use of chemical synthesis and enzyme-mediated reactions, all these isomers were prepared [54,55] and submitted to the olfactory evaluation. Two different synthetic approaches to this class of compounds are based on the enantioselective lipase-mediated acetylation of suitable hydroxy-derivatives. The first lies on the resolution of a compound showing two stereogenic centers, one due to the presence of the sp^3 C6 carbon atom and the other due to the presence of a secondary hydroxyl group. The separation of the enantioenriched isomeric forms following to the transformation (or removal) of the hydroxyl group to each of the isolated enantiomers affords the single enantiomeric forms of ionone or irone derivatives. The second is based on the resolution of a racemic hydroxy-derivative of a compound showing the sp^2 C6 carbon atom followed by transformation of the sp^2 center in a sp^3 center with concomitant transfer of chirality. Regarding the first approach, since the enzyme-mediated acetylation of 9-hydroxy-derivatives proceeds with high enantioselectivity and no diastereoselectivity, the racemic ionols and irols **11** (R = H and CH_3, respectively), their corresponding α-epoxy-derivatives **12**, and the diols **13** were prepared in a diastereoisomerically pure form and then resolved by lipase-mediated acetylation [54,55]. The obtained enantioenriched isomers were then oxidized to the suitable ketone derivatives. Using this approach, the isomeric forms of the artificial violet odorant α-iralia [56] of the passion fruit flavor component 7,11-epoxymegastigma-5(6)-en-9-ol [57] and of the artificial woody odorant timberol [58] were also prepared starting from racemic alcohol **14, 15**, and **16**, respectively. Similarly, the enantioselective lipase PS-mediated acetylation of alcohol of type **17** and **18** were the key steps in the preparation of the 3,4-dehydro-γ-ionone [59], γ-iralia isomers [60] and in the synthesis of 13-alkyl-α-ionone isomers [61], respectively. Otherwise, lipases catalyzed the regio- and enantioselective acetylation of diastereoisomerically pure diols **19–22** whose isolated enantiomeric forms were used in the preparation of damascone [62] (rose odorants), dehydrotheaspirone [63] (nectarine flavor), karahana lactone (hops flavor), and crocusatin C (saffron flavor) [64], respectively.

Concerning the second general approach described above, it is worth to note that the enantiopure isomers obtained by lipase-mediated resolution of the racemic alcohol derivatives **23–25** were used as starting material in the synthesis of irone [65], damascone [66], and karahana lactone [67], respectively.

9.3.1.4 Ambergris and Jasmine.

Similar biotechnological approaches have been studied for ambergris and jasmine fragrances. The first class of compounds derives its name from the compound ambergris [68], which is a secretion found in the intestinal tract of the sperm whale. The latter material contains the odorless triterpene ambreine (Figure 9.12), which on exposure to sunlight, air, and seawater, undergoes a degradative process. The deriving flavor compounds are responsible for the complex odor of ambergris.

The most appreciated one is the tricyclic ether (−)-ambrox, which is currently produced by semisynthesis from sclareol, a diterpene present in clary sage. Recently, ambrox was obtained in few chemical transformations from enantiopure (+)-albicanol **26** and diol (+)-**27**. The latter compounds were prepared by kinetic resolution of the corresponding racemic materials mediated by lipase PL-266 (lipase from *Alcaligenes* sp.) [69] and lipase PS [70], respectively. In addition, acylase from *Aspergillus melleus* catalyzes enantioselectively [71] the hydrolysis of the phenolic acetate **28**, whose single isomeric form was used in (−)-ambrox synthesis.

(+)-γ-Dihydroionone and (+)-γ-coronal are also relevant flavor components of ambergris. The first one was prepared in an optically pure form by regioselective reduction of (+)-γ-ionone, which was obtained by lipase PS-mediated resolution of the racemic γ-ionol [72]. Otherwise, the kinetic acetylation of

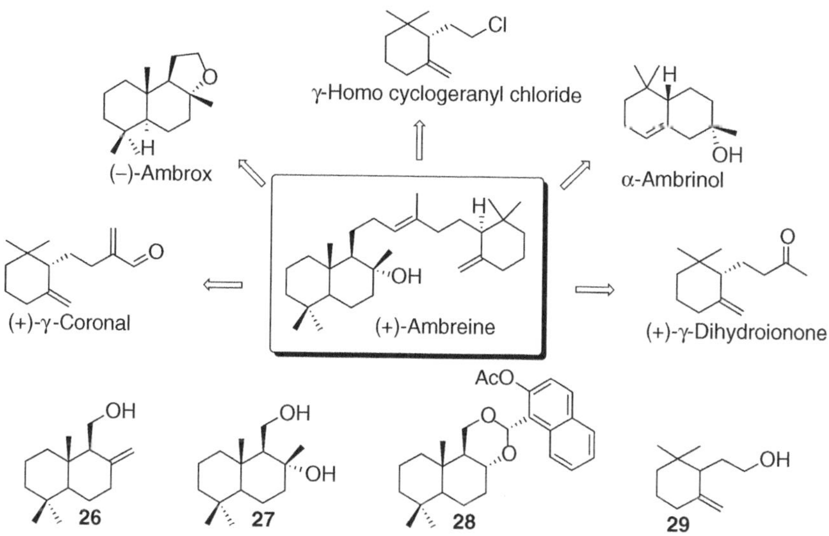

Figure 9.12 Ambergris fragrance: preparation of enantiopure amber odorants by the use of lipase-mediated resolution. Compounds **26–29** are key intermediates in their synthesis.

racemic γ-cyclohomogeraniol **29** was catalyzed by lipase AK (lipase from *Pseudomonas* AK) [73]. The (*S*) enantiomer obtained was used as the building block in the (+)-γ-coronal synthesis.

(−)-*Trans*-jasmonate and (+)-*cis*-jasmonate (Figure 9.13) are the relevant components of jasmine oil fragrance that occur in an enantiopure form in a diastereoisomeric ratio of 93:3. The synthetic *trans* ester is commercially available in a racemic form and its resolution has been recently reported [74]. The key steps were the reduction of ketone functionality, the separation of the alcohol **30** as a single diastereoisomer, and its resolution by lipase PS-mediated acetylation.

Methyl dihydrojasmonate is also a valuable compound for the manufacture of perfumes (trade names: Cepionate®, Hedione®). The *cis* diastereoisomer is much more powerful than the *trans* diastereoisomer and its odor is mainly due to the (1*R*,2*S*) enantiomer. Racemic allylic alcohol **31** was submitted to enzymatic kinetic resolution [75], by treatment with dimethyl malonate and using *Candida antartica* as catalyst. The trimethylsilyl ketene acetal derivative of the enantioenriched malonate **32** obtained was submitted to Claisen rearrangement and the product was converted into enantiopure (1*R*,2*S*)-*cis* dihydrojasmonate by a number of reactions. The unconverted alcohol derivatives were racemized under acidic conditions and recycled, thus improving the efficiency of the enzymatic process.

9.3.1.5 Further Use of Lipases in the Synthesis of Odor Active Components of Artificial Fragrances.

A great number of artificial fragrances are currently prepared and used as a mixture of isomers though the specific contribution of their components to the overall odor remains unknown. All the

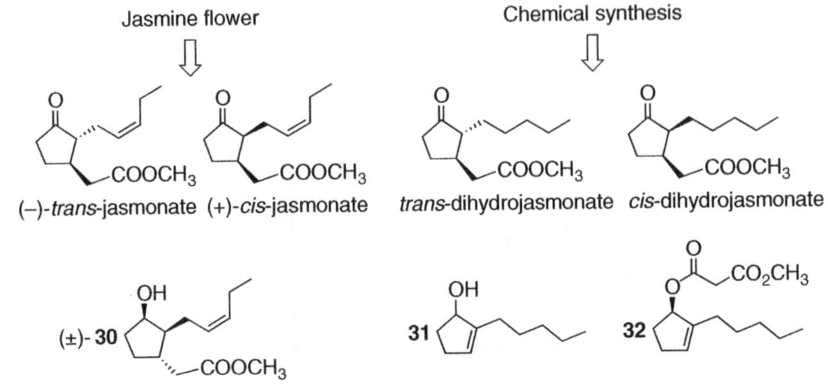

Figure 9.13 Jasmine fragrance: preparation of enantiopure jasmine odorants by the use of lipase-mediated resolution. Compounds **30–32** are key intermediates in their synthesis.

possible stereoisomers of a certain chiral odorant have to be prepared in an enantiopure form in order to be evaluated and the corresponding odor threshold and quality determined.

Since lipase-catalyzed kinetic resolution is a general and efficient method to prepare all the enantiomers of a certain fragrance in high optical purity, a number of works on this topic have been recently reported [38]. In addition to the fragrances illustrated in the preceding paragraphs, described here are a number of artificial fragrances the preparation of whose isomeric forms has been accomplished using the lipase-mediated resolution approach.

Owing to the great versatility of the hydroxyl functional group, many fragrances with different structures could be obtained by simple chemical manipulations of the enantioenriched alcohol derivatives. For instance, the enantiomeric forms of the alcohols muguesia®, pamplefleur® [76], and mugetanol® [77] (Figure 9.14) were prepared starting from racemic alcohols **33, 34**, and **35**, respectively. The latter compounds were resolved by means of lipase catalysis and their enantiomeric forms were used as chiral building blocks. The reductive removal of the carbonylic functionality of **33**, the C3 homologation of **34** followed by a second resolution step, and the aromatic ring reduction of **35** afforded the wanted fragrances. Similarly, cyclic ethers clarycet®, florol®, and pelargene® [78,79] were obtained by ring closure of the corresponding enantioenriched 1,5 diol derivatives, obtained in turn by enzymatic resolution of the hydroxy-derivatives **36** and **37** followed by few chemical transformations. Acetal and ketone derivatives are other relevant classes of fragrances obtainable by diols and alcohols manipulations, respectively. In this context, the enantiomeric forms of the 1,3-dioxane odorants floropal® and magnolan® [80,81] have been prepared using the lipase-mediated resolution of the hydroxy-ketone **38** and of the diol **39**, respectively, as a key step. Otherwise, the diastereoselective reduction of the fragrance nectaryl® [82] affords a mixture of two separable racemic diastereoisomers **40**. The enzyme-mediated resolution of each diastereoisomers followed by oxidation

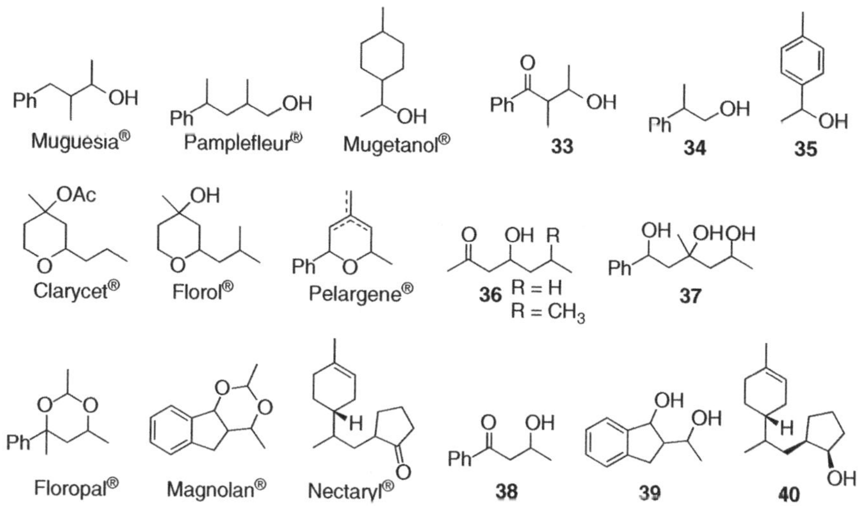

Figure 9.14 Examples of artificial fragrances that can be prepared in enantiopure form by aid of the lipase-mediated resolution. Compounds **33–40** are the racemic starting materials.

of the hydroxyl group allowed the synthesis and the evaluation of all the isomeric forms of the commercial compound.

9.3.2 Use of Baker's Yeast in Flavors and Fragrances Synthesis

The use of enzymes or microorganisms is compulsory when natural flavors would be produced starting from a natural precursor. Some instances on the use of baker's yeast for this kind of transformations are previously reported in the section describing natural routes for flavor production. It is worth noting that the latter microorganism is able to perform a great number of chemical transformations, often with high chemical selectivity. In addition, it is very cheap, safe, and easily available. For these reasons, baker's yeast has been used in the preparation of a number of chemical intermediated, although the employment of the latter compounds in a multistep synthesis affords only nature-identical or artificial flavors and fragrances.

In this context, two kinds of reaction are especially significant: the biohydrogenation of the double bonds and the reduction carbonyl functionality. Both transformations received renewed interest in the context of the preparation of chiral synthetic intermediates [83]. Indeed, the baker's yeast-mediated biohydrogenation of activated olefins allows the creation of up to two stereogenic centers, whereas the reduction of the carbonyl functionality affords chiral alcohol. Moreover, for olefins of type **41–43** (Figure 9.15), the stereochemical course of the reduction is by now predictable, affording invariably saturated compounds of type **44–46**, respectively.

By these means, a large number of enantioenriched compounds have been thus prepared some of which have been used in stereoselective syntheses of

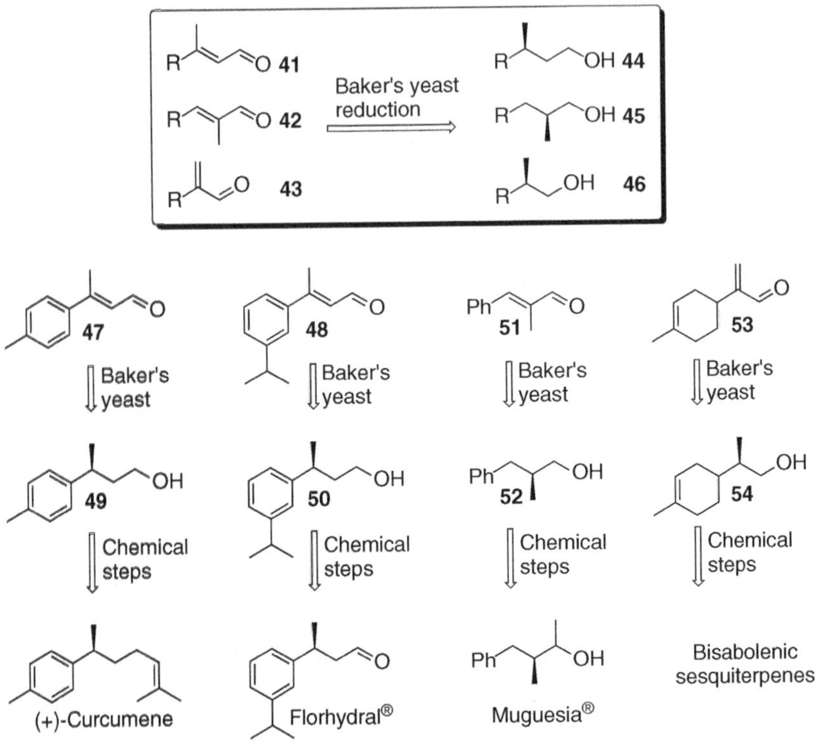

Figure 9.15 Baker's yeast-mediated biohydrogenation of prochiral double bonds as a key step in the enantioselective preparation of flavors and fragrances.

flavors and fragrances. Some relevant items are those involving the reduction of substituted methyl cinnamaldehydes. For instance, the baker's-yeast-mediated reduction of aldehydes **47** and **48** proceed with high enantioselectivity affording saturated alcohols **49** and **50**, respectively, which were used as chiral building blocks in the synthesis of natural bisabolanic sesquiterpenes curcumene [84] and for the preparation of the olfactory active isomer of the artificial fragrance florhydral [85], respectively. Similarly, enantioenriched alcohol **52** was obtained by reduction of α-methyl cinnamaldehydes **51** and was employed in the synthesis of artificial fragrance muguesia [86]. In addition, baker's yeast reduced α-substituted acroleins with high enantioselectivity as demonstrated [87] by the transformation of the *p*-menthanic aldehydes **53** into saturated alcohol **54** that is a useful synthon in bisabolenic sesquiterpenes synthesis.

Regarding the reduction of carbonyl functionality, the yeast-mediated reduction of keto-acids of type **55** and **56** (Figure 9.16) and of the 1,3-diketones of type **57** gives chiral synthons of type **58**–**60**, respectively that have been used in synthesis of flavors and fragrances.

Indeed, baker's yeast reduction of a compound of type **55**, where R is a *n*-butyl or *n*-pentyl aliphatic chain, affords directly the corresponding *trans*-lactones

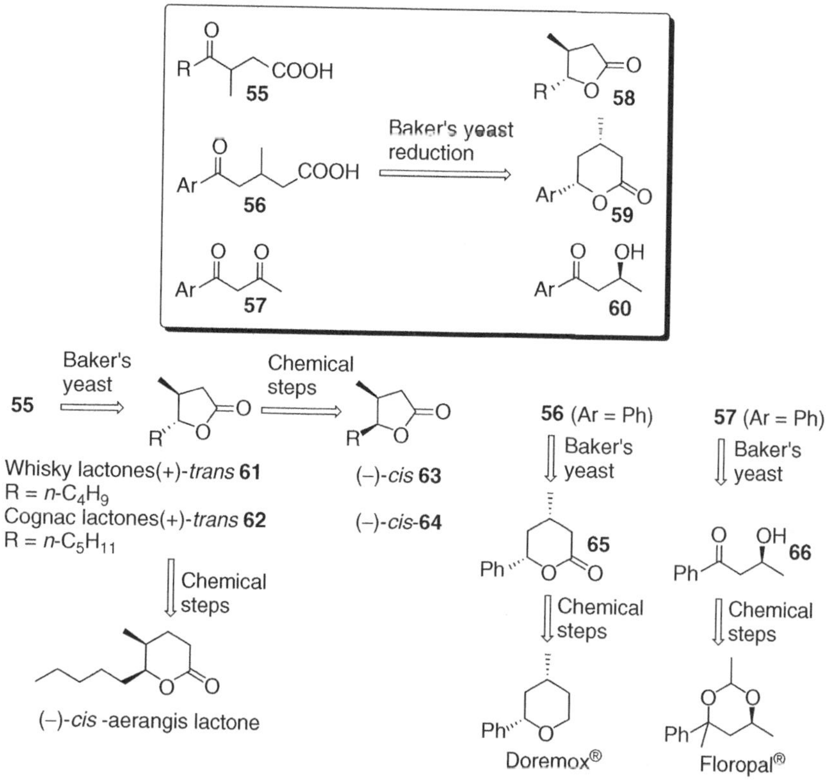

Figure 9.16 Baker's yeast-mediated reduction of keto-acid and diketone derivatives as key steps in the enantioselective preparation of flavors and fragrances.

61 and **62** [88,89], respectively with complete enantio- and diastereoselectivity. These substances, which are natural flavors known as whisky and cognac lactones, have also been converted in the corresponding *cis* derivatives **63** and **64**. The latter ester has been used as starting material in the synthesis of the natural floral fragrance *cis*-aerangis lactone [89]. Similarly, reduction of compounds of type **56** and **57**, where R is a phenyl group, gives lactone **65** and hydroxy ketone **66**, respectively, which are converted into the artificial fragrances doremox [90] and floropal [80,81], respectively.

9.3.3 Use of Other Enzymes or Biosynthetic Tools

Similar to biosynthetic preparations of flavors and fragrances described in the previous and in the following paragraphs, further odorous compounds have been prepared by means of the biocatalysis. Although a plethora of transformations based on the use of bacteria, fungi, plant cell cultures, basidiomycetes, and isolated enzymes is reported in the literature [91–94], few approaches hold preparative value. Among these, some processes that use natural and renewable starting compounds are noteworthy. Terpenes [95–98] and especially monoterpenes and carotenoids are easily obtained by extraction from plants and their bio-oxidation is a source of a great number of flavors and fragrances. For instance, limonene, pinene, and β-carotene are suitable substrates in the production of a number of oxygenated monoterpenes and ionone [99] derivatives, respectively. In addition, many valuable odorous compounds can be prepared by microbial degradation of amino acids and fatty acids, including substances that provide roasted note, green notes, mushroom flavors, and methylketones [91–94]. An interesting example is the fungi-mediated reduction of the natural hexanoic and hexenoic acids isomers to the corresponding alcohol derivatives (green notes) [100,101].

9.4 BIOCATALYSIS FOR THE PRODUCTION OF COSMETIC INGREDIENTS, FUNCTIONAL PERFUMERY

9.4.1 General Considerations, Biocatalyzed Preparation of Cosmetic Ingredients, and Use of Renewable Sources

A large number of the fragrances described above are used in cosmetic formulations. In spite of this fact, the main part of the cosmetic ingredients is odorless having different biological activity, for instance, as emulsifiers, emollient, moisturizing, antioxidants, thickening, or refatting agents. These compounds are usually prepared by extraction from natural sources or by conventional chemical synthesis. Otherwise, their preparation processes should meet a number of demanding requirements that, in some instances, are well met by biocatalytic methods. The most relevant case concerns the fatty acid esters, which are widespread in cosmetic formulation. The manufacture of these compounds requires that the esterification process does not afford unwanted by-products (especially if they posses an unpleasant odor), does not give degradation of the reagents or of the ester, and does not need solvents whose residues are undesired in cosmetics. The biocatalyzed methods work at room temperature with high selectivity and offer an evident advantage over conventional chemical processes involving harsh reaction conditions. In addition, the possibility of reusing the biocatalyst has minimized the catalyst cost to an acceptable fraction of the total production cost. In this context, the instance of cetyl ricinoleate [102] is of outstanding relevance. This ester is a wax compound that causes a silky but nonoily feeling on the skin and it is widely used in cosmetic formulations. It is obtainable by condensation of ricinoleic acid and cetyl alcohol but conventional chemical processes afford products close to the suitable ester, polyesters of ricinoleic acid. Otherwise, the use of the enzyme-mediated esterification (usually by lipases) dramatically reduces the polyester formation greatly increasing the final quality of the product. Similarly, a number of other fatty acid ester derivatives have been prepared by biocatalytic methods and some of these are currently manufactured by the same. For example, isopropyl myristate, isopropyl palmitate [103], cinnamyl laurate [104], and a number of retinyl esters [105] have been obtained by lipase-mediated synthesis, whereas ceramides and hydrid ceramides are produced by lipase-mediated selective N-acylation of glycosphingolipids [106]. Another relevant class of cosmetic ingredients is based on the use of glucoside derivatives. Indeed, fatty acid glucoside esters are characterized by an amphiphilic structure that makes them suitable nonionic skin detergents and emulsifiers. In addition, the specific modification of the sugar or of the fatty acid moiety changes the biological activity of the obtained ester. Accordingly, lactose and sucrose monolaurate [107] are used as skin detergents, whereas unsaturated and polyunsaturated fatty acid glucoside esters [108] have been employed as suitable "carrier molecules" capable of transporting the fatty acid into the epidermis. Similarly, since the antioxidant properties of the ascorbic acid are limited in the hydrophobic environment, L-ascorbyl laurate [109] has been used as antioxidant in both lipid-containing food and cosmetics. All these esters were prepared by lipases-mediated transesterification processes that afforded the products in higher yields and under milder experimental conditions compared with the classical chemical catalysis. Other enzymes such as glycosidases and phospholipases have been employed in the preparation of cosmetic ingredients. A relevant example is that of β-glucosylglicerol [110], an important moisturizing agent, which was prepared by glycosidase-mediated transglucosylation.

A further consideration is noteworthy: all the above-described processes are intrinsically "green," since they are well suited to the use of renewable raw materials and produce an inferior amount of dangerous waste materials. Indeed, the chemical transformations occurring in the natural environment are based on the same starting compounds and are catalyzed by the same kind of enzymes thus ensuring high efficiency and selectivity to the industrial processes based on these biotechnological tools. The biocatalyst itself is "biodegradable" and does not show the toxic features of metal-based catalysts. This aspect is very relevant in the synthesis of cosmetic ingredients, since biocatalytic methods cause only reduced pollution and allow preparation of safer final products. Moreover, a large number of chemicals used directly or indirectly in cosmetic preparations are available from industries that process renewable materials [111]. For instance, fatty acids, some of their ester, and glycerol are prepared from natural (vegetal or animal) oils and fats. In addition, other chemicals such as acetic acid, lactic acid, ascorbic acid, and phenylalanine are currently produced by fermentation, whereas sugars and polysaccharides are obtained by extraction from vegetal sources.

9.4.2 Functional Perfumery: Enzyme-Triggered Profragrances

Fragrances are often high volatile compounds. Their diffusion on air and thus their perception, for example, after application of a product for body care or household applications, are not linear over a long period of time. Therefore, many attempts to control the release of fragrances in order to increase their performance have been intensively carried out. The design of selective and efficient delivery systems in functional perfumery products has become an important research topic in the flavor and fragrance industry. Moreover, it is necessary to distinguish between physical delivery systems (e.g., microencapsulation) and chemical delivery systems, in which an odorless profragrance generates the fragrance molecule by means of a chemical reaction in which a covalent bond is selectively cleaved. The latter approach is the most relevant one and different kinds of chemical reactions have been used to trigger the release of perfumes from their corresponding precursors [112]. Considering the case of cosmetic ingredients, a number of enzyme-triggered profragrance have been prepared and used in commercial products. Indeed, human skin and hair are places of intense enzyme activity mainly because of the presence of different microorganisms. The knowledge of the biochemical transformations that take place on the skin helps to create a number of specific profragrances that are able to liberate the fragrances slowly only after contact with the human body. This approach is conceptually the opposite of that described before, where the flavors, the fragrances, or the cosmetic ingredients are prepared by means of enzyme catalysis.

Two kinds of enzyme-triggered profragrances have been developed. The first is based on the simple derivatization of the fragrance itself (Figure 9.17).

In this case, the biotransformation of the profragrance directly releases the fragrance. For instance, a number of glycosides of fragrance alcohols that are odorless stable compounds can be hydrolyzed by means of the glycosidase from the skin or skin bacteria as illustrated by the formation of phenylethanol after application of its β-glucosyl derivative **67** on the skin. Moreover, since skin bacteria can transform odorless proteinaceous secretions into malodours, the study of this peptide degradation has underlined the presence of amino acid β-lyases and of amino–acylase activities. Accordingly, the profragrances **68** and **69** were synthesized. The exposition of these compounds to axillary bacteria (e.g., by use the same in a deodorant formulation) produces the release of phenylethanol and of *cis*-3-hexenol, respectively, also reducing the formation of malodours from protein-containing secretions.

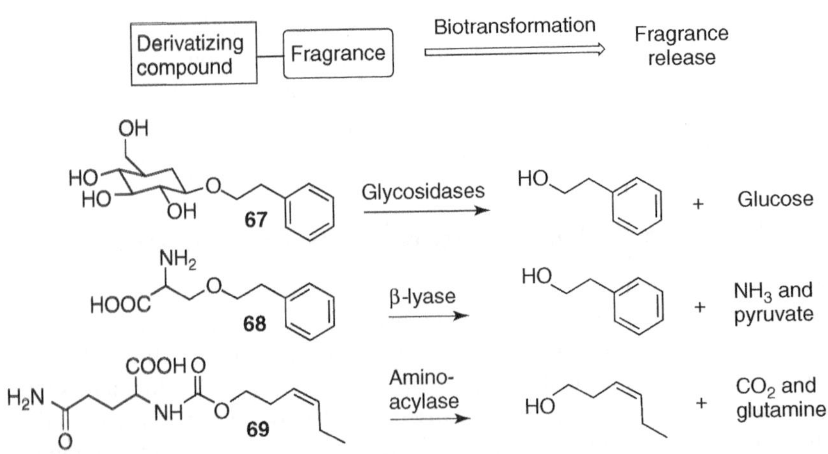

Figure 9.17 Examples of enzyme-triggered profragrances based on the derivatization of the fragrance itself.

A second relevant enzyme-triggered class of transformations is based on a two-step process (Figure 9.18).

By this way, the enzyme-catalyzed transformation of the starting precursor gives an intermediated compound(s) that undertakes further chemical transformations responsible for the final fragrance release. A domino-cleavage mechanism of this kind is involved in the degradation of carbonates of type **70** [113], which are compounds with considerable commercial value. These *o*-coumarate derivatives are used especially for laundry applications where the profragrance is added to the fabric softener. Since the detergent usually contain lipases, during the washing cycle both the hydrolytic enzyme and **70** are deposited together onto the laundry. Thanks to the normal humidity, the carbonate moiety is slowly hydrolyzed releasing a first fragrant alcohol molecule. Then the exposure of the fabric to the light produces the E/Z isomerization of the unsaturated ester **71** and the so-formed (Z)-*o*-hydroxycinnamate **72** undertakes ring closure to give a further molecule of alcohol and a molecule of the fragrance coumarin. By means of this technology when the clothes are worn, the fragrances are slowly released.

Overall, a great number of these kinds of compounds have been used in cosmetic and household applications as described in the patent literature [112]. In addition, the ability of certain enzymes to act on polymers afforded the preparation of polymeric and dendritic compounds [114] able to act as carrier substrate for the enzyme-triggered generation of fragrances. The number of the studies on all the above-mentioned topics is continuously growing as justified by their industrial and academic interest.

9.5 SUMMARY AND OUTLOOK

The examples presented here point out the power of biocatalysis in the production of flavors, fragrances, and cosmetic ingredients. Although there is considerable variability in the different methods used, the outstanding possibilities offered by biocatalysis are certainly a new opportunity for the industry as illustrated by the description of some methods of industrial and academic interest. Indeed,

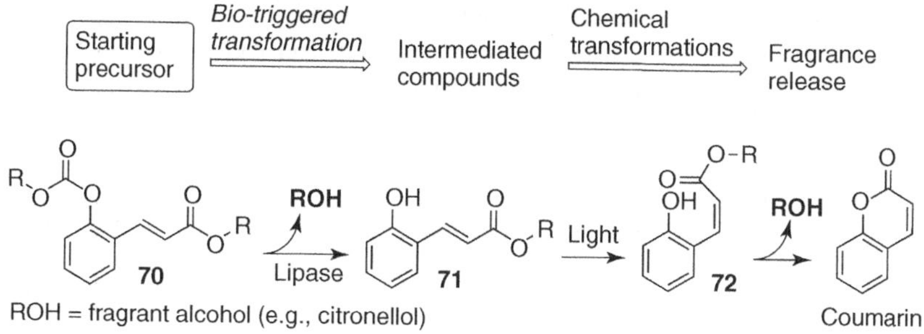

Figure 9.18 An example of enzyme-triggered profragrance based on the domino-cleavage mechanism.

the intrinsic "green" aspect of the enzyme-mediated transformation close to the use of renewable resources will take advantage of consumers' preference for natural compounds and for products obtainable from renewable raw materials. In addition, the current studies of biotechnology, biochemical pathways, and microbiology will support new strategies for natural flavor biogeneration, whereas the preparation of nature-identical flavors using biocatalysis will enhance the possibilities offered by chemical syntheses rather than compete with them.

REFERENCES

1. Pybus DH. The history of aroma chemistry and perfume. In: Pybus DH, Sell CS, editors. The chemistry of fragrances. Cambridge (UK): RSC; 1999. pp. 3–23.
2. Surburg H, Panten J. Common fragrance and flavor materials: preparation, properties and uses. Weinheim (Germany): Wiley-VCH; 2006.
3. Fráter G, Bajgrowicz JA, Kraft P. Fragrance chemistry. Tetrahedron 1998;54: 7633–7703.
4. http://www.leffingwell.com/top_10.htm
5. Brenna E, Fuganti C, Serra S. Enantioselective perception of chiral odorants. Tetrahedron: Asymmetry 2003;14:1–42.
6. Serra S. Chiral chemistry and food flavourings. In: Taylor A, Hort J, editor. Modifying flavour in food. Cambridge (UK): Woodhead Publishing Limited; 2007. pp. 107–130.
7. Serra S, Fuganti C, Brenna E. Biocatalytic preparation of natural flavours and fragrances. Trends Biotechnol 2005;23:193–198.
8. Berger RG. Biotechnology of flavours-the next generation. Biotechnol Lett 2009;31:1651–1659.
9. Schrader J, Etschmann MMW, Sell D, Hilmer J-M, Rabenhorst J. Applied biocatalysis for the synthesis of natural flavour compounds-current industrial processes and future prospects. Biotechnol Lett 2004;26:436–472.
10. Vandamme EJ, Soetaert W. Bioflavours and fragrances via fermentation and biocatalysis. J Chem Technol Biotechnol 2002;77:1323–1332.
11. Krings U, Berger RG. Biotechnological production of flavours and fragrances. Appl Microbiol Biotechnol 1998;49:1–8.
12. Cheetham PSJ. Combining the technical push and the business pull for natural flavours. Adv Biochem Eng Biotechnol 1997;55:1–49.
13. Baines D. Flavouring substances: from chemistry and carriers to legislation. In: Taylor A, Hort J, editors. Modifying flavour in food. Cambridge (UK): Woodhead Publishing Limited; 2007. pp. 10–40.
14. US Code of Federal Regulations. *21*, 101.22a.3. Washington (DC): Food and Drug Administration; 1985
15. Regulation (EC) No 1334/2008 of the European Parliament and of the Council of 16 December; 2008.
16. Walton NJ, Mayer MJ, Narbad A. Molecules of interest—Vanillin. Phytochemistry 2003;63:505–515.
17. Priefert H, Rabenhorst J, Steinbüchel A. Biotechnological production of vanillin. Appl Microbiol Biotechnol 2001;56:296–314.

18. Xu P, Hua D, Ma C. Microbial transformation of propenylbenzenes for natural flavour production. Trends Biotechnol 2007;25:571–576.
19. Etschmann MMW, Bluemke W, Sell D, Schrader J. Biotechnological production of 2-phenylethanol. Appl Microbiol Biotechnol 2002;59:1–8.
20. Kosjek B, Stampfer W, van Deursen R, Faber K, Kroutil W. Efficient production of raspberry ketone via 'green' biocatalytic oxidation. Tetrahedron 2003; 59:9517–9521.
21. Fuganti C, Zucchi G. Product distribution in the microbial biogeneration of raspberry ketone from 4-hydroxybenzalacetone. J Mol Catal B Enzym 1998;4:289–293.
22. Brenna E, Fronza G, Fuganti C, Gatti FG, Grande V, Serra S, Guillou C, Reniero F, Serra F. Stable isotope characterization of the *ortho*-oxygenated phenylpropanoids: coumarin and melilotol. J Agric Food Chem 2005;53:9383–9388.
23. Häser K, Wenk HH, Schwab W. Biocatalytic production of dihydrocoumarin from coumarin by *Saccharomyces cerevisiae*. J Agric Food Chem 2006;54:6236–6240.
24. Schwab W, Davidovich-Rikanati R, Lewinsohn E. Biosynthesis of plant-derived flavor compounds. Plant J 2008;54:712–732.
25. Waché Y, Aguedo M, Nicaud J-M, Belin J-M. Catabolism of hydroxyacids and biotechnological production of lactones by *Yarrowia lipolytica*. Appl Microbiol Biotechnol 2003;61:393–404.
26. Gill I, Valivety R. Polyunsaturated fatty acids, part 2: Biotransformations and biotechnological applications. Trends Biotechnol 1997;15:470–478.
27. Villa M, Fuganti C, Zucchi G, Allegrone G, Barbeni M, Cabella P. Process for the production of gamma nonalactones in natural form. US 5646022. 1997.
28. Fronza G, Fuganti C, Mendozza M, Rigoni R, Servi S, Zucchi G. Stereochemical aspects of flavour biogeneration through baker's yeast mediated reduction of carbonyl-activated double bonds. Pure Appl Chem 1996;68:2065–2071.
29. Muller B, Gautier A, Dean C, Kuhn J-C. Process for the enzymatic preparation of aliphatic alcohols and aldehydes from linoleic acid, linolenic acid, or a natural precursor. US 5464761. 1995.
30. Dennis MJ. Establishing food authenticity. Chem Ind 1997; 997–1000.
31. Fronza G, Fuganti C, Serra S, Burke A, Guillou C, Reniero F. The positional $\delta(^{18}O)$ values of extracted and synthetic vanillin. Helv Chim Acta 2001;84:351–359.
32. Fronza G, Fuganti C, Guillou C, Reniero F, Joulain D. Natural abundance ^{2}H nuclear magnetic resonance study of the origin of raspberry ketone. J Agric Food Chem 1998;46:248–254.
33. Fronza G, Fuganti C, Pedrocchi-Fantoni G, Serra S, Zucchi G, Fauhl C, Guillou C, Reniero F. Stable isotope characterization of raspberry ketone extracted from Taxus baccata and obtained by oxidation of the accompanying alcohol (Betuligenol). J Agric Food Chem 1999;47:1150–1155.
34. Aleu J, Fronza G, Fuganti C, Serra S, Fauhl C, Guillou C, Reniero F. Differentiation of natural and synthetic phenylacetic acids by ^{2}H NMR of the derived benzoic acids. Eur Food Res Technol 2002;214:63–66.
35. Barbeni M, Cisero M, Fuganti C. Natural abundance ^{2}H nuclear magnetic resonance study of the origin of (*Z*)-3-hexenol. J Agric Food Chem 1997;45:237–241.
36. Del Mar Caja M, Preston C, Kempf M, Schreier P. Flavor authentication studies of α-ionone, β-ionone, and α-ionol from various sources. J Agric Food Chem 2007;55:6700–6704.

37. Schmid RD, Verger R. Lipases: interfacial enzymes with attractive applications. Angew Chem Int Ed 1998;37:1608–1633.
38. Abate A, Brenna E, Fuganti C, Gatti FG, Serra S. Lipase-catalysed preparation of enantiomerically enriched odorants. J Mol Catal B Enzym 2004;32:33–51.
39. Sell C. Ingredients for modern perfumery industry. In: Pybus DH, Sell CS, editors. The chemistry of fragrances. Cambridge (UK): RSC; 1999. pp. 70–76.
40. Akutagawa S. A pratical synthesis of (-)-menthol with the Rh-BINAP catalyst. In: Collins AN, Sheldrake GN, Crosby J, editors. Chirality in industry; Chichester: John Wiley & Sons Ltd; 1992. pp. 313–323.
41. Gatfield I-L, Hilmer J-M, Bornscheuer UT, Schmid RD, Vorlová S. Verfahren zur herstellung von D- oder L-menthol. EP1223223 A1. 2002.
42. Vorlová S, Bornscheuer UT, Gatfield I, Hilmer J-M, Bertram H-J, Schmid RD. Enantioselective hydrolysis of D,L-menthyl benzoate to L-(-)-menthol by recombinant Candida rugosa lipase LIP1. Adv Synth Catal 2002;344:1152–1155.
43. Chaplin JA, Gardiner N, Mitra RK, Parkinson CJ, Portwig M, Mboniswa BA, Dickson MDE, Brady D, Marais SF, Reddy S. Process for preparing (-)-menthol and similar compounds. WO 02/36795 A2. 2002.
44. Yamamoto T. Method for purifying (-)-n-isopulegol and citrus perfume compositions containing purified (-)-n-isopulegol obtained by the method. US 5773410. 1998.
45. Gaudin J-M. Synthesis and organoleptic properties of *p*-menthane lactones. Tetrahedron 2000;56:4769–4776.
46. Ferraz HMC, Longo LS, Grazini MVA. Syntheses of mintlactone and isomintlactone. Synthesis 2002; 2155–2164.
47. Guth H. Determination of the configuration of wine lactone. Helv Chim Acta 1996;79:1559–1571.
48. Wüst M, Mosandl A. Important chiral monoterpenoid ethers in flavours and essential oil-enantioselective analysis and biogenesis. Eur Food Res Technol 1999;209:3–11.
49. Serra S, Brenna E, Fuganti C, Maggioni F. Lipase-catalyzed resolution of *p*-menthan-3-ols monoterpenes: preparation of the enantiomer-enriched forms of menthol, isopulegol, *trans*- and *cis*-piperitol, and *cis*-isopiperitenol. Tetrahedron: Asymmetry 2003;14:3313–3319.
50. Serra S, Fuganti C. Enzyme-mediated preparation of enantiomerically pure *p*-menthan-3,9-diols and their use for the synthesis of natural *p*-menthane lactones and ethers. Helv Chim Acta 2002;85:2489–2502.
51. Serra S, Fuganti C. Natural *p*-menthene monoterpenes: synthesis of the enantiomeric forms of wine lactone, epi-wine lactone, dill ether and epi-dill ether starting from a common intermediate. Helv Chim Acta 2004;87:2100–2109.
52. Shimizu M, Kamikubo T, Ogasawara K. An enantio- and diastereocontrolled route to mintlactone and isomintlactone using a chiral cyclohexadienone equivalent. Synlett 1998;655–657.
53. Forzato C, Nitti P, Pitacco G, Valentin E. Synthesis of optically active condensed γ-lactones from 2-(2-nitroethyl)cyclohexanols. Part II. Gazz Chim Ital 1995; 125:223–231.
54. Brenna E, Fuganti C, Serra S, Kraft P. Optically active ionones and derivatives: preparation and olfactory properties. Eur J Org Chem 2002;967–978.
55. Brenna E, Fuganti C, Serra S. Applications of biocatalysis in fragrance chemistry: the enantiomers of α-, β-, and γ-irones. Chem Soc Rev 2008;37:2443–2451.

56. Abate A, Brenna E, Fuganti C, Malpezzi L, Serra S. The enantiomers of Iralia®: preparation and odour evaluation. Tetrahedron: Asymmetry 2007;18:1145–1153.

57. Brenna E, Fuganti C, Serra S. Synthesis and olfactory evaluation of the enantiomerically enriched forms of 7,11-epoxymegastigma-5(6)-en-9-one and 7,11-epoxymegastigma-5(6)-en-9-ols isomers, identified in *Passiflora edulis*. Tetrahedron: Asymmetry 2005;16:1699–1704.

58. Brenna E, Fronza G, Fuganti C, Righetti A, Serra S. Enzyme-mediated preparation of the single enantiomers of the olfactory active components of the woody odorant Timberol®. Helv Chim Acta 1999;82:1762–1773.

59. Serra S, Fuganti C, Brenna E. Synthesis, olfactory evaluation, and determination of the absolute configuration of the 3,4-didehydroionone stereoisomers. Helv Chim Acta 2006;89:1110–1122.

60. Barakat A, Brenna E, Fuganti C, Serra S. Synthesis, olfactory evaluation and determination of the absolute configuration of the β- and γ-Iralia® isomers. Tetrahedron: Asymmetry 2008;19:2316–2322.

61. Luparia M, Boschetti P, Piccinini F, Porta A, Zanoni G, Vidari G. Enantioselective synthesis and olfactory evaluation of 13-alkyl-substituted α-ionones. Chem Biodivers 2008;5:1045–1057.

62. Serra S, Fuganti C. Synthesis of the enantiomeric forms of α- and γ-damascone starting from commercial racemic α-ionone. Tetrahedron: Asymmetry 2006;17:1573–1580.

63. Serra S, Barakat A, Fuganti C. Chemoenzymatic resolution of cis- and trans-3,6-dihydroxy-α-ionone. Synthesis of the enantiomeric forms of dehydrovomifoliol and 8,9-dehydrotheaspirone. Tetrahedron: Asymmetry 2007;18:2573–2580.

64. Serra S, Gatti FG, Fuganti C. Lipase-mediated resolution of the hydroxy-cyclogeraniol isomers: application to the synthesis of the enantiomers of karahana lactone, karahana ether, crocusatin C and γ-cyclogeraniol. Tetrahedron: Asymmetry 2009;29:1319–1329.

65. Inoue T, Kiyota H, Oritani T. Synthesis of both enantiomers of cis-α-irone and cis-γ-irone, principal constituents of iris oil, via resolution of (±)-2,2,4-trimethyl-3-cyclohexene-1-carboxylic acid. Tetrahedron: Asymmetry 2000;11:3807–3818.

66. Mori K, Amaike M, Itou M. Synthesis of (S)-alpha-damascone. Tetrahedron 1993;49:1871–1878.

67. Galano J-M, Audran G, Monti H. Straightforward enantioselective synthesis of both enantiomers of karahana lactone using a domino ring-closure sequence. Tetrahedron 2000;56:7477–7481.

68. Sell C. The chemistry of ambergris. Chem Ind 1990;516–520.

69. Akita H, Nozawa M, Mitsuda A, Ohsawa H. A convenient synthesis of (+)-albicanol based on enzymatic function: total syntheses of (+)-albicanyl acetate, (-)-albicanyl 3,4-dihydroxycinnamate, (-)-drimenol, (-)-drimenin and (-)-ambrox. Tetrahedron: Asymmetry 2000;11:1375–1388.

70. Tanimoto H, Oritani T. Pratical synthesis of ambrox® from farnesyl acetate involving lipase catalyzed resolution. Tetrahedron: Asymmetry 1996;7:1695–1704.

71. Akita H, Nozawa M, Shimizu H. Synthesis of decalin type chiral synthons based on enzymatic functionalisation and their application to the synthesis of (-)-ambrox and (+)-zonarol. Tetrahedron: Asymmetry 1998;9:1789–1799.

72. Fuganti C, Serra S, Zenoni A. Synthesis and olfactory evaluation of (+)- and (-)-γ-ionone. Helv Chim Acta 2000;83:2761–2768.
73. Horiuchi S, Takikawa H, Mori K. Enzymatic resolution of (±)-γ-cyclohomogeraniol and conversion of its (*S*)-isomer to (*S*)-γ-coronal, the ambergris odorant. Bioorg Med Chem 1999;7:723–726.
74. Kiyota H, Higashi E, Koike T, Oritani T. Lipase-catalyzed preparation of both enantiomers of methyl jasmonate. Tetrahedron: Asymmetry 2001;12:1035–1038.
75. Fehr C, Galindo J. A new variant of the Claisen rearrangement from malonate-derived allylic trimethylsilyl ketene acetals: efficient, highly enantio- and diastereoselective syntheses of (+)-methyl dihydroepijasmonate and (+)-methyl epijasmonate. Angew Chem Int Ed 2000;39:569–573.
76. Abate A, Brenna E, Fuganti C, Gatti FG, Giovenzana T, Malpezzi L, Serra S. Chirality and fragrance chemistry: stereoisomers of the commercial chiral odorants Muguesia and Pamplefleur. J Org Chem 2005;70:1281–1290.
77. Vieira GAB, Lemos TLG, de Mattos MC, de Oliveira M, Melo VMMM, de Gonzalo G, Gotor-Fernandez V, Gotor V. Chemoenzymatic synthesis of optically active mugetanol isomers: use of lipases and oxidoreductases in fragrance chemistry. Tetrahedron: Asymmetry 2009;20:214–219.
78. Abate A, Brenna E, Fronza G, Fuganti C, Gatti FG, Serra S, Zardoni E. Preparation of the enantiomerically enriched isomers of the odorous cyclic ethers Clarycet®, Florol® and Rhubafuran® by enzymatic catalsysis. Helv Chim Acta 2004;87:765–779.
79. Abate A, Allievi M, Brenna E, Fuganti C, Gatti FG, Serra S. Enzymatic approach to and odor description of the twelve enantiomerically pure isomers of Pelargene®. Helv Chim Acta 2006;89:177–189.
80. Abate A, Brenna E, Fronza G, Fuganti C, Ronzani S, Serra S. Enzyme-mediated preparation of chiral 1,3-dioxane odorants. Helv Chim Acta 2003;86:592–606.
81. Abate A, Brenna E, Fuganti C, Serra S. Enzyme-mediated synthesis of new 1,3-dioxane odorants related to Floropal®. Flavour Fragrance J 2004;19:382–393.
82. Brenna E, Fuganti C, Gatti FG, Malpezzi L, Serra S. Synthesis and olfactory evaluation of all stereoisomers of the fragrance Nectaryl®. Tetrahedron: Asymmetry 2008;19:800–807.
83. Csuk R, Glänzer BI. Baker's yeast mediated transformations in organic chemistry. Chem Rev 1991;91:49–97.
84. Fuganti C, Serra S, Dulio A. Baker's yeast mediated enantioselective synthesis of the bisabolane sesquiterpenes curcumene, turmerone, dehydrocurcumene and nuciferal. J Chem Soc Perkin Trans 1 1999;279–282.
85. Abate A, Brenna E, Dei Negri C, Fuganti C, Serra S. Biocatalysed synthesis of the enantiomers of the floral odorant Florhydral®. Tetrahedron: Asymmetry 2002;13:899–904.
86. Abate A, Brenna E, Costantini A, Fuganti C, Gatti FG, Malpezzi L, Serra S. Enzymatic approach to enantiomerically pure 5-alken-2,4-diols and 4-hydroxy-5-alken-2-ones: application to the synthesis of chiral synthons. J Org Chem 2006;71:5228–5240.
87. Serra S, Fuganti C, Gatti FG. A chemoenzymatic, preparative synthesis of the isomeric forms of p-menth-1-en-9-ol: application to the synthesis of the isomeric forms of the cooling agent 1-hydroxy-2,9-cineole. Eur J Org Chem 2008;1031–1037.

88. Benedetti F, Forzato C, Nitti P, Pitacco G, Valentin E, Vicario M. Synthesis of all stereoisomers of cognac lactones via microbial reduction and enzymatic resolution strategies. Tetrahedron: Asymmetry 2001;12:505–511.

89. Brenna E, Dei Negri C, Fuganti C, Serra S. Baker's yeast-mediated approach to (-)-*cis*- and (+)-*trans*-Aerangis lactones. Tetrahedron: Asymmetry 2001;12: 1871–1879.

90. Brenna E, Fuganti C, Ronzani S, Serra S. Biocatalyzed preparation of the optically enriched stereoisomers of 4-methyl-2-phenyl-tetrahydro–2*H*-pyran (Doremox®. Can J Chem 2002;80:714–723.

91. Longo MA, Sanromán MA. Production of food aroma compounds: microbial and enzymatic methodologies. Food Technol Biotechnol 2006;44:335–353.

92. Giri A, Dhingra V, Giri CC, Singh A, Ward OP, Narasu ML. Biotransformations using plant cells, organ cultures and enzyme systems: current trends and future prospects. Biotechnol Adv 2001;19:175–199.

93. Vandamme EJ. Bioflavours and fragrances via fungi and their enzymes. Fungal Divers 2003;13:153–166.

94. Lomascolo A, Stentelaire C, Asther M, Lesage-Meessen L. Basidiomycetes as new biotechnological tools to generate natural aromatic flavours for the food industry. Trends Biotechnol 1999;17:282–289.

95. Da Carvalho CCCR, Da Fonseca MMR. Biotransformation of terpenes. Biotechnol Adv 2006;24:134–142.

96. Bicas JL, Dionisio AP, Pastore GM. Bio-oxidation of terpenes: an approach for the flavor industry. Chem Rev 2009;109:4518–4531.

97. Duetz WA, Bouwmeester H, van Beilen JB, Witholt B. Biotransformation of limonene by bacteria, fungi, yeasts, and plants. Appl Microbiol Biotechnol 2003;61: 269–277.

98. Fraatz MA, Berger RG, Zorn H. Nootkatone-a biotechnological challenge. Appl Microbiol Biotechnol 2009;83:35–41.

99. Belin J-M, Dumont B, Ropert F. Enzymatic process for the preparation of flavours, in particular the ionones and C_6 to C_{10} aldehydes. US 5705372. 1998.

100. Fronza G, Fuganti C, Grasselli P, Servi S, Zucchi G, Barbeni M, Villa M. Reduction of carboxylates to alkanols catalysed by *Colletotrichum gloeosporoides*. Chem Commun 1995;439–440.

101. Dulio A, Fuganti C, Zucchi G. Biogeneration of (E,E)-2,4-hexadienal and (E)-4-hexen-1-ol from sorbic acid by *Colletotrichum gloeosporoides* cultures. Flavour Fragrance J 1999;14:79–81.

102. Veit T. Biocatalysis for the production of cosmetic ingredients. Eng Life Sci 2004;4:508–511.

103. Hasan F, Shah AA, Hameed A. Industrial applications of microbial lipases. Enzyme Microb Technol 2006;39:235–251.

104. Yadav GD, Dhoot SB. Immobilized lipase-catalysed synthesis of cinnamyl laurate in non-aqueous media. J Mol Catal B Enzym 2009;57:34–39.

105. Maugard T, Legoy MD. Enzymatic synthesis of derivatives of vitamin A in organic media. J Mol Catal B Enzym 2000;8:275–280.

106. Smeets JWH, De Pater RM, Lambers JWJ. Enzymatic synthesis of ceramides and hybrid ceramides. US 5610040.

107. Walsh MK, Bombyk RA, Wagh A, Bingham A, Berreau LM. Synthesis of lactose monolaurate as influenced by various lipases and solvents. J Mol Catal B Enzym 2009;60:171–177.

108. Bousquet M-P, Willemot R-M, Monsan P, Boures E. Enzymatic synthesis of unsaturated fatty acid glucoside esters for dermo-cosmetic applications. Biotechnol Bioeng 1999;63:730–736.

109. Chang S-W, Yang C-J, Chen F-Y, Akoh CC, Shieh C-J. Optimized synthesis of lipase-catalyzed L-ascorbyl laurate by Novozym® 435. J Mol Catal B Enzym 2009;56:7–12.

110. Schwarz A, Thomsen MS, Nidetzky B. Enzymatic synthesis of β-glucosylglycerol using a continuous-flow microreactor containing thermostable β-glycoside hydrolase celB immobilized on coated microchannel walls. Biotechnol Bioeng 2009;103:865–872.

111. Sheldon RA, Arends I, Hanefeld U. Chemicals from renewable raw materials. Green chemistry and catalysis. Weinheim (Germany): Wiley-VCH Verlag GmbH & Co. KGaA; 2007. pp. 329–387.

112. Herrmann A. Controlled release of volatiles under mild reaction condition: from nature to everyday products. Angew Chem Int Ed 2007;46:5836–5863.

113. Flachsmann F, Gautschi M, Bachmann J-P, Brunner G. Enzyme-triggered and self-cleaving fragrant alcohol precursors. Chem Biodivers 2008;5:1115–1136.

114. Aulenta F, Drew MGB, Foster A, Hayes W, Rannard S, Thornthwaite DW, Youngs TGA. Fragrance Release from the surface of branched poly(amide)s. Molecules 2005;10:81–97.

CHAPTER 10

APPLICATION OF BIOCATALYSIS IN THE AGROCHEMICAL INDUSTRY

DAVID A. JACKSON
Syngenta Crop Protection Muenchwilen AG., Muenchwilen, Switzerland

10.1 INTRODUCTION

Chemical manufacture and the use of chemical products have had an environmental impact for centuries—for just as long as there has been an ever growing realization and learning of the need to improve processes, to minimize not only cost but also the consequences of unconstrained release of waste to land and into sea and air. This has led to increasingly demanding regulatory requirements that have driven the typical issues and imperatives of waste management in the form of avoidance, abatement, recovery, recycle, and second use. Expressions such as "telescoping" of reactions or "domino reactions" are modern terms for process organization that have been undertaken in industry for decades, usually to reduce costs. The term "Green Chemistry" is a similar recent idiom that is used to capture a host of concepts and technologies that seek to minimize the environmental impact of a given transformation, but do not always serve to address the wider issues of either process organization or sustainability [1]. Nevertheless, the term "Green Chemistry" has spawned some useful concepts [2,3] and tools [4,5] for assessing the environmental impact of transformations and processes [6,7].

The population of the world will increase to possibly eight billion by 2030, with a calorific intake 50% higher than today. Land and water resources are under increasing pressure. The environmental load has never been greater than today. The issue of producing sufficient food for the world in a sustainable way is currently a major global concern that warrants examination of all the aspects of

Biocatalysis for Green Chemistry and Chemical Process Development, First Edition.
Edited by Junhua (Alex) Tao and Romas Kazlauskas.

food, feed, farming, and production, including their biology and chemistry. As a consequence of this, there is a demand for new improved products and better ways to make them. The agribusinesses are rising to these challenges in multiple ways that include minimum-input farming practices, improved seeds, increased efficacy of products, increased selectivity over nontarget organisms, new products with reduced environmental impact, and withdrawal of older products with undesirable properties.

Biotransformations, in the sense of fermentation or biocatalysis, are being actively used in research to identify biologically active molecules and in the large-scale production of active ingredients. However, the use of biotransformations is still a relatively small component of the overall production technologies applied by industry. Biological systems are used to address specific needs that are not easily met by abiotic technologies. Biotransformations have proved useful in producing natural products by fermentation, enantioselective manufacture of chiral active substances, or in functinalization of relatively unactivated positions. Biotransformations have most commonly been used for either high-value low-volume products such as pharmaceuticals or for low-value high-volume products such as sugar transformations. It is important to recognize that for the agricultural business, cost of goods is a critical business driver.

This review exemplifies the use of biotransformations in the manufacture of active substances, highlights the alternative approaches that have been applied, examines and contrasts a real example of abiotic and a biotransformation, and draws conclusions about the future applications of biotransformations to manufacture of agrochemicals. Out of scope of the review is the use of biotransformations in active substance discovery, in animal health products, or in pheromone synthesis. A general review of the use of biotransformations in the synthesis of agrochemicals appeared in 2006 [8], and in the synthesis of pheromones in 2007 [9]. Many monographs also contain additional reviews relevant to manufacture of agrochemicals [10].

10.2 NATURAL PRODUCT PRODUCTION BY FERMENTATION

Natural products have been a source of inspiration for people searching for interesting biological properties. Microorganisms and plants are used on a small scale in a variety of pest control applications [11]. The crushed flowers of *Chrysanthemum cinerariaefolium* contain natural pyrethroid compounds and have been used as insecticides since the nineteenth century. *Bacillus thuringiensis* (Bt), a soil bacteria, has insecticidal properties and has been used since the 1920s as a spray of bacterial spores or proteins. The use of the toxin produced by Bt has now increased significantly because of the transfer of appropriate genes for toxin production into crops such as corn and cotton.

Fermentation broths have yielded many compounds of interest to the agricultural industry. Often, the complexity of the molecules obviates a total abiotic synthesis. Fermentation is used to manufacture such molecules using either wild or, more usually, engineered organisms optimized for fermentation productivity and improved robust product isolations. Examples of the more important of these are given below.

10.2.1 Abamectin

A family of macrolides was isolated from fermentation of *Streptomyces avermitilis*, a naturally occurring soil Actinomycete. The discovery and development of Abamectin by workers at Merck has been well described [12]. The product, introduced by Merck in 1985 as an acaricide and insecticide, is a mixture of two compounds, avermectin B1a and avermectin B1b (ratio about 90:10) (Figure 10.1). Abamectin is now also produced in China. The fermentation over 7–14 days produces Abermectin at 5–10 g/L. The product remains in the cells and is extracted either directly from the fermentation broth or from the filtered biomass using hydrocarbon or alcohol solvents. There are ongoing efforts to further improve the production technology in strain development [13] and extraction methods [14]. Notably, the waste fermentation broth is extremely toxic and requires treatment at about 150°C before it can undergo normal biodigestion.

10.2.2 Bilanafos and L-Phosphinothricin

Bilanafos, also known as Bialaphos, is a tripeptide with herbicidal properties. It was first identified as a fermentation product of *Streptomyces hygoscopicus* and introduced as a product by Meiji Seika Kaisha in 1984. A description of the fermentation scale-up has been given [15]. During investigations into the structure, biosynthesis, and laboratory synthesis of Bilanafos, the left part of the molecule was prepared independently by workers in Germany [16] and Japan [17] (Figure 10.2). Notably, the new compound Phosphinothricin, also known as Gluphosinate, was found to have remarkable herbicide properties. It has become a significant product as the racemic mixture manufactured by total synthesis. The single enantiomer has also been evaluated by fermentation [18]. Meiji is still evaluating improvements to both the fermentation and synthesis of Phosphinothricin [19–21].

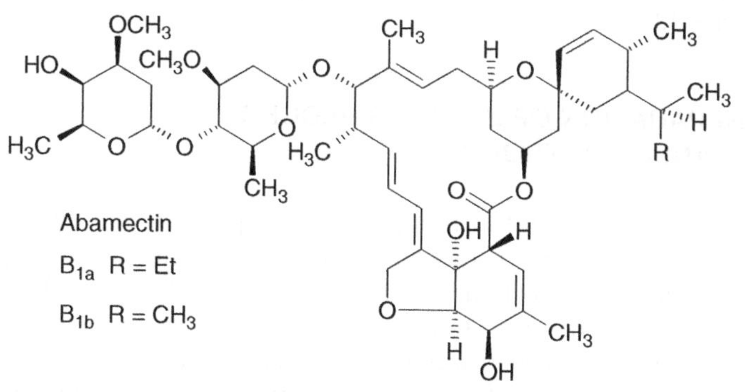

Figure 10.1 Abamectin produced by fermentation is a mixture of two components.

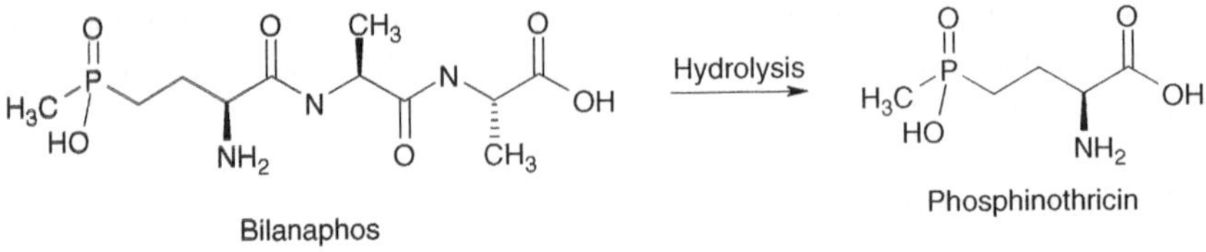

Figure 10.2 Tripeptide Bilanaphos, a fermentation product, is the origin of Phophinothricin.

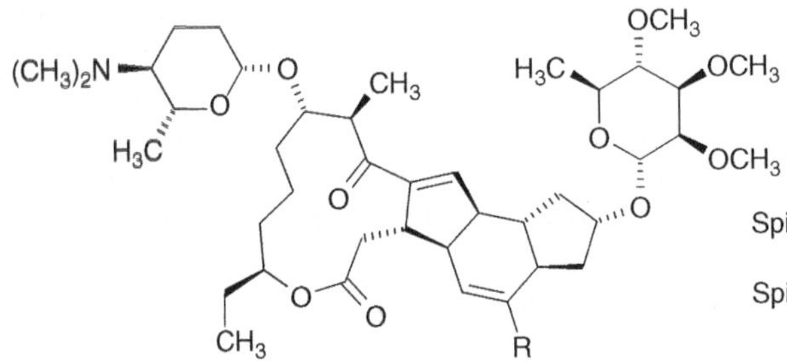

Figure 10.3 The soil bacteria *Saccharopolyspora spinosa* produces the insecticidal Spinosyns.

10.2.3 Spinosad

During the 1980s, joint investigation by Eli Lilly and Co. and Dow Chemical Co. of fermentation broths identified a soil bacterium, *Saccharopolyspora spinosa*, that produces metabolites with interesting insecticidal properties. The organism produced a variety of related molecules characterized as macrolides that were named spinosyns. The two most active molecules against insects were found to be spinosyn A and spinosyn D (Figure 10.3). Dow AgroSciences subsequently developed the product Spinosad, which contains A and D in a ratio of approximately 5:1 to 6:1. The product was brought to market in 1997. The discovery, microbiology, fermentation, and manufacture have been described by Racke [22]. The use of biotechnology to improve strain productivity has also been reported [23].

10.3 PREPARATION OF BUILDING BLOCKS BY BIOTRANSFORMATIONS

Active substances are identified from both natural products and libraries of synthetic molecules. However, in order to exploit the interesting biological properties as sales products, it is usually necessary to modify the structures. Typically, the synthetic analogs are designed to have higher efficacy, greater biological selectivity, reduced toxicity, and reduced environmental persistence. Inevitably, the resultant new structures are all prepared industrially using chemical transformations. Only a minority of these transformations are performed with some sort of biological assistance such as live cells, dead cells, or isolated enzymes. This section describes examples of the biggest applications of biotransformations to prepare building blocks specific to the production of agrochemicals. Biotransformations for the preparation of general building blocks used in a range of other product sectors are not covered, for example, amino acids.

10.3.1 Preparing Single Enantiomers

10.3.1.1 Approaches to the Phenoxypropionoates. There is a group of selective herbicides that are active against the target plant species by enzyme inhibition of fatty acid biosynthesis. Within this group are a number of active compounds that posses a common structural moiety; an aryloxy-substituted phenoxypropionoate. Representative examples are displayed in Figure 10.4. Notably, the molecules with the *R*-configuration at position 2 are the most active.

Several synthetic approaches have been made to this group of molecules as single enantiomers. These are mostly based on some form of resolution technology on an early-stage intermediate. The different synthetic paths are described in

Clodinafop-propagyl

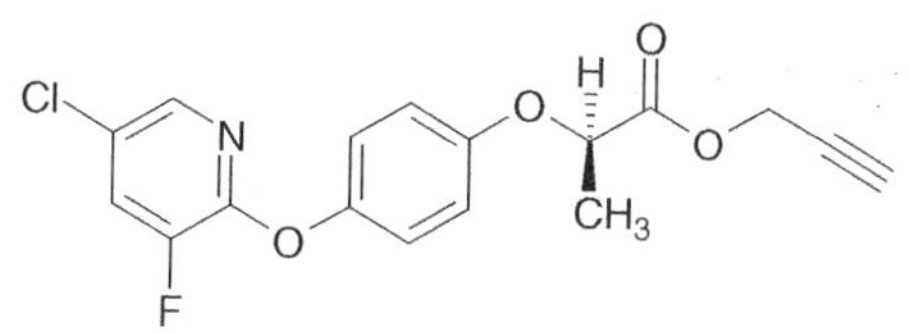

Cyhalofop-butyl

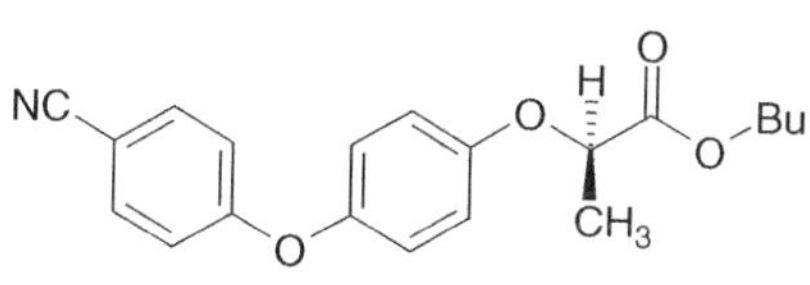

Fenoxaprop-*P*-ethyl

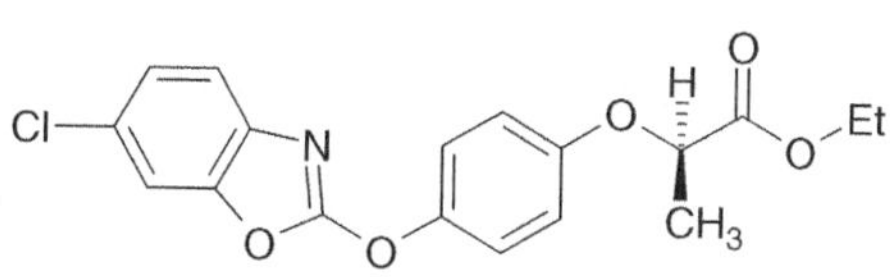

Propaquinizafop

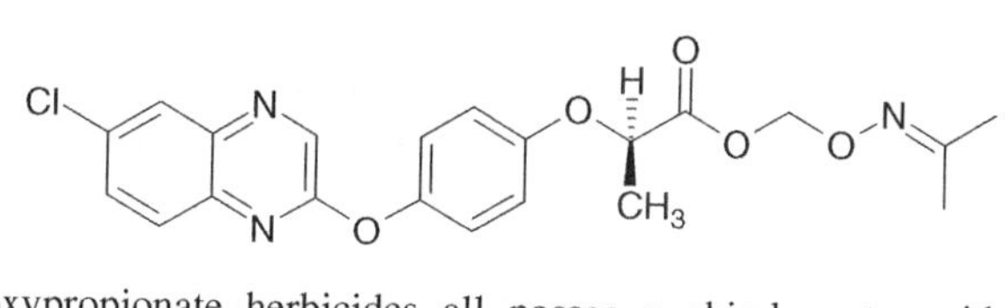

Figure 10.4 The phenoxypropionate herbicides all posses a chiral centre with the *R*-configuration. Here are some representative examples.

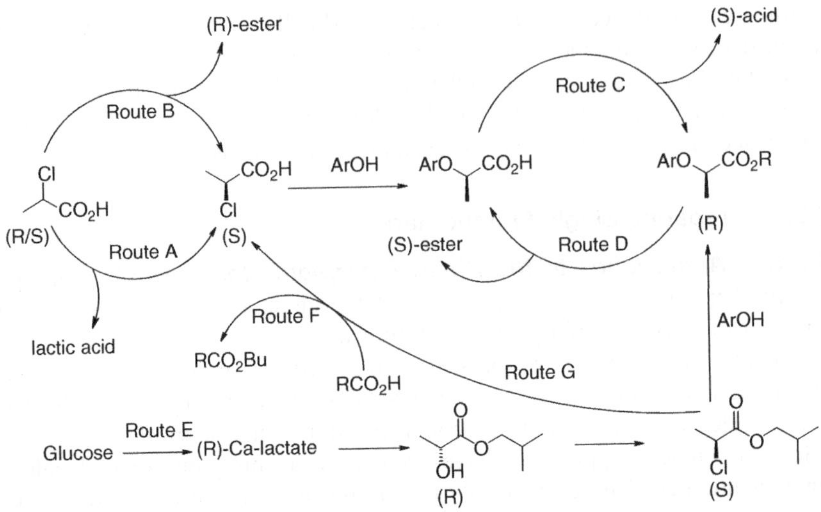

Figure 10.5 The (*R*)-2-aryloxpropionate herbicides can be prepared by different routes. Several have been operated at industrial scale. The details of the bio transformations for each route A–G are given in Table 10.1 below.

TABLE 10.1 (*R*)-Aryloxypropionate Routes and References

Route	Information	References
A	Halidohydrolase catalyzed hydrolysis of the unwanted enantiomer. Suspended dead cells of *Pseudomonas putida*. Substrate about 11 gm/L, pH 7.4, 30°C. 99% *ee*	[10a,24,25]
B	Lipase-catalyzed esterification of the unwanted enantiomer in hexane-containing butanol	[26]
C	Hydrolase on celite-catalyzed esterification of the unwanted enantiomer	[27]
D	Desired enantiomer is hydrolyzed using esterase from *Bacillus subtilis*	[28]
E	Fermentation of glucose using a *lactobacillus*	[29]
F	Transesterification using an additional carboxylic acid receiver molecule. 98% *ee*.	[29]
G	Hydrolysis of the single enantiomer catalyzed by lipase from *Pseudomonas* at pH 5–7, 20–40°C, 98% *ee*.	[30]

Figure 10.5. The route information and references are included in the accompanying Table 10.1. The steps in routes A, E, and F have certainly been operated on a large scale for production of (*R*)-2-aryloxpropionate herbicides.

10.3.1.2 Approaches to (S)-1-Methoxy-2-Aminopropane. The chloroacetanilides are a class of grass herbicides used in crops such as corn. Two particular

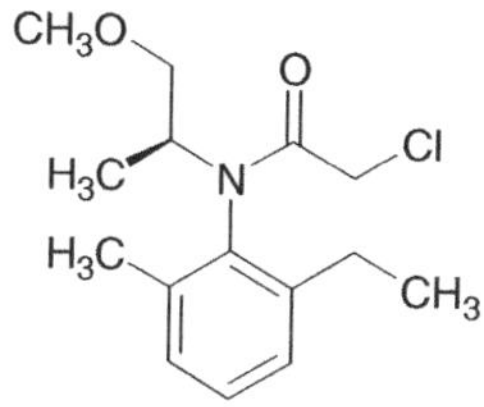

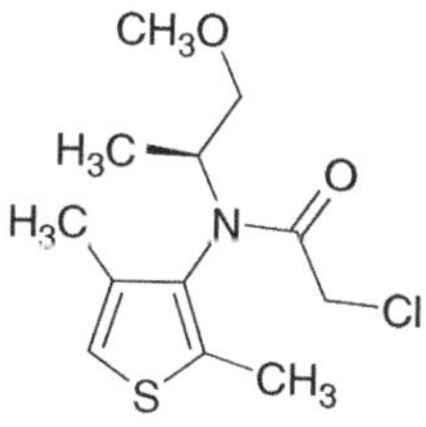

Figure 10.6 The chloroacetanilide herbicides typically contain a chiral centre alpha to the amine. Where steric hindrance restricts rotation about the anilide bond atropisomers exist, for example, in (*S*)-Metolachlor.

products, Metalochlor (commercialized 1976) and Dimethenamid (discovered mid 1980s), were originally developed as racemic mixtures. The chirality of the molecules lies in the recognizable 1-methoxy-2-aminopropane part of the structures. Metolachlor also contains an additional chiral element because of restricted rotation around the C_{Ar}–N bond. Subsequent to the discovery of Metolachlor and Dimethanamid, it was further identified that a major part of the herbicidal activity resides in particular enantiomers (Figure 10.6).

Changing from the racemic products to the single enantiomers is very desirable because of reduced costs for producers and farmers and reduced environmental load in production and application. Chemical and biotransformation technologies have been evaluated for the preparation of both molecules. In the case of (*S*)-Metolachlor, the method of manufacture utilizes hydrogenation of an imine catalyzed by an iridium catalyst system utilizing the enantiomerically pure ligand "xyliphos" to produce the desired enantiomers. The development and implementation of this abiotic technology is now a well told story [31].

Racemic Dimethenamid was developed by Sandoz, but as a result of industrial reorganizations, the product was sold to BASF in 1996. The original route to Dimethenamid [32] is described in Figure 10.7.

BASF implemented a switch to the single enantiomer Dimethenamid-P. The existing synthesis route provided an obvious opportunity; substituting racemic 1-methoxy-2-aminopropane with the (*S*)-enantiomer. BASF synthesize (*S*)-1-methoxy-2-aminopropane, using a resolution procedure [29,33]. This possibly involves selective acylation with ethyl 2-methoxyacetate catalyzed by a lipase [34], or a longer chain ester [35], as shown in Figure 10.8. The reaction takes place without a solvent using neat components at 20–60°C and achieves, for the decyl-ester (R = n-$C_{10}H_{21}$), conversions of >60% with (*S*)-amine *ee's* of >99%. The desired (*S*)-amine can be recovered from the unwanted (*R*)-amide and alcohol by-product by either extraction or distillation.

A significant drawback of the procedure is typical of a resolution process; the unwanted enantiomer, in this case the (*R*)-amide, requires an outlet by recycle, alternative use, or disposal. To this end, BASF [36] and others [37] have developed processes for recycle of the undesired (*R*)-amide by liberation of

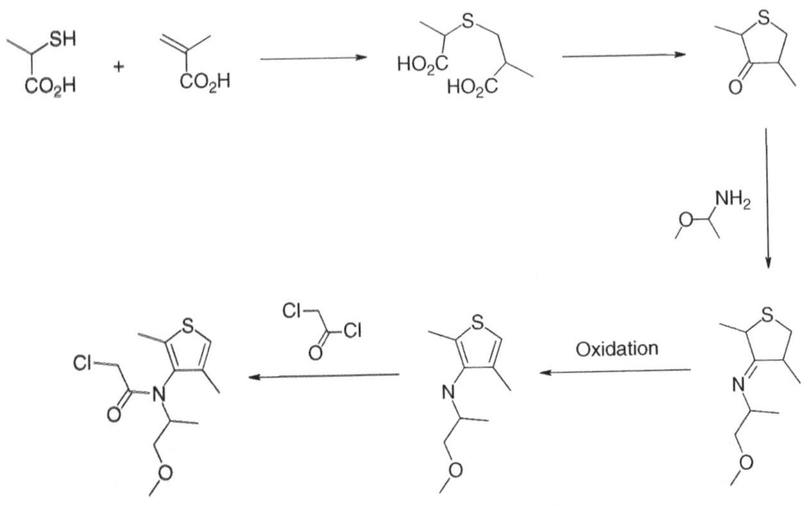

Figure 10.7 The original sandoz route to dimethenamid utilized racemic 1-methoxy-2-aminopropane.

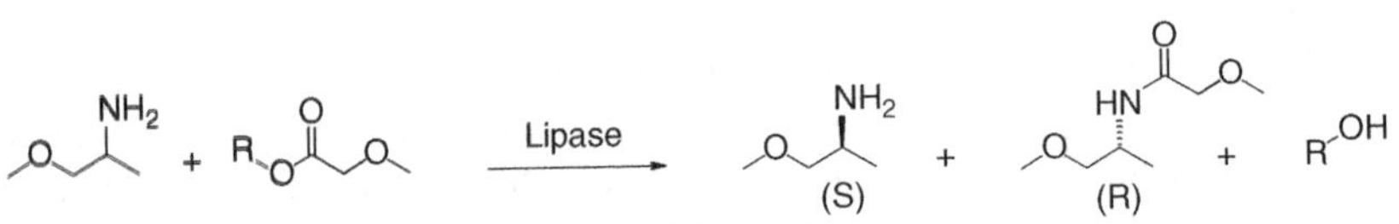

Figure 10.8 (S)-1-methoxy-2-aminopropane is resolved by a bio transformation based on acylation of the amino group of the unwanted enantiomer.

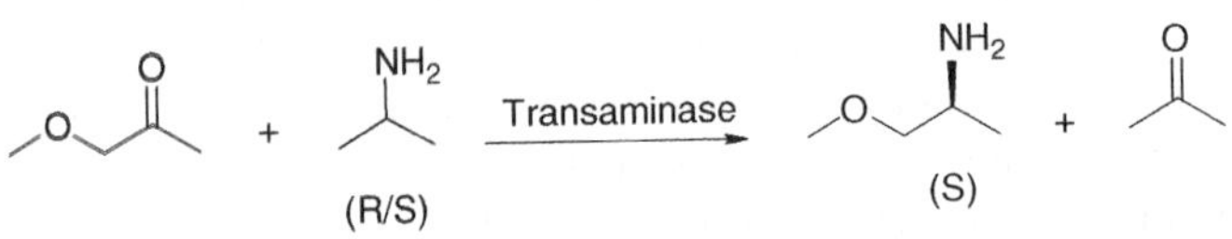

Figure 10.9 Transamination produces (S)-1-methoxy-2-aminopropane directly without the need for recovery, racemization, and recycle of the unwanted enantiomer.

the (R)-amine and racemization by, for example, heating the amine at a high temperature in the gas phase in the presence of hydrogen and metal catalysts (Cu, ZnO, Ni, and Cu).

Celgro has taken a significantly different strategy based upon transamination (Figure 10.9). Beginning from enzyme activity in a soil bacterium, the gene responsible was transferred into *Escherichia coli*. Enzyme improvement was undertaken by mutagenesis and selection to further improve the productivity; concentration, time, and activity. The steps taken and the final outcome have been reported [38].

The reaction runs in water at 2 M concentration, pH 7.5, 50°C over 7 h and requires a small amount of pyridoxal phosphate (0.2 mM). The actual biocatalyst

(5 g/L) is a powder obtained after spray drying the fermentation broth. (*S*)-1-methoxy-2-aminopropane is obtained in a 99% *ee*. An important process feature to achieve acceptable conversions, >90%, involves continuous removal of acetone from the reaction mass under reduced pressure (0.1 Atm).

A somewhat more complicated transamination procedure has also been demonstrated using alanine as the amino source [39]. The sequence involves a first-step kinetic resolution followed by reductive amination. Notably, the procedure requires two transaminases with opposite stereo preferences. A 99% yield with a 99% *ee* is obtained in a 93% chemical yield from methoxyacetone.

Celgro has also reported the synthesis of (*S*)-Metolachlor from (*S*)-1-methoxy-2-aminopropane [40]. The key reaction is a Pd-catalyzed homogeneous coupling of the (*S*)-amine to an activated aromatic electrophile (Figure 10.10). It is interesting to compare the Celgro biotransformation approach with the conventional (*S*)-Metolachlor synthesis that involves catalytic enantioselective hydrogenation [31] using a homogeneous catalyst (Xyliphos/Iridium) Figure 10.11.

Both routes probably start from the aniline and pass through the intermediate (*S*)-NAA. Notably, the conventional sequence produces only 30 kg of waste per 1000 kg of (*S*)-NAA. Extremely low catalyst (Xyliphos/Iridium) loading is

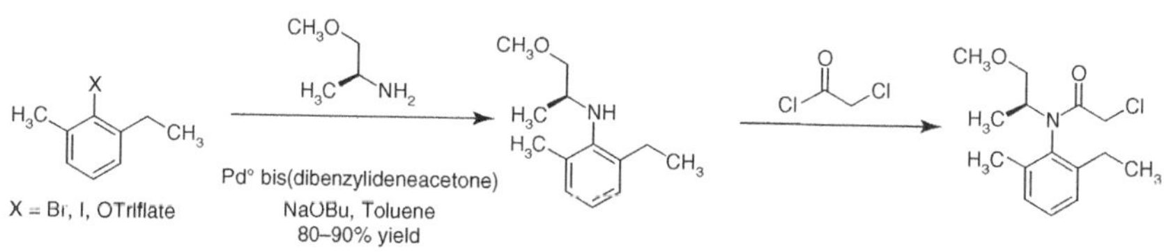

Figure 10.10 The celgro route to (*S*)-metolachlor required the development of palladium catalysed coupling of (*S*)-1-methoxy-2-aminopropane with an aryl derivative.

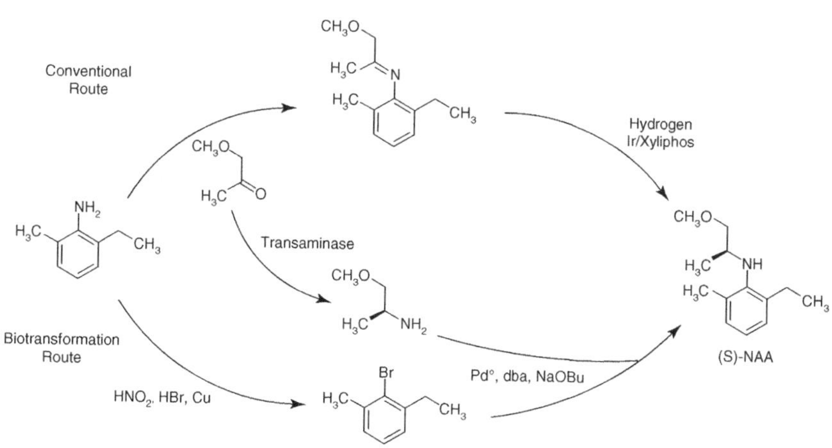

Figure 10.11 The stereoselctive route to (*S*)-NAA an intermediate for (*S*)-metolachlor is short and very efficient.

critical in achieving a commercially viable process substrate : catalyst ratio of 1,000,000:1. The origin of the chirality in the ligand Xyliphos is a biotransformation! A lipase-catalyzed esterificationis is used to resolve 1-ferrocenylethanol in an early step of the ligand synthesis [41].

10.3.1.3 Approaches to Cyanohydrins. The insecticidal properties of ground flowers of Chrysanthemum were recognized in the nineteenth century and led to the preparation and sale of various products. The worldwide production of dried flowers increased to over 20,000 tones per annum in the 1970s [42]. The active molecules are typically based on structures described in Figure 10.12.

To broaden the activity and potency, synthetic pyrethroids began to be developed in the 1940s and work continues to the present day. Typically, the synthetic pyrethroids are mixtures of geometrical and diastereoisomers. Notable additional structural elements are the introduction of halogenated vinyl substituents and the acyl-cyanohydrin group typified by the structures in Figure 10.13.

This class of molecules is often prepared by the acylation of an appropriate cyanohydrin. The recognition that a significant proportion of the insecticidal activity resides in one of the diastereoisomers and/or enantiomers has led to considerable attention to the preparation of (*S*)-cyanohydrins and, in particular, (*S*)-3-phenoxybenzaldehyde cyanohydrins [43]. Biocatalyltic methods have played a leading role in the synthesis of single enantiomers. The principal methods are

1. Reaction of an aldehyde with HCN mediated by an isolated oxynitrilase [43].
2. Kinetic resolution of a racemic cyanohydrins using an oxynitrilase [44].
3. Resolution of racemic cyanohydrins by a selective acylation [45].
4. Kinetic resolution of the racemic ester of a cyanohydrins by a selective deacylation [46–50].

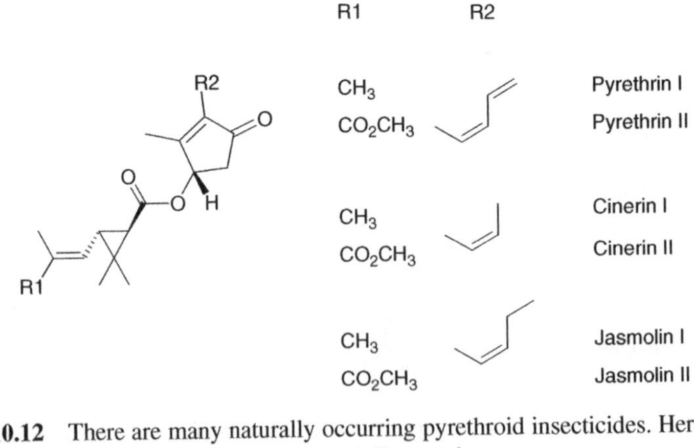

Figure 10.12 There are many naturally occurring pyrethroid insecticides. Here are some examples from those found in the flowers of Chrysantheum.

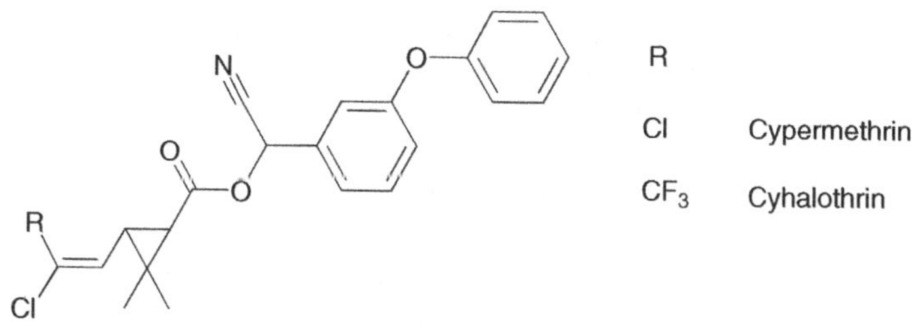

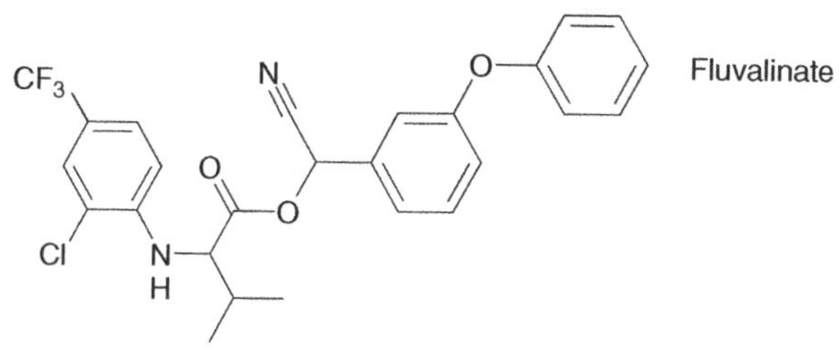

Figure 10.13 The synthetic pyrethroids notably have halogenated vinyl substituents and the acyl-cyanohydrin group to broaden activity, potency, and stability. Here are some examples.

Method (1) is the most efficient and has been applied in an industrial process by DSM [51]. The (*S*)-oxynitrilase enzyme derived from a plant was cloned and expressed in a microbial host [52]. Further steps that have been evaluated to improve this approach include identification of enzymes with wider pH tolerance [53] and immobilization of the enzyme [54]. Resolution processes inevitably require attention to recycle of the unwanted enantiomer. Although this can be achieved simply by hydrolysis of the undesired cyanohydrin to the starting aldehyde and recycle, there have been some attempts to be more efficient. The Effenberger group [49] has demonstrated the steps for enantioselective synthesis of (1*R*,*cis*,α*S*)-cypermethrine via a lipase resolution of racemic 3-phenoxybenzaldehyde cyanohydrin acetate [44] and application of the Mitsuda method for the racemization of the undesired isomer by heating with triethylamine [55].

10.3.2 Functional Group Introduction

There are two significant examples of production-scale biotransformations for the introduction of functionality to enable potentially more efficient construction of agrochemicals. Both are regiospecific, oxidative hydroxylations of aromatic carbon–hydrogen bonds: a benzenoid to generate a phenol and a pyridine to form a pyridone. Both are impressive and now well-known processes.

10.3.2.1 Preparation of (R)-2-(4-Hydroxyphenoxy)propionoic Acid. The more widely applied routes to the (*R*)-2-(4-hydroxyphenoxy) propionates were described earlier in the chapter (Section 10.3.1.1, Preparation of (*R*)-2-(4-Hydroxyphenoxy)propionoic Acid). BASF developed [56] what they claim to be a more environmentally friendly route, described in Figure 10.14.

Both phenol and 1,4-hydroquinone (HQ) are usually derived from oxidation of the isopropenyl alkylation products of benzene. HQ is more expensive than phenol and is sometimes obtained via chemical oxidation of phenol. The route from BASF has potential advantages; it uses cheaper phenol compared to HQ and the O-alkylation with l-chloropropionates does not suffer a selectivity problem that afflicts O-monoalkylation of HQ, resulting in about 10% yield loss due to bis-O-alkylation, which requires a difficult purification process involving crystallization. The regiospecific hydroxylation of the phenoxypropionate in the 4-position by conventional chemistry is not currently feasible. However, BASF has made efforts to develop a biotransformation using live cells of *Beauveria bassiana* that will convert a range of similar substrates at concentrations of >5 g/L. We have no information about consumption of feed, fermentation efficiency, or the disposal of biomass, so it is difficult to evaluate where the environmental benefit lies in the transformation of benzene through four steps to (*R*)-2-(4-hydroxyphenoxy)propionates. This is a good example of where care should be taken in framing the boundaries when asserting that a given technology is "greener" or more environmentally friendly when compared to another technology.

In order to evaluate which route and technology has the smallest environmental impact, one must examine the technology of all the steps involved, which is beyond the scope of this discussion. Manufacturers choose the route and technology best suited to their needs and capabilities using criteria such as cost, equipment availability, capital investment, expertise, waste disposal, resource, and timing.

10.3.2.2 Preparation of 6-Hydroxynicotinic Acid and Related Structures. There is an important class of insecticides of which the following

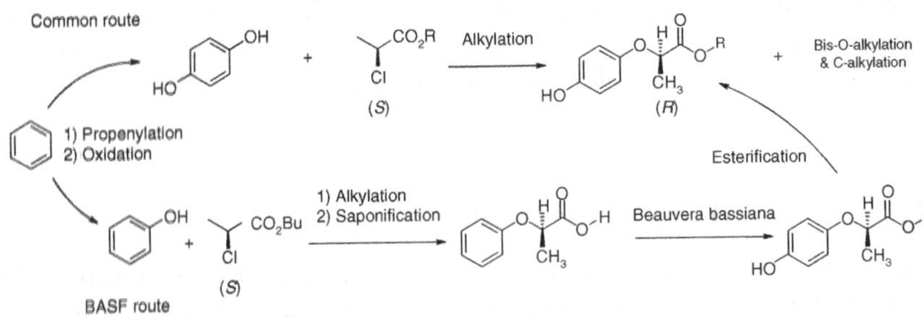

Figure 10.14 A schematic comparison of the common route and the BASF route to (*R*)-2-(4-hydroxyphenoxy)propionoic acid. It is difficult to superficially determine the relative environmental merits of the two technologies.

are examples: Imidacloprid (Bayer), Nitenpyram (Takeda), and Acetamiprid (Nippon). Imidacloprid is the biggest product of this type.

The compounds 2-chloro-5-chloromethylpyridine and 5-aminomethyl-2-chloropyridine in Figure 10.15 are potential key building blocks in the manufacture of these molecules. For this reason, there are several published syntheses of the two compounds. There are many syntheses in which a pyridone is the precursor of a chlorination step. 3-Methylpyridine is a very cheap building block but direct chlorination is not regiospecific at the 6-position; it is accompanied by substitution at the 2- and the 4-positions and polychlorinated products. The regiospecific synthesis is possible via alicyclic ring construction [57]. The Bayer syntheses are described [58,59] in Figure 10.16.

Lonza has developed [60] a regiospecific hydroxylation of aromatic *N*-hetrocycles, compounds that are useful for a range of crop protection structural types.

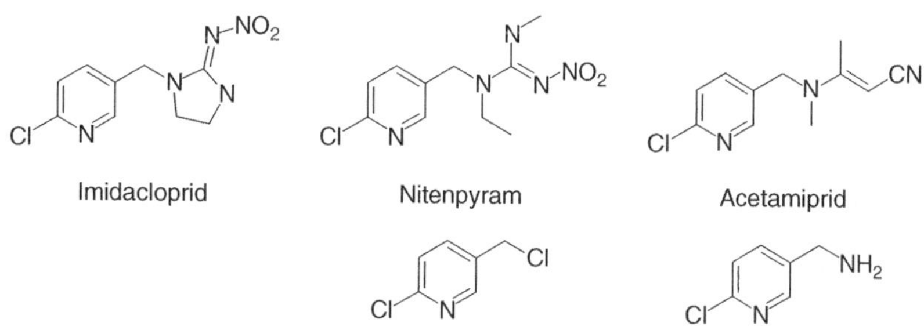

Figure 10.15 Examples of nicotinoid insecticides. The methyl-chloro and methyl-amino building blocks are key components in their industrial production.

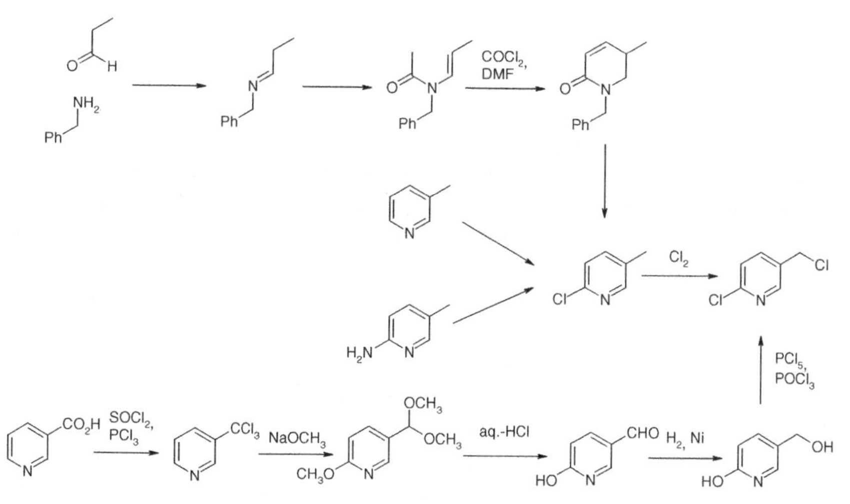

Figure 10.16 There are several demonstrated routes to 2-chloro-5-chloromethylpyridine that have also been operated on an industrial scale.

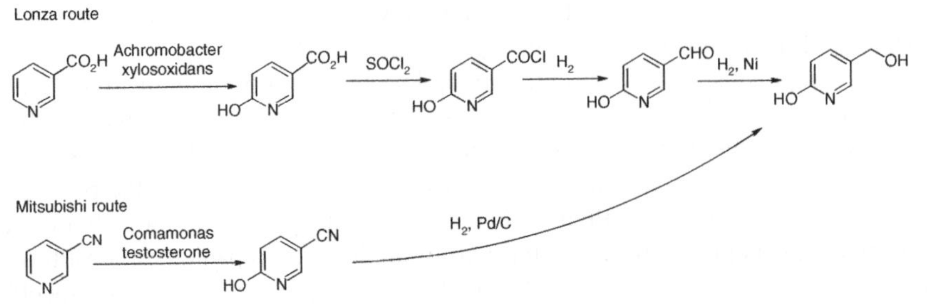

Figure 10.17 Lonza and Mitsubishi have developed alternative bio transformation routes to nicotinyl building blocks.

They have further elaborated the concept [61,62] as shown in Figure 10.17. The biotransformation utilizes suspended whole cells of *Achromobacter xylosoxidans* at pH 7, 30°C, and a substrate concentration of >60 g/L. Conversions of >90% are obtained in a >90% yield [10a].

A similar approach has also been reported by Mitsubishi [63–65] beginning from nicotinamide, oxidized using *Comamonas testosteroni*, and combined with hydrogenation of a cyano group directly to the methyl alcohol.

Overall, this regiospecific functionalization of a relatively unactivated C–H bond using a biotransformation enables more efficient routes to the key building blocks for the manufacture of this class of insecticides.

10.4 EMAMECTIN CASE—CAN BIOCATALYSIS PLAY A ROLE?

Emamectin benzoate **1** is used in the control of insects with application rates of 5–25 g/ha. It was discovered by Merck & Co. Inc. in 1988 through evaluation of synthetic analogs of Abamectin [66], **2**. Emamectin differs from Abamectin at position $C_{4''}$, where an equatorial hydroxyl group is replaced by an axial methylamino group. Emamectin **1** is synthesized from Abamectin **2** using the route described [67] in Figure 10.18. The methylamino group is introduced by stereoselective reductive amination of ketone **4**. In order to achieve the regioselective oxidation of the hydroxyl attached to $C_{4''}$, protection at the hydroxyl attached to C_5 is required. Protected compound **3** is oxidized with dichlorophosphate, DMSO, and triethylamine to give **4**, which is then followed by reductive amination using methylamine and borohydride to give compound **5**. The protecting group is then removed.

Alternative technologies and synthesis strategies for the conversion of Abamectin to Emamectin have been considered. These include electrochemical reductive amination of the oxo intermediates [68] and glycosidation [69].

Syngenta has evaluated whether an oxidoreductase biological system could be used selectively at position $C_{4''}$ to obviate the need for protection and deprotection steps [70], thus saving the materials and avoiding waste and costs of two steps in the manufacture. To this end, over 3300 microorganisms were screened,

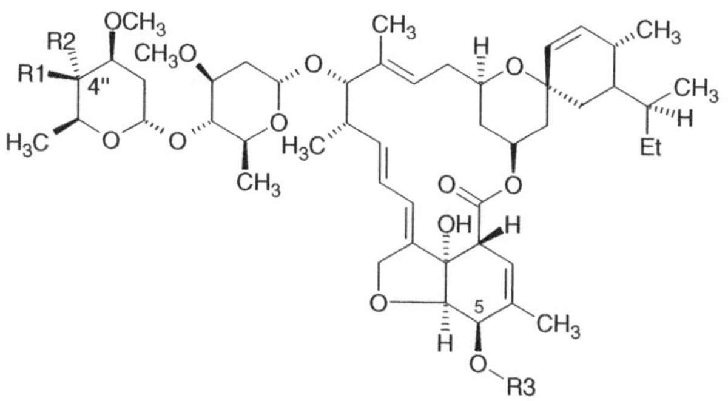

		R1	R2	R3
Emamectin	**1**	CH_3NH	H	H
$Abamectin_{B1}$	**2**	H	OH	H
	3	H	OH	Protection
	4	=O		Protection
	5	CH_3NH	H	Protection

2 —Protection→ 3 —Oxidation→ 4 —Reductive amination→ 5 —Deprotection→ 1

Figure 10.18 Emamectin is prepared by chemical modification of the fermentation product Abamectin. Different protection strategies have been applied.

from which 17 strains of Streptomycetes with promising biocatalytic activity were identified. In order to justify replacement of the existing manufacturing technology with a biological transformation, we found that the following biological performance would be necessary.

- Conversion yield **2** to **4** of >70% after isolation.
- Transformation conditions of 50 g/L and >80% conversion in 24 h.
- No new by-products of regulatory significance.

Our studies revealed that side reactions during the transformation were significantly reducing the yield. Most notably, the desired oxidation was accompanied by demethylation at position $C_{3''}$ (Figure 10.19) in a ratio of $k_1/k_2 = 1.35$.

To achieve our targets, we clearly had to improve the performance of the biological system. We worked with an external commercial organization with expertise in the technology of enzyme improvement, host selection,

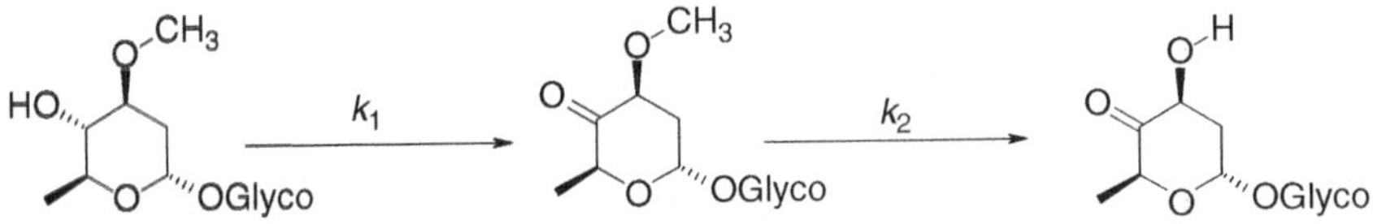

2 4"-oxo-abamectin

Figure 10.19 The biological oxidation of Abamectin obviates the need for protection in the Emamectin synthesis. However, the yield of the transformation is lowered by a competitive demethylation.

and fermentation optimization. Our studies included identification and characterization of the P450 enzyme by X-ray crystallography, gene reassembly techniques, and selective mutagenesis. Over 130,000 clones were created and tested in a library of five generations. The best performing clone met our targets.

Five host organisms were tested. They exhibited different levels of selectivity. For example, *Pseudomonas putida* had $k_1/k_2 = 2.35$ and *Streptomycetes tubercidicus* $k_1/k_2 = 1.14$. *S. tubercidicus R922*, a native producer of the enzyme, possessed the best growth characteristics and overall performance.

The low solubility of Abamectin in water, 1.2 mg/L, was a significant hurdle to overcome in the fermentation development and scale-up. The concentration could be increased to 4 g/L by addition of DMSO and detergents. However, agitation and aeration caused severe foaming that necessitated compromises in cell activity, and the detergents caused problems and complexity in the product recovery, which involved removal of the cells by centrifugation of a multiphase mixture. The crude 4″-oxo-abamectin could be isolated by continuous extraction and solvent distillation at a 20% w/w concentration.

After several millions of dollars spent on development, we have achieved a new process with the following characteristics.

- Reaction concentration <50 g/L, too low.
- Conversion rate, too low.
- Expensive solvents with complicated recycles and significant losses to aqueous phases.
- Waste load (kilogram waste/kilogram product)
    - Cells 50
    - Water 1000
    - Solvent 90
- Additional waste water treatment to remove recalcitrant toxic mectin byproducts.
- Difficult purification to meet the regulatory specifications.
- Equipment investment costs estimated to be >$10 million.

In parallel, we had also continued to improve the efficiency of the conventional technology described in Figure 10.18. The business case for the switch from conventional to a biotransformation was in the end too small to justify the necessary resources. At this point we abandoned further study of this approach.

10.5 CONCLUSIONS AND OUTLOOK

For the manufacture of complex natural products, fermentation will remain a key technology. The continuing progress in genetics, site-specific mutation, high-throughput screening, and host selection will improve the productivity and robustness of fermentations and the speed with which development and scale-up can be achieved. However, annually there are unlikely to be many new agrochemical active substances brought to the market that can be obtained by fermentation.

The application of biotransformations to the manufacture of agrochemical products has been evaluated by the industry over the last 20 years. Until now the number of transformations that have been used on a large scale remains small in comparison to normal abiotic technology. This is not due to a lack of interest on the part of manufacturers. There are clearly some notable successes of biotransformations that were developed in the 1990s, for example, phenoxypropionates and hydroxylations. The low rate at which the technology has been applied is also apparent in the number of commercial organizations that have closed internal groups dedicated to studying biotransformations or refocused them to other areas such as biofuels. For the manufacture of agrochemicals, biocatalysis is an important tool for application in some niche areas. The current strategy of Syngenta is to collaborate with manufacturers of fine chemicals whose business model includes expertise in fermentation and biotransformations as a core competence. There have been some great improvements in the development and application of biotransformation technology and in the speed it can be deployed [71]. However, the features that still limit the application of biotransformation technology to agrochemical manufacture include:

- the limited range of the types of generic transformations available.
- high cost of live cell transformations.
- avoidance of chirality in active substance design.

New active substances are being brought to market. Fermentation and biotransformations will have a role to play where they support delivery of acceptable cost of goods.

ACKNOWLEDGMENTS

The author would like to thank Dr. B. Antelmann and Dr. T. Schuez for assistance in the preparation of the Emamectin case study.

REFERENCES

1. Graedel T, Allenby BR. Industrial ecology. London: Prentis Hall International; 1995.
2. Paul TA, Warner JC. Green chemistry: theory and practice. New York: Oxford University Press; 1998.
3. Anastas P, Williamson TC. Green chemistry: an overview. ACS Symp Ser 1996; 626: 1–17.
4. Sheldon RA. Consider the environmental quotient. Chemtech 1994; 24(3): 38–47.
5. Sheldon RA. The E factor: fifteen years on. Green Chem 2007; 9: 1273–1283.
6. Douglas JM. Process synthesis for waste minimization. Ind Eng Chem Res 1992; 31(1): 238–243.
7. Lapkin A, Crittenden B. Framework for evaluating the "Greenness" of Chemical Processes: Case Studies for a Novel VOC Recovery technology. Environ Sci Technol 2004; 38(21): 5815–5823.
8. Aleu J, Bustillo AJ, Hernàndez-Galàn R, Collado IG. Biocatalysis applied to the synthesis of agrochemicals. Curr Org Chem 2006; 10: 2037–2054.
9. Aleu J, Bustillo AJ, Hernàndez-Galàn R, Collado IG. Biocatalysis applied to the synthesis of pheromones. Curr Org Chem 2007; 11(8): 693–705.
10. For general reviews containing applications relevant to Agrochemicals see: (a) Liese A, Seelbach K, Wandrey C. Industrial biotransformations. Weinheim: Wiley-VCH Verlag; 2000; (b) Liese A. Replacing chemical steps by biotransformations: industrial application and processes using biocatalysis. In: Drauz K, Waldmann H, editors. Volume III, Enzyme catalysis in organic synthesis. 2nd ed. Weinheim: Wiley-VCH Verlag; 2002. pp. 1419–1459; (c) Faber K, editor. Biotransformation in organic chemistry. 5th ed. Berlin: Springer; 2004; (d) Drauz K, Waldmann H, editor. Volume I-III, Enzyme catalysis in organic synthesis. 2nd ed. Weinheim: Wiley-VCH Verlag; 2002.
11. Rosskopf EN, Charudattan K, Kadir JB. In: Bellows TS, Fischer TW, editors. Handbook of biological control: principles and applications of biological control. New York: Academic Press; 1999. pp. 891–918.
12. Campbell WC, editor. Ivermectin and abamectin. New York: Springer; 1989.
13. (a) Fan Y, Min Y, Xiong W, Chen F, Liang Y, Lu H, Xing R, Zheng Y. Application of integrative expression of orfX gene in industrial breeding of Streptomyces avermitilis. Nongye Shengwu Jishu Xuebao 2007; 15(6): 1048–1052 CAN 151:146169; (b) Yang Q, Xu Y, Yu Z, Wu Y, Zheng Z. Method for increasing abamectin productivity by reducing fermentation period. CN 1928108 A, CAN 146:378327. 2007; (c) Wu X, Shen D, Yang Y, Chu X, Guo Z. Application of Streptomyces avermitilis X-28 capable of producing abamectin B. CN 1928075 A, CAN 146:374188. 2007.
14. Liu S, Liu S, Wang D. Method and apparatus for extracting abamectin. CN 101362785 A, CAN 150:276858. 2009.
15. Takebe H, Takane N, Hiruta O, Satoh A, Kataoka H, Tanaka H. Scale-up of bialaphos production. J Ferment Bioeng 1994; 78(1): 93–99.
16. Bayer E, Gugel KH, Haegele K, Hagenmeier H, Jessipow S, Koenig WA, Zaehner H. Metabolic products of microorganisms. 98. Phosphinothricin and phosphinothricylalanylanine. Helv Chim Acta 1972; 55: 224–239.

17. Kondo Y, Shomura T, Ogawa Y, Tsuruoka T, Watanabe H, Totsukawa K, Suzuki T, Moriyama C, Yoshida J, Inouye S, Niida T. New antibiotic SF-1293. I. Isolation and physicochemical and biological characterization of SF-1293 substance. Sci Rep Meiji Seika Kaisha 1973; 13: 34–41.

18. Takebe H, Namiki A, Tanioku K, Ooshima N, Endo T. Manufacture of L-2-amino-4-(hydroxymethylphosphinyl)butyric acid with transaminase-producing microorganisms. JP 06245780 A, CAN 122:8186. 1994.

19. Minowa N, Mitomi M. Preparation of phosphorus-containing. alpha.-keto acids and their intermediates. JP 2008230992 A, CAN 149:402504. 2008.

20. Minowa N, Nakanishi N, Mitomi M. Preparation of alkoxymethylphosphinylalkyl cyanide and preparation of L-2-amino-4-(hydroxymethylphosphinyl)butanoic acid. WO 2008035687 A1, CAN 148:403567.

21. Minowa N, Nakanishi N, Mitomi M. Method for producing L-2-amino-4-(hydroxymethylphosphinyl)- butanoic acid. WO 2006104120 A1, CAN 145:397784.

22. Racke KD. A reduced risk insecticide for organic agriculture: spinosad case study. ACS Symposium Series (2007), 947(Crop Protection Products for Organic Agriculture). pp. 92–108.

23. Baltz RH, Broughton MC, Crawford KP, Madduri K, Merlo DJ, Treadway PJ, Turner JR, Waldron C. Cloning and expression of genes for the enzymes of biosynthesis of insecticidal spinosyns of Saccharopolyspora spinosa. (Dow Agrosciences Llc, USA). PCT Int. Appl. WO 9946387 A1. 1999.

24. Taylor SC. Imperial Chemical Industries PLC. US 4758518. 1988.

25. Taylor SC. (S)-2-Chloropropionic acid: developments in its industrial manufacture. Chirality in industry II. UK: John Wiley & Sons, Inc.; 1997. pp. 207–223.

26. Klibanov AM, Kirchner G. Enzymic production of optical isomers of 2-halopropionic acids. US 4601987. 1986.

27. Pan S, Kawamoto T, Fukui T, Sonomoto K, Tanaka A. Stereoselective esterification of halogen-containing carboxylic acids by lipase in organic solvent: effects of alcohol chain length. Appl Microbiol Biotechnol 1990; 34(1): 47–51.

28. (a) Mutsaers JHGM, Kooreman HJ. Preparation of optically pure 2-aryl- and 2-aryloxypropionates by selective enzymic hydrolysis. Rec des Trav Chimiques des Pays-Bas 1991; 110(05): 185–188; (b) Bertola MA, De Smet MJ, Marx AF, Phillips GT. Preparation of specific stereoisomers of aryloxypropionic acids and esters by microbial or enzymatic action. Eur. Pat. Appl. EP 299559 A1. 1989.

29. Ladner WE, Ditrich K. Biocatalytic production of chiral intermediates. Cimica Oggi 1999; 7–8, 51–54.

30. Ladner W, Rettenmaier H, Zipperer B, Hansen H. Manufacture of L-2-chloropropionic acid and its salts. DE 4328231. 1994.

31. Blaser H, Buser H, Coers K, Hanreich R, Jalett H, Jelsch E, Pugin B, Schneider H, Spindler F, Wegmann A. The chiral switch of metolachlor. The development of a large-scale enantioselective catalytic process. Chimia 1999; 53(6): 275–280.

32. Schneider H. Substituted aminothiophenes. CH 657129 A5, CAN 106:32825. 1986.

33. Balkenhohl F, Ditrich K, Hauer B, Ladner W. Optically active amines via lipase-catalyzed methoxyacetylation. J Fuer Prakt Chemie/Chemiker-Zeitung 1997; 339(4): 381–384.

34. Balkenhohl F, Ditrich K, Nuebling C. Racemate separation of primary and secondary heteroatom-substituted amines by enzyme-catalyzed acylation. PCT Int. Appl. WO 9623894. 1996.
35. Nuebling C, Ditrich K, Dully C. Enzyme-catalyzed separation of racemic primary amines. Ger. Offen., DE 19837745. 2000.
36. Riechers H, Simon J, Hoehn A, Kramer A, Funke F, Siegel W, Nuebling C. Process and catalysts for the gas-phase racemizations of optically active amines. Ger. Offen., DE 19852282 A1. 2000.
37. Hayes KS, Lutz EG, Turcotte MG. Process and catalysts for the racemization of optically active alkoxyamines in the presence of hydrogen and ammonia. US 6060624 A. 2000.
38. Matcham G, Bhatia M, Lang W, Lewis C, Nelson R, Wang A, Wu W. Enzyme and reaction engineering in biocatalysis. synthesis of (S)-methoxyisopropylamine (= (S)-1-methoxypropan-2-amine). Chimia 1999; 53(12): 584–589.
39. Koszelewski D, Clay D, Rozzell D, Kroutil W. Deracemisation of. alpha.-chiral primary amines by a one-pot, two-step cascade reaction catalysed by omega-transaminases. Eur J Org Chem 2009; 14: 2289–2292.
40. Lin SL, Zhou WM, Matcham GW. Process and catalysts for the preparation of secondary and tertiary amines. WO 2001044162 A1, CAN 135:45977. 2001.
41. Wang YF, Lalonde JJ, Momongan M, Bergbreiter DE, Wong CH. Lipase-catalyzed irreversible transesterifications using enol esters as acylating reagents: preparative enantio- and regioselective syntheses of alcohols, glycerol derivatives, sugars and organometallics. J Am Chem Soc 1988; 110(21): 7200–7205.
42. Metcalf R. Insect control. Ullmann's Encyclopedia of Industrial Chemistry. Weinheim: Wiley-VCH Verlag GmbH & Co. KGaA; 2005.
43. Effenberger F. Synthesis and reactions of optically active cyanohydrins. Angew Chem Int Ed Engl 1994; 33: 1555–1564.
44. Zhang T, Yang L, Zhu Z, Wu J. The kinetic study on lipase-catalyzed transesterification of. alpha.-cyano-3-phenoxybenzyl alcohol in organic media. J Mol Catal B Enzym 2002; 18(4–6): 315–323.
45. Effenberger F, Schwämmle A. Preparation of (*S*)-cyanohydrins by enantioselective cleavage of racemic cyanohydrins with (*R*)-hydroxynitrile lyase. Biocatal Biotransformation 1997; 14: 167–179.
46. Chrisochoou AA, Schaber K, Stephan K. Phase equilibria with supercritical carbon dioxide for the enzymic production of an enantiopure pyrethroid component. Part 1. Binary systems. J Chem Eng Data 1997; 42(3): 551–557.
47. Mitsuda S, Yamamoto H, Umemura T, Hirohara H, Nabeshima S. Biological preparation of an optically active alcohol. Part III. Enantioselective hydrolysis of. alpha.-cyano-3-phenoxybenzyl acetate with Arthrobacter lipase. Agric Biol Chem 1990; 54(11): 2907–2912.
48. Fadnavis NW, Babu RL, Sheelu G, Deshpande A. 'Gelozymes' in organic synthesis. Part 2. Candida rugosa lipase mediated synthesis of enantiomerically pure (S)-cyano(3-phenoxyphenyl)methyl butyrate. Tetrahedron: Asymmetry 2001; 12(12): 1695–1699.
49. Roos J, Stelzer U, Effenberger F. Enzyme-catalyzed reactions. 34. Synthesis of (1R,cis,.alpha.S)-cypermethrine via lipase catalyzed kinetic resolution of racemic m-phenoxybenzaldehyde cyanohydrin acetate. Tetrahedron: Asymmetry 1998; 9(6): 1043–1049.

50. Fishman A, Zviely M. Chemo-enzymatic synthesis of (S)-.alpha.-cyano-3- phenoxybenzyl alcohol. Tetrahedron: Asymmetry 1998; 9(1): 107–118.
51. Enzymatic process for the preparation of (S)-cyanohydrins. Pochlauer P, Schmidt M, Wirth I, Neuhofer R, Zabelinskaja-Mackova AA, Griengl H, Van Den Broek C, Reintjens RWEG, Wories HJ. (DSM Fine Chemicals Austria G.m.b.H., Austria). EP 927766 A1, CAN 131:72766. 1999.
52. Aehle W, Waldmann H, Schultz C, Gröger H, Dinkel C, Drauz K. Enzymes in organic synthesis. Ullmanns Encyclopedia of Chemical Technology. Weinheim: Wiley-VCH Verlag; 2008.
53. von Langermann J, Guterl J, Pohl M, Wajant H, Kragl U. Hydroxynitrile lyase catalyzed cyanohydrin synthesis at high pH-values. Bioprocess Biosyst Eng 2008; 31(3): 155–161.
54. Cabirol FL, Hanefeld U, Sheldon RA. Immobilized hydroxynitrile lyases for enantioselective synthesis of cyanohydrins: sol-gels and cross-linked enzyme aggregates. Adv Synth Catal 2006; 348(12+13): 1645–1654.
55. Mitsuda S, Yamamoto H, Umemura T, Hirohara H, Shigeyasu N. Biological preparation of an optically active alcohol. Part III. Enantioselective hydrolysis of. alpha.-cyano-3-phenoxybenzyl acetate with Arthrobacter lipase. Agric Biol Chem 1990; 54(11): 2907–2912.
56. Dingler C, Ladner W, Krei GA, Cooper B, Hauer B. Preparation of (R)-2-(4- hydroxyphenoxy) propionic acid by biotransformation. Pestic Sci 1996; 46(1): 33–16.
57. Shaw NM, Robins KT, Kiener A. Lonza: 20 years of biotransformations. Adv Synth Catal 2003; 345(4): 425–435.
58. (a) Lantzsch R. Preparation of N-(cis-1-propenyl)acetamides. EP 748798 A1, CAN 126:131194. 1996; (b) Kraus H, Klausener A, Fuerstenwerth H. Process for the prepa ration of 2-substituted 5-alkylpyridines. EP 592896 A1, CAN 121:35346. 1994.
59. Jelich K. Preparation of 2-chloro-5-(chloromethyl)pyridine as an insecticide intermediate. EP 373463 A2, CAN 113:211856. 1990.
60. Gallenkamp B. Preparation of 5-(chloromethyl)pyridines as insecticide intermediates. DE 3630046 A1, CAN 109:73345. 1988.
61. Werbitzky O, Studer P. Preparation of 2-chloro-5-chloromethylpyridine. EP 569947 A1, CAN 120:134302. 1993.
62. Kulla HG. Enzymic hydroxylations. Chimia 1991; 45: 81–85.
63. Yasuda M, Ohkishi H, Sato K, Morimoto Y, Nagasawa T. Preparation of 6- hydroxy pyridines or pyrazines by microbial hydroxylation. EP 558022 A2, CAN 119:248182. 1993.
64. Takuma J, Sekine M, Kasuga J, Sakaguchi H, Tanaka T. Method for producing 3-hydroxymethyl-6-hydroxypyridine. JP 07224037 A, CAN 123:339761. 1995.
65. Yasuda M, Sakamoto T, Sashida R, Ueda M, Morimoto Y, Nagasawa T. Microbial hydroxylation of 3-cyanopyridine to 3-cyano-6-hydroxypyridine. Biosci Biotechnol Biochem 1995; 59(4): 572–575.
66. Mrozik H, Eskola P, Linn BO, Lusi A, Shih TL, Tischler M, Waksmunski FS, Wyvratt MJ, Hilton NJ, Anderson TE, Babu RJ, Bybas RA, Preiser FA, Fischer MH. Discovery of novel avermectins with unprecedented insecticidal activity. Experientia 1989; 45(3): 315–316.
67. Cvetovich RJ, Kelly DH, DiMichele LM, Shuman RF, Grabowski EJJ. Syntheses of 4-epi-Amino-4-deoxyavermectins B1. J Org Chem 1994; 59(25): 7704–7708.

68. Cvetovich RJ, Amato JS, DiMichele L, Weinstock L, Hazen G. Novel electrochemical reductive amination of 4-Oxo-5-O-(tert-butyl- dimethylsilyl)avermectin B1. J Org Chem 1997; 62(19): 6697–6698.
69. Loewe MF, Cvetovich RJ, DiMichele LM, Shuman RF, Grabowski EJJ. Glycosidation Route to 4-epi-(Methylamino)-4-Deoxyavermectin B1 (MK-244, Emamectin Benzoate). J Org Chem 1994; 59(25): 7870–7875.
70. Pachlatko JP, Pitterna T, Jungmann V. Preparation of a macrocyclic lactone. WO 2001036434 A1. 2001.
71. (a) Schoemaker HE, Mink D, Wubbolts MG. Dispelling the myths-biocatalysis in industrial synthesis. Science 2003; 299(5613): 1694–1698; (b) Lorenz P, Eck J. Screening for novel industrial biocatalysts. Eng Life Sci 2004; 4(6): 501–504; (c) Robertson DE, Steer BA. Recent progress in biocatalyst discovery and optimization. Curr Opin Chem Biol 2004; 8(2): 141–149.

CHAPTER 11

GREEN PRODUCTION OF FINE CHEMICALS BY ISOLATED ENZYMES

ROLAND WOHLGEMUTH
Sigma-Aldrich, Buchs, Switzerland

11.1 INTRODUCTION

Many great achievements in the enhancement of the quality of life in modern societies have been built on the supply of intermediates, the fine chemicals, which need to be produced in one way or other from easily accessible and inexpensive raw materials. Green production considerations provide a framework [1] for the design of synthetic routes, transformations, and process intensifications, which require less raw materials, solvents, and reagents, and generate less waste. Considering all the materials involved and not just the product is a molecular economy perspective with finite elements on our planet and the economic incentives and boundary conditions are also shaping the progress in enzymatic production methods. Selective and convenient synthetic tools with high performance are needed to build up fine chemicals, since side-reactions, lengthy synthetic schemes, and the use of protecting groups may require a massive redesign of synthesis. The limited resources and crowded supply chains on one side contrasts with the increasing energy/product demands and waste accumulation along the synthetic routes on the other side, leading to closer interactions between the industrial and living environment for humans, animals, and plants. Scientific interest in these interactions of the living and nonliving world, biodegradation and biotransformation of synthetic chemicals, and more than a century of research has unraveled the central importance of enzymes in the biosphere. Therefore, our global biological environment with an increasing world population and an increasing global fine chemical production emphasizes the harmonization of ecology and economy of enzymes.

Biocatalysis for Green Chemistry and Chemical Process Development, First Edition.
Edited by Junhua (Alex) Tao and Romas Kazlauskas.

The prime attention given to the principles of green chemistry, engineering, and production represents the global importance of safety, health, and environmental aspects of industrial activities in a densely populated world [2–4]. Various concepts and strategies, such as the E-factor [5] for the amount of waste per unit of product, the atom economy [6], step economy [7] and redox economy [8] and their combination [9,10], have been developed and are complementary to the classic principles of organic synthesis design. A mass flow perspective in a synthetic scheme, as shown in Figure 11.1, illustrates the significant amount of waste, which is accumulated during the product formation. Efforts to improve the molecular economy of the overall transformation from starting compound to product aim at both green production processes and cost reductions.

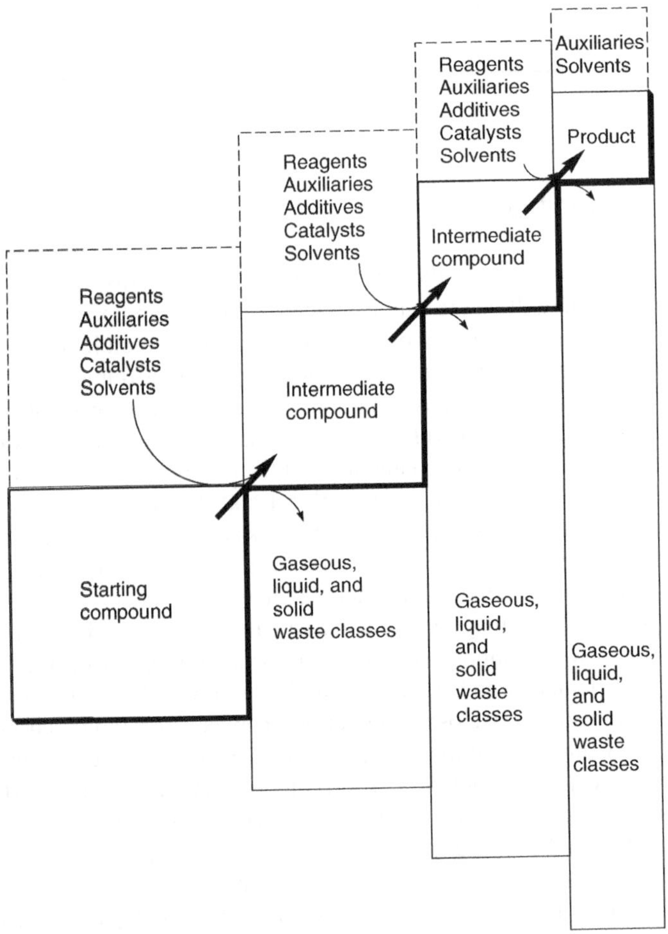

Figure 11.1 Molecular economy of a fine chemical production from starting compound to product. Green production considerations involve the atom economy of each reaction step, the optimization of the changes of redox states over the synthetic sequence as well as the optimization of the total number of steps from the starting compound to the product.

A key to green productions are catalytic process technologies instead of classic organic syntheses, which require auxiliary reagents in stoichiometric amounts, minimize the amount of waste per kilogram of product (E-factor), and improve environment, health and environment issues (EHS-issues) by avoiding problematic auxiliary reagents [11–13]. The development of catalytic processes replacing stoichiometric conversions has made tremendous contributions and 85%–90% of the products of chemical industry have been estimated to be manufactured by catalytic processes [14].

This minimization of waste per kilogram of product has to be balanced with catalyst toxicity and sustainability. The history of enzymes as being easily degradable and nontoxic catalysts is intricately linked to the history of catalysis and the isolation and purification of enzymes from whole-cell biomass. Whether the required biocatalysts are utilized as whole-cell biocatalysts or as isolated enzymes may depend on a variety of factors like the availability, knowledge, and complexity of the enzyme, and operational stability as well as costs of the bioprocess. The use of isolated enzymes in comparison with whole-cell biocatalysts is fundamentally advantageous if (i) substrate access and product removal from the enzyme are hindered by the corresponding cell membrane of the whole-cell biocatalysts or (ii) substrate or product is transformed to side products or degradation products by other enzymes of the whole-cell biocatalysts. The application of enzymes in chemistry started in the nineteenth century and is now experiencing a renaissance as a result of much scientific progress. The enzyme's inherent nature of chirality due to the chirality of its amino acid building blocks has led to higher efficiency in terms of enantioselectivity and yield than most synthetic asymmetric catalysts. The rapid progress in life sciences has led to a continued growth of newly discovered, tailor-made, and evolved enzymes together with increased knowledge on structure–function relationships.

The increased availability of isolated enzymes through large-scale production, the concomitant preparation of auxiliary chemicals, and the corresponding enzyme reaction engineering for fine chemical production have brought the application of isolated enzymes from a specialty research area to the main stream production of a wide variety of industrial products [15–17].

The enzyme classification system with the four-digit number is unique among the catalysts and based on the nature and direction of the catalyzed reaction. This modular catalyst nomenclature system is open for extension and useful for keeping the global overview while novel enzymes are discovered. The numbers and types of large-scale enzyme applications are growing due to the exquisite chemo-, regio-, and stereoselectivity of suitable enzymes, the continuing improvements in process design and product workup and purification.

Practical and schematic guidelines [18] for reporting effectively the experimental data on biocatalytic reactions creates value for different scientific and technological fields. The implementation of the guidelines and the simple availability of the reported experimental data in the required and sufficient detail enable other scientists to reproduce the experiments. Standardized reporting combined with the global communication and search tools helps to bring forward global innovation in enzyme technology as well as its outreach into related scientific areas.

Isolated enzymes represent the ideal biocatalysts because they work efficiently and selectively under mild conditions [19–23] and can be produced in a stable form. The organic chemistry of many enzyme-catalyzed reactions has been worked out to molecular details and reaction mechanisms [24,25]. They are increasingly and successfully being used in single reaction steps [26] and contribute an enormous variety of methodologies to green production [27]. In the following sections, the enzymatic reaction steps for the production of fine chemicals are discussed with respect to the reaction classes of the enzyme classification system. Apart from the single reaction step, it is also important to identify and overcome the bottlenecks and challenges in the adaptation and interfacing with the preceding and subsequent reaction steps of the whole synthetic sequence [28]. The use of two and more isolated enzymes in a one-pot cascade-, domino-, multistep, or multicomponent reaction can be developed without intermediate product recovery steps, similar to that taking place in biological cells [29]. The ultimate goal of enzyme reaction engineering is the one-pot enzymatic reaction cascade with optimized mass and energy economy. It is therefore no wonder that multistep enzyme reactions in the same reactor provide the gold standard for cascade reactions in organic chemistry and for mixing catalysts for a series of subsequent reaction steps in one pot [30].

11.2 ENZYMATIC OXIDATIONS AND REDUCTIONS FOR THE GREEN PRODUCTION OF FINE CHEMICALS

Organic synthesis would not be possible without a variety of reducing agents like metal hydrides or molecular hydrogen and without oxidizing agents like inorganic or organic oxidants, molecular oxygen, or ozone, which have been made available. These methods are standard practice in organic chemistry and fine chemical production for the electron transfer reactions involved in reductions and oxidations of organic compounds. Oxidants and reducing agents, which are toxic or critical to exothermic events, fire and explosion risks, are however a major issue at production scale. Their removal from the final product or their replacement by safe, healthy, and environmentally compatible reagents is of prime importance. A comparison of enzymatic and chemical processes for the oxidation of primary alcohol groups to aldehyde groups and for the oxidative deamination of primary donor groups to keto groups is shown in Figure 11.2. In addition to solving these green aspects, the development of selective and orthogonal redox methodology, which does not introduce additional protection–deprotection loops for other labile functional groups, continues to be a topic for catalytic asymmetric synthesis.

Oxidizing enzymes have therefore been of much interest for modern biooxidation methods [26,30,31]. D- and L-Specific amino acid oxidases have been used to prepare both the chiral amino acid as well as the α-ketoacid from the racemic amino acid substrate. The combination of an enantiospecific oxidation of a racemic amino acid substrate catalyzed by D-amino acid oxidase with a chemical

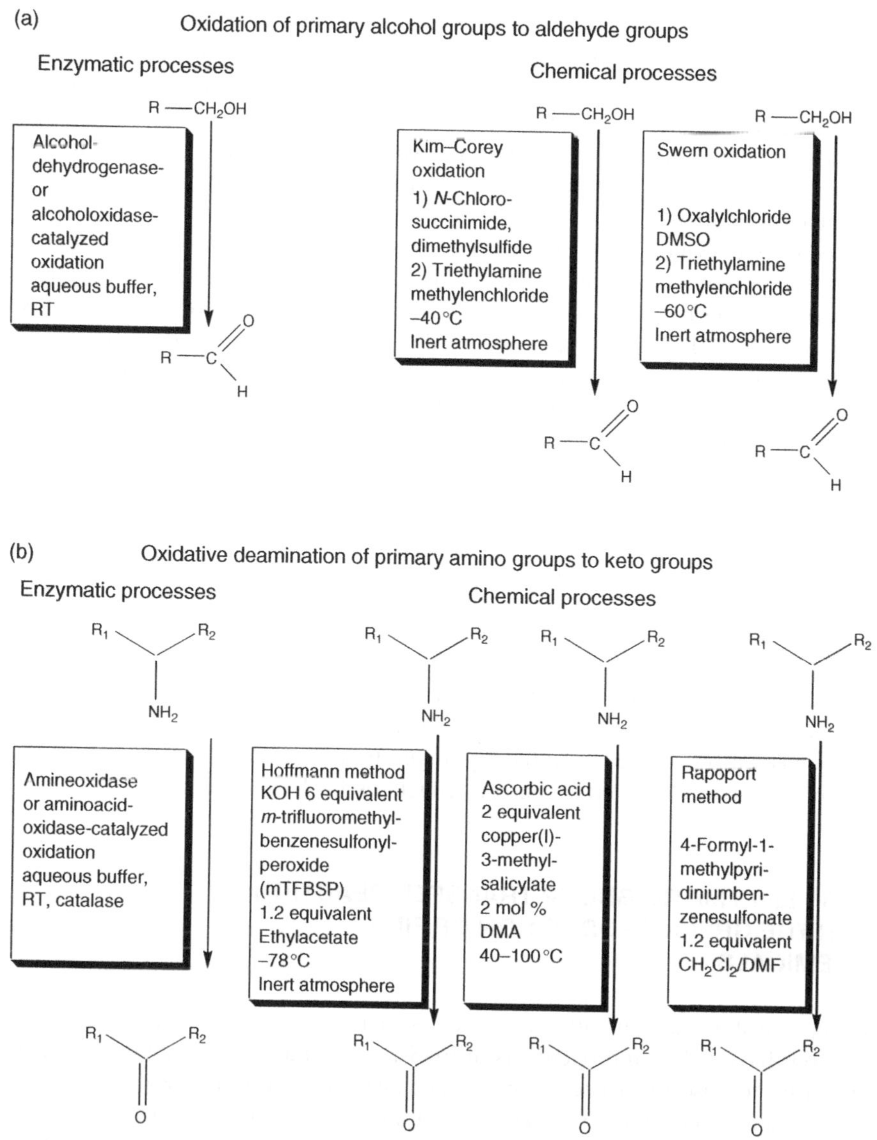

Figure 11.2 Green enzymatic oxidation processes in comparison with classical chemical oxidation processes.

nonenantiospecific reduction of the product with sodium borohydride in one pot has been used for the production of L-amino acids in high enantiomeric purity and yield. The enzymatic oxidation of residual peroxides by isolated catalases is utilized in many chemical large-scale oxidations whereby peroxides are used as oxidants. Since peroxides are also the by-products of enzymatic oxidation reactions, like, for example, the oxidation of amino acids by amino acid oxidases

[32], the analogous catalase methods can be used for their selective removal. Biocatalytic asymmetric oxidations have been achieved not only with oxidases, but also with other isolated enzymes like peroxidases, dehydrogenases, and mono- and dioxygenases [26,33–35], which can compete with the best available classical chemical oxidation methods. The further exploration of the catalytic potential of oxidizing enzymes for highly selective and green synthetic applications is very promising [26,30,36].

Reducing enzymes like the dehydrogenases and reductases have become excellent tools for modern bioreduction methods which in the case of ketone reductions outperform other ketone reduction chemistries and make them the general method of choice [37–40]. Biphasic reaction media, which are advantageous for poorly water-soluble ketones and for reactions at higher substrate concentrations, have been developed for the asymmetric reduction of ketones with *in situ* cofactor regeneration, whereby both the alcohol dehydrogenase and the formate dehydrogenase remain stable [41]. Since many oxidoreductase reactions depend on the nicotinamide cofactors NAD^+/NADH and $NADP^+$/NADPH, efficient *in situ* cofactor regeneration systems have been engineered [42], which can be scaled up. Catalytic asymmetric reductions of carbon–carbon double bonds can be done in a *cis*- or *trans*-fashion and generate up to two new chiral centers. The biocatalytic asymmetric *trans*-reduction of carbon–carbon double bonds is complementary to the high standard of transition-metal catalyzed *cis*-hydrogenation and has been achieved using cloned enoate reductases [43,44]. An interesting novel nicotinamide-independent asymmetric reduction of activated carbon–carbon double bonds was developed by direct hydrogen transfer from a sacrificial enone as hydrogen donor, catalyzed by enoate reductases [45].

11.3 ENZYMATIC GROUP TRANSFER REACTIONS FOR THE GREEN PRODUCTION OF FINE CHEMICALS

Group transfer reactions are another class of reactions where green production methods have led to improvements in EHS-issues, and stoichiometric waste generated by the use of protecting group chemistry can be minimized through the use of selective enzymes. Transketolases, transaminases, and glycosyltransferases have been developed for enzymatic ketol-, amino-, and glycosyl-group transfer reactions with complete conversions.

The ketol group transfer reaction catalyzed by transketolase corresponds to the central reaction in metabolic pathways, which is a blueprint for the green production of fine chemicals by transketolase-catalyzed C2-elongation bioprocesses. A variety of carbohydrates, chiral intermediates, stable-isotope-labeled compounds, and metabolites can be built up from nonphosphorylated 2-hydroxyaldehydes and phosphorylated aldehydes [46–49].

Although transaminases have already been studied and applied for many years, the biocatalytic asymmetric conversion of keto functional groups to an amino functional group of the desired *R*- or *S*-configuration continues to be an important research topic [50–54]. Since a variety of ketone substrates like epoxy- and hydroxyketones, keto- and hydroxyketoacids can be made by chemical synthesis, biocatalytic transamination opens up an easy access to a wide spectrum of more complex amino group-containing compounds. Interfacing the preceding chemical synthesis with the biocatalytic transamination is therefore an attractive synthetic route to chiral amines.

The application of glycosyltransferases for the synthesis of complex and highly pure carbohydrates and glycosylated compounds is currently recognized as being the most effective [55–57]. The synthesis of a trisaccharide [58] shows the advantage of a direct one-step enzymatic process in comparison with the seven-step chemical process (Figure 11.3).

11.4 HYDROLASES IN THE GREEN PRODUCTION OF FINE CHEMICALS: THE WORKHORSES

Hydrolytic reactions and their reverse reactions in organic solvents have been the classic domain where the introduction of isolated enzymes in the synthesis of fine chemicals has had the largest number of successful industrial applications [15–17,19–23]. On the basis of the availability of proteases, esterases, and lipases at industrial scale and the vast knowledge in this area, the development of new reactions has become much more rapid and is now part of standard synthetic methodology. The development of reliable enzymatic reaction platforms like the enzymatic hydrolysis tools with the large range of hydrolases available [59,60] to catalyze forward or backward reactions depending on solvent and auxiliary reagents has paved the way for the expansion of reaction methodologies. The elimination of one or more symmetry elements precluding chirality of a starting compound has been an attractive approach to prepare chiral products in theoretically 100% yield [61].

Isolated enzymes have successfully been used in such enantioselective desymmetrizations and Figure 11.4 shows some examples using hydrolases and oxidoreductases. Besides the ever increasing number of protease-, esterase-, and lipase-catalyzed reactions, other hydrolases like epoxide hydrolases, lactone hydrolases, and nitrilases have been developed for kinetic hydrolytic resolutions.

The asymmetric ring opening reaction of epoxides catalyzed by epoxide hydrolase has been developed for the synthesis of chiral epoxides, epoxyalcohols, tetrahydrofuran derivatives, and vicinal diols [62–64]. Since enantioselective microbial epoxide hydrolases can be produced at large scale [65], the preparation of enantiopure epoxides and diols by enzymatic kinetic resolution is of much interest [66].

Lactone hydrolases have been of interest for the preparation of D- and L-pantolactone from DL-pantolactone by kinetic resolution [67,68]. Although whole-cell biocatalysis saves costs by avoiding enzyme isolation, the use of isolated lactone hydrolases has the advantage of avoiding side-reactions of functionalized lactones.

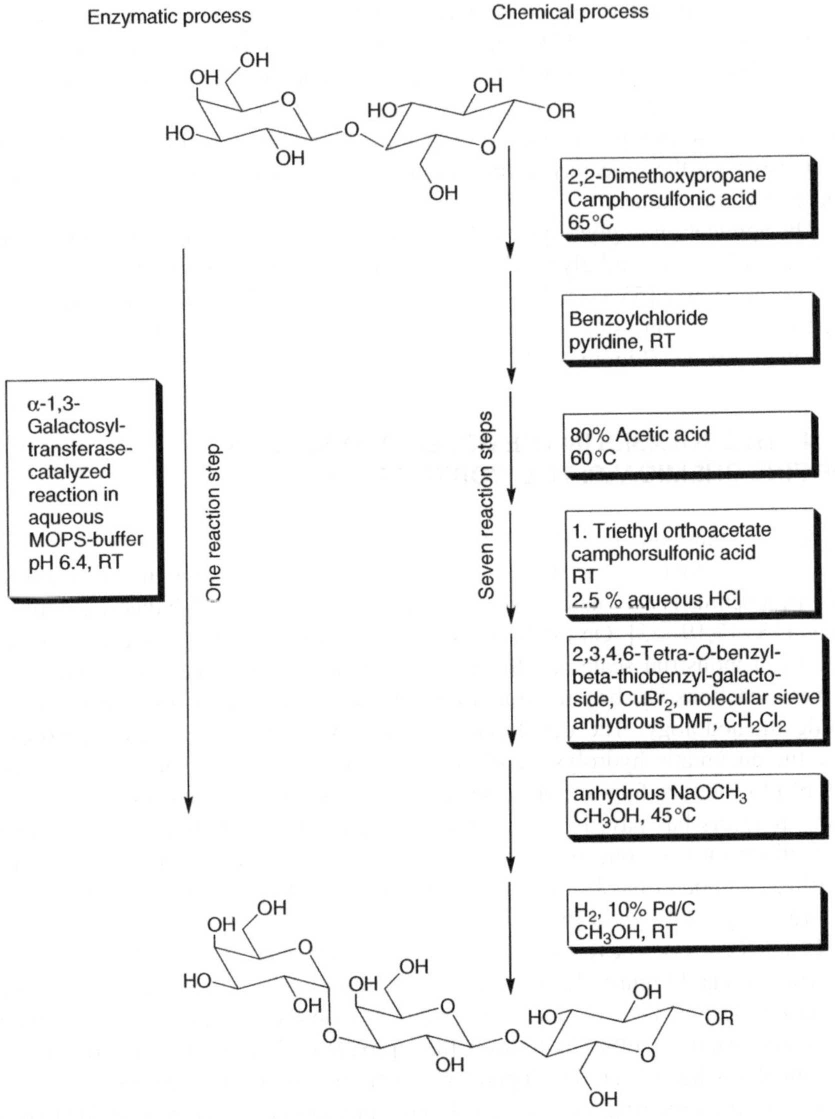

Figure 11.3 Green enzymatic glycosylation process in comparison with classical chemical carbohydrate synthesis using protecting groups.

The direct conversion of nitriles to carboxylic acids without applying the harsh reaction conditions of strong acids or bases at high temperature is of high synthetic interest for selective conversions, avoiding damage to labile functional groups. The selective hydrolysis of nitriles catalyzed by nitrilases has been developed for the desymmetrization of dinitriles [69–72] as well as for the asymmetric hydrolysis of mononitriles [73].

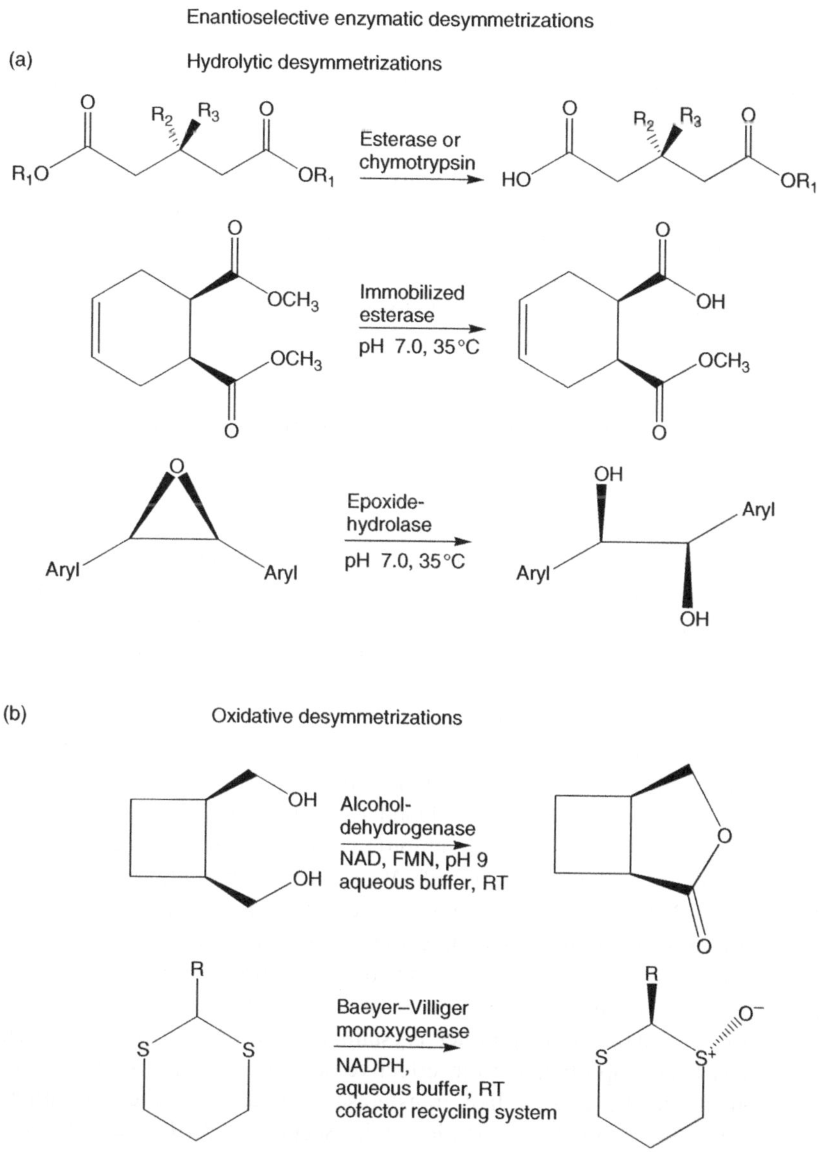

Figure 11.4 Green enantioselective enzymatic desymmetrization processes.

As many proteases are produced on a large industrial scale for detergent applications, the availability of inexpensive, highly stable, and active proteases has been a driving force for the development of production processes for fine chemicals. A great variety of unnatural amino acids and carboxylic acids are produced by acylase-, hydantoinase-, protease-, and amidase-catalyzed selective hydrolysis reactions of amides and esters and resolutions of racemates. In the synthetic direction, enzyme-catalyzed selective coupling of N-protected amino acids and esters has been developed to production scale [74].

Halohydrin dehalogenase-catalyzed ring opening of epoxides in the presence of anions like halides, cyanide, or azide proceed with excellent regio- and enantioselectivity and an industrial process for the one-pot enzymatic conversion of ethyl (*S*)-4-chloro-3-hydroxy-butyrate (ECHB) to ethyl (*R*)-4-cyano-3-hydroxybutyrate, involving an optimized bacterial halohydrin dehalogenase with 4000-fold improved volumetric activity, has been developed [75].

The application of glycosidases in the synthesis of glycosides has been advanced by the discovery of glycosynthases, which no longer catalyzed product hydrolysis [76,77].

11.5 LYASES IN GREEN PRODUCTION OF FINE CHEMICALS

A great variety of enzymes for the formation of carbon backbones in natural products are reflected by the different families of carbon–carbon lyases. Catalytic asymmetric synthesis of new carbon–carbon bonds with a high degree of stereochemical control remains one of the most challenging areas in synthetic organic chemistry and is of prime importance in building the carbon backbone of fine chemicals. Aldol reactions catalyzed by aldolases have been described in research for constructing C–C bonds with complete stereochemical control of the reaction course and the configurations of the newly formed stereogenic centers. Aldolases depending on various donors like dihydroxyacetone-phosphate, pyruvate, acetaldehyde, and dihydroxyacetone have been used for the synthesis of carbohydrates and analogs [78–84]. Donor-specific bottlenecks include narrow donor specificity, donor stability, inhibition phenomena, and thermodynamic equilibria [85,86]. The synthesis of *N*-acetyl-D-neuraminic acid (NANA), catalyzed by NANA-aldolase, from *N*-acetyl-D-mannosamine and pyruvate on a multiton scale represents a milestone in industrial aldolase applications [87]. Integrated process development including recombinant enzyme expression and immobilization, use of very high *N*-acetyl-D-mannosamine concentrations up to 20% w/v and simple product crystallization directly from the reaction mixture has been important for the effective scale-up. Besides carbon–carbon lyases, other enzyme classes like transferases can catalyze the formation of new carbon–carbon bonds as mentioned in Section 11.3. Transketolases and transaldolases have found increased use for synthetic ketol- and aldol-transfer reactions, making C_2- and C_3-chain extensions an attractive methodology, which can be integrated in synthetic planning. Benzaldehyde lyases and a variety of thiamine diphosphate-dependent enzymes have been successfully used for a variety of carbon–carbon bond formation reactions like asymmetric benzoin condensations, the synthesis of 2-hydroxyketones and bis(2-hydroxyketones), and asymmetric homo- and cross-coupling of aldehydes [88]. An important high-yield tyrosine phenol-lyase-catalyzed synthesis of L-3,4-dihydroxyphenylalanine (L-DOPA) has been industrialized in the last few decades [89].

The synthesis of α-hydroxynitriles by enzymatic condensation has been known since 1908, but it was not until novel (*R*)- and (*S*)-selective hydroxynitrile lyases have been discovered and made available for further studies by large-scale production that a variety of procedures for the synthesis of chiral cyanohydrins by enzymatic carbon–carbon bond formation have been developed [90,91]. An important aspect of the scale-up development was the tailoring of the process to existing equipment which has been dedicated to the exclusive use of HCN (hydrogen cyanide) instead of investing money in the realization of new vessels and workup equipment [91]. In the scale-up from the laboratory scale to industrial production, the use of a biphasic system to achieve a high enantiomeric excess has been an interesting aspect among more than 10 variables of the multidimensional process optimization. The process requires optimized stirring to achieve high reaction rate and enantiomeric excess, but at the same time avoiding both the formation of stable emulsions and enzyme denaturation due to shear stress [92].

The enzymatic carbon–carbon bond formation utilizing CoA (coenzyme A) thioesters requires cofactor recycling due to the high costs of CoA cofactors [93].

The enzyme-catalyzed addition of water to carbon–carbon double bonds by hydrolyases or carbon–oxygen lyases is not only atom-efficient, but also creating new chiral centers [20], which is not possible with the corresponding chemical addition reaction under harsh conditions.

Although isolated hydratases have been used at the laboratory scale [94], the present industrial applications have so far favored the use of whole-cell biocatalysts and the immobilization of the biocatalysts due to cost considerations [95]. While the addition of water to alkenoic acids is not common, the addition to alkenedioic acids like fumaric acid is performed on a multi-ten tonne scale [96].

The enzyme-catalyzed addition of water to carbon–carbon or carbon–nitrogen triple bonds enables further interesting hydration reactions. The nitrile hydratase-catalyzed addition of water to the nitrile triple bond is of interest for the conversion of nitriles into higher value amides and the example of the acrylnitrile to acrylamide bioconversion has been a success story, now in large-scale industrial production [97]. Although this process also uses whole-cell biocatalysts, the use of isolated nitrile hydratases for the mild and selective synthesis of more complex fine chemicals at laboratory scale is highly interesting.

The enzyme-catalyzed addition of ammonia to carbon–carbon double bonds is another class of addition reactions, which correspond ideally to green production principles. The continuous production of L-aspartic acid by immobilized aspartate ammonia-lyase has been an industrial large-scale application for decades [98]. The use of phenylalanine ammonia-lyase for catalyzing the reaction of cinnamic acid derivatives with an amino group donor to produce optically active phenylalanine compounds is another attractive process [99]. A recombinant aspartate ammonia-lyase from *Bacillus* sp. has been found to catalyze the Michael addition of hydroxylamine, hydrazine, methoxylamine, and methylamine to fumarate [100], thus giving the corresponding *N*-substituted aspartic acids with >97% enantioselectivity.

11.6 ISOMERASES IN THE GREEN PRODUCTION FINE CHEMICALS

Although an advantage of enzymatic methods is the absence of reaction conditions which would isomerize the substrate or the product, selective enzymatic isomerization is of great synthetic and economic interest, if the product is otherwise not easily accessible or if an inexpensive raw material can be isomerized in one step to the product. As isomerases catalyze intramolecular structural changes from one isomer to another isomer within a single molecule, no other substrates are involved and only activators and cofactors are required to reach maximal enzyme activity. To limit the by-product formation, short reaction times by using high activities of immobilized isomerases are needed, which gives isolated enzymes an advantage over whole cells.

The range of synthetic routes is increased by epimerases, *cis–trans* isomerases, mutases, racemases, and other isomerases. This creates more opportunities for value creation by step economy from inexpensive raw materials.

A variety of aldose isomerases catalyze the conversion of ketohexoses to their corresponding aldohexoses [101]. Immobilized L-rhamnose isomerase enabled the large-scale conversion of D-psicose to D-allose [102]. The enzymatic epimerization of the *N*-acetyl group in the 2-position of *N*-acetyl-D-glucosamine to *N*-acetyl-D-mannosamine by *N*-acyl-D-glucosamine 2-epimerase has been essential for the synthesis of *N*-acetyl-neuraminic acid [103].

The phenylalanine aminomutase-catalyzed addition reaction of ammonia to substituted cinnamic acids, which are easily accessible from the corresponding aldehydes, yields a mixture of two regioisomeric, but enantiopure α- and β-amino acids for most of the substrates [104,105]. Cinnamic acid derivatives with electron-donating groups at the para position yielded the corresponding β-amino acids with >90% yield and >99% *ee* [105].

11.7 LIGASES IN THE GREEN PRODUCTION OF FINE CHEMICALS

Ligases are not yet used in the production of fine chemicals, but there is increasing interest to develop new ligases and to extend the application of ligases from the analytical to the preparative scale, both for the synthesis of high- and low-molecular weight compounds.

Acyl-activating enzymes such as acyl-CoA ligases, have the potential to produce varied starter CoA precursors. As mycobacterial fatty acyl-CoA ligases have shown remarkable substrate flexibility, activate a variety of fatty acids with modifications at the α, β, ω, and ω−ν positions, and also utilize *N*-acetyl-cysteine, the shorter acceptor terminal portion of CoA, new routes for metabolite syntheses are possible [106]. Dipeptide synthesis from unprotected amino acids has been shown to be catalyzed by an L-amino acid ligase from *Ralstonia solanacearum* in an ATP-dependent manner [107].

The four cytoplasmic peptidoglycan cell wall precursors, UDP-*N*-MurNAc-L-Ala, UDP-*N*-MurNAc-L-Ala-γ-D-Glu, UDP-*N*-MurNAc-L-Ala-γ-D-Glu-*m*-diaminopimelic acid, and UDP-*N*-MurNAc-L-ala-γ-D-glu-*m*-diaminopimelic acid-D-ala-D-ala, have been synthesized from UDP-*N*-acetylmuramic acid (UDP-*N*-MurNAc). The attachment of L-Ala, D-Glu, and *m*-diaminopimelic acid is done enzymatically by the action of the murein ligases MurC, MurD, MurE, and MurF [108]. Each murein ligase catalyzes the ATP-dependent addition of an amino acid to the free carboxy end of its substrate, thereby enabling the first high-yield synthesis of these murein substrates [108].

The ATP-dependent coupling of amino acid monomers, elongation, and cyclization is an interesting way that peptides can be assembled independent of ribosomal-based mRNA or the nonribosomal peptide synthetase machinery [109].

11.8 ENZYME REACTION ENGINEERING

The synthetic routes to fine chemicals usually consist of a series of reactions, intermediate recovery, and purification steps. Since both chemical and enzymatic reaction steps may be involved, the development of the right interface between these steps is essential [27], if preceding or subsequent intermediates are produced by established chemical synthesis routes. The optimized green production route from substrate via intermediate(s) to product has to be selected from a three-dimensional reaction space defined by the dimensions of the classical chemical reaction steps, the enzymatic reaction steps, and the recovery/purification steps as shown in Figure 11.5.

Once the basic enzyme reaction is defined in molecular terms and the starting material, reagents, biocatalysts are available, the kinetic, thermodynamic, and reaction engineering parameters can be characterized [16]. The attention to parameters like selectivity, enantiomeric excess, yield of the reaction, and overall isolated yield, which determine the enzyme reaction in the research phase, shifts more to complete conversion, space–time yield, efficiency, and stability of the biocatalyst in the process development phase. In a similar way, the criteria by which an enzymatic reaction step is judged against the corresponding chemical reaction step depend on the different requirements of the research versus the scale-up and production phase [17,21]. In addition to the scientific and technological criteria, safety, health, renewable resources, energy conservation, and waste minimization are of increasing global and local influence on the routes and reactions selected [12]. The reaction reengineering toward green production involves attention to many details including the replacement of hazardous reagents by environment-friendly reagents and favors enzymatic reactions in general. For a single reaction step with recovery and purification of the reaction, product attention to the total effort in obtaining the pure product-in-the-bottle is essential as most often the downstream part of the process contributes the major costs.

Whether a single-step enzymatic reaction or a complete sequence of enzymatic reactions is the ultimate goal, the design of both the molecular aspects like

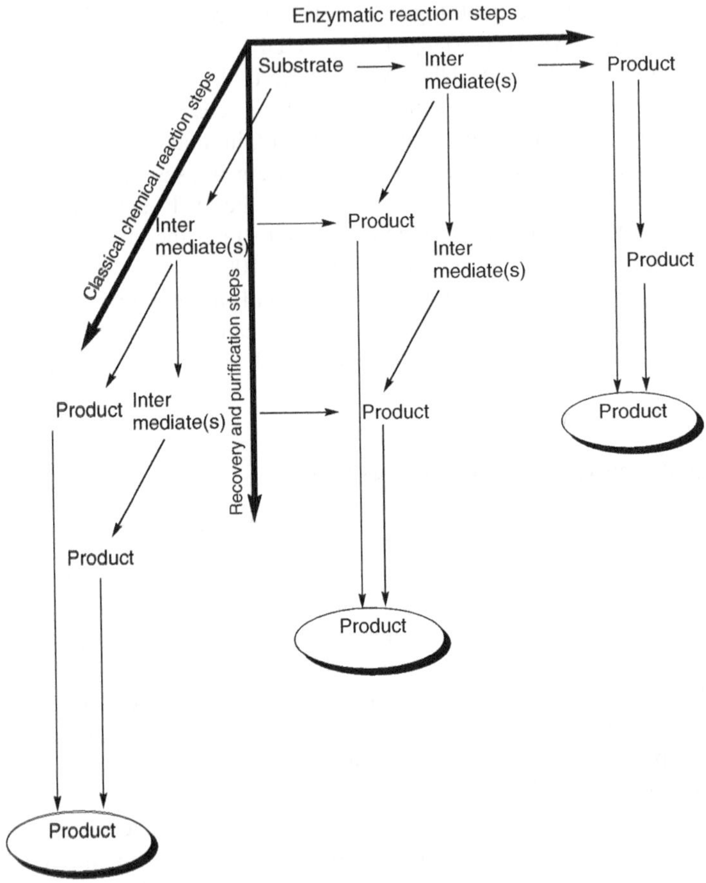

Figure 11.5 Interfacing green reaction steps with classical chemical reaction steps and recovery/purification steps in the molecular, thermodynamic, and kinetic landscape of feasible and practically available transformation processes for the selection of the optimized green production route.

the synthetic route or the performance optimization of the required enzymes as well as the engineering aspects like bioprocess design or downstream processing is key for obtaining the fine chemical not just as a laboratory curiosity, but in a viable industrial large-scale process. As an example of the change toward the economically and environmentally attractive production of fine chemicals, the two-step enzymatic bioprocess for an atorvastatin intermediate illustrates the different options for enzymatic route selection and enzyme performance optimization under large-scale reaction conditions [110,111].

The methods of directed evolution of enantioselective enzymes provide a toolbox for engineering the enzyme to the desired properties for a given green process like broader substrate acceptance, enantioselectivity, activity, and stability [112].

11.9 SUMMARY AND OUTLOOK

The use of isolated enzymes from all six classes has been constantly increasing over the last decades in the production of fine chemicals. The development of new reaction platforms and their scale-up has been and continues to be a success story. Despite the progress achieved, enzymatic reactions still cover only a small fraction of the chemical reaction space developed over the past centuries. It is therefore both of fundamental and practical interest for the science of synthesis that novel enzymes and novel synthetic reactions are discovered and developed in order to replace nasty and nonselective chemical reactions or to shorten the number of reaction steps in a synthetic sequence. The development of strategies for new enzymatic reaction chemistry will benefit from a close interaction between chemistry and biology [113]. The search for novel enzymes involves a variety of approaches like sequence- and structure-based search and optimization, protein design by computational methods and directed evolution, metagenome- and classical microbiological screening for enzyme activities. The discovery of novel enzyme functions requires meaningful, robust, and sensitive analytical methodologies. The standardization of quantitative and reproducible measurements of substrate-to-product conversions and their reporting in publications [18] will be beneficial for the further global growth of this field. As the availability of isolated enzymes is essential for their applications in organic synthesis, the development and production of new isolated enzymes in a form which shows high operational and storage stability is a prerequisite.

Reaction engineering and product recovery are equally important and green chemistry will continue to be a useful central design framework for the translation of this new knowledge into daily industrial practice of fine chemical production [114]. The key element for the success of enzymatic reactions in green production methods is the implementation of fine chemical production and the intensified inclusion in organic chemistry, catalysis, and green chemistry [115]. Since the chemical production is highly complex, diverse, and based on a variety of scientific and technological disciplines, detailed analyses of existing process challenges in a certain industrial production area and research perspectives for green production methodologies are useful [116]. The introduction of novel enzymatic tools for key catalytic asymmetric transformations will change chemical manufacturing in the twenty-first century toward green production methodology.

REFERENCES

1. Anastas P, Eghbali N. Green chemistry: principles and practice. Chem Soc Rev 2010;39:301–312.
2. Li CJ, Trost BM. Green chemistry for chemical synthesis. Proc Natl Acad Sci USA 2008;105:13197–13202.
3. Noyori R. Synthesizing our future. Nat Chem 2009;1:5–6.
4. Sheldon RA, Arends I, Hanefeld U. Green chemistry and catalysis. Weinheim: Wiley-VCH; 2007.

5. Sheldon RA. The E factor: fifteen years on. Green Chem 2007;9:1273–1283.
6. Trost BM. The atom economy: a search for synthetic efficiency. Science 1991;254: 1471–1477.
7. Wender PA, Verma VA, Paxton TJ, Pillow TH. Function-oriented synthesis, step economy, and drug design. Acc Chem Res 2008;41:40–49.
8. Burns NZ, Baran PS, Hoffmann RW. Redox economy in organic synthesis. Angew Chem Int Ed 2009;48:2854–2867.
9. Sheldon RA. E factors, green chemistry and catalysis: an odyssey. Chem Commun 2008; 3352–3365.
10. Clarke PA, Santos S, Martin WHC. Combining pot, atom and step economy in organic synthesis. Synthesis of tetrahydropyran-4-ones. Green Chem 2007;9:438–440.
11. Ran N, Zhao L, Chen Z, Tao J. Recent applications of biocatalysis in developing green chemistry for chemical synthesis at the industrial scale. Green Chem 2008;10:361–372.
12. Wohlgemuth R. Modular and scalable biocatalytic tools for practical safety, health and environmental improvements in the production of speciality chemicals. Biocatal Biotransformation 2007;25:178–185.
13. Wohlgemuth R. Tools for selective enzyme reaction steps in the synthesis of laboratory chemicals. Eng In Life Sci 2006;6:577–583.
14. Chorkendorff I, Niemandsverdriet JW. Concepts of modern catalysis and kinetics. Weinheim: Wiley-VCH; 2003.
15. Breuer M, Ditrich K, Habicher T, Hauer B, Kesseler M, Stuermer R, Zelinski T. Industrial methods for the production of optically active intermediates. Angew Chem Int Ed 2004;43:788–824.
16. Liese A, Seelbach K, Wandrey C, editors. Industrial biotransformations. 2nd, completely revised and extended ed. Weinheim: Wiley-VCH; 2006.
17. Ghisalba O, Meyer HP, Wohlgemuth R. Industrial biotransformation. In: Flickinger MC, editor. Encyclopedia of industrial biotechnology. Hoboken (NJ): Wiley; 2010.
18. Gardossi L, Poulsen P, Ballesteros A, Hult K, Svedas V, Vasic-Racki D, Carrea G, Magnusson A, Schmid A, Wohlgemuth R, Halling P. Guidelines for reporting biocatalytic reactions. Trends Biotechnol 2010;28:171–180.
19. Wong CH, Whitesides GM. Volume 12, Enzymes in synthetic organic chemistry, Tetrahedron organic chemistry series. Oxford: Elsevier; 1994.
20. Drauz KH, Waldmann H. Volume I-III, Enzyme catalysis in organic synthesis: a comprehensive handbook. 2nd, completely revised and enlarged ed. Weinheim: Wiley-VCH; 2002.
21. Bommarius AS, Riebel BR. Biocatalysis. Weinheim: Wiley-VCH; 2004.
22. Roberts S. Preparative biotransformations: whole cell and isolated enzymes in organic synthesis. Chichester: John Wiley & Sons, Inc.; 1998.
23. Faber K. Biotransformations in organic chemistry. 5th revised & corr. ed.; Berlin: Springer-Verlag; 2004.
24. Silverman RB. The organic chemistry of enzyme-catalyzed reactions. San Diego (CA): Academic Press; 2000.
25. Walsh C. Enzymatic reaction mechanisms. San Francisco (CA): W.H.Freeman; 1979.

26. Fessner WD, Anthonsen T, editors. Modern biocatalysis. Weinheim: Wiley-VCH; 2009.
27. Wohlgemuth R. Interfacing biocatalysis and organic synthesis. J Chem Technol Biotechnol 2007;82:1055–1062.
28. Garcia-Junceda E, editor. Multi-step enzyme catalysis. Weinheim: Wiley-VCH; 2008.
29. Wohlgemuth R. The locks and keys to industrial biotechnology. New Biotechnol 2009;25:204–213.
30. Schmid RD, Urlacher V, editors. Modern biooxidation: enzymes, reactions and applications. Weinheim: Wiley-VCH; 2007.
31. Burton SG. Oxidizing enzymes as biocatalysts. Trends Biotechnol 2003;21:243–249.
32. Benz P, Wohlgemuth R. Amino acid oxidase-catalysed resolution and Pictet-Spengler reaction towards chiral and rigid unnatural amino acids. J Chem Technol Biotechnol 2007;87:1082–1086.
33. Kroutil W, Mang H, Edegger K, Faber K. Biocatalytic oxidation of primary and secondary alcohols. Adv Synth Catal 2004;346:125–142.
34. Alphand V, Wohlgemuth R. Applications of baeyer-villiger monoxygenases in organic synthesis. Curr Org Chem 2010;14:1928–1965.
35. Richter N, Neumann M, Liese A, Wohlgemuth R, Eggert T, Hummel W. Characterisation of a recombinant NADP-dependent glycerol dehydrogenase from gluconobacter oxydans and its application in the production of L-glyceraldehyde. Chembiochem 2009;10:1888–1896.
36. Witayakran S, Ragauskas AJ. Synthetic applications of laccase in green chemistry. Adv Synth Catal 2009;351:1187–1209.
37. Kroutil W, Mang H, Edegger K, Faber K. Recent advances in the biocatalytic reduction of ketones and oxidation of sec-alcohols. Curr Opin Chem Biol 2004;8:120–126.
38. Goldberg K, Schroer K, Lütz S, Liese A. Biocatalytic ketone reduction - a powerful tool for the production of chiral alcohols - part I: processes with isolated enzymes. Appl Microbiol Biotechnol 2007;76:237–248.
39. De Wildeman SMA, Sonke T, Schoemaker HE, May O. Biocatalytic reductions: from lab curiosity to "First Choice". Acc Chem Res 2007;40:1260–1266.
40. Moore JC, Pollard DJ, Kosjek B, Devine PN. Advances in the enzymatic reduction of ketones. Acc Chem Res 2007;40:1412–1419.
41. Gröger H, Hummel W, Buchholz S, Drauz K, Van Nguyen T, Rollmann C, Hüsken H, Abokitse K. Practical asymmetric enzymatic reduction through discovery of a dehydrogenase-compatible biphasic reaction media. Org Lett 2003;5:173–176.
42. Wichmann R, Vasic-Racki D. Cofactor regeneration at the lab scale. Adv Biochem Eng Biotechnol 2005;92:225–260.
43. Hall M, Stueckler C, Kroutil W, Macheroux P, Faber K. Asymmetric bioreduction of activated alkenes using cloned 12-oxophytodienoate reductase isoenzymes OPR-1 and OPR-3 from lycopersicon esculentum (Tomato): a striking change of stereoselectivity. Angew Chem Int Ed 2007;46:3934–3937.
44. Stuermer R, Hauer B, Hall M, Faber K. Asymmetric bioreduction of activated C=C bonds using enoate reductases from the old yellow enzyme family. Curr Opin Chem Biol 2007;11:203–213.

45. Stueckler C, Reiter TC, Baudendistel N, Faber K. Nicotinamide-independent asymmetric bioreduction of C=C-bonds via disproportionation of enones by enoate reducta-ses. Tetrahedron 2010;66:663–667.

46. Wohlgemuth R. Tools and ingredients for the biocatalytic synthesis of carbohydrates and glycoconjugates. Biocatal Biotransformation 2008;26:42–48.

47. Wohlgemuth R. C_2-Ketol elongation by transketolase-catalyzed asymmetric synthesis. J Mol Catal B Enzym 2009;61:23–29.

48. Shaeri J, Wright I, Rathbone EB, Wohlgemuth R, Woodley JM. Characterization of enzymatic D-xylulose 5-phosphate synthesis. Biotechnol Bioeng 2008;101:761–767.

49. Wohlgemuth R, Smith MEB, Dalby PA, Woodley JM. In: Flickinger MC, editor. Encyclopedia of industrial biotechnology. Hoboken (NJ): Wiley; 2010.

50. Ward J, Wohlgemuth R. High-yield biocatalytic amination reactions in organic synthesis. Curr Org Chem 2010;14:1914–1927.

51. Zhu D, Hua L. Biocatalytic asymmetric amination of carbonyl functional groups—a synthetic biology approach to organic chemistry. Biotechnol J 2009;4:1420–1431.

52. Höhne M, Bornscheuer UT. Biocatalytic routes to optically active amines. ChemCatChem 2009;1:1–11.

53. Schell U, Wohlgemuth R, Ward JM. Synthesis of pyridoxamine 5′-phosphate using an MBA:pyruvate transaminase as biocatalyst. J Mol Catal B Enzym 2009;59:279–285.

54. Hanson RL, Davis BL, Chen Y, Goldberg SL, Parker WL, Tully TP, Montana MA, Patel RN. Preparation of (R)-amines from racemic amines with an (S)-amine transaminase from bacillus megaterium. Adv Synth Catal 2008;350:1367–1375.

55. Oehrlein R. Glycosyltransferase-catalyzed synthesis of non-natural oligosaccharides. Top Curr Chem 1999;200:227–254.

56. Wohlgemuth R. Development, production, and application of recombinant yeast biocatalysts in organic synthesis. Chimia 2005;59:735–740.

57. Weijers CAGM, Franssen MCR, Visser GM. Glycosyltransferase-catalyzed synthesis of bioactive oligosaccharides. Biotechnol Adv 2008;26:436–456.

58. Ratcliffe RM, Kamath VP, Yeske RE, Gregson JM, Fang YR, Palcic MM. Large-scale chemical and enzymatic synthesis of a *Clostridium difficile* toxin a binding trisaccharide. Synthesis 2004;14:2293–2296.

59. Bornscheuer UT, Kazlauskas RJ. Hydrolases in organic synthesis. Weinheim: Wiley-VCH; 2006.

60. Wohlgemuth R. Large-scale applications of hydrolases in biocatalytic asymmetric synthesis. In: Blaser HU, Federsel HJ, editors. Large-scale asymmetric catalysis. Weinheim: Wiley-VCH; 2010.

61. Garcia-Urdales E, Alfonso I, Gotor V. Enantioselective enzymatic desymmetrizations in organic synthesis. Chem Rev 2005;105:313–354.

62. Choi WJ. Biotechnological production of enantiopure epoxides by enzymatic kinetic resolution. Appl Microbiol Biotechnol 2009;84:239–247.

63. Archelas A, Furstoss R. Synthetic applications of epoxide hydrolases. Curr Opin Chem Biol 2001;5:112–119.

64. Lee EY. Epoxide hydrolase-mediated enantioconvergent bioconversions to prepare chiral epoxides and alcohols. Biotechnol Lett 2008;30:1509–1514.

65. Naundorf A, Melzer G, Archelas A, Furstoss R, Wohlgemuth R. Influence of pH on the expression of a recombinant epoxide hydrolase in Aspergillus niger. Biotechnol J 2009;4:756–765.

66. Deregnaucourt J, Archelas A, Barbirato F, Paris JM, Furstoss R. Enzymatic transformations; 61. preparation of enantiopure trifluoromethyl-substituted aromatic epoxides and vicinal diols using the aspergillus niger epoxide hydrolasecatalysed resolution. Adv Synth Catal 2006;348:1165–1169.

67. Zhang X, Xu JH, Pan J. Isolation and properties of a levo-lactonase from Fusarium proliferatum ECU2002: a robust biocatalyst for production of chiral lactones. Appl Microbiol Biotechnol 2007;75:1087–1094.

68. Kataoka M, Honda K, Sakamoto K, Shimizu S. Microbial enzymes involved in lactone compound metabolism and their biotechnological applications. Appl Microbiol Biotechnol 2007;75:257–266.

69. DeSantis G, Wong K, Farwell B, Chatman K, Zhu Z, Tomlinson G, Huang H, Tan X, Bibbs L, Chen P, Kretz K, Burk MJ. Creation of a productive, highly enantioselective nitrilase through gene site saturation mutagenesis (GSSM). J Am Chem Soc 2003;125:11476–11477.

70. DeSantis G, Zhu Z, Greenberg WA, Wong K, Chaplin J, Hanson SR, Farwell B, Nicholson LW, Rand CL, Weiner DP, Robertson DE, Burk MJ. An enzyme library approach to biocatalysis: development of nitrilases for enantioselective production of carboxylic acid derivatives. J Am Chem Soc 2003;125:11476–11477.

71. Wang MX. Progress of enantioselective nitrile biotransformations in organic synthesis. Chimia 2009;63:331–333.

72. Bergeron S, Chaplin DA, Edwards JH, Ellis BSW, Hill CL, Holt-Tiffin K, Knight JR, Mahoney T, Osborne AP, Ruecroft G. Nitrilase-catalyzed desymmetriza-tion of 3-hydroxyglutaronitrile: preparation of a statin side-chain intermediate. Org Proc Res Dev 2006;10:661–665.

73. Benz P, Muntwyler R, Wohlgemuth R. Chemoenzymatic synthesis of chiral carboxylic acids via nitriles. J Chem Technol Biotechnol 2007;82:1087–1098.

74. Oyama K. The industrial production of aspartame. In: Collins AN, Sheldrake G, Crosby J, editors. Chirality in industry. Chichester: John Wiley & Sons, Ltd.; 1995.

75. Fox RJ, Davis SC, Mundorff EC, Newman LM, Gavrilovic V, Ma SK, Chung LM, Ching C, Tam S, Muley S, Grate J, Gruber J, Whitman JC, Sheldon RA, Huisman GW. Improving catalytic function by ProSAR-driven enzyme evolution. Nat Biotechnol 2007;25:338–344.

76. Jahn M, Whithers SG. New approaches to enzymatic oligosaccharide synthesis: glycosynthases and thioglycoligases. Biocatal Biotransformation 2003;21:159–166.

77. Perugino G, Cobuzzi-Ponzano B, Rossi M, Moracci M. Recent advances in the oligosaccharide synthesis promoted by engineered glycosidases. Adv Synth Catal 2005;347:941–950.

78. Fessner WD. In: Mahrwald R, editor. Volume 1, Modern aldol reactions. Weinheim; Wiley-VCH; 2004.

79. Dean SM, Greenberg WA, Wong CH. Recent advances in aldolase-catalyzed asymmetric synthesis. Adv Synth Catal 2007;349:1308–1320.

80. Breuer M, Hauer B. Carbon–carbon coupling in biotransformation. Curr Opin Biotechnol 2003;14:570–576.

81. Samland AK, Sprenger GA. Microbial aldolases as C–C bonding enzymes—unknown treasures and new developments. Appl Microbiol Biotechnol 2006;71: 253–264.
82. Clapes P, Sprenger GA, Joglar J. In: Fessner WD, Anthonsen T, editors. Modern biocatalysis. Weinheim: Wiley-VCH; 2009. pp. 299–366.
83. Wolberg M, Dassen BHN, Schürmann M, Jennewein S, Wubbolts MG, Schoemaker HE, Mink D. Large-scale synthesis of new pyranoid building blocks based on aldolase-catalysed carbon-carbon bond formation. Adv Synth Catal 2008;350:1751–1759.
84. Schürmann M, Sprenger GA. Fructose-6-phosphate aldolase is a novel class i aldolase from escherichia coli and is related to a novel group of bacterial transaldolases. J Biol Chem 2001;276:11055–11061.
85. Schneider S, Gutierez M, Sandalova T, Schneider G, Clapes P, Sprenger GA, Samland AK. Redesigning the active site of transaldolase TalB from *Escherichia* coli: new variants with improved affinity towards non-phosphorylated substrates. Chembiochem 2010. in press.
86. Sanchez-Moreno I, Iturrate L, Doyagüez E, Martinez JA, Fernadez-Mayoralas A, Garcia-Junceda E. Activated α,β-unsaturated aldehydes as substrate of dihydroxyace-tone phosphate (DHAP)-dependent aldolases in the context of a multienzyme system. Adv Synth Catal 2009;351:2967–2975.
87. Mahmoudian M, Noble D, Drake CS, Middleton RF, Montgomery DS, Piercey JE, Ramlakhan D, Todd M, Dawson MJ. An efficient process for production of N-acetylneuraminic acid using N-acetylneuraminic acid aldolase. Enzyme Microb Technol 1997;20:393–400.
88. Müller M, Gocke D, Pohl M. Thiamin diphosphate in biological chemistry: exploitation of diverse thiamin diphosphate-dependent enzymes for asymmetric chemoenzymatic synthesis. FEBS J 2009;276:2894–2904.
89. Koyanagi T, Katayama T, Suzuki H, Onishi A, Yokozeki K, Kumagai H. Hyperproduction of 3,4-dihydroxyphenyl-L-alanine (L-DOPA) using *Erwinia herbicola* cells carrying a mutant transcriptional regulator TyrR. Biosci Biotechnol Biochem 2009;73:1221–1223.
90. Griengl H, Schwab H, Fechter M. The synthesis of chiral cyanohydrins by oxynitrilases. Trends Biotechnol 2000;18:252–256.
91. Purkarthofer T, Skranc W, Schuster C, Griengl H. Potential and capabilities of hydroxynitrile lyases as biocatalysts in the chemical industry. Appl Microbiol Biotechnol 2007;76:309–320.
92. Poechlauer P, Skranc W, Wubbolts M. The large-scale biocatalytic synthesis of enantiopure cyanohydrins. In: Blaser HU, Schmidt E, editors. Asymmetric catalysis on industrial scale: challenges, approaches and solutions. Weinheim: Wiley-VCH; 2004. pp. 151–164.
93. Billhardt UM, Stein P, Whitesides GM. Enzymatic methods for the preparation of acetyl-CoA and analogs. Bioorg Chem 1989;17:1–12.
94. Findeis MA, Whitesides GM. Fumarase-catalyzed synthesis of L-threo-chloromalic acid and its conversion to 2-deoxy-D-ribose and D-erythro-sphingosine. J Org Chem 1987;52:2838–2848.
95. Tanaka A, Tosa T, Kobayashi T, editors. Industrial application of immobilized biocatalysts. New York: Marcel Dekker; 1993.

96. Chibata I, Tosa T, Takata I. Continuous production of L-malic acid by immobilized cells. Trends Biotechnol 1983;1:9–11.

97. Yamada H, Kobayashi M. Nitrile hydratase and its application to industrial production of acrylamide. Biosci Biotechnol Biochem 1996;60:1391–1400.

98. Tosa T, Sato T, Mori T, Matuo Y, Chibata I. Continuous production of L-aspartic acid by immobilized aspartase. Biotechnol Bioeng 1973;XV:69–84.

99. van Assema FBJ, Sereinig N. Method for producing optically active phenylalanine compounds from cinnamic acid derivatives employing a phenylalanine ammonia-lyase derived from idiomarina loihiensis. WO/2008/031578, PCT/EP2007/007945.

100. Weiner B, Poelarends GJ, Janssen DB, Feringa BL. Biocatalytic enantio-selective synthesis of N-substituted aspartic acids by aspartate ammonia lyase. Chem Eur J 2008;14:10094–10100.

101. Izumori K. Bioproduction strategies for rare hexose sugars. Naturwissenschaften 2002;89:120–124.

102. Morimoto K, Park CS, Ozaki M, Takeshita K, Shimonishi T, Granström TB, Takata G, Tokuda M, Izumori K. Large scale production of d-allose from d-psicose using continuous bioreactor and separation system. Enzyme Microb Technol 2006;38:855–859.

103. Maru I, Ohnishi J, Ohta Y, Tsukuda Y. Simple and large-scale production of N-acetylneuraminic acid from N-acetyl-D-glucosamine and pyruvate using N-acyl-D-gluco-samine 2-epimerase and N-acetylneuraminate lyase. Carbohydr Res 1998;306:575–578.

104. Wu B, Szymanski W, Wietzes P, de Wildeman S, Poelarends GJ, Feringa BL, Janssen DB. Enzymatic synthesis of enantiopure α- and β-amino acids by phenylalanine aminomutase-catalyzed amination of cinnamic acid derivatives. Chembiochem 2009;10:338–344.

105. Szymanski W, Wu B, Weiner B, de Wildeman S, Feringa BL, Janssen DB. Phenylalanine aminomutase-catalyzed addition of ammonia to substituted cinnamic acids a route to enantiopure α- and β-amino acids. J Org Chem 2009;74:9152–9157.

106. Arora P, Vats A, Saxena P, Mohanty D, Gokhale RS. Promiscuous fatty Acyl CoA ligases produce Acyl-CoA and Acyl-SNAC precursors for polyketide biosynthesis. J Am Chem Soc 2005;127:9388–9389.

107. Kino K, Nakoto YN, Yagasaki M. Dipeptide synthesis by L-amino acid ligase from *Ralstonia solanacearum*. Biochem Biophys Res Commun 2008;371:536–540.

108. Reddy SG, Waddell ST, Kuo DW, Wong KK, Pompliano DL. Preparative enzymatic synthesis and characterization of the cytoplasmic intermediates of murein biosynthesis. J Am Chem Soc 1999;121:1175–1178.

109. Hollenhorst MA, Clardy J, Walsh CT. The ATP-dependent amide ligases DdaG and DdaF assemble the fumaroyl-dipeptide scaffold of the dapdiamide antibiotics. Biochemistry 2009;48:10467–10472.

110. Ma SK, Gruber J, Davis C, Newman L, Gray D, Wang A, Grate J, Huisman GW, Sheldon RA. A green-by-design Biocatalytic process for atorvastatin intermediate. Green Chem 2010;12:81–86.

111. Patel JM. Biocatalytic synthesis of atorvastatin intermediates. J Mol Catal B Enzym 2009;61:123–128.

112. Reetz MT. Directed evolution of enantioselective enzymes: an unconventional approach to asymmetric catalysis in organic chemistry. J Org Chem 2009;74:5767–5778.
113. Wohlgemuth R. Biocatalysis—key to sustainable industrial chemistry. Current Opinion in Biotechnology 2010;21:713–724.
114. Anastas PT. The transformative innovations needed by green chemistry for sustainability. ChemSusChem 2009;2:391–392.
115. Anastas PT, Crabtree RH, editors. Volume 3, Handbook of green chemistry, biocatalysis. Weinheim: Wiley-VCH: 2009.
116. Constable DJC, Dunn PJ, Hayler JD, Humphrey GR, Leazer JL Jr, Linder-man RJ, Lorenz K, Manley J, Pearlman BA, Wells A, Zaks A, Zhang TY. Key green chemistry research areas - a perspective from pharmaceutical manufacturers. Green Chem 2007;9:411–420.

CHAPTER 12

WHOLE CELL PRODUCTION OF FINE CHEMICALS AND INTERMEDIATES

KAREN ROBINS
Senior Research Associate, Lonza AG, Visp, Switzerland

JOHN GORDON
LES R&D Evaluations and Business Process Excellence, Lonza AG, Visp, Switzerland

12.1 INTRODUCTION

Stoichiometric reagents, hazardous compounds, solvents, and energy inefficiency are out. Catalysis, efficient utilization of raw materials, improved safety, and waste prevention are in, now and for the future. This could be the opening sentence of a newspaper article but it is, in fact, the reality in the current fine chemical industry. For some time now, there has been a strong push to implement the 12 principles of green chemistry because of the growing awareness of the pressing need for environmental protection and sustainable development [1]. It also makes strong economic sense because of the huge cost savings that can be achieved with efficient processes in the chemical industry. Utilization of energy and raw materials is greatly reduced and large savings are also made as waste treatment costs are reduced.

Since the 1960s, Lonza has developed its business based on the high-temperature cracking of liquid petroleum gas (LPG) to produce valuable feedstocks such as ethylene, acetylene, hydrogen, methane, and carbon dioxide. To remain competitive, it has always been imperative that highly integrated processes be developed that avoid waste and create high value-added compounds from these simple building blocks (Figure 12.1) [2]. This has led to a strong culture of process integration and recycling of waste streams where necessary.

Biocatalysis for Green Chemistry and Chemical Process Development, First Edition.
Edited by Junhua (Alex) Tao and Romas Kazlauskas.

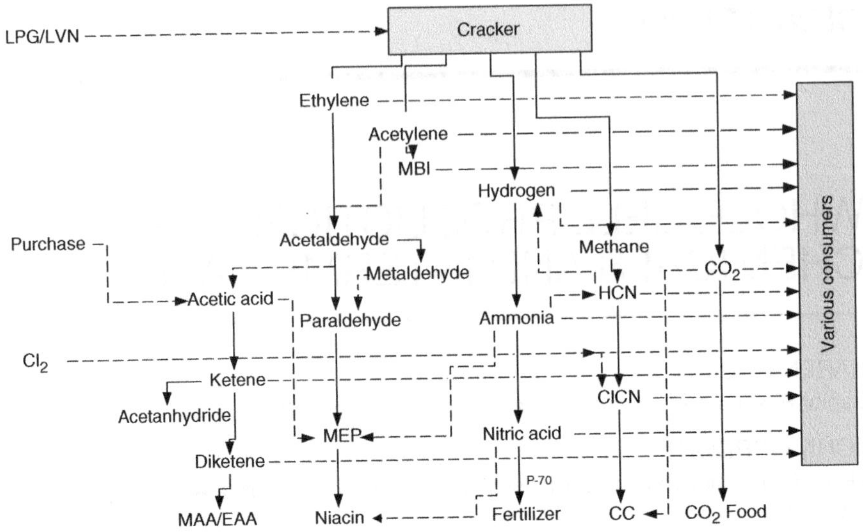

Figure 12.1 The interconnected product lines resulting from the cracker. LVN, light vergin naptha; MBI, 2-methyl-butin-2-ol; MAA, methyl acetoacetate; EAA, ethyl acetoacetate; MEP, 3-methyl-5-ethylpyridine; CC, cyanuric chloride; ClCN, chlorocyanide; HCN, hydrogen cyanide.

The production of L-carnitine, nicotinamide, (S)-(−)-2,2-dimethylcyclopropane carboxamide, 6-hydroxynicotinic acid, 5-hydroxy-2-pyrazine carboxylic acid, 5-methylpyrazinecarboxylic acid, and ethyl (R)-4-cyano-3-hydroxybutyrate are more recent examples of the shift in Lonza from traditional processes using stoichiometric amounts of reagents and large amounts of solvent in the downstream processing to processes that successfully integrate a number of principles of green chemistry and engineering. These processes use catalytic steps, membrane technologies where possible in the downstream processing to avoid the use of solvents, recycling of by-products and heat, and in two cases one-pot biotransformations where two or three biocatalytic reactions are carried out avoiding the necessity of isolating intermediates and reducing the number of unit operations.

12.2 L-CARNITINE

Lonza produces L-carnitine for feed, food, and pharmaceutical applications. L-Carnitine was first isolated from muscle tissue in 1905 [3,4]. In the early 1950s, L-carnitine was found to be an essential growth factor for the meal worm, *Tenebrio molitor* and hence became known as *vitamin* B_T [5]. For human and animal nutrition, L-carnitine is classified as a vitamin-like nutrient. L-Carnitine is synthesized in the liver from two amino acids, lysine and methionine, and plays a very important role in the energy metabolism of heart and muscle tissue.

Long-chain fatty acids are unable to pass the inner mitochondrial membrane. L-Carnitine transports activated fatty acids from the cytoplasm across this membrane into the mitochondria. Inside the mitochondria, the fatty acids undergo β-oxidation producing energy in the form of ATP [6]. The two natural sources of L carnitine in humans are from endogenous synthesis and dietary intake. The highest concentration of L-carnitine is found in meat products, whereas the concentration in plant foods is very low. In humans, L-carnitine has been shown to provide beneficial effects for the elderly, pregnant women, infants on milk formulation, and for vegetarians. It also supports cardiovascular health [7].

About 25 years ago, Lonza developed a four-step chemical process for the synthesis of L-carnitine starting from epichlorohydrin (Figure 12.2) [8].

The first step was the reaction of epichlorohydrin with trimethylamine and L-tartaric acid in methanol to produce crude (*S*)-bis[(3-chloro-2-hydroxypropyl) trimethylammonium]L-tartrate, which was crystallized out of this mixture by the addition of acetone. The pure diastereomeric salt was obtained after recrystallization with methanol and acetone. Next, water, calcium chloride, and sodium cyanide were added to remove the tartaric component as an insoluble salt of calcium and soluble (*R*)-(3-cyano-2-hydroxypropyl)trimethylammonium chloride was produced by cyanide nucleophilic displacement. The calcium tartrate was removed by filtration and the filtrate dried and then recrystallized in ethanol. The (*R*)-(3-cyano-2-hydroxypropyl)trimethylammonium chloride was converted to L-carnitine ((*R*)-3-hydroxy-4-(trimethylammonium)butanoate) by acid hydrolysis with hydrochloric acid. This mixture was passed over active charcoal and then over an ion exchange column to purify the product further. The solution from the ion exchange column was concentrated and recrystallized in isobutanol. The overall yield was 22%.

The disadvantages of this process were low yield, the need to recycle the tartaric acid (2.3 kg/kg L-carnitine), the solvent waste (about 4.5 kg/kg L-carnitine), and the impurity D-carnitine (0.5%). It is desirable to produce L-carnitine free of D-carnitine because it competes with L-carnitine for active transport into the eukaryotic cell [9].

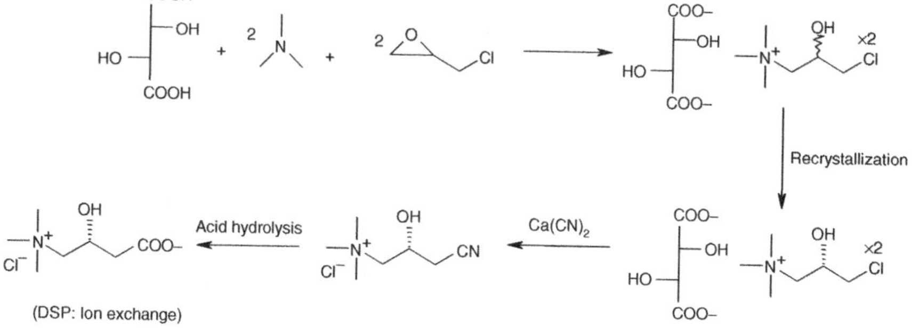

Figure 12.2 The synthetic route for the production of L-carnitine starting from epichlorohydrin.

The second-generation process developed by Lonza for the production of L-carnitine addressed the above problems by employing a biocatalytic asymmetric synthesis starting from γ-butyrobetaine. In this way, L-carnitine could be produced in high yield, with a high enantiomeric excess (*ee*), completely free of D-carnitine while applying the principles of green chemistry.

The γ-butyrobetaine costs were a critical factor. Intensive developmental work concentrated on reducing the number of unit operations, which avoided as much as possible, the isolation of intermediates and implemented recycling of solvents wherever possible. Although the reaction steps remained the same, they led to a faster cycle time, nearly a 10-fold increase in throughput, reduction in the number of unit operations by half, and also a 50% reduction in the amount of waste produced (Figure 12.3). The synthesis starts with the reaction of γ-butyrolactone and hydrochloric acid to form chlorobutyric acid, which is then esterified with methanol. This methyl ester is reacted with trimethylamine to form γ-butyrobetaine methyl ester. The ester is hydrolyzed in the presence of a base to produce the raw γ-butyrobetaine which is then purified by passing the solution over active charcoal. The solution is further purified with electrodialysis to produce a 40% solution of γ-butyrobetaine (pure) [10]. The reaction has a 65% molar efficiency with regard to the starting material, γ-butyrolactone. Excess solvent usage is avoided by employing electrodialysis to purify the γ butyrobetaine, and the methanol (0.4 kg/kg product) used in the process is recycled and reused in the third step.

From a chemical viewpoint, γ-butyrobetaine is then converted in a one-pot multicatalytic cascade using water as the solvent and under mild conditions of temperature and pressure to produce L-carnitine in high yield and high enantiomeric excess. In biological terms, this would be described as a fed-batch process using *Agrobacterium/Rhizobium* HK1349. This microorganism contains three enzymes that carry out this conversion (Figure 12.4). After the γ-butyrobetaine is taken up by the cell, it is activated by the enzyme γ-butyrobetaine-CoA-synthetase to form γ-butyrobetainyl-CoA. A double bond is then introduced to this intermediate by the enzyme γ-butyrobetainyl-CoA-dehydrogenase to form crotonyl-betainyl-CoA. The last step is the introduction

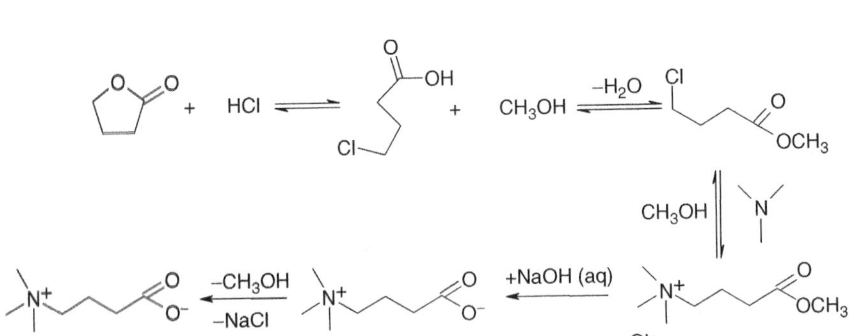

Figure 12.3 The Lonza improved γ-butyrobetaine synthesis.

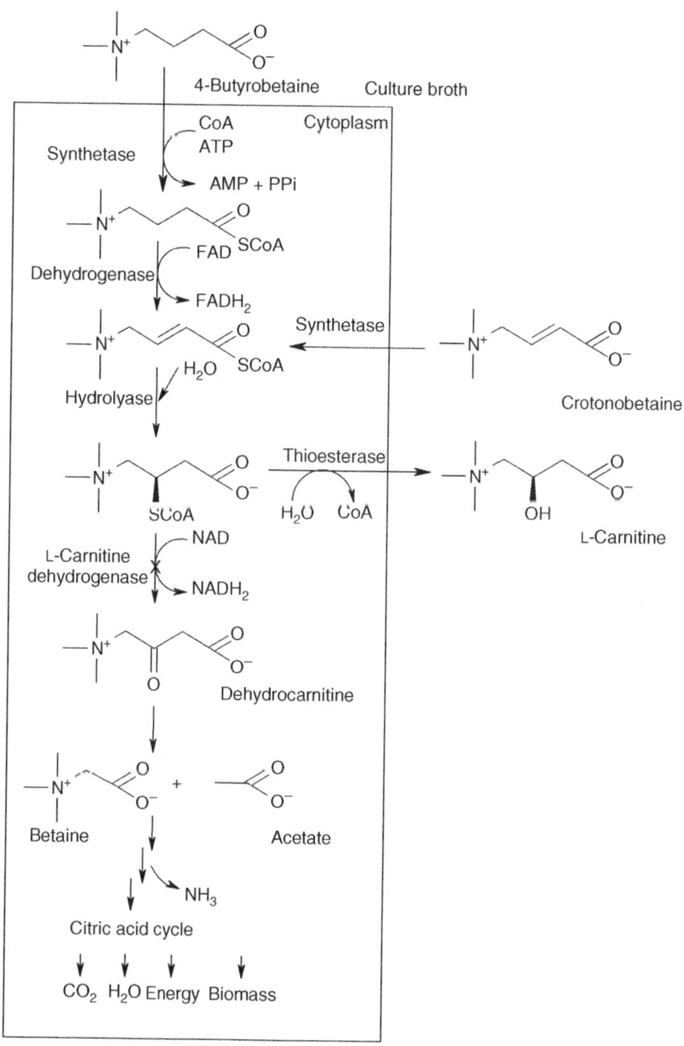

Figure 12.4 The degradation of butyrobetaine and crotonobetaine in the wild-type strain, *Agrobacterium/Rhizobium* HK4. The L-carnitine dehydrogenase is inactivated in the production strain (X), which leads to the accumulation of L-carnitine in the medium.

of the hydroxyl group using water as the O-atom donor to form L-carnitine-CoA [11]. L-Carnitine with an enantiomeric excess of more than 99.9% is excreted into the medium. It is not known if a thioesterase or spontaneous cleavage removes the CoA from the L-carnitine. This is a very energy- and cofactor-intensive process that requires living cells so that the energy requirements of the reaction and the cofactor recycling are fulfilled.

After the fermentation, the biomass is removed by ultrafiltration. The salt is removed from the filtrate by electrodialysis and the solution is passed over active charcoal to remove color and organic compounds from the solution. This solution is concentrated by evaporation and sold for animal feed formulations. L-Carnitine is recrystallized from *iso*-butanol/methanol mixture only for food and pharmaceutical applications where the specifications for γ-butyrobetaine are very low [12].

The development of this process started with the isolation of the wild-type strain, *Agrobacterium/Rhizobium* HK4. A mutation was required to inactivate the L-carnitine dehydrogenase enzyme that would otherwise further degrade L-carnitine in the cell (Figure 12.4). This enabled L-carnitine accumulation but at very low levels. Hans Kulla submitted the strain to an elegant strain development program that introduced high product tolerance, at the same time preserving γ-butyrobetaine affinity so that nearly quantitative conversion of high concentrations of γ-butyrobetaine could be achieved and a commercially viable process could be developed [11]. Initially, a continuous process with cell recycling was developed with very high productivity, exploiting the maintenance state of the cells where high metabolic activities are associated with low growth rates ensuring that the energy required for the process is not wasted on biomass production [13–15]. The disadvantage of this process was that the product solution contained about 92% L-carnitine and 8% γ-butyrobetaine. L-Carnitine and γ-butyrobetaine have very similar physiochemical properties, which makes them very difficult to separate from an aqueous solution. L-Carnitine was isolated from the biosolution by ultrafiltration, electrodialysis, active charcoal treatment, and drying but two additional recrystallizations in *iso*-butanol were required to reduce γ-butyrobetaine to <0.1%, which is required in order to fulfill the product specifications of L-carnitine (Figure 12.5)

The disadvantages of the continuous process could have been solved with a cascade reaction but this would have required high investment in equipment necessary to run the process. The economic alternative proved to be the development of a fed-batch process that had four times less volumetric productivity than the continuous process but the bioconversion was >99% [15–17]. The process design also integrated the upstream and downstream aspects of the process optimally, which was another important aspect of the development of an economic and environmentally friendly process. The phosphoric acid used for pH control was replaced by H_2SO_4. This resulted in an increase in efficiency of the electrodialysis step because the sulfate is removed 30% faster than the phosphate salts. The color of the isolated L-carnitine was also improved by replacing the technical grade betaine and glycerol, which are important components of the fermentation medium, with more expensive pure grades of these components. This led to more effective color removal in the active charcoal step. As a consequence of the increase in conversion achieved in the fed-batch process and the changes in the quality of the raw materials used in the process, both of the solvent-requiring crystallization steps could also be omitted from the workup for feed grade product and the process costs were also reduced considerably (Table 12.1) [15,16].

The process improvements addressed the process as a whole and included optimization of the chemical process for γ-butyrobetaine synthesis, strain

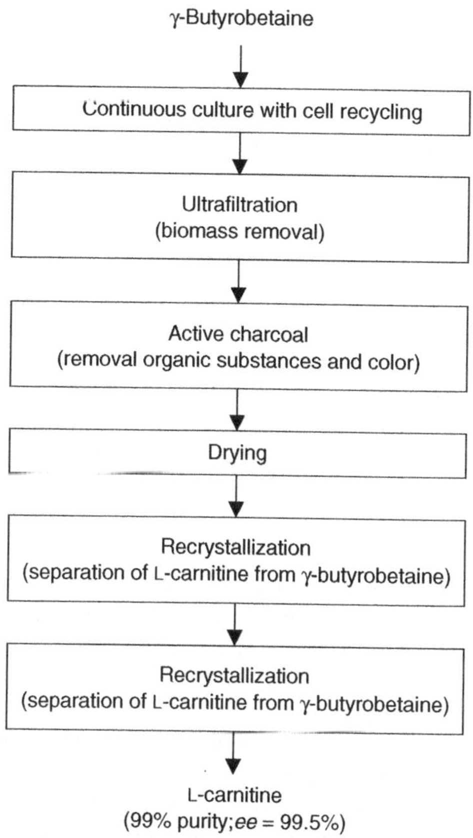

Figure 12.5 Unit operations required for the production of L-carnitine from γ-butyrobetaine [16].

TABLE 12.1 Comparison of the Continuous Process with Cell Recycling with the Fed-Batch Process for the Conversion of γ-Butyrobetaine to L-Carnitine

Parameter	Continuous Process with Cell Recycling	Fed-Batch Process
Volumetric productivity (g/l/d)	130	30
Bioconversion yield (molar % of substrate)	92	99.5
ee (%)	>99.9%	>99.9%
D-Carnitine (%)	n.d.	n.d.
Cost estimation of biotransformation and isolation of L-carnitine (arbitrary units)	100	60

n.d., not detected.

improvement, an upgrade in the quality of the critical components of the medium, a change from continuous to fed-batch process, and the optimization of the downstream processing. The biotransformation combined three biocatalytical steps in an extremely atom-efficient one-pot cascade which led to high volumetric and space time yields, reduced the unit operations, and generated less waste, which translated into substantial environmental and economical benefits.

12.3 NICOTINAMIDE

Lonza is the world's largest producer of niacin (nicotinic acid and nicotinamide). Niacin is otherwise known as vitamin B_3 or vitamin PP (pellagra preventing factor). Niacin is an important building block of NADH or NADPH, which are essential for all living cells and play a central role in metabolism. Nicotinic acid is used as an additive in animal feed and nicotinamide is used for food applications. Although nicotinic acid and nicotinamide are essentially equivalent in the human body, nicotinic acid is not used as a food additive because it causes flushing in humans.

The first synthesis of nicotinic acid was carried out in 1867. Interestingly enough, the synthesis involves the oxidation of nicotine, a renewable resource with chromic acid, which is carcinogenic and environmentally undesirable! Ignorance or schizophrenia? On the basis of the data from R. Chuck which assume 100% yield for this reaction and does not include the product isolation, the E Factor is 8.6 (Figure 12.6) [18,19].

Lonza began the production of nicotinic acid in 1952 [20]. It was produced from 2-methyl-5-ethyl-pyridine (MEP) via pyridine-2,5-dicarboxylic acid with a mixture of sulfuric acid and nitric acid. This process was changed and optimized several times before a continuous process was introduced in 1965 starting from paraldehyde via 2-methyl-5-ethyl-pyridine [21].

Chemically, 3-picoline is more suited than 2-methyl-5-ethyl-pyridine in the production of nicotinic acid because the methyl group can be directly and easily oxidized to nicotinic acid with high regiospecificity, which yields few side-products. The disadvantage of 3-picoline as a starting product and the reason that Lonza based their original process on 2-methyl-5-ethyl-pyridine lies in the traditional process meant for 3-picoline production. When formaldehyde, acetaldehyde, and ammonia are reacted, a mixture of pyridine, as the main product, and 3-picoline are produced. The ratio is 2:1. This means that the pyridine has always determined the price and the availability of 3-picoline. 2-Methyl-5-ethyl-pyridine was chosen by Lonza as the precursor to nicotinic acid because it can

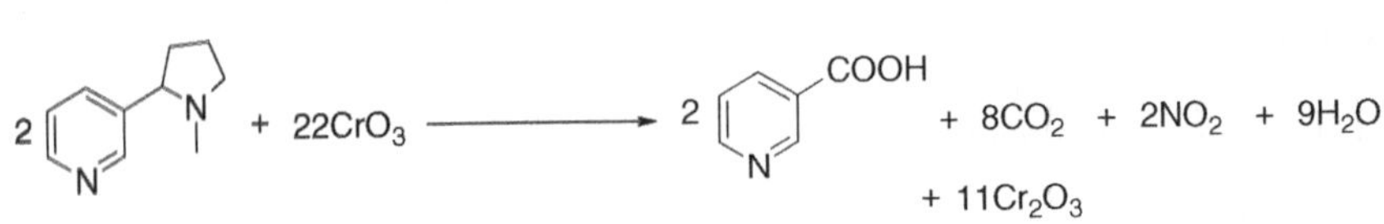

Figure 12.6 Chromic acid oxidation of nicotine to produce nicotinic acid.

be produced by the liquid-phase reaction of paraldehyde and ammonium with a selectivity of about 70% and is considerably cheaper than 3-picoline to produce (Figure 12.7). The 2-methyl-5-ethyl-pyridine is then oxidized with nitric acid under conditions of high temperature and pressure to produce nicotinic acid. More than 15,000 ton/yr of nicotinic acid is produced with this process. The next step is the reaction of nicotinic acid with ammonia to produce nicotinamide. The selectivity of this process is high (>80%), the nitric oxides (NO_x) produced are recycled, and all other waste streams are either recycled or rendered neutral. There are, however, several disadvantages. [22]

- Safety issues arise when nitric acid is used at high temperature and pressure;
- More than 1 ton of CO_2 is produced per ton of niacin;
- Nitric oxides are produced. Although these nitric oxides are recycled, they require additional processing and it is not possible to remove all the nitric oxide from the waste gases. This requires an additional step to catalytically remove the remaining nitric oxides;
- Recrystallization and decolorization are required to reach product specifications.

Carbon efficiency is at best 0.75 for the conversion of 2-methyl-5-ethyl-pyridine to niacin and two carbon atoms are lost. The reactant to product ratio is <15% if recycling is neglected.

Lonza worked for years to develop an alternative process for the production of nicotinamide. The congruent pathway that was used to produce nicotinic acid and nicotinamide had the disadvantage that part of the nicotinic acid produced, which could otherwise be sold directly, had to be used for the production of nicotinamide. It made sense to develop a process that was independent of nicotinic acid and at the same time solved some of the problems of the process mentioned above.

The four catalytic steps involved in the production of nicotinamide are shown in Table 12.2 [23].

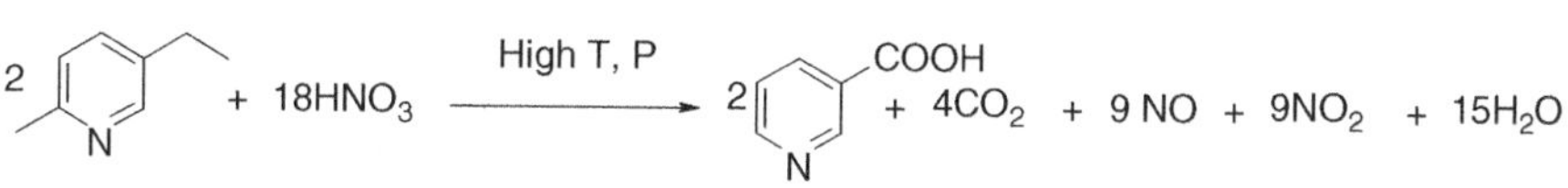

Regeneration of nitric acid

$$NO + [O] \longrightarrow NO_2$$

$$3NO_2 + H_2O \longrightarrow 2HNO_3 + NO$$

Figure 12.7 The oxidation of 2-methyl-5-ethyl-pyridine with nitric acid to nicotinic acid [22].

TABLE 12.2 The Chemo-Biocatalytic Lonza Process for the Production of Nicotinamide in China

Step 1

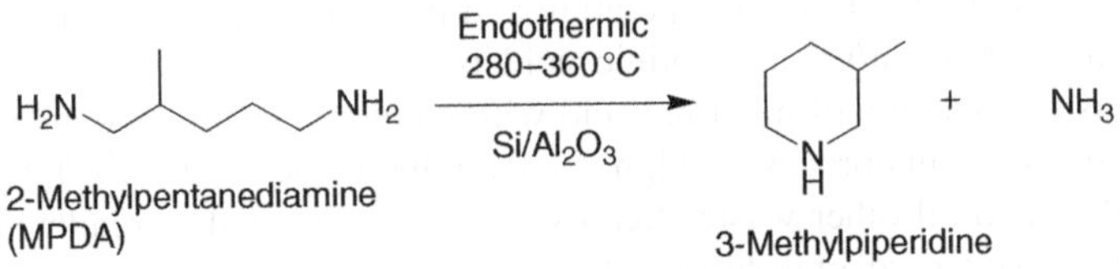

3-Methylpiperidine	Starting product for step 2
1,5-Diamino-2-methylpentane	Waste product from the production of nylon
Si/Al_2O_3	Catalytic amounts
NH_3	Recycled in step 3
Endothermic	Heat generated in step 2 used in this step

Step 2

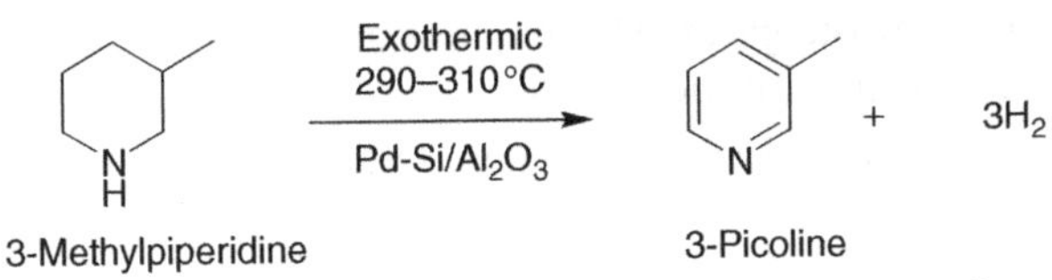

3-Picoline	Starting product for step 3
3-Methylpiperidine	Product from step 1
$Pd–Si/Al_2O_3$	Catalytic amounts
H_2	Used to generate heat for step 1
Exothermic	Heat produced used in step 1

Step 3

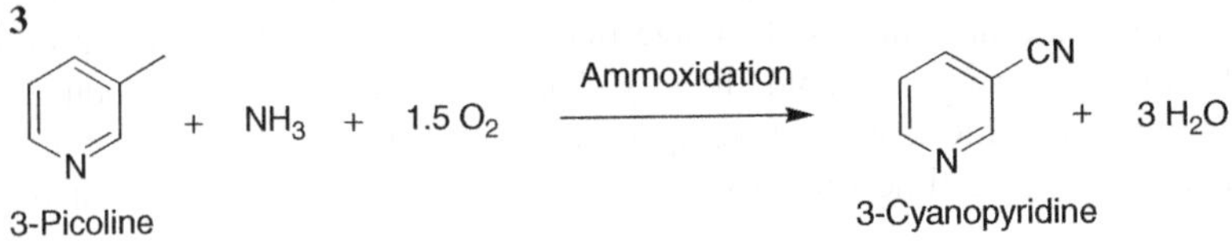

3-Cyanopyridine	Starting product for step 4
3-Picoline	Product from step 2
NH_3	Liberated in step 1
O_2	Sustainable resource
H_2O	Sustainable resource

Step 4

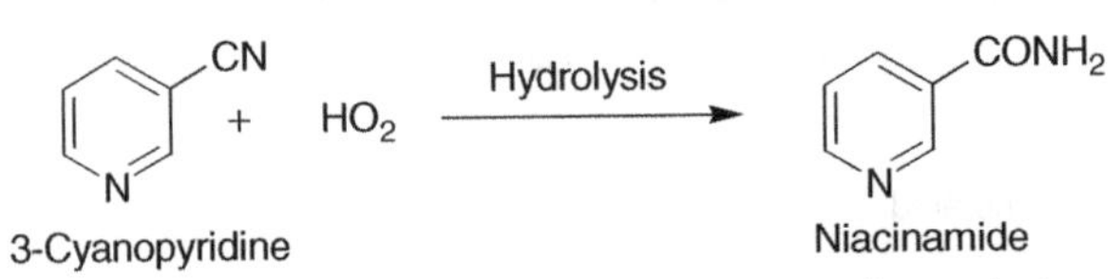

Nicotinamide	Highly pure product solution
3-Cyanopyridine	Product from step 3
Immobilized *Rhodococcus rhodochrous* J1 (nitrile hydratase)	Catalytic amounts
H_2O	Sustainable resource

The first step involves the catalytic ring closure of 2-methylpentanediamine (MPDA), a waste from the nylon industry, to 3-methylpiperidine and liberates ammonia. The 3-methylpiperidine is very pure and the ammonia generated in this step is used in the third step of the reaction, which is the ammoxidation of 3-picoline to 3-cyanopyridine. The next step is the exothermic dehydrogenation of 3-methyl piperidine to produce 3-picoline and hydrogen. The hydrogen is used directly to generate heat for the first reaction. The production of 3-picoline is approximately energy neutral as the energy liberated in the second step is used in the first step, the endothermic ring closure.

Subsequently, the 3-picoline undergoes gas-phase ammoxidation, which is the selective oxidation of 3-picoline with air in the presence of ammonia to form 3-cyanopyridine. The ammonia that is liberated from the ring closure in the first step is the source of the ammonia for the ammoxidation. The 3-cyanopyridine is extracted into toluene and distilled. The 3-cyanopyridine is fed directly into the next step, where it is hydrolyzed by a nitrile hydratase producing nicotinamide. Immobilized wild-type cells of *Rhodococcus rhodochrous* J1 from DiaNitrix are used to carry out this hydrolysis [24]. The reaction takes place at ambient temperature; no pressure is required; and the solvent is a water-based buffer with minimal salt concentrations. The conversion of 3-cyanopyridine to nicotinamide is 100% and unlike the chemical conversion of 3-cyanopyiridine to nicotinamide, no nicotinic acid is produced as a side-product of this reaction. Owing to the high purity of the product, the following steps required to isolate the nicotinamide are very simple and cheap. The filtrate is passed over an active charcoal column and then through a nanofiltration unit. The resulting solution is then concentrated and the process water is recycled back into the biohydrolysis step and the concentrated solution is dried directly in the spray dryer. The product meets all of the customer specifications on purity, color (white powder), particle size, and flow characteristics, which are so important for products used in the food industry. A schematic representation of the process can be found in Figure 12.8. It shows the inputs, outputs, and the streams that are recycled.

For the biohydrolysis step to work so efficiently, a strain containing a stable nitrile hydratase with high activity and no amidase activity, that would otherwise convert the nicotinamide to the unwanted nicotinic acid, was chosen. Any trace of nicotinic acid would complicate the workup and lead to added costs and more unit operations. Another important feature of the strain was that a color-negative mutant was used. This improved the color of the isolated product negating the necessity of additional purification steps. Detailed kinetic studies of the enzyme were also required. This kinetic data was important for the development of an efficient process which functioned optimally and minimized the amount of biocatalyst used in the process. The K_m for nitrile hydratase catalyzed hydrolysis of 3-cyanopyridine is 200 mM, which means that substrate concentrations need to be high for high reaction rates [25]. This is contradicted by the fact that 3-cyanopyridine is one of the major factors leading to the inactivation of the enzyme. To avoid this problem, substrate concentration should be kept low. This

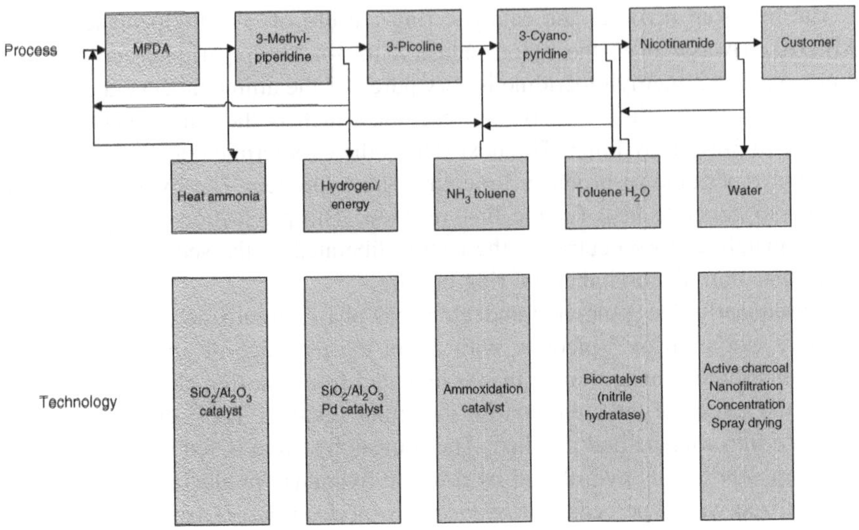

Figure 12.8 The new Lonza process in Guangzhou and Nansha, China.

problem was solved with a technical solution by developing a continuous three-step cascade reaction. In this way, the 3-cyanopyridine concentration could be kept low and the slow reaction rates at low substrate concentrations could be compensated for by imposing the same retention times in each tank. Approximately 99% of the 3-cyanopyridine is converted in the first tank and the rest in the second tank. The end concentration of 3-cyanopyridine in the second tank should be <0.01%. The third tank functions as a watch dog in case there is an increase in the concentration of 3-cyanopyridine above specifications as the biocatalyst becomes exhausted and just prior to recharging the tanks with fresh biocatalyst, which is after about three months.

There are many aspects of this process that are in accordance with the 12 principles of green chemistry [1,22]:

- Starting material, MPDA, is a waste product of nylon industry and it is used to create a higher value product;
- Four highly selective catalytic reactions;
- Air replaces chemical oxidants in step 3;
- Wastes are treated with two additional catalytic steps. They oxidize and remove the nitric oxides from the waste gases to remove side-products (overall <10%) and produce N_2, H_2O, and CO_2 (about 200 kg CO_2 is produced per ton of nicotinamide);
- The first three reactions are carried out in the gas phase and, therefore, chemical catalysts do not have to be recovered from solution;
- Energy from the exothermic reaction (step 2) can be recovered and reused in the first step;

- Liberated H_2 in step 2 is also used to generate heat for the first step;
- Low number of unit operations;
- The only solvents are water and toluene. Toluene is used for the extraction of 3-cyanopyridine and is recycled (nearly 100%);
- Ammonia is recycled in step 3;
- Biohydrolysis is carried out at ambient temperature and atmospheric pressure and is completely selective;
- Minimal waste due to highly selective reactions;
- High conversion (overall yield from MPDA is 90%) and carbon efficiencies (ratio of carbon in product to reactants) is 90%.

Roger Sheldon, Isabel Arends, and Ulf Hanefeld wrote that "the ultimate green process involves the integration of several catalytic steps into catalytic cascade processes, thus emulating the multienzyme processes which are occurring in the living cell" [26]. This new process demonstrates how effectively the combination of chemo- and biocatalysis in such a system creates a highly efficient and selective multistep reaction. Side-products are avoided by the high selectivity of the reactions and there is a high carbon efficiency, which minimizes CO_2 emissions. By involving process engineers early in the process development, the steps are also highly interconnected with the waste stream from one step being used as a starting material or energy source for another step. This further minimizes the waste streams. The complete selectivity of the last step, the biohydrolysis of 3-cyanopyridin to nicotinamide means that the product isolation is very simple and does not require the use of solvents to remove unwanted side-products.

The success of this "green" process can be measured by the fact that we successfully compete in the very competitive environment found in China. Our first plant, which was started in 1997 in Guangzhou, has a capacity of 5000 mt/year. Since then, a second plant has been built in Nansha with a capacity of 12,000 mt/year.

12.4 (S)-(–)-2,2-DIMETHYLCYCLOPROPANE CARBOXAMIDE

Lonza has developed several processes for the production of chiral acids and amides. Figure 12.9 shows some examples of these processes. One example, the production of (*S*)-(–)-2,2-dimethylcyclopropane carboxamide on a large scale, is described in more detail below.

(*S*)-(–)-2,2-Dimethylcyclopropane carboxamide is a chiral intermediate for the production of cilastatin, a renal dehydropeptidase inhibitor. Cilastatin is used in conjunction with imipenem and carbapenem antibiotics. When cilastatin is used in combinations with these antibiotics, this peptidase is inhibited so that it is unable to inactivate these antibiotics *in vivo*. The dosage of antibiotic can therefore be reduced because the half-life is extended considerably.

During the development of the process for the production of (*S*)-(–)-2,2-dimethylcyclopropane carboxamide, both chemical and chemoenzymatic routes

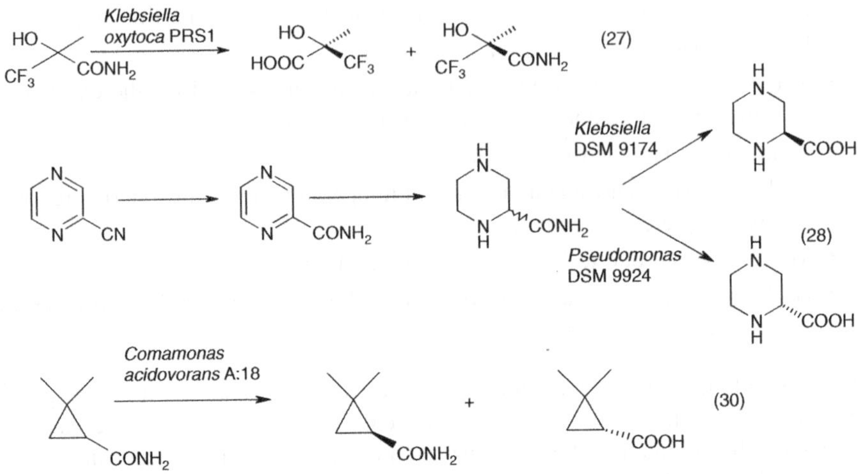

Figure 12.9 Some Lonza examples of the application of stereoselective amidases.

were evaluated. Both routes chemically produced the racemic amide via the nitrile or acid (Figure 12.10) [29]. The enantiomeric excess percentage and the yields were lower with the racemic resolution of the amide using classical chemical resolution compared to the enzymatic resolution. The racemic resolution of the amide was carried out with the wild-type strain, *Comamonas acidovorans* A:18 containing a (*R*)-specific amidase. The pH and temperature profile of this enzyme revealed that by running the biotransformation at 37°C and at a pH of 8.0, the rate and selectivity for the (*R*)-amide could be increased considerably without effecting the stability of the enzyme (Figure 12.11) (Personal communication Nick Shaw)

A conversion of 45.2% (not isolated) with a 99.3% *ee* was achieved with the enzymatic racemic resolution of the amide [30]. The biomass was removed by ultrafiltration, the filtrate was concentrated, and the (*S*)-(−)-2,2-dimethylcyclopropane carboxamide crystallized from the solution.

An unexpected problem that arose with this process was caused by the low solubility of the chemically synthesized racemic amide (the racemic amide had a solubility of 0.5%, whereas the pure isomer had a solubility of 2.5%). The undissolved crystals proved difficult to mix and partly floated on the surface of the reaction mixture. This problem was solved in the second-generation process by replacing the chemical conversion of the nitrile to the racemic amide with the enzymatic hydrolysis using a nitrile hydratase containing strain of *Rhodococcus equi* TG328-2. The needle crystals of the amide produced by the enzymatic reaction could be easily and evenly dispersed in the reaction mixture. The nitrile hydratase was combined with the amidase step in a one-pot two-step biocatalytic reaction. The two enzymatic reactions were run sequentially because the nitrile otherwise inactivated the amidase in the subsequent step. This innovation reduced the number of unit operations in the process and avoided the necessity

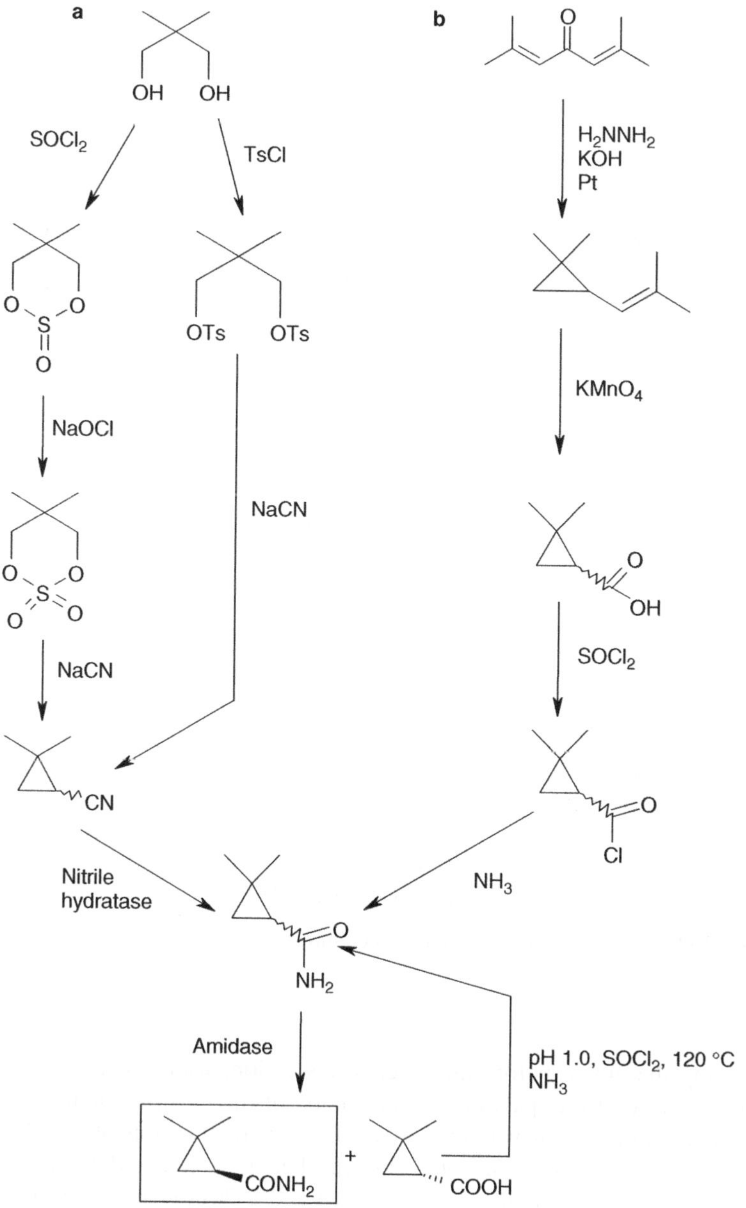

Figure 12.10 Synthesis of *S*-(+)-2,2-dimethylcyclopropane carboxamide via the racemic nitrile (route a) and synthesis of *S*-(+)-2,2-dimethylcyclopropane carboxamide via the racemic acid (route b).

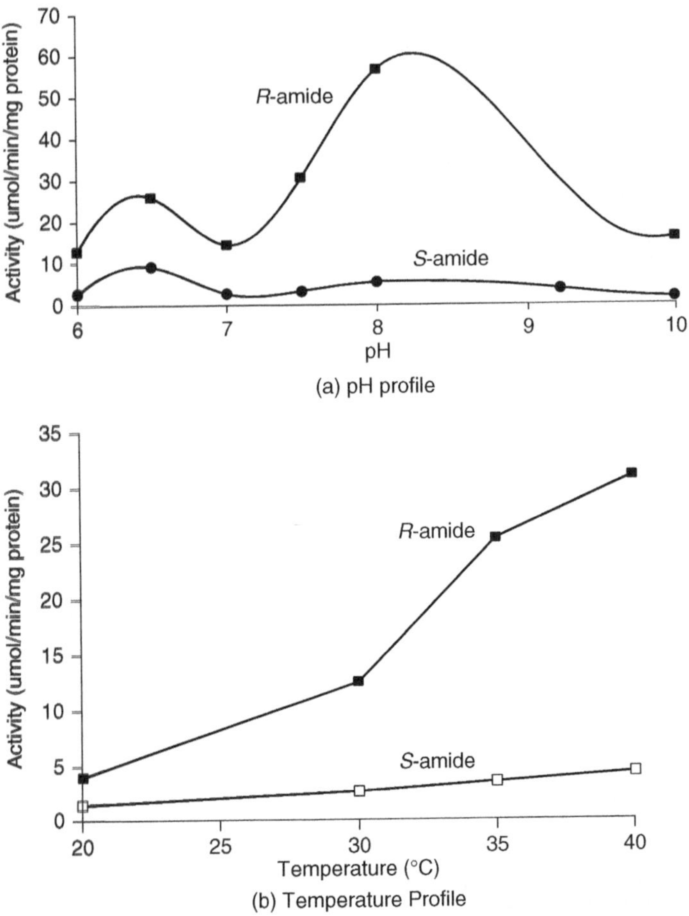

Figure 12.11 The pH profile (a) and temperature profile (b) of the *S*-specific amidase from *Comamonas acidovorans* A:18.

of isolating the amide before the racemic resolution, which increased the yield by 20% because of the quantitative conversion of the nitrile to amide [31].

Another improvement was the cloning and overexpression of the stereospecific amidase in *Escherichia coli*. This resulted in a fourfold increase in the specific activity of the amidase (Table 12.3) and a 20-fold increase in the total activity after the fermentation [32]. In a campaign carried out in the production plant in Kourim (Czech Republic) for the production of approximately 500 kg of (*S*)-(−)-2,2-dimethylcyclopropane carboxamide, the molar yield (not isolated) calculated from the nitrile after the two biotransformation steps was 44.5% with a 99.3%–99.8% *ee*. The downstream processing was also improved by introducing electrodialysis, which removed most of the (*R*)-acid and the salts from the filtrate. Less than 10% of product was lost in this step. The remaining traces of (*R*)-acid

TABLE 12.3 A Comparison of the Cloned Stereospecific Amidase with the Wild-Type Amidase

	C. acidovorans A:18	*E. coli* DH5/pCAR6
Inducer	Inducible (substrate)	Constitutive
Specific activity (g/L/h/OD_{650nm})	0.5	2.1
$OD_{650nm,max}$	6	30
Total activity (g/L/h)	3	63

were removed with ion exchange. The pure solution of the (*S*)-amide was concentrated with reverse osmosis at 60°C, allowing the filtrate to be concentrated before crystallization. The crystallization was carried out at 2–4°C without solvents and the product was then dried. The isolated molar yield (calculated from the nitrile) from this 500 kg campaign was 36.6%. The option of recycling the 2,2-dimethylcyclopropane carboxylic acid which was isolated after the ion exchange step (49.6% molar yield from the nitrile) was also evaluated (Figure 12.9). The acid can be converted to the racemic amide with thionyl chloride and ammonia, which can be reintroduced into the enzymatic racemic resolution step. If an 80% yield is assumed for this conversion of waste acid to racemic amide then, by means of repeated, multifold processing of the acid a maximal molar yield of the (*S*)-amide based on the nitrile of 85% could, in theory, be reached.

This process was developed in the early 1990s. At that time, no stereospecific nitrilases were known but the combination of a highly efficient nitrile hydratase with the stereospecific amidase was also effective. The whole process was improved by implementing a new synthetic pathway for the production of the racemic nitrile as a starting product for the two-step biocatalytic cascade and cloning of the amidase to increase productivity. The product isolation was also improved by incorporating electrodialysis and reverse osmosis in the isolation steps, which allowed the crystallization of the product directly from the aqueous filtrate avoiding the use of solvents and led to the production of a highly pure product that met specifications.

12.5 REGIOSPECIFIC HYDROXYLATION OF HETEROCYCLIC CARBOXYLIC ACIDS

Production of regiospecifically hydroxy-substituted aromatic *N*-heterocyclic carboxylic acids enables the effective synthesis of many important intermediates for the synthesis of agrochemicals and pharmaceuticals. The first process developed in the biotransformation group in Lonza was the production of 6-hydroxynicotinic acid [33]. This process used a microorganism that catabolizes nicotinic acid via 6-hydroxynicotinic acid. This process was further adapted for the large-scale production of 5-hydroxy-2-pyrazine carboxylic acid. Both of these processes are discussed below in more detail. This hydroxylation platform was expanded further

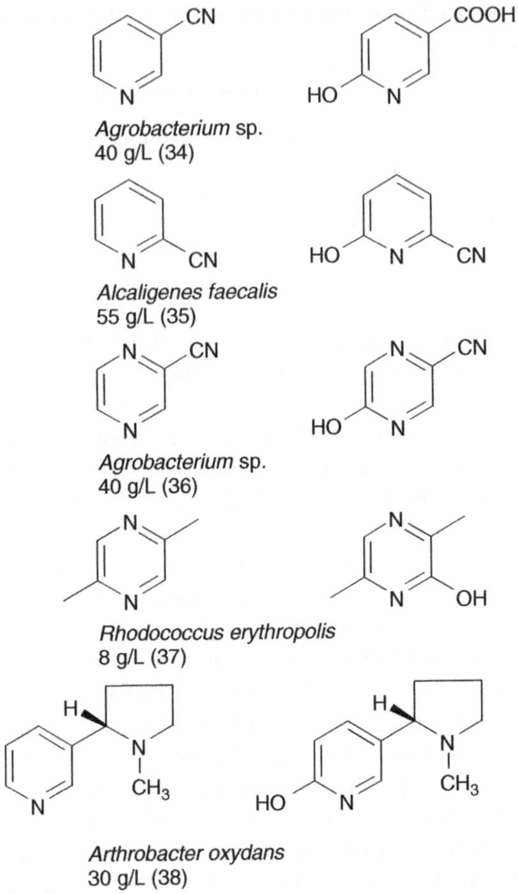

Figure 12.12 Further examples of the microbial hydroxylation of various pyridine derivatives.

by Andreas Kiener so that a wide range of hydroxylated heterocycles could be produced using highly regiospecific enzymes (Figure 12.12).

12.5.1 6-Hydroxy-Nicotinic Acid

6-Hydroxynicotinic acid is a versatile building block for the synthesis of some neonicotinoid insecticides. It can be chlorinated to form, 6-chloronicotinic acid, a precursor of the insecticide, imidacloprid [39]. The first process in Lonza was the chemical synthesis of 6-chloronicotinic acid starting from isocinchomeronic acid-*N*-oxide (Figure 12.13) [40]. The problem with the synthesis was the product mix of 6-chloronicotinic acid, 6-hydroxynicotinic acid (53.5:43.7), and the other by-products which required a complicated workup to obtain pure product.

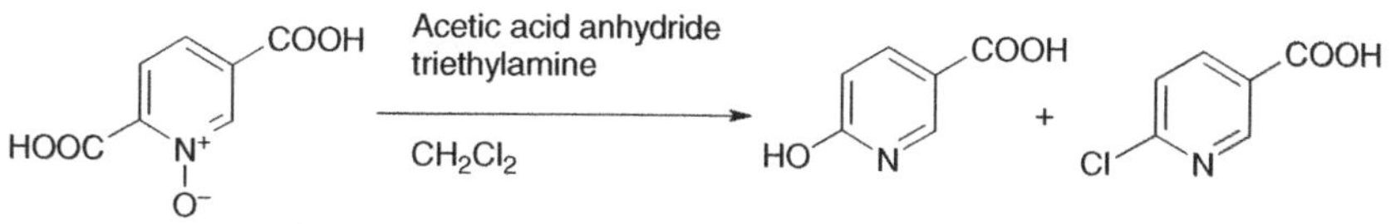

Figure 12.13 Chemical synthesis of 6-chloronicotinic acid.

The bioprocess was developed after the serendipitous discovery of pure 6-hydroxynicotinic acid crystals in a sample from the Lonza nicotinic acid production plant [11]. The microorganism isolated from this sample was *Achromobacter xylosoxydans* LK1, which was growing on the nicotinic acid as sole carbon and nitrogen source. The pathway is shown in Figure 12.14. The process depended on the fact that the enzyme responsible for the further degradation of 6-hydroxynicotinic acid, 6-hydroxynicotinate hydroxylase, was inhibited by more than 10 g/L nicotinic acid but the nicotinic acid hydroxylase was unaffected. This allowed the process to be segregated into an initial growth phase where active biomass was produced when a limiting amount of nicotinic acid was present and then a biotransformation phase when an excess of nicotinic acid was added. A concentration of 74 g/L could be achieved in 25 h with this process. The isolation of the product was carried out by an acid pH shift which resulted in the precipitation of the product. This process was carried out in multiton scale at our fermentation plant in Kourim (Czech Republic) and an isolated yield of 85%–91% was achieved.

12.5.2 5-Hydroxy-Pyrazine Carboxylic Acid

Substituted pyrazinecarboxylic acids are used for the synthesis of several active pharmaceutical ingredients such as a potassium sparing diuretic, Amiloride, an antidiabetic, Glipicide and Pyrazinamide, a tuberculostatic [41a,b,c].

The production of 5-hydroxy-2-pyrazine carboxylic acid is carried out analogous to the 6-hydroxynicotinic acid process for the biomass production phase

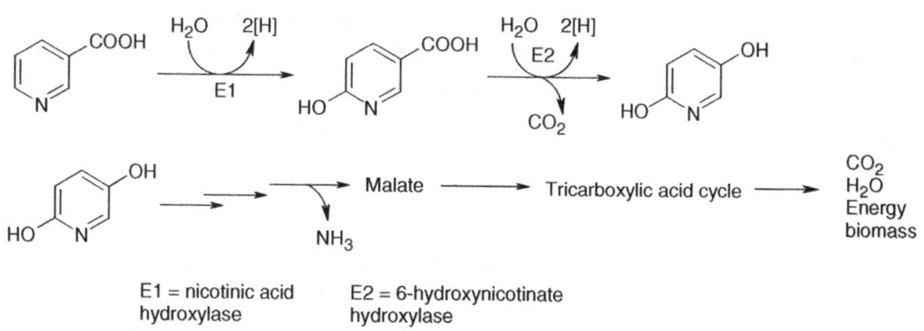

Figure 12.14 The degradation pathway of nicotinic acid by *Achromobacter xylosoxydans* LK1.

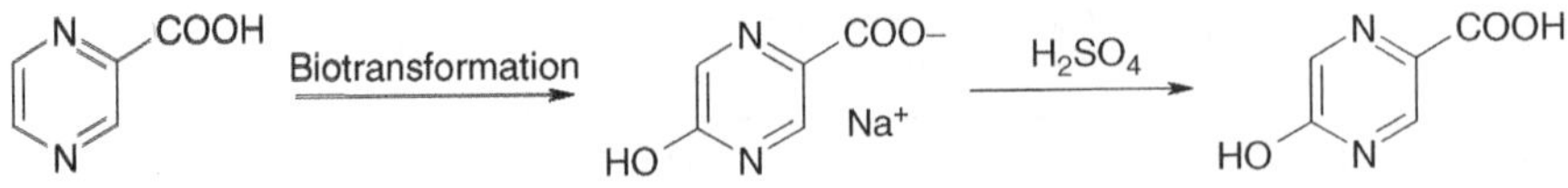

Figure 12.15 The production of 5-hydroxy-2-pyrazine carboxylic acid.

and instead of nicotinic acid, pyrazine carboxylic acid is used as the starting material for the biotransformation step [42]. After the biotransformation, the biomass is removed by ultrafiltration at 50°C and pH 8.2. The pH of the filtrate is then adjusted to pH 1.8 with H_2SO_4 and the temperature reduced to about 5°C which causes the product to precipitate out of the solution. The product crystals are then harvested by centrifugation, washed with water, and dried. An isolated molar yield of 75.3% is achieved with this process (Figure 12.15). The only waste products are biomass and salts from the precipitation step. Process improvements are on-going to increase the end concentration of the product substantially.

12.6 METHYL AND ETHYL GROUP OXIDATION OF AROMATIC HETEROCYCLES

The selective oxidation of ethyl and methyl substituted aromatic *N*-heterocycles for the preparation of the corresponding monocarboxylic acids is extremely difficult with classical chemistry. It usually leads to the formation of unwanted side-products that are difficult to remove without extensive isolation steps usually requiring large amounts of solvent.

Lonza developed two strategies to solve this problem giving them access to either the monocarboxylic acid or the acetic acid derivative in high purity.

The first strategy is based on the degradation of xylene by *Pseudomonas putida*. It is known that *P. putida* degrades xylene first by oxidation to 4-methylbenzylalcohol by xylene monooxygenase, which is then oxidized to 4-methylbenzoic acid by benzylalcohol and benzaldehyde dehydrogenase. Further, it is degraded to methylcatechol before the aromatic ring is cleaved. After cleavage, the product enters the tricarboxylic acid cycle. Andreas Kiener observed that when the wild-type strain of *P. putida* ATCC 33015 was grown on xylene as the sole carbon and energy source it was able to oxidize the methyl group on five- and six-membered heterocycles to the corresponding monocarboxylic acid (Figure 12.16). These monocarboxylic acids were not further degraded and accumulated in the medium [43,44].

The second strategy is based on the degradation of octane by *Pseudomonas oleovorans* ATCC 29347. It is known that during the growth of *P. oleovorans* on octane the first three enzymes (alkane monoxygenase, alcohol dehydrogenase, and aldehyde dehydrogenase) degrade octane to the corresponding carboxylic acid, analogous to the previously mentioned xylene degradation in *P. putida*. It was observed that this strain when grown on octane catalyzed the regioselective oxidation of the ethyl group on heteroaromatic compounds to the corresponding

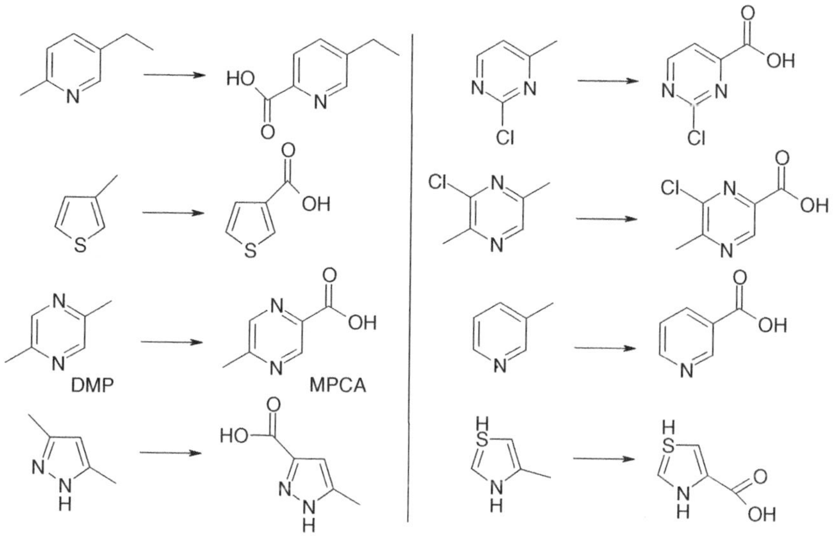

Figure 12.16 Some examples of the regioselective oxidation of methylated heterocycles by *Pseudomonas putida* [43,44]. *Note*: DMP, 2,5-Dimethylpyrazine; MPCA, 5-Methylpyrazine-2-carboxylic acid.

acetic acid derivative. The selectivity of the reaction is demonstrated by the very selective oxidation of the ethyl group of 5-ethyl-2-methylpyridine. Only (5-methyl-pyridyl-3-)acetic acid is found in the medium and absolutely no trace of the monocarboxylic acid is detected [43]. Some examples of this oxidation system can be seen in Figure 12.17.

12.6.1 5-Methylpyrazine-2-Carboxylic Acid

5-Methylpyrazine-2-carboxylic acid (MPCA) is an intermediate for the production of an active pharmaceutical ingredient with antilipolytic activity, acipimox [45].

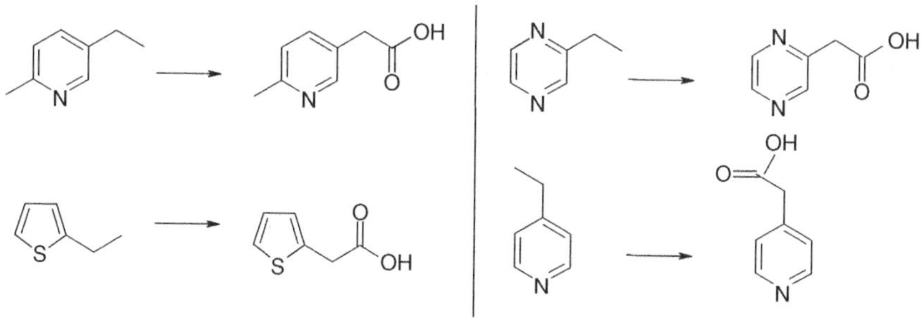

Figure 12.17 Some examples of the regioselective oxidation of ethylated heterocycles by *Pseudomonas oleovorans*.

Lonza developed a process for the production of this compound based on the methyl group oxidation technology and scaled the process to tonne scale. 2,5-Dimethylpyrazine (DMP) was oxidized to MPCA by *P. putida* while growing on xylene (Figure 12.16). On-line measurement of the xylene concentration in the exhaust gases and on-line HPLC (high-performance liquid chromatography) measurement of the DMP concentrations were important parameters for process control. A product concentration of 24 g/L was produced, which corresponds to a yield of more than 95%. Less than 0.1 g/L of DMP remained in the medium and no trace of xylene could be detected [43]. The biomass was removed by ultrafiltration and the filtrate passed through a nanofiltration unit to remove chromophoric substances and larger molecules like protein from the solution. The filtrate was treated with active charcoal to further decolorize the solution which was then concentrated using electrodialysis. The MPCA was then precipitated out of the filtrate by acidifying the solution.

12.7 ETHYL (*R*)-4-CYANO-3-HYDROXYBUTYRATE

Ethyl (*R*)-4-cyano-3-hydroxybutyrate (hydroxynitrile, HN) (**4**) is the key chiral building block for the production of Pfizer's Atorvastatin (Lipitor). At the time of the work described below, Lipitor was the world's biggest selling drug with peak annual sales of $12.4 billion in 2008 [46].

Large-scale HN processes practiced before the start of this work involved the nucleophilic substitution of the halogen atom by cyanide in a 4-halo-3-hydroxybutyrate in classical batch manner. The reaction was low yielding, provided a product which required a complicated fractional distillation to remove the side products produced in the process [47]. Considering the size of the market for Atorvastatin and the technical problems with the available production processes, it was an obvious target for process improvement.

Codexis Inc. identified a two-step enzymatic route from ethyl 4-chloroacetoacetate (ECAA) which provided HN in high yield and high purity (Figure 12.18). The first step is a ketoreductase (KRED)-mediated enantioselective reduction of ethyl 4-chloroacetoacetate [48,49].

The second step uses a halohydrin dehalogenase (HHDH) to catalyze the displacement of the 4-chloro group in Ethyl (*S*)-4-chloro-3-hydroxybutyrate (ECHB) [2] with cyanide [50]. HHDHs catalyze the formation of epoxides from halohydrins, the epoxide thus formed may be opened with various nucleophiles among them, namely, cyanide, azide, hydroxyl [49]. This step is the key to the new process and allows the introduction of the cyano group under such mild and selective conditions that the "crude" product obtained after extraction and evaporation is to all intents and purposes pure.

The proof of concept had been demonstrated but there were two problems to be solved: (i) the activities of the wild-type strain enzymes were not sufficient for an economically viable process, and (ii) a suitable partner was required for the development of a large scale process.

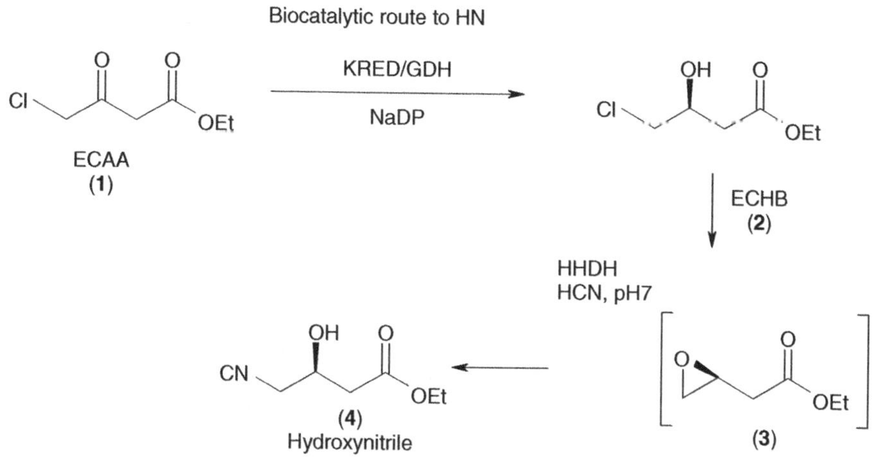

Figure 12.18 Biocatalytic route to hydroxynitrile.

Codexis' proprietary directed evolution (shuffling) technology was perfectly suited for optimizing the enzyme characteristics (activity, selectivity, stability, etc.). Lonza, a large-scale producer and user of both ethyl 4-chloroacetoacetate (**1**) and hydrogen cyanide (HCN) with a large experience in the development of industrial scale biocatalytic processes was the ideal partner for the project. Targets (Table 12.4) were set for both enzyme activity and process performance which would allow the target sales price to be achieved.

The optimization of the KRED and glucose dehydrogenase (GDH) required for the reduction step was achieved within a few rounds of shuffling. The HHDH proved somewhat more challenging; this enzyme system was much less well known than the KRED/GDH and it was also being asked to perform a transformation nature did not design it for. Various problems had to be overcome (temperature stability, enzyme inhibition by ECCA among others). After a few extra rounds of shuffling, the target activities were achieved.

The enzyme optimization was performed in close cooperation with chemical process optimization groups at Lonza and Codexis.

The main hurdles to overcome during process development were the following:

- removing phosphate buffer system (high waste treatment costs);

TABLE 12.4 Process Targets Necessary to Achieve Target Sales Price

	Step 1		Step 2
Target	KRED	GDH	HHDH
Substrate loading	160 g/L	160 g/L	160 g/L
Enzyme loading	160 mg/L	160 mg/L	160 mg/L
Reaction time	16 h	16 h	16 h
Improvement versus wild type	140	75	4000
Step yield (isolated)	95%		85%

- improving the process throughout;
- streamlining of work-up;
- recycling of the extraction solvent;
- safe handling of HCN.

Both steps were simplified so that they could be performed simply in a standard multipurpose plant. The same extraction solvent (*n*-butylacetate, *n*-BuOAc) was used in both steps and recycled directly into the process.

Step 1: A mixture of KRED, GDH, NADP, *n*-BuOAc, water, and ECAA (**1**) was held at pH 6.75 at 30°C by the addition of NaOH solution. After in-process control showed the reaction to be complete (about 8 h), celite was added and the solids removed via a filter. The filtered mixture was extracted with *n*-BuOAc and the organic extracts concentrated by distillation to give ECHB (**2**), which was used directly in the next step.

Step 2: HHDH and ECHB (**2**) were added to a solution of HCN at pH 7.0–7.5. The pH was held at 7.5 by the addition of NaCN solution. When in-process control showed the reaction to be complete, the pH was reduced to 6.5 and the residual HCN stripped via a washer (the recycling of this HCN is foreseen at later date). Celite was added as a filter aid and the solids removed using a centrifuge. The aqueous phase was extracted with *n*-BuOAc and after removal of the solvents HN was obtained which met the required specifications provided. The unwanted enantiomer could not be detected by GC (gas chromatography) analysis and the only impurity above 0.1 area% in the GC-chromatogram was residual *n*-BuOAc. If desired, the product could be distilled over a wiped film evaporator to remove residual *n*-BuOAc and color.

Codexis was awarded the U.S. Environmental Protection Agency's Presidential Green Chemistry Challenge Award in 2006 for this work [51]. The process was run at 10 m^3 scale at Lonza's Visp site to produce a total of 20 tons of HN. The process is currently (2009) being performed for Codexis at Arch Pharmalabs in India.

12.8 CONCLUSION

Inclusion of biocatalytical steps in chemical syntheses and the trend of carrying out complete *de novo* synthesis will lead to continual improvement in the fine chemical industry leading to processes that are extremely environmentally friendly and responsible. It should not be forgotten that for this development to be completely successful a holistic approach has to be taken. Taking a holistic approach means that there is an awareness of all aspects of the process, such as substrate synthesis, strain design, biomass production, and biotransformation which will lead to a minimal number of steps required to isolate the product from the culture broth or reaction mixture. Also, alternative isolation steps such as membrane technologies should be implemented when possible to replace or minimize solvent use.

This chapter has attempted to describe examples where this holistic approach was successfully applied. Most of these processes include steps to minimize or eliminate side-product production already during the strain selection and design phase. Our experience has also shown that sometimes the appropriate fermentation or biotransformation process leading to high product purity is not always accompanied by the highest productivity. An appropriately designed process leading to high product purity using highly selective enzyme systems always proves to be the best process in the end because of the large positive environmental and financial impact that simple product isolation entails.

ACKNOWLEDGMENT

I thank Mariel Von der Gruen, Felix Previdoli, Detlef Gerritzen, and Katrina Kulova for their support in supplying essential process information, and Manuela Avi and Antonio Osorio for proofreading the manuscript.

REFERENCES

1. Paul TA, Warner JC. Green chemistry: theory and practice. Oxford, Oxford University Press; 1998.
2. Gerritzen D. Oekonomie und Oekologie im Einklang—nachhaltige chemische Production am Beispiel des Lonza-Produktionsverbundes in Visp. 7. Freiburger Symposium der Division Industrielle Chemie/Swiss Chemical Society, Nachhaltige chemische Produktion "22nd & 23rd September; 2005.
3. Gulewitsch W, Krimberg R. Zur kenntnis der extraktivstoff der muskeln. Hoppe-Seyler's Z Physiol Chem 1905;45:326–330.
4. Tomita M, Sendju Y. Ueber die Oxyaminoverbindungen, welche die Biuretreaktion zeigen. III. Spaltung der γ-Amino-β-oxy-buttersaeure in die optischaktiven Komponenten. Hoppe-Seyler's Z Physiol Chem 1927;169:263–277.
5. Friedman S, Fraenkel GS. New and unidentified growth factors. III. Carnitine. In: Sebrell WH Jr, Harris RS, editor. The Vitamins. New York. Academic Press; 1972. pp. 329–355.
6. Barlett K, Eaton S. Mitochondrial β-oxidation. Eur J Biochem 2004;271:462–469.
7. Mueller DM, Seim H, Kiess W, Loester H, Richter T. Effects of oral L-carnitine supplementation on *in vivo* long chain fatty acid oxidation in healthy adults. Metabolism 2002;51(11):1389–1391.
8. Voeffray R, Perlberger J-C, Tenud L. L-carnitine. Novel synthesis and determination of the optical purity. Helv Chim Acta 1987;70:2058–2064.
9. Scholte HR, De Jonge PC. Metabolism, function and transport of carnitine in health and disease. In: Gitzelmann R, Baerlocher K, Steinmann B, editors. Carnitin in der medizin. Stuttgart: Schattauer; 1987. pp. 21–59.

10. Hardt P, Eyer M. Preparation of betaines from lactones and tertiary amines, Lonza. CH 687984. 1993.
11. Kulla H. Enzymatic hydroxylations in industrial application. Chimia 1991;45:81–85.
12. Hoeks FWJMM. Process for the microbiological discontinual preparation of L-carnitine Lonza. EP 0410430. 1991.
13. Kulla H, Lehky P. Process for the production of L-carnitine in a microbiological way. Lonza. EP 0158194. 1985.
14. Kulla H, Lehky P, Squaratti A. Process for the continuous production of L-carnitine Lonza. EP 0195944. 1986.
15. Hoeks FWJMM, Kulla H, Meyer H-P. Continuous process cell-recycling with for L-carnitine production: performance, engineering and downstream process aspects compared with discontinuous processes. J Biotechnol 1992;22:117–128.
16. Zimmermann TP, Robins KT, Hoeks FWJM. Bio-transformation in the Production of L-Carnitine. In: Collins AN, Sheldrake GN, Crosby J, editors. Chirality in industry. Chichester. John Wiley & Sons, Ltd.; 1997. pp. 287–305.
17. Meyer H-P, Robins K. Large scale bioprocess for the production of optically pure L-carnitine. Chem Mon 2005;136:1269–1277.
18. Chuck R. A catalytic green process for the production of niacin. Chimia 2000; 54:508–513.
19. Sheldon RA. The E factor: fifteen years on. Green Chem 2007;9:1273–1283.
20. Bergamin RA. Produktionsintegrierter umweltschutz in der praxis am fall-beispiel der nicotinsäureproduktion CONCEPT-symposiums. Produktionstechnik unter Umweltaspekten; 1991; Mannheim.
21. Chuck R. Technology development in nicotinate production. Appl Catal A Gen 2005;280:75–82.
22. Chuck R. Green sustainable chemistry in the production of nicotinates. Ist International IUPAC Conference on Green-Sustainable Chemistry; 2006.
23. Heveling J, Ambruster E, Utiger L, Rohner M, Dettwiler H-R, Chuck RJ. Process for the preparation of nicotinic acid amide Lonza AG, Switzerland. EP 0770687. 1997.
24. Nagasawa T, Mathew CD, Mauger J, Yamada H. Nitrile hydratase-catalyzed production of nicotinamide from 3-cyanopyridine in *Rhodococcus rhodochrous* J1. Appl Environ Microbiol 1988;54:1766–1769.
25. Nagasawa T, Takeuchi K, Yamada H. Characterization of a new cobalt-containing nitrile hydratase purified from urea-induced cells of *Rhodococcus rhodochrous* J1. Eur J Biochem 1991;196:581–589.
26. Sheldon RA, Arends I, Hanefeld U. Preface from the authors. Green chemistry and catalysis. Weinheim: Wiley-VCH; 2007.
27. Shaw NM, Naughton A, Robins K, Tinschert A, Schmid E, Hischier M-L, Venetz V, Werlen J, Zimmermann T, Brieden W, de Riedmatten P, Roduit J-P, Zimmermann B, Neumueller R. Selection, purification, characterisation, and cloning of a novel heat-stable stereo-specific amidase from Klaebsiella oxytoca, and its application in the synthesis of enantiomerically pure (R)- and (S)-3,3,3-trifluoro-2-hydroxy-2-methylpropionic acids and (S)-3,3,3-trifluoro-2-hydroxy-2-methylpropionamide. Org Process Res Dev 2002;6:497–504.

28. Eichhorn E, Roduit J-P, Shaw N, Heinzmann K, Kiener A. Preparation of (S)-piperazine—2-carboxylic acid, (R)-piperazine-2-carboxylic acid, and (S)-piperidine-2-carboxylic acid by kinetic resolution of the corresponding racemic carboxamides with stereoselectiv amidases in whole bacterial cells. Tetrahedron: Asymmetry 1997;8(15):2533–2536.
29. (a) Hanselmann P. Process for the preparation of cyclopropanenitrile derivatives. EP 0491323. 1992; (b) Meul T, Kaempfen U. Process for the preparation of dimethylcyclopropancarboxylic acid. EP 0491330. 1992.
30. Robins KT, Gilligan T. Biotechnological process for the production of (*S*)-(+)-2,2-dimethylcyclopropane carboxamide and (*R*)-(-)-2,2—dimethylcyclopropane carboxylic acid. EP 0502525. 1992.
31. Birch OM, Boehlen E, Gilligan T, Hoeks FWJMM, Malanik V, Robins K, Shaw NM, Zimmermann T. Development of an industrial process for S-2,2-dimethylcyclopropanecarboxamide. Chiral Europe '94; 1994.
32. Zimmermann T, Robins K, Birch O, Boehlen L. Process for the preparation of (*S*)-(+)-2,2-dimethyl-cyclopropane carboxamide by genetically engineered microorganisms. EP 0524 604. 1993.
33. Lehky P, Kulla H, Mischler S. Process for the preparation of 6-hydroxynicotinic acid. EP0152948. 1985.
34. Kiener A. Microbiological process for the preparation of 6-hydroxy nicotinic acid from 3-cyanopyridine. EP 504819. 1992.
35. (a) Kiener A. Microbiological process for the preparation of 6-hydroxy picolinic acid from 2-cyanopyridine. EP 504818. 1992; (b) Kiener A, Gloeckler R, Heinzmann K. Preparation of 6-oxo-1,6-dihydropyridine-2-carboxylic acid by microbial hydroxylation of pyridine-2-carboxylic acid. J Chem Soc [Perkin Trans 1] Organic and Bio-Organic Chemistry (1072–1999) 1993;11:1201–1202.
36. Wieser M, Heinzmann K, Kiener A. Bioconversion of 2-cyanopyrazine to 5-hydroxypyrazine-2-carboxylic acid with Agrobacterium sp. DSM 6336. Appl Microbiol Biotechnol 1997;48(2):174–176.
37. Kiener A, Van Gameren Y, Bokel M. US 5284767. 1994.
38. Roduit JP, Wellig A, Kiener A. Renewable functionalised pyridines derived from microbial metabolites of the alkaloid (S)-nicotine. Heterocycles 1997;45(9): 1687–1702.
39. Buckingham S, Lapied B, Corronc H, Sattelle F. Imidocloprid actions on insect neuronal acetylcholine receptors. J Exp Biol 1997;200:2685–2692.
40. Quarroz D. Process for the preparation of 2-halogen-pyridine derivatives. EP 0084118. 1983.
41. (a) 406. Amiloride. The Merck Index, 10th Edition Merck & Co, USA. 1983. (b) 4302. Glipicide. The Merck Index, 10th Edition Merck & Co, USA. 1983. (c) 7858. Pyrazinamide. The Merck Index, 10th Edition Merck & Co, USA. 1983.
42. Kiener A. Microbiological manufactures of 5-hydroxypyrazine carboxylic acid. EP 519512. 1992.
43. Kiener A. Biosynthesis of functionalized aromatic N-heterocycles. Chemtech 1995;25:31–35.
44. Kiener A. Enzymatic oxidation of methyl groups on aromatic heterocycles: a versatile method for the preparation of heteroaromatic carboxylic acids. Angew Chem Int Ed Engl 1992;31(6):774–775.

45. Orsini G, Lovisolo PP, Chiari A, Branzoli U. Effect of the antilipolytic compound acipimox on peroxisome marker enzymes, lipid pattern. 1985.
46. Jarvis LM. Pfizer buys Wyeth. Chem Eng News 2009;87(5):7.
47. Matsuda H, Shibata T, Hashimoto H, Kitai M. Method for producing (R)-4-cyano-3-hydroxybutyric acid lower alkyl esters. US 5908953. 1999.
48. Davis SC, Grate JH, Gray DR, Gruber JM, Huisman GW, Steven K, Newman LM, Sheldon R, Wang Li A. Enzymatic processes for the production of 4-substituted 3-hydroxybutyric acid derivatives. WO015132. 2004.
49. Davis SC, Grate JH, Gray DR, Gruber JM, Huisman GW, Steven K, Newman LM, Sheldon R, Wang Li A. Enzymatic processes for the production of 4-substituted 3-hydroxybutyric acid derivatives and vicinal cyano, hydroxy substituted carboxylic acid esters. WO018579. 2005.
50. Davis CS, Fox RJ, Gavrilovic V, Huisman GW, Newman LM. Improved halohydrin dehalogenases and related polynucleotides. WO2005017141. 2005.
51. Ritter SK. Going green keeps getting easier. Chem Eng News 2006;84(28):24–27.

PART IV

APPLICATION AND CASE STUDIES—POLYMERS AND RENEWABLE CHEMICALS

CHAPTER 13

GREEN CHEMISTRY FOR THE PRODUCTION OF BIODEGRADABLE, BIORENEWABLE, BIOCOMPATIBLE, AND POLYMERS

JOHN S. F. BARRETT
Department of Chemical Engineering and Material Science, University of Minnesota, Minneapolis, Minnesota

FRIEDRICH SRIENC
Department of Chemical Engineering and Material Science, University of Minnesota, Minneapolis, Minnesota

13.1 INTRODUCTION

One particularly exciting topic at the intersection of biocatalysis and green chemistry is that of biopolymers. Such compounds are considered desirable for their biodegradability, biocompatibility, and capacity for synthesis from renewable feedstocks. Furthermore, biopolymers are aptly named because a biocatalysis step is usually implemented at some point in their production whether it is in the synthesis of feedstocks or their polymerization process. Biopolymers are ubiquitous in nature and include compounds such as nucleic acids, polypeptides, and cellulose. Biopolymers for commercial applications as plastics include starch-based plastics, polylactic acid (PLA), polyhydroxyalkanoates (PHAs), polyurethanes from vegetable oil-derived polyols, and polyethylene synthesized from bio-derived ethanol. In this chapter, we focus mainly on the PHAs and their applications. We also give a short discussion about the synthesis of PLA.

Biocatalysis for Green Chemistry and Chemical Process Development, First Edition.
Edited by Junhua (Alex) Tao and Romas Kazlauskas.

13.2 POLYHYDROXYALKANOATES

PHAs occur naturally and are found in a wide range of microorganisms which use them to store excess carbon and energy in preparation for times of nutrient limitations [1]. The polymers are formed within the cell as large insoluble inclusions surrounded by a lipid monolayer [2]. In the case of PHA, polymerization occurs by the action of a polymerase enzyme, which catalytically adds monomers to a growing polymer chain. Commercial production of PHA is carried out by microbial fermentation using a variety of feedstocks.

PHA is a polyester because the monomers that comprise PHA are linked by oxoester bonds. The key enzyme for PHA synthesis is the PHA synthase. This enzyme polymerizes (*R*)-3-hydroxyacyl-CoA (coenzyme A) thioester monomers into PHA with the release of CoA [3]. A broad diversity of monomer types exists in nature, and more than 120 have been discovered [4]. Within the broad class of PHAs, the polymers are further classified according to the number of carbon atoms within monomer units that comprise them. PHA made of monomers containing 3–5 carbon atoms are referred to as PHA_{scl} (short chain length (scl) while those made of monomers containing 6–14 carbon atoms are referred to as PHA_{mcl} (medium chain length (mcl)). PHA_{scl} polymers include polyhydroxybutyrate (PHB) and polyhydroxyvalerate (PHV). PHA_{mcl} polymers include polyhydroxyhexanoate (PHH) and polyhydroxyoctanoate (PHO) (Figure 13.1).

The division of PHA into these two subclasses of PHA_{scl} and PHA_{mcl} arises from the substrate specificity of the polymerase enzymes used to synthesize the polymers. For example, the organism *Ralstonia eutropha* contains a polymerase enzyme (*phbC*) that accepts (*R*)-3-hydroxyacyl-CoA substrates that contain from three to five carbons only [5]. In contrast, the organism *Pseudomonas oleovorans* contains a different polymerase enzyme (*phaC1*) that has specificity for longer (*R*)-3-hydroxyacyl-CoA substrates containing 6–14 carbon atoms [6]. Furthermore, this distinction between monomer specificity is so prevalent that whole classes of synthase genes from various organisms have been catalogued accordingly [3]. While many polymerases fall within these two categories, some polymerases have been identified, which demonstrate specificity for both scl and mcl-type monomers [7–12].

13.3 ENGINEERING POLYMER PROPERTIES

Within the field of polymer engineering scientists have created a wide variety of materials exhibiting a range of physical properties, which, in turn, dictate the types of technological applications for which a polymer is suited. Important physical properties include polymer strength, stiffness, toughness, elasticity, and thermoplasticity. Strong materials can withstand heavy loads before failure. Stiff materials support this load with little or no strain, that is, distortion of shape. Steel is a strong and stiff material. If on loading the distortion occurs

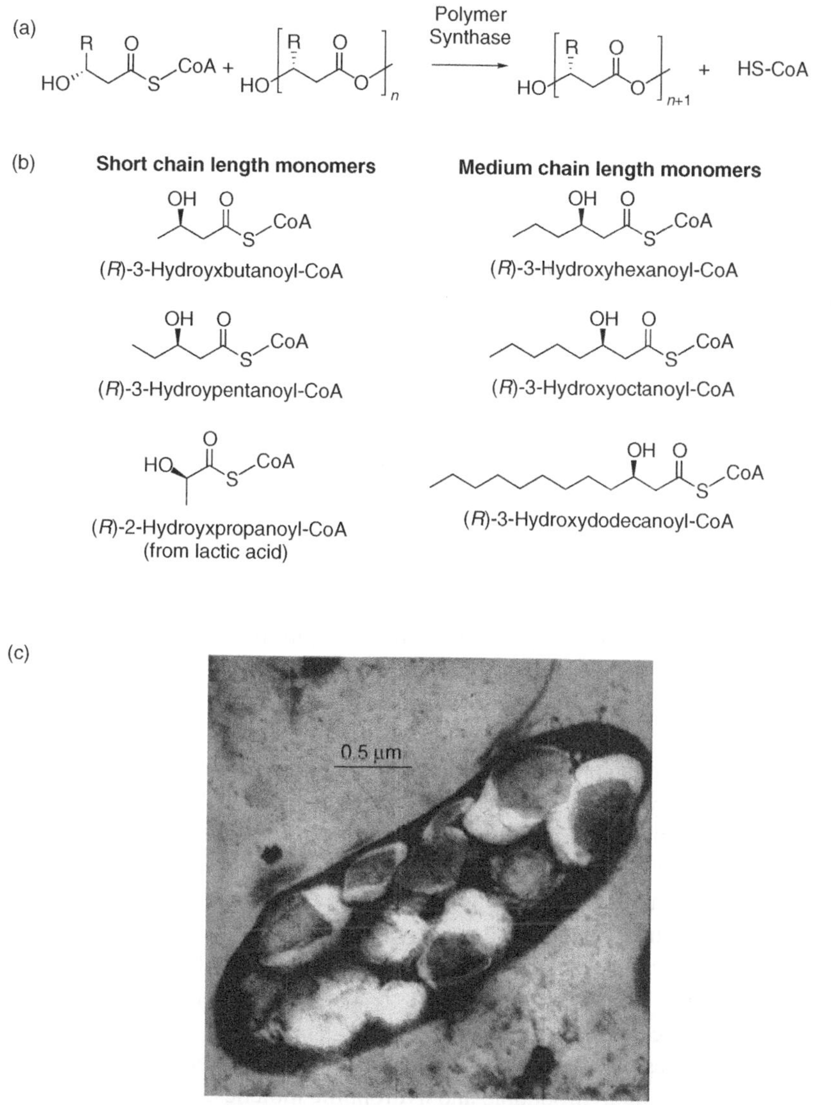

Polyhydroxybutyrate-co-polyhydroxyvalerate granules in *Ralstonia eutropha*. TEM image by Eric N. Pederson

Figure 13.1 (a) Polyhydroxyalkanoate synthesis. (b) Various hydroxyalkanoate monomers. (c) Polyhydroxyalkanoate granules in *Ralstonia eutropha*.

suddenly and causes catastrophic fracture, such materials are considered brittle. Glass is a strong but brittle material. Tough is the opposite of brittle. If sufficient load is applied to a tough material permanent deformation will occur, but the material can undergo significantly more strain before catastrophic failure. Kevlar and Nylon are tough materials. Elastomeric materials can undergo a large strain without permanent deformation. Rubber is an elastomer. Thermoplastic materials easily undergo plastic deformation at high temperature but regain stiffness on cooling.

For practical applications the distinction between PHA_{scl} and PHA_{mcl} is important, as monomer composition can greatly affect the material properties of the resulting polymer. The properties of PHAs are similar to those of conventional petroleum-derived polyesters. In general, PHA_{scl} polymers are strong and stiff thermoplastics that are brittle with a high-melting temperature. In contrast, PHA_{mcl} polymers are more elastomeric, have a lower melting temperature, and are tougher when subjected to shock and shearing stresses. See Table 13.1 for a comparison of polymer properties.

One factor that greatly affects a polymer's physical performance is molecular weight, which correlates directly with polymer chain length. Increasing the molecular weight of a polymer chain usually results in a higher softening temperature and increased toughness [25]. In 1997, Sim *et al*. observed, in recombinant *Escherichia coli* producing PHA, that the expression of additional pha synthase molecules was inversely correlated with polymer molecular weight [26], thus providing a tool for controlling polymer properties. In 2002, researchers produced PHB samples of varying molecular weights by controlling the aeration rate to a culture of *Azotobacter chroococcum* 6B. Testing of the polymer samples showed that tensile strength increased with increasing polymer molecular weight [15].

Another common approach for enhancing polymer performance is by blending two different polymers to make a novel compound with hybrid mechanical properties, for example, a polymer that is stiff and strong yet tough against impact and tearing. Polymer blending is a suitable practice when the types and amounts of polymer used readily form a homogeneous solution or are combined in a stable dispersion. Polymer blends containing PHAs have been explored with various materials including starch [27], rubber [28], and polylactide [29,30]. Composite plastics utilizing PHAs have also been created using montmorillonite clay [31] and single-wall carbon nanotubes [32].

In some cases, however, the polymers may be chemically dissimilar, for example, containing hydrophobic and hydrophilic moieties, which results in phase separation of the two polymers [33]. One way to prevent phase separation is by linking the two different polymers using a covalent bond. These types of molecules are called *copolymers*. The copolymer may consist of short- or long repeats of each monomer which may be ordered periodically or randomly, for example, DNA is a copolymer of the nucleotides A, G, C, and T in an irregular order. When a single monomer type occurs in a long repeat followed by long repeat of the alternate polymer such molecules are called *block copolymers*. Depending on the size of the homogeneous domains within the chain, the copolymer may undergo microphase separation to form periodic mesostructures. This additional level of structural order can be useful for a variety of

TABLE 13.1 Polymer Properties

Polymer	Melting Temperature (°C)	Young's Modulus (GPa)	Tensile Strength (MPa)	Elongation at Break (%)	Reference
PHB	175–180	3.5–4	40	3–8	[13,14]
PHB					[15]
MW = 445 kDa	—	—	38	—	
MW = 1000 kDa	—	—	113	—	
MW = 1240 kDa	—	—	215	—	
Poly(3HB-co-3HV)					[16]
0 mol% 3HV	179	3.5	40	—	
3 mol%	170	2.9	38	—	
9 mol%	162	1.9	37	—	
14 mol%	150	1.5	35	—	
20 mol%	145	1.2	32	—	
25 mol%	137	0.7	30	—	
Poly(3HB-co-3HV)					[17,18]
8 mol% 3HV (random)	151	0.260	18	15	
29 mol% 3HV (random)	101	0.210	18	15	
15 mol% 3HV (block)	164	0.210	28	90	
23 mol% 3HV (block)	161	0.150	15	175	
Poly(3HB-co-4HB)					[19,20]
3 mol% 4HB	166	—	28	45	
10 mol% 4HB	159	—	24	242	
16 mol% 4HB	—	—	26	444	
64 mol% 4HB	50	30	17	591	
90 mol% 4HB	50	100	65	1080	
100 mol% 4HB	53	149	104	1000	
Polyhydroxyoctanoate (PHO)	61	—	6–10	300–450	[21]
Polylactic acid (PLA)					[22]
L-PLA	145–186	1.2–3.0	28–50	2–6	
L,D-PLA	—	1.9–2.4	29–35	5–6	
Conventional polymers					[23]
Polypropylene (PP)	160–175	1.1–1.6	31–41	100–600	
High density polyethylene (HDPE)	130–137	1.1	22–31	10–1200	
Low density polyethylene (LDPE)	98–115	0.2–0.3	8.2–31	100–650	
Polyethylene terephthalate (PET)	212–265	2.8–4.1	48–72	30–300	
Polycarbonate (PC)	150 (T_g)	2.4	63–72	110–120	
Polystyrene (PS)	74–104 (T_g)	2.3–3.3	36–52	1.2–2.5	
Nylon-66	255–265	1.6–3.4	75.8	150–300	

Source: Adopted from van der Walle *et al*. [24].

technical applications. For more common synthetic polymers, numerous structures have been observed experimentally and include micelles, vesicles, cylinder arrays, sphere arrays, continuous phase gyroids, and lamellae [34]. Micelles and vesicles are useful for applications in drug delivery and particle encapsulation. Sphere arrays are useful for enhancing polymer toughness or strength. Cylinder arrays and lamellae can serve as templates for patterning other materials with ordered mesostructures. Continuous phase gyroids, cylinder arrays, and lamellae offer useful structures for designing polymer materials with unique transport properties.

13.4 PHA COPOLYMERS

In light of the potential benefits of copolymer architecture on material properties and applications, much work in the field of PHAs has focused on the production of copolymer molecules [6,35–37]. Such polymers can arise when an organism simultaneously consumes two separate substrates [38] or when successive rounds of metabolic cycles, for example, fatty acid β-oxidation or fatty acid *de novo* biosynthesis generate multiple monomer types from the same starting material [6]. One example of a PHA copolymer is poly(3-hydroxybutyrate-*co*-3-hydroxyvalerate)) (PHBV) which can be synthesized from *R. eutropha* via the simultaneous feeding of metabolic precursors fructose and valerate [38]. Synthesis of this particular polymer is possible because the *phbC* synthase accepts both the (*R*)-3-hydroxybutyrate (HB) and (*R*)-3-hydroxyvalerate substrates which are both scl type monomers. The polymer synthase of *P. oleovorans* which is known to have specificity for mcl-type monomers produces poly(3-hydroxyhexanoate-*co*-3-hydroxyoctanoate)) (PHBHO) when cultured on octane as the soul carbon source [6]. For the synthesis of copolymers containing both scl and mcl monomer types, researchers have found some enzymes with broader specificity capable of accepting both monomer types and producing polymers such as poly(3-hydroxybutyrate-*co*-3-hydroxyhexanoate). PHA synthase enzymes with broad substrate specificity have been found in *Aeromonas caviae, Aeromonas hydrophila, Pseudomonas* sp. 61-3, *Pseudomonas stuzteri, Nocardia coralline*, and *Bacillus* sp. 88D [7,8,11,12]. Furthermore, using the techniques of protein engineering, for example, random mutagenesis and site-directed mutagenesis, researchers have been able to create novel broad specificity polymerases showing enhanced affinity toward scl type monomers [9,10].

For the types of copolymers described above, the sequence of monomers is random; however, as discussed earlier, block copolymers with their ordered microstructure offer new possibilities for unique technical applications. To achieve the desired block organization, some type of sequential substrate feeding is usually employed. In 2006, Pederson *et al*. synthesized block copolymers of PHB and PHV (PHB-*b*-PHV) by periodic feeding of pentanoic acid to *R. eutropha* cultured on fructose as a primary carbon source [17]. Additionally, they demonstrated that measurements of oxygen uptake rate and carbon dioxide evolution rate could be used to indicate the consumption of specific substrates in the batch, thereby allowing for feedback control of substrate addition. Later, McChalicher and Srienc performed extensive physical testing of these PHB-*b*-PHV block copolymers and compared them to random copolymers of the same monomers. On aging, the block copolymers showed superior toughness versus the random copolymers [18]. A similar demonstration of block copolymer synthesis via alternate substrate feeding was also made in 2008 by Pereira *et al.* [39]. Recognizing the importance of process control strategies for manipulating polydispersity and copolymer organization, Mantzaris *et al.* utilized population balance modeling to predict the optimal period for substrate feeding [40]. Later Jurasek and Marchessault created a computer simulation based on a highly mechanistic model of PHB granule assembly. The model was able to predict such statistics of granule size, number, polymer molecular weight, and polydispersity as a function of gene expression levels of polymerase, depolymerases, and phasins [41]. More recently, Iadevaia and Mantzaris presented modeling results suggesting that block copolymer organization could be controlled by the use of synthetic gene networks designed to control transient expression of genes necessary for monomer synthesis. In this way, both monomers could be transiently produced using a single carbon source. In one network design, monomer switching was controlled via the addition of external inducer molecules [42], while an alternate device coupled monomer switching to a genetic device capable of generating stable oscillations [43].

13.5 METABOLIC PATHWAYS TO POLYMER PRECURSORS IN NATIVE PHA PRODUCING ORGANISMS

In addition to studies focused on the polymerization process, considerable effort has also been made to uncover the enzymatic pathways involved in the conversion of feedstocks to polymer precursors. These pathways are linked to both catabolic and anabolic processes within the cell. For the synthesis of mcl (*R*)-3-hydroxyacyl-CoA substrates, the most direct route is through the catabolism of fatty acids and alkanes. This is accomplished via the β-oxidation pathway which is a primary supply of PHA_{mcl} monomers in *Pseudomonas* spp. Various intermediates of this pathway can be converted to PHA monomers [44]. For the production of mcl monomers from unrelated carbon sources such as glucose and gluconate, a *phaG* gene coding for an (*R*)-3-hydroxyacyl transferase has been identified in *Pseudomonas* spp. [45,46]. This enzyme allows the organism to convert (*R*)-3-hydroxyacyl-ACP, an intermediate of fatty acid synthesis, directly to the polymer precursor, (*R*)-3-hydroxyacyl-CoA. For the production of the scl substrate, (*R*)-3-hydroxybutyrl-CoA, the organism, *R. eutropha*, has developed a particularly direct pathway [47]. This route involves the expression of two additional genes *phbA*, a β-ketothiolase, and *phbB*, an acetoacetyl-CoA reductase, which converts acetyl-CoA to PHB monomers in only two steps.

13.6 METABOLIC PATHWAYS IN RECOMBINANT *ESCHERICHIA COLI*

In addition to production of PHAs in native host strains, for example, *P. oleovorans* and *R. eutropha*, researchers have also investigated PHA synthesis using recombinant systems. The industrial workhorse, *E. coli*, is an ideal candidate for metabolic engineering because of its sequenced genome, well-studied metabolism, and rapid growth rate. The capacity to synthesize PHAs requires two specific activities: (i) the ability to synthesize monomer precursors from either related or unrelated carbon sources and (ii) the ability to catalyze the formation of oxoester linkages between monomers. Wild-type *E. coli* does not naturally produce PHAs and so attempts at recombinant expression have required the inclusion of a polymerase gene as well as additional metabolic engineering to increase monomer substrate production. A thorough review of PHA production in recombinant *E. coli* is provided by Li *et al*. [11]. Here, we provide a summary of the topic along with relevant figures.

13.6.1 Production of PHB

In the production of PHB, researchers have achieved success by cloning the PHB biosynthesis genes from both *R. eutropha* and *Alcaligenes latus* among others [48,49]. In each case, three specific genes are essential: *phbA*, a β-ketothiolase enzyme for the conversion of two acetyl-CoA to a single acetoacetyl-CoA molecule; *phbB*, a reductase enzyme for the conversion of acetoacetyl-CoA to (*R*)-3-hydroxybutyryl-CoA; and *phbC*, a polymer synthase enzyme. Alternatively, Taguchi *et al*. produced PHB from glucose using a *phaC* gene from *A. caviae* along with overexpression of the *E. coli* protein, 3-ketoacyl-ACP synthase III (*FabH*). The authors proposed that the transacylating activity of the enzyme could convert (*R*)-3-hydroxybutryl-ACP, an intermediate of fatty acid *de novo* synthesis, to the proper (*R*)-3-hydroxybutryl-CoA substrate [50]. In an alternate synthesis route, Taguchi *et al*. showed that the native *E. coli* gene, 3-ketoacyl-ACP reductase (*FabG*), also had activity toward analogous CoA-tagged substrates thus allowing the conversion of the β-oxidation intermediate 3-ketoacyl-CoA to the polymer precursors (*R*)-3-hydroxyacyl-CoA, including scl monomer, (*R*)-3-hydroxybutryl-CoA [51]. Figure 13.2 shows some of the main metabolic pathways through which PHB can be synthesized.

13.6.2 Production of PHA_{mcl}

For the production of PHA_{mcl} in *E. coli*, researchers inserted a type II mcl synthase from *Pseudomonas aeruginosa. E. coli* cells harboring only the synthase gene produced minimal PHA_{mcl} [52]; however, when the gene was inserted into an *fadB* mutant deficient in β-oxidation, PHA synthesis increased significantly [53,54]. Later, the same authors demonstrated that addition of acrylic acid, a known inhibitor of fatty acid β-oxidation, increased the production of PHA_{mcl}

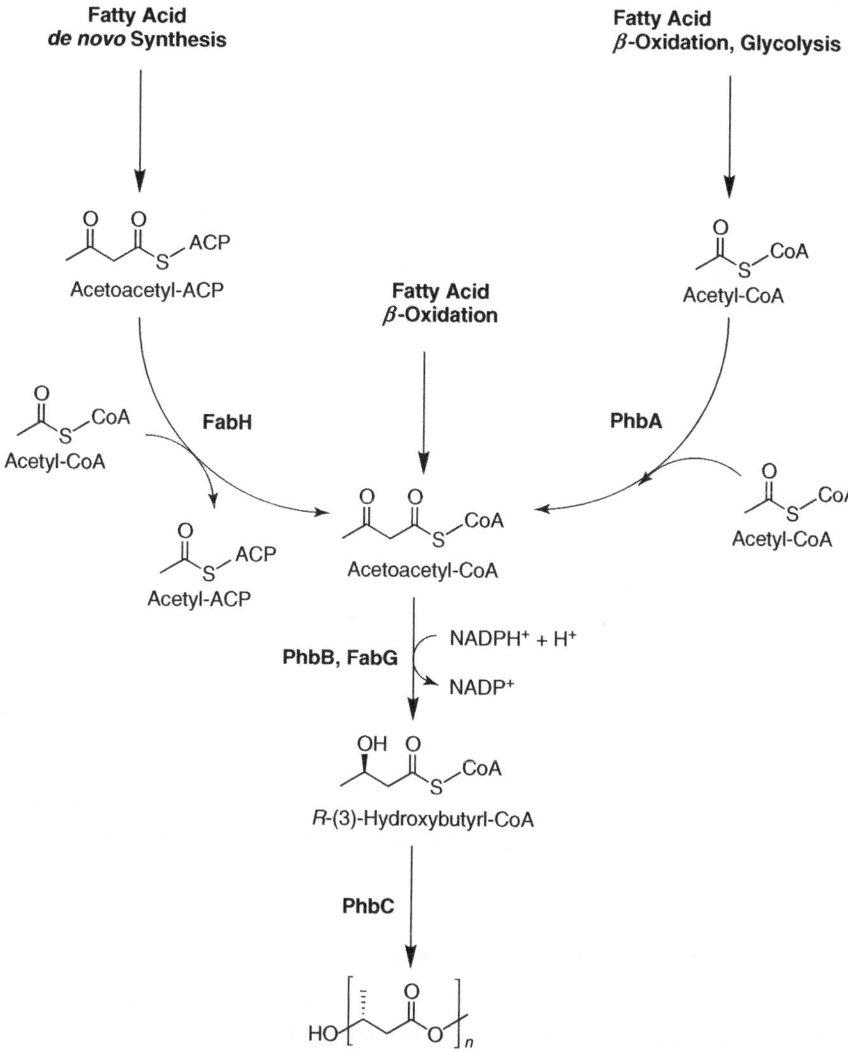

Figure 13.2 Metabolic pathways involved in PHB synthesis. Enzymes shown include those that have been cloned from foreign organisms as well as others that are native to *E. coli*. Enzymes from *R. eutropha*: PhbA, β-ketothiolase, PhbB, acetoacetyl-CoA reductase, PhbC, type I polymer synthase. Enzymes from *E. coli*: FabH, 3-ketoacyl-ACP synthase III, FabG, 3-ketoacyl-ACP reductase.

[55]. In *E. coli*, the *fadB* gene codes for enoyl-CoA hydratase, which converts enoyl-CoA to (*S*)-3-hydroxyacyl-CoA, another intermediate of the β-oxidation cycle; however, the polymer synthase enzyme accepts only (*R*)-3-hydroxyacyl-CoA substrates. By knocking out this pathway via the *fadB* deletion, the conversion of enoyl-CoA was shunted away from the unusable *S*-enantiomer. In

P. aeruginosa, the production of the usable, *R*-configuration 3-hydroxyacyl-CoA monomers has been attributed to an *R*-specific enoyl-CoA hydratase (PhaJ1) [56,57]. In *E. coli*, a homologous enzyme, MaoC, has been linked to PHA synthesis, and is believed to be a new enoyl-CoA hydratase with increased selectivity for producing (*R*)-3-hydroxyacl-CoA [58]. Furthermore, other researchers showed that *P. aeruginosa* contained at least four different PhaJ proteins with unique specificities for various enoyl-Co substrates. Individually, cloning each *phaJ* gene in recombinant *E. coli* with the *Pseudomonas* sp. 61-3 synthase produced PHA with varying proportions of even number mcl monomers when the cells were fed with sodium dodecanoate [59]. Additionally, Taguchi *et al*. demonstrated that the *E. coli* gene for 3-ketoacyl-ACP reductase, *fabG*, also showed activity in the conversion of 3-ketoacyl-CoA, another β-oxidation intermediate, to (*R*)-3-hydroxyacyl-CoA monomers. Overexpression of this gene along with a synthase gene, *phaC1*, from *Pseudomonas* sp. 61-3 in *E. coli* strain HB101 resulted in the accumulation of PHA_{mcl} [51]. In 2001, Rehm *et al*. engineered *E. coli* to produce PHA_{mcl} via the fatty acid *de novo* biosynthesis pathway from the unrelated carbon source gluconate. This was done by the recombinant expression of the *phaG* gene from *Pseudomonas putida* coding for a transacylase enzyme capable of converting (*R*)-3-hydroxyacyl-ACP, a fatty acid *de novo* synthesis intermediate, to (*R*)-3-hydroxyacyl-CoA monomers [60]. In another attempt to derive mcl monomers from unrelated carbon sources, Rehm *et al*. inserted an acyl-ACP thioesterase from *Umbellularia californica* into *E. coli* mutants deficient in the β-oxidation pathway. The enzyme, which shows enhanced selectivity for ACP thioesters over CoA thioesters, was able to convert lauryl-ACP, a 12-carbon intermediate of the fatty acid *de novo* synthesis, directly to lauric acid, which was subsequently converted to PHA_{mcl} via truncated fatty acid β-oxidation [61]. Figure 13.3 shows some of the main metabolic pathways through which PHA_{mcl} can be synthesized.

13.6.3 Production of PHA_{scl}-co-PHA_{mcl}

As discussed previously, copolymers comprising both PHA_{scl} and PHA_{mcl} have distinct advantages over conventional homopolymers. Generation of such compounds requires synthase enzymes with broad substrate specificity for both scl- and mcl-type monomers as well as metabolic pathways for producing each monomer. Synthesis of both scl and mcl precursors may be obtained from the same fatty acid starting material through a truncated β-oxidation cycle. As β-oxidation progresses, mcl enoyl-CoA molecules are reduced by two carbons every time a cycle is completed, eventually creating scl enoyl-CoA molecules. Then, the action of an (*R*)-specific enoyl-CoA hydratase (PhaJ) catalyzes the formation of (*R*)-3-hydroxyacyl-CoA monomers from the enoyl-CoA precursors. In one case, a single PhaJ may have specificity for both scl and mcl enoyl-CoA molecules [62,63]. In another case, multiple PhaJs may act uniquely on enoyl-CoA molecules of different lengths [59]. Utilization of fatty acids with an even number of carbons produces (*R*)-3-hydroxyacyl-CoA monomers with an even

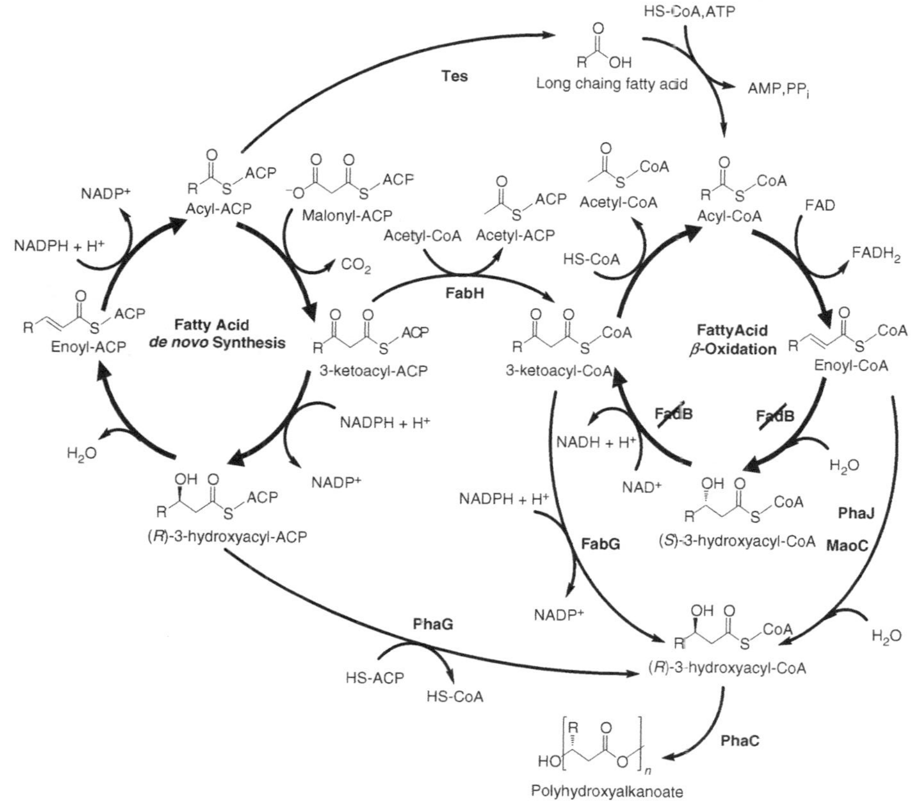

Figure 13.3 Metabolic pathways involved in PHA_{mcl} synthesis. Enzymes shown include those that have been cloned from foreign organisms as well as others are native to *E. coli*. Enzyme from *P. aeruginosa*: PhaJ, *R*-specific enoyl-CoA hydratase. Enzyme from *P. putida*: PhaG, (*R*)-3-hydroxyacyl-ACP to (*R*)-3-hydroxyacyl-CoA transacylase. Enzyme from *U. californica*: Tes, acyl-ACP thioesterase. Enzyme from *Pseudomonas*. sp. 61-3: PhaC, type II polymer synthase. Enzymes from *E. coli*: FadB, enoyl-CoA hydratase, 3-hydroxyacyl-CoA dehydrogenase, FabH, 3-ketoacyl-ACP synthase III, FabG, 3-ketoacyl-ACP reductase, MaoC, *R*-specific enoyl-CoA hydratase.

number of carbons, while utilization of fatty acids with an odd number of carbons produces odd-numbered PHA monomers. Owing to the cyclic nature of β-oxidation and fatty acid synthesis which produces successively smaller and larger substrates, respectively, the copolymers produced often contain a broad range of monomer types (Table 13.2) [6,64].

In 2004, Nomura *et al.* demonstrated that coexpression of a mutated 3-ketoacyl-ACP synthase III (*fabH*), with expanded substrate specificity, along with a PHA synthase, *phaC1*, gene of *Pseudomonas* 61-3 resulted in the production of PHA_{scl}–PHA_{mcl} copolymers utilizing glucose as a substrate [65]. Continuing with their investigation they then showed that the additional

TABLE 13.2 Polyhydroxyalkanoate Monomers Produced from Successive Rounds of Fatty Acid β-Oxidation Beginning With Dodecanoic Acid as the Starting Material.

β-Oxidation cycle number	β-Oxidation Intermediate	(*R*)-3-Hydroyacyl-CoA Monomer
1	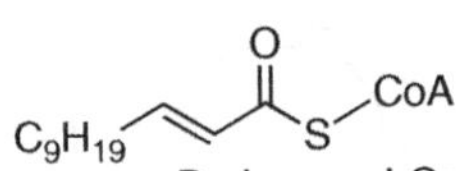 Dodecanoyl-CoA	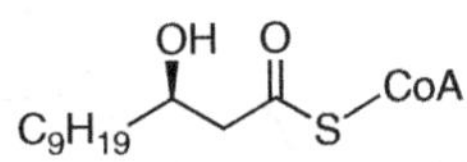 (*R*)-3-Hydroxylauryl-CoA
2	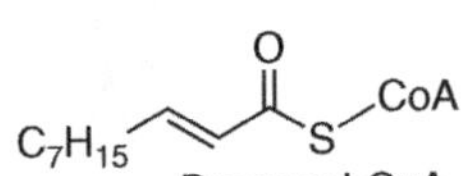 Decanoyl-CoA	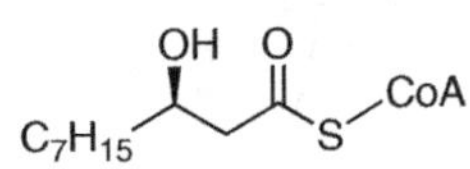 (*R*)-3-Hydroxycapryl-CoA
3	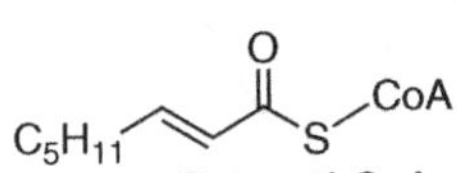 Octanoyl-CoA	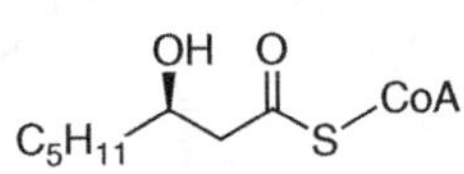 (*R*)-3-Hydroxycaprylyl-CoA
4	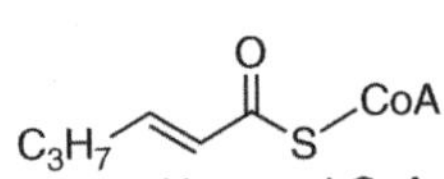 Hexanoyl-CoA	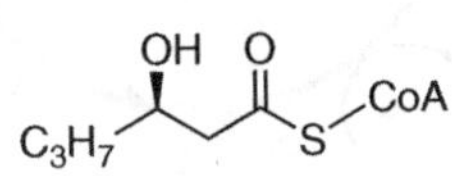 (*R*)-3-Hydroxyhexanoyl-CoA
5	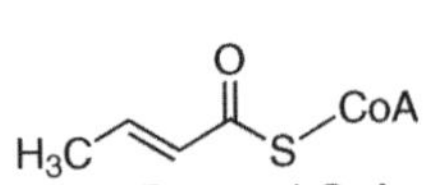 Butanoyl-CoA	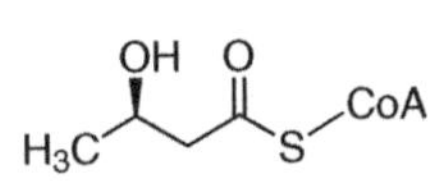 (*R*)-3-Hydroxybutyryl-CoA

The action of an *R*-specific enoyl-CoA hydratase is required to convert the b-oxidation intermediate to the *R*-3-hydroxyalkanoate monomer. In addition, each round of β-oxidation also produces one molecule of acetyl-CoA, which is a precursor for the formation of PHB.

overexpression of 3-ketoacyl-ACP reductase (*fabG*) from either *E. coli* or *Pseudomonas* sp. 61-3 could be used to control the composition of PHA_{mcl} within the copolymer [66,67]. In an alternate approach to copolymer synthesis, Park and Lee combined the *phaAB* genes of *R. eutropha* with the PHA synthase gene of *Pseudomonas* sp. 61-3 in mutant *E. coli* strains deficient in the β-oxidation pathway. When the cells were cultured on a mixture of gluconate and decanoate, expression of the *phaAB* genes allowed the cells to more effectively produce 3-hydroxybutyrate monomers from gluconate while deficiency in the β-oxidation pathway enhanced production of mcl monomers from decanoate [68].

13.7 PHA PRODUCTION IN EUKARYOTIC CELLS

Saccharomyces cerevisiae, also known as *common yeast*, is another organism that has been investigated for recombinant expression of PHAs, because yeast are already widely used in industries for the production of ethanol as biofuel. Additionally, producing both renewable fuels and plastics in the same organism is in line with the recent interest in biorefining technologies, that is, the production of multiple value-added products from a single complex feedstock. Yeast are also an interesting host for PHA production because production of specialty plastics from a eukaryotic organism may offer enhanced biocompatibility for biomedical applications in eukaryotic human cells. In 1996, Leaf *et al*. demonstrated that PHB production in *S. cerevisiae* was possible by expressing the PHB synthase gene of the bacterium *R. eutropha* [69]. The production of PHB using only the synthase gene without additional synthesis enzymes indicated that *S. cerevisiae* already contained the necessary metabolic routes for PHB monomer synthesis. In 2006, Carlson and Srienc showed that by recombinantly expressing the *phaAB* genes of *R. eutropha* they could boost PHA production by 45× when compared to cells expressing only the *phaC* synthase gene [70]. Furthermore, they also demonstrated that the yeast were capable of simultaneously producing PHA and ethanol under anaerobic conditions. In the same year Zhang *et al*. showed that PHA_{mcl} could be expressed in the yeast cytosol using a recombinant, cytosolically expressed *P. oleovorans* synthase in a *pex5* mutant [71]. β-Oxidation enzymes utilized for the production of PHA_{mcl} precursors are typically expressed in yeast peroxisomes. The small size of these cellular compartments limits the available volume for polymer production. Cells deficient in the Pex5p receptor (*pex5*) are unable to target these β-oxidation enzymes to their native peroxisomal location and instead accumulate them in the cytosol, thereby creating a cytosolic PHA_{mcl} production pathway.

Another alternative for the economical production of PHAs is using transgenic plants. Researchers first achieved this feat in 1992 by the transformation of *phbB* and *phbC* genes from *R. eutropha* into *Arabadopsis thaliana*, a model organism. Cloning of the *phbA* gene was not included as the organism already produced its own β-ketothiolase within the cytoplasm. Production of PHB was low, about

0.1% dry weight [72]. Later, researchers boosted this level to 40% dry weight by transforming all three *R. eutropha* genes, *phbABC*, into *A. thaliana* and targeting them for expression in the chloroplasts via plastidial targeting sequences [73]. With an interest in the potential for biorefining, researchers have also turned their attention to plants already in use or being considered for food or energy production. Plants investigated include maize, sugar cane, rapeseed, and switchgrass [74–77]. In these efforts, targeting polymer production to plastids or peroxisomes is a common approach. Mooney provides a comprehensive review on this topic [78].

13.8 RAW MATERIAL SOURCES FOR PHA PRODUCTION

Currently, the vast majority of plastics throughout the world are synthesized from petroleum-based feedstocks, a source which is not sustainable and likely to be depleted within the current century. Furthermore, the production of petroleum-derived plastics also results in the net increase of atmospheric carbon dioxide, a significant contributor to global warming. With these concerns in mind, PHA-based plastics offer an attractive alternative because of their capacity to be synthesized from renewable feedstocks. Common feedstocks are purified sugars, molasses, and propionate. However, such feedstocks are more costly and less abundant than petroleum which in part contributes to the significantly higher price of PHAs when compared to conventional plastics [79]. In 2008, the price of Mirel brand PHA plastic was $\sim$5 kg^{-1}, three times higher than the price of polypropylene [80]. While the use of renewable feedstocks may lessen the amount of associated greenhouse gas emissions, energy, often from fossil fuel sources, is still required for polymer production. Nevertheless, a life-cycle analysis in 2008 by Kim and Dale suggested that PHB production provided significant environmental advantages to conventional petroleum-derived plastics. The authors estimated that PHB production from corn grain required 2.5 MJ/kg PHB of nonrenewable energy compared to 69–100 MJ/kg plastic for petroleum-derived plastics. Much of the energy savings was accrued by burning leftover corn stover for electricity and steam generation. With respect to their impact on global warming, they estimated that PHB provided greenhouse gas credits of −2.3 kg CO_2 equivalents/kg of polymer versus a greenhouse gas debit of 1.9–5.4 kg CO_2 equivalents/kg for petroleum-derived plastics [81].

Recently, much research has been focused on the testing of other renewable feedstocks for which supplies may be greater, less costly, and require less refining before conversion to PHA. The most ideal of these are waste streams from other industrial or bioprocesses. Carbon sources explored include beet molasses, sugarcane liquor, hydrolysates from wheat, bagasse hydrolysates, corn stover hydrolysates, dairy whey, glycerol (a by-product of biodiesel production), plant oils including corn, palm, and soybean, CO_2 emissions from fossil fuel consumption or ethanol production, sewer sludge, waste solvent streams, and solid waste products that remain after oil extraction from biomass [82].

Processes utilizing CO_2 are particularly exciting as they directly sequester greenhouse gasses. In addition to CO_2, these processes require the feeding of H_2 and O_2 as a source of energy for carbon fixation. Interestingly, the native PHB producer, *R. eutropha*, is a natural chemolithoautotroph and is adept at survival on these gaseous substrates. Attempts at PHB production utilizing *R. eutropha* and CO_2 have been successful compared to conventional cultivation with organic carbon sources [82]; however, such efforts have been encumbered by the risk of hydrogen explosion. That is, the need to keep oxygen level below the flammability limit reduces the capacity of the bioreactor for effective oxygen mass transfer [83].

13.9 PHA PRODUCTION USING MIXED MICROBIAL CULTURES

Two particularly relevant features of renewable feedstocks are that they often contain a broad variety of nutrients and they may experience significant variations in quality depending on seasonal environmental factors. Isolation of a single nutrient source from a renewable mixture, for example, glucose from corn grain, requires additional energy for pretreatment and separation processes. If a monoculture is used for conversion of a mixed feed source, it is likely that considerable work would be needed to optimize the strain for production and still this organism might be selective for just one specific nutrient within the mixture, for example, *E. coli*'s preferential consumption of glucose over xylose in cellulosic biomass. Thus, a complete conversion of all the available biomass may not be achieved. One potential solution to this problem is through the use of a mixed microbial culture such as the ones used for the treatment of industrial and municipal waste streams. By process of natural selection, those organisms that are better at utilizing the substrates available within the waste stream will become enriched within the population. If multiple carbon sources are present in the waste stream, a unique community of organism will evolve to consume the waste. If an upset in nutrient composition occurs, natural selection will again produce an optimized population for the new conditions. Because of the competitive advantage that PHA production confers to microorganisms for survival during times of nutrient scarcity, this trait has been readily adopted by numerous organisms [84]. When mixed populations of microorganisms are subjected to repeated cycles of nutrient excess (feast) followed by nutrient limitation (famine), those organisms that can survive the feast and famine regimen by utilizing their stores of accumulated PHA become enriched within the population. Such conditions of feast and famine occur routinely in waste treatment processes, and in fact the production of PHAs by activated sludge has been well documented [85–87]. In this way, mixed microbial populations can be optimized for both multiple substrate utilization and PHA productivity. PHA content in mixed microbial communities has been observed as high as 50–85% of cell dry weight [88–91], and specific productivity (g of PHA/g of substrate/h) of these cultures is approximately 10-fold higher than for PHA producing

E. coli cultures [90]. The highest levels of polymer accumulation are obtained by deliberate feast/famine cycling to enrich the culture in high PHA producing species. Furthermore, mixed microbial communities offer the advantage of not requiring sterilization, a significant energy requirement, and may be conducted in simple concrete basins as opposed to the expensive steel tanks required for monoculture fermentations requiring a significant capital investment.

Despite the many advantages to mixed microbial cultures, volumetric productivity (g of PHA/L/h) remains below that of pure cultures because of low cell densities which are approximately 10-fold lower than for pure cultures [90]. This low-volumetric productivity leads to additional costs for purification and recovery. Furthermore, the polymer material produced from mixed microbial cultivation is likely to have a higher variability in monomer composition and may be less pure than polymer produced from conventional monoculture processes thus targeting it toward applications with lower product quality requirements.

13.10 POLYLACTIC ACID

Within the field of renewable polymers, another significant product is polylactic acid, otherwise known as polylactide or PLA. Production of this polymer is considered renewable because the main feedstock, lactic acid, is produced by the fermentation of sugars, often corn-derived dextrose. Polylactide degrades naturally to carbon dioxide and water first by hydrolysis into lower molecular weight fragments and then via enzymatic degradation by microbes [92]. The lactic acid molecule has one chiral center allowing for the existence of two stereoisomers, L-lactic acid and D-lactic acid. The chirality of the lactic acid produced depends on the type of microorganisms used for fermentation with different organisms producing pure D or L or a mixture of the two [93]. After the lactic acid has been purified from the microbial culture, it is then catalytically converted to polylatctic acid; however, the polymerization also releases water which thermodynamically drives the reaction in the reverse direction. This prevents the formation of long chains. To bypass this limitation, the mixture of low-molecular weight oligomers is then subjected to a catalytic degradation reaction to produce lactide, the cyclic dimer of two lactic acid molecules [94]. By vacuum distillation, this product is then purified to remove the water. Depending on the starting quantities of L-lactic acid and D-lactic acid from the fermentation, the purified product contains L-lactide (two L-lactic acid molecules), D-lactide (two D-lactic acid molecules), and DL-lactide (a molecule of both L-lactide and D-lactide) [95]. The lactide is then converted to long chain PLA by either a ring-opening polymerization typically using an anionic initiator (e.g., alkali metal oxides) or a coordinated insertion polymerization using metal carboxylates as catalysts (e.g., stannous octanoate) (Figure 13.4).

Isotactic PLA composed of only L-lactide (PLLA) is straight and regular; however, the inclusion of occasional D-lactide or DL-lactide impurities creates kinks

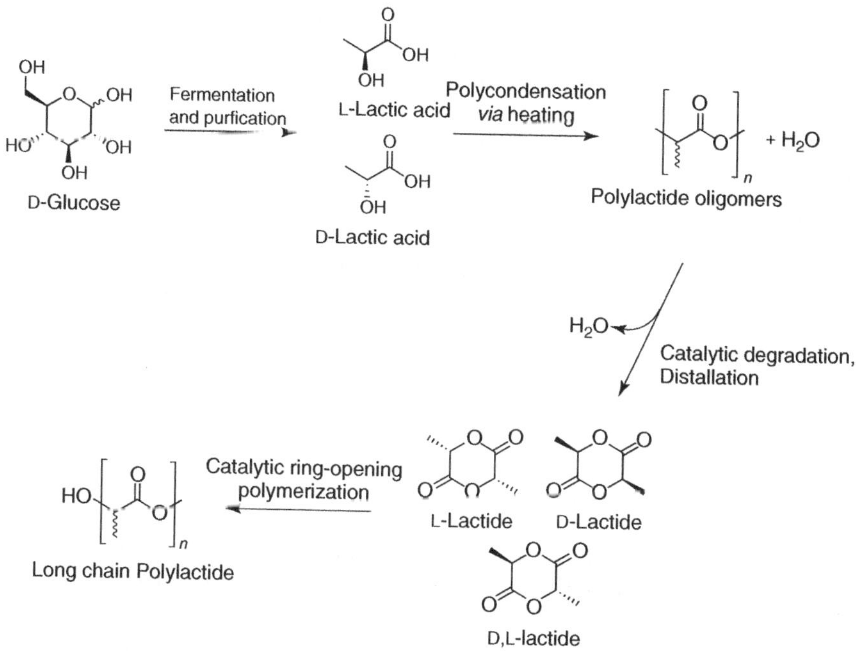

Figure 13.4 The most commonly employed process for synthesis of polylactide includes fermentation, purification, and three organic synthesis steps.

in the chain which reduce the degree of crystallization in the polymer. Tacticity affects polymer crystallinity (i.e., amorphous vs semicrystalline structure), melting temperature, and degradation rate [96]. Control of tacticity may be achieved by varying the ratio of stereoisomers in the mixture or by the use of catalysts with a high degree of selectivity for one stereoisomer [97]. Using these selective catalysts, research has demonstrated the synthesis of "stereo-block" copolymers from racemic lactide [98]. In a phenomenon known as *stereocomplexation*, polymer chains with opposing tacticity display a higher affinity for one another than they do for chains of the same tacticity. For polylactide, stereocomplexation of PLLA with PDLA produces a material with a melting temperature of 220–230°C, significantly higher than for PLLA alone, 170–190°C [99]. In another route to the creation of renewable block copolymers utilizing PLA, researchers are investigating the addition of polymenthide (PM) derived from menthol found in mint plants. The menthol is first converted to menthone which is subsequently converted to menthide [100]. The menthide is then used as monomer for the synthesis of PM by ring-opening polymerization. PM is then joined to polylactide to form a triblock copolymer, PLA-*b*-PM-*b*-PLA, with enhanced elongation and elasticity over PLA homopolymer [101].

Unlike the production of PHA which can be accomplished entirely through biosynthesis in a single organism, the commercial production of PLA is a multistep process requiring (i) microbial fermentation to produce lactic acid,

(ii) purification of lactic acid, (iii) chemical synthesis of the lactide dimer, (iv) purification of lactide, and (v) *in vitro* ring-opening polymerization to create PLA. Recognizing the advantage of a more direct synthesis route, Taguchi *et al.* recently engineered recombinant *E. coli* to produce copolymers of PLA and PHA enzymatically [102]. Noting that the chemical structure of lactic acid (2-hydroxypropanoic acid) is similar to the structure of 3-hydroxybutyric acid, they tested seven unique polymerase enzymes for their ability to incorporate the LA-CoA monomer. Only one of these enzymes was successful, a mutant of the *Pseudomonas* 61-3 polymerase with two amino acid substitution. However, the enzyme would incorporate the LA-CoA only in the presence of HB-CoA monomers. To generate an ample supply of LA-CoA monomers, they inserted a *pct* gene from *Megasphaera elsdenii* coding for a propionyl-CoA transferase which had also been shown to have transferase activity toward lactic acid. For the production of HB-CoA monomers, they included *phbAB* genes from *R. eutropha*. The recombinant *E. coli* produced polymer to 19 wt% of cell dry weight with a lactate fraction of 6 mol%. Since then similar investigations have increased both polymer content and lactate fraction by means of improved polymerases, improved propionate CoA transferase enzymes, and mutant *E. coli* strains with enhanced lactate production pathways [103–105]. The most successful attempt boasts 46 wt% polymer production with 70 mol% lactate synthesized from glucose alone. This area of research is particularly exciting as PLA is already widely used and production by complete biosynthesis could result in a significant cost savings.

13.11 *IN VITRO* SYNTHESIS OF PHAs

In addition to synthesis from fermentation, PHAs have also been synthesized *in vitro*. Although such synthesis routes may be less direct than *in vivo* synthesis, in theory they offer the advantage of allowing greater control over monomer compositions and product purity. Such qualities may be particularly important for biomedical applications where quality requirements are more stringent. In 1995, Gerngross and Martin reported the successful synthesis of PHB *in vitro* by combining purified polymer synthase from *R. eutropha* with synthetically produced (*R*)-3-hydroxybutyryl CoA in a suitable buffer. By varying the initial concentration of polymer synthase, they could control the molecular weight of the polymer molecules [106]. Beyond this early success, researchers realized that any large scale application of *in vitro* PHA synthesis would require the ability to recycle the costly CoA cofactor. In 2005, Satoh *et al.* devised a solution to this problem by incorporating an acyl-CoA synthetase cloned from *P. oleovorans* into their reaction mixture along with CoA cofactor in addition to purified polymer synthase from *R. eutropha* and either the 3-hydroxybutyrate or 4-hydroxybutyrate substrates in a suitable buffer. This mixture allowed for the continuous regeneration of the HB-CoA complex. Additionally, they showed that the monomer composition of P(3HB-*co*-4HB) copolymer could be strictly controlled by varying the ratios of the monomers in the reaction mixture [107].

Another *in vitro* route for the synthesis of PHB is through ring-opening polymerization of β-butyrolactone [108]. This substrate can be synthesized from 3-hydroxybutyric acid, a product which is easily obtained from microbial fermentation [109]. This route is similar to the way in which PLA is obtained through the ring-opening polymerization of lactide. While conventional ring-opening catalysts may be employed, these are often toxic and may compromise the biocompatibility of the polymer. A greener alternative is the use of purified lipase enzymes [110] Such reactions must be performed under anhydrous conditions as lipases also catalyze lactone hydrolysis in the presence of water. Thus, neat organic solvents must be used to dissolve the monomers and associated polymer.

One alternative to the use of organic solvents is the use of ionic liquids with low-temperature melting points. A significant advantage of these solvents is their nonvolatility which eliminates a significant pathway for environmental release. While some ionic liquids can be toxic others have limited toxicity. In 2007, Gorke *et al*. tested several ionic liquid and lipase combinations for the synthesis of PHAs from various lactone precursors. Using lipase B from *Candida antarctica* dissolved in the ionic liquid, 1-butyl-3-methylimidazolium bis(trifluoromethane)-sulfonimide, they synthesized polymers from β-propiolactone, δ-valerolactone, and ε-caprolactone with degrees of polymerization as high as 170, 25, and 85, respectively. Activity of the lipase toward β-butyrolactone and γ-butyrolactone was not as high, producing only five-length oligomers [111].

13.12 APPLICATIONS OF PHAs

Over the years, PHA and PLA have been available as commercial products from several manufacturers. Trade names include NatureWorks, Ingeo (PLA) Biopol, Biomer, Nodax, Biogreen, and Mirel (all PHA) [82]. Bulk applications include biodegradable plastic bags, drinking cups, plastic cutlery, fibers for clothing, packaging materials for dairy and vegetables, film overwraps, and blister packages [82,112]. The application of PHA particles as an aqueous dispersion has been investigated for the creation of polymer coatings on paper and cardboard, biodegradable cheese rinds, and paints [24]. Because PHA polymers are synthesized from only the *R* stereoisomer of hydroxyalkanoic acids, they have also been considered as a source of enantiomerically pure precursors for other fine chemical and pharmaceuticals [79,109]. PHA monomers have been produced by *in vitro* chemical digestion of PHA [113] as well as by *in vivo* enzymatic digestion using depolymerase enzymes [114,115]. Interest has also been centered on the use of PHAs for biomedical purposes because of their biodegradability and biocompatibility. Applications that have been investigated include tissue patches, orthopedic implants, tissue scaffolding, sutures, drug release medium, nerve regeneration, and wound healing [116].

More recently, PHA granules, the native cellular bodies that contain the polymer, have also attracted attention. When PHA is produced within the cell it forms small, spherical, water-insoluble inclusions. These granules are surrounded by a phospholipid monolayer [2,117,118]. Embedded within this layer are several proteins, including PHA synthase, PHA depolymerase, and PHA phasins. Depolymerase enzymes are responsible for degrading the polymer during times of nutrient scarcity while phasins interfere with granule coalescence which affects granule size [119]. Using these "granule associated proteins" experimentalists have been able to attach various nonnative proteins to the surface of the granules creating functionalized "biobeads." One application of these PHA biobeads is for protein purification. This is done by making a fusion protein in which the protein of interest is linked to the C terminal of a PhaP phasin protein. The phasin acts as an affinity tag by which the protein of interest becomes linked to the PHA granule. The PHA granules may be produced within the same cell as the protein of interest [120,121] or may be produced in separate cells and added *in vitro* to cell lysates containing the phasin fusion proteins [122]. Centrifugation is used to separate the insoluble biobeads along with the protein of interest attached to the surface. Within the protein linker segment connecting the phasin to the protein of interest a protease cleavage site is used to release the protein from the biobead. Additionally, fusion proteins have also been constructed using the PhaC polymerase. The synthesis of bifunctional biobeads containing both a Green Fluorescent Protein (GFP) fusion protein and a mouse antigen demonstrated that the technology could also be useful as cell markers for flow cytometry [123]. Recently, Yao *et al*. demonstrated that PHA biobeads loaded with the dye molecule rhodamine B isothiocyanate and modified with cell-specific ligands could be used to target delivery of the model drug to two unique tissue types in *in vivo* mouse experiments [124].

13.13 RECOVERY AND PURIFICATION OF PHAs FROM MICROBIAL CULTURE

Another important factor impacting PHA production is polymer recovery and purification, which affects both the cost of the product and its value as an environmentally friendly substitute to petroleum-derived plastids. One common extraction technique is with the use of organic solvents which serve to both permeate the cell membrane and dissolve the polymer [125]. Solvents investigated include chloroform, 1,2-dichloroethane, and methylene chloride [126]; ethylene carbonate and propylene carbonate [127,128]; 1,2-propandiol, glycerol, diethyl succinate, and butyrolactone [129]. Of these solvents, propylene carbonate has most recently experienced renewed interest for its advantageous environmental properties, for example, extraction at atmospheric pressure, low volatility, and low toxicity [128]. Dissolved PHA is precipitated by the addition of a nonsolvent, for example, methanol or water [130,131]. Drawbacks of this type of purification include the large quantities of solvent required, the environmental hazard of such solvents, and the cost of solvent recovery operations. Advantages of organic solvent extraction are high product purity (>99%) and minimal reduction in polymer molecular weight [132].

Other methods of cell digestion include treatment with surfactants, for example, palmitoyl carnitine [133] and sodium dodecyl sulfate [134], that are less harmful to the environment but still require treatment of process wastewater due to the large amounts of surfactants used. Another additive which is very effective at cell lysis is sodium hypochlorite and purities as high as 99% have been obtained using this method; however, hypochlorite can in some cases cause a significant reduction in polymer molecular weight by ~50% [135]. Researchers have also investigated the use of lytic enzymes for cellular digestion [136–139]. These methods are desirable because they are much more environmentally friendly and produce good recovery yields; however, they are hampered by the high cost of enzymes [132]. In addition, various mechanical methods have been investigated for the recovery of PHA, including bead mill disruption, high-pressure homogenization, and ultrasonication [140–142]. Finally, researchers have investigated the potential for self-lysing cells producing PHAs. Such a functionality is engineered usually by the recombinant expression lytic enzymes cloned from viruses [143].

13.14 SUMMARY

Petroleum-derived plastics have become ubiquitous in human society; however, their production from nonrenewable sources, contribution to greenhouse gas emissions, and permanent presence in the environment after discard demand an alternative solution that is both environmentally friendly and sustainable. Both PHAs and polylactide show significant potential for meeting this demand. These materials are advantageous for their biodegradability and production from renewable feedstocks, and both are already available commercially. Applications of these materials include packaging, fibers for textiles, biomedical materials, and protein-functionalized biobeads.

While the homopolymer forms of these materials may lack toughness, polymer blending and block copolymer synthesis have been successful in improving material properties. Furthermore, the large diversity of monomer units available for PHAs results in a spectrum of material properties ranging from stiff and strong to elastomeric. The ability to combine different monomers into copolymers has improved with the creation of mutant polymer synthase proteins with expanded monomer specificity.

Despite their environmental benefits, the most significant obstacle to the proliferation of PHAs and polylactide is their cost which is approximately three to five times higher than the cost of petroleum-derived plastics. Much of this price disparity should decrease in the future as the cost of oil will surely rise with diminishing supplies. Other factors affecting the price of these bioplastics are raw materials cost and energy needed for sterilization of feedstocks. Typical fermentations to produce these plastics require the use of refined, that is, expensive, feedstocks such as glucose. To address this issue, researchers are investigating the use of less refined feed sources including biomass and waste streams from other processes. Ideally, PHAs could be produced as part of a larger biorefining operation, for example, producing PHAs using glycerol waste from biodiesel production or CO_2 from ethanol production, thereby offsetting the overhead costs of utilities and infrastructure.

Furthermore, research to optimize the metabolic pathways of microorganisms will allow higher conversion of feedstocks. In some instances, mixed microbial cultures may be used to produce plastics without the need for feedstock sterilization.

Another way to reduce costs may be through process simplification. Researchers have already demonstrated that polylactide can be synthesized using an *in vivo* polymerase in the same way that PHAs are created, thereby eliminating several purification and chemical synthesis steps. Eventually, plant-based technologies may allow direct production of plastics from carbon dioxide and sunlight in the same way that natural rubber is produced (Figure 13.5).

Lastly, for bioplastics to retain their image as environmentally friendly alternatives that are truly superior to petroleum-derived plastics, processes should be designed in such a way that the need for toxic solvents in product recovery and purification is eliminated.

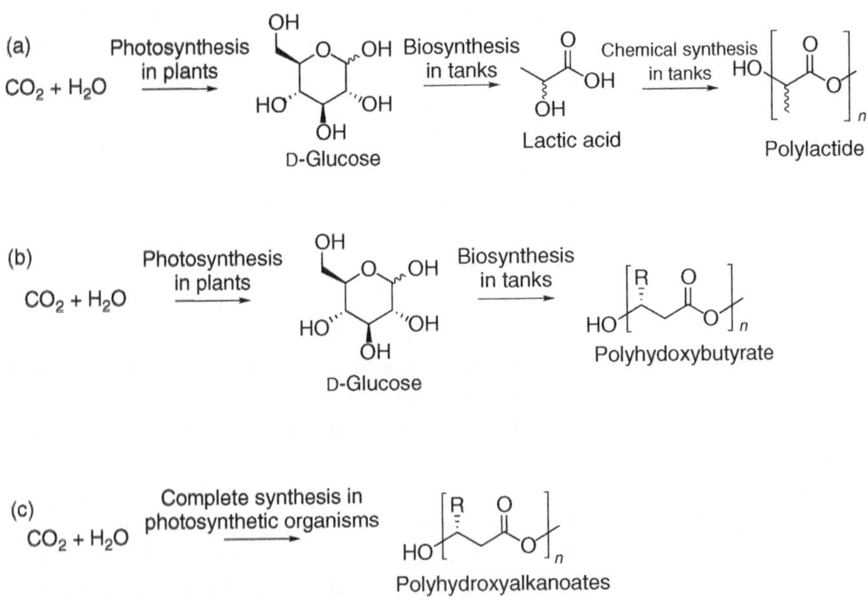

Figure 13.5 Simplification of bioplastic production. (a) The current process for polylactide production. (b) The current process for polyhydroxybutyrate production. Researchers have also demonstrated the ability to synthesize polylactide via a similar biosynthesis. (c) Synthesis of bioplastics in plants or other photosynthetic organisms could further simplify processing. This technology has already been demonstrated in the laboratory.

REFERENCES

1. Reis MAM, Serafim LS, Lemos PC, Ramos AM, Aguiar FR, Van Loosdrecht MCM. Production of polyhydroxyalkanoates by mixed microbial cultures. Bioprocess Biosyst Eng 2003;25:377–385.
2. Mayer F, Hoppert M. Determination of the thickness of the boundary layer surrounding bacterial pha inclusion bodies, and implications for models describing the molecular architecture of this layer. J Basic Microbiol 1997;37:45–52.
3. Rehm BHA. Polyester synthases: natural catalysts for plastics. Biochem J 2003; 376:15–33.
4. Steinbüchel A, Valentin HE. Diversity of bacterial polyhydroxyalkanoic acids. FEMS Microbiol Lett 1995;128:219–228.
5. Slater S, Gallaher T, Dennis D. Production of poly-(3-Hydroxybutyrate-Co-3-Hydroxyvalerate) in a recombinant *Escherichia Coli* strain. Appl Environ Microbiol 1992;58:1089–1094.
6. Lageveen RG, Huisman GW, Preusting H, Ketelaar P, Eggink G, Witholt B. Formation of polyesters by *pseudomonas oleovorans*: effect of substrates on formation and composition of poly-(R)-3-hydroxyalkanoates and poly-(R)-3-hydroxyalkenoates. Appl Environ Microbiol 1988;54:2924–2932.
7. Fukui T, Doi Y. Cloning and analysis of the poly(3-Hydroxybutyrate-Co-3-Hydroxyhexanoate) biosynthesis genes of *Aeromonas caviae*. J Bacteriol 1997; 179:4821–4830.
8. Matsusaki H, Manji S, Taguchi K, Kato M, Fukui T, Doi Y. Cloning and molecular analysis of the poly(3-Hydroxybutyrate) and poly(3-Hydroxybutyrate-Co-3-Hydroxyalkanoate) biosynthesis genes in *Pseudomonas sp.* Strain 61-3. J Bacteriol 1998;180:6459–6467.
9. Takase K, Taguchi S, Doi Y. Enhanced synthesis of poly(3-Hydroxybutyrate) in recombinant *Escherichia Coli* by means of error-prone Pcr mutagenesis, saturation mutagenesis, and *in vitro* recombination of the type II polyhydroxyalkanoate synthase gene. J Biochem 2003;133:139–145.
10. Sheu D-S, Lee C-Y. Altering the substrate specificity of polyhydroxyalkanoate synthase 1 derived from *Pseudomonas putida* Gpo1 by localized semirandom mutagenesis. J Bacteriol 2004;186:4177–4184.
11. Li R, Zhang H, Qi Q. The production of polyhydroxyalkanoates in recombinant *Escherichia Coli*. Bioresour Technol 2007;98:2313–2320.
12. Reddy S, Thirumala M, Mahmood S. A novel *Bacillus sp.* Accumulating poly (3-Hydroxybutyrate-Co-3-Hydroxyvalerate) from a single carbon substrate. J Ind Microbiol Biotechnol 2009;36:837–843.
13. Doi Y. Microbial polyesters. Wiley-VCH, New York: 1990.
14. de Koning GJM, Scheeren AHC, Lemstra PJ, Peeters M, Reynaers H. Crystallization phenomena in bacterial poly[(R)-3-Hydroxybutyrate]: 3. Toughening via texture changes. Polymer 1994;35:4598–4605.
15. Quagliano Javier C, Miyazaki Silvia S. Microbial production of the biodegradable polyester poly-3-hydroxybutyrate (Phb) from Azotobacter chroococcum 6b: relation between Phb molecular weight, thermal stability and tensile strength. In: Alain Durieux and Jean Paul Simon, editors. Appl Microbiol, *Focus on Biotechnology*. Springer, Netherlands; 2002; Vols 2 and 3:135–140.

16. Hocking PJ, Marchessault RH. Polyhydroxyalkanoates. In: Kaplan DL, editor. Biopolymers from renewable resources. Springer, New York; 1998. pp. 220–248.
17. Pederson EN, McChalicher CWJ, Srienc F. Bacterial synthesis of Pha block copolymers. Biomacromolecules 2006;7:1904–1911.
18. McChalicher CWJ, Srienc F. Investigating the structure-property relationship of bacterial Pha block copolymers. J Biotechnol 2007;132:296–302.
19. Holmes P. Biologically produced (R)-3-Hydroxyalkanoate polymers and copolymers. In: Basset CD, editor. Development in crystalline polymers. Elsevier, London; 1988. pp. 1–65.
20. Verhoogt H, Ramsay BA, Favis BD. Polymer blends containing poly(3-Hydroxyalkanoate)s. Polymer 1994;35:5155–5169.
21. Gagnon KD, Lenz RW, Farris RJ, Fuller RC. Crystallization behavior and its influence on the mechanical properties of a thermoplastic elastomer produced by *Pseudomonas oleovorans*. Macromolecules 1992;25:3723–3728.
22. Mark JE, editor. Polymer Data Handbook. 2nd ed. Oxford University Press, USA; 2009.
23. Kaplan WA, editor. Modern plastics Encyclopedia '99. McGraw-Hill, New York; 1998.
24. van der Walle G, de Koning G, Weusthuis R, Eggink G. Properties, modifications and applications of biopolyesters. In: Wolfgang Babel and Alexander Steinbüchel, editors. Biopolyesters, *Advances in Biochemical Engineering / Biotechnology*. Springer Berlin / Heidelberg; 2001;71:263–291.
25. Ram A. Fundamentals of polymer engineering. Plenum Press, New York; 1997.
26. Sim SJ, Snell KD, Hogan SA, Stubbe J, Rha C, Sinskey AJ. Pha synthase activity controls the molecular weight and polydispersity of polyhydroxybutyrate *in Vivo*. Nat Biotechnol 1997;15:63–67.
27. Parulekar Y, Mohanty AK. Extruded biodegradable cast films from polyhydroxyalkanoate and thermoplastic starch blends: fabrication and characterization. Macromol Mater Eng 2007;292:1218–1228.
28. Bhatt R, Shah D, Patel KC, Trivedi U. Pha-rubber blends: synthesis, characterization and biodegradation. Bioresour Technol 2008;99:4615–4620.
29. Noda I, Satkowski MM, Dowrey AE, Marcott C. Polymer alloys of Nodax copolymers and poly(Lactic Acid). Macromol Biosci 2004;4:269–275.
30. Schreck KM, Hillmyer MA. Block copolymers and melt blends of polylactide with Nodax(Tm) microbial polyesters: preparation and mechanical properties. J Biotechnol 2007;132:287–295.
31. Chen GX, Hao GJ, Guo TY, Song MD, Zhang BH. Structure and mechanical properties of poly(3-hydroxybutyrate-Co-3-hydroxyvalerate) (Phbv)/Clay nanocomposites. J Mater Sci Lett 2002;21:1587–1589.
32. Yun S, Lo V, Noorman J, Davis J, Russell R, Holden P, Gadd G. Morphology of composite particles of single wall carbon nanotubes/biodegradable polyhydroxyalkanoates prepared by spray drying. Polym Bull 2010;64:99–106.
33. Yu L, Dean K, Li L. Polymer blends and composites from renewable resources. Prog Polym Sci 2006;31:576–602.
34. Meuler AJ, Hillmyer MA, Bates FS. Ordered network mesostructures in block polymer materials. Macromolecules 2009;42:7221–7250.

35. Eggink G, van der Wal H, Huijberts GNM, de Waard P. Oleic acid as a substrate for poly-3-hydroxyalkanoate formation in *Alcaligenes eutrophus* and *Pseudomonas putida*. Ind Crops Prod 1992;1:157–163.

36. Matsusaki H, Abe H, Doi Y. Biosynthesis and properties of poly(3-Hydroxybutyrate-Co 3 Hydroxyalkanoates) by recombinant strains of *Pseudomonas sp.* 61-3. Biomacromolecules 2000;1:17–22.

37. Qiu Y-Z, Han J, Chen G-Q. Metabolic engineering of Aeromonas hydrophila for the enhanced production of poly(3-Hydroxybutyrate- Co-3-Hydroxyhexanoate). Appl Microbiol Biotechnol 2006;69:537–542.

38. Volova TG, Kalacheva GS. The synthesis of hydroxybutyrate and hydroxyvalerate copolymers by the bacterium *Ralstonia eutropha*. Microbiology 2005;74:54–59.

39. Pereira SMF, Sánchez RJ, Rieumont J, Cabrera JG. Synthesis of biodegradable polyhydroxyalcanoate copolymer from a renewable source by alternate feeding. Polym Eng Sci 2008;48:2051–2059.

40. Mantzaris NV, Kelley AS, Daoutidis P, Srienc F. A population balance model describing the dynamics of molecular weight distributions and the structure of pha copolymer chains. Chem Eng Sci 2002;57:4643–4663.

41. Jurasek L, Marchessault RH. Polyhydroxyalkanoate (Pha) granule formation in Ralstonia eutropha cells: a computer simulation. Appl Microbiol Biotechnol 2004;64:611–617.

42. Iadevaia S, Mantzaris NV. Genetic network driven control of Phbv copolymer composition. J Biotechnol 2006;122:99–121.

43. Iadevaia S, Mantzaris NV. Synthesis of Phbv block copolymers driven by an oscillatory genetic network. J Biotechnol 2007;128:615–637.

44. Witholt B, Kessler B. Perspectives of medium chain length poly(Hydroxyalkanoates), a versatile set of bacterial bioplastics. Curr Opin Biotechnol 1999; 10:279–285.

45. Rehm BHA, Kruger N, Steinbüchel A. A new metabolic link between fatty acid de novosynthesis and polyhydroxyalkanoic acid synthesis. J Biol Chem 1998;273:24044–24051.

46. Matsumoto Ki, Nakae S, Taguchi K, Matsusaki H, Seki M, Doi Y. Biosynthesis of poly(3-Hydroxybutyrate-Co-3-Hydroxyalkanoates) copolymer from sugars by recombinant Ralstonia eutropha harboring the Phac1ps and the Phagps genes of pseudomonas Sp. 61-3. Biomacromolecules 2001;2:934–939.

47. Schubert P, Steinbüchel A, Schlegel HG. Cloning of the Alcaligenes eutrophus genes for synthesis of poly-beta-hydroxybutyric acid (Phb) and synthesis of Phb in *Escherichia Coli*. J Bacteriol 1988;170:5837–5847.

48. Peoples OP, Sinskey AJ. Poly-beta-hydroxybutyrate (Phb) biosynthesis in Alcaligenes eutrophus H16. Identification and characterization of the Phb polymerase gene (Phbc). J Biol Chem 1989;264:15298–15303.

49. Choi J-I, Lee SY, Han K. Cloning of the Alcaligenes latus polyhydroxyalkanoate biosynthesis genes and use of these genes for enhanced production of poly(3-Hydroxybutyrate) in *Escherichia Coli*. Appl Environ Microbiol 1998;64:4897–4903.

50. Taguchi K, Aoyagi Y, Matsusaki H, Fukui T, Doi Y. Over-expression of 3-ketoacyl-Acp synthase Iii or malonyl-CoA-Acp transacylase gene induces monomer supply for polyhydroxybutyrate production in *Escherichia Coli* Hb101. Biotechnol Lett 1999;21:579–584.

51. Taguchi K, Aoyagi Y, Matsusaki H, Fukui T, Doi Y. Co-expression of 3-Ketoacyl-Acp reductase and polyhydroxyalkanoate synthase genes induces Pha production in *Escherichia Coli* Hb101 strain. FEMS Microbiol Lett 1999;176:183–190.
52. Timm A, Steinbüchel A. Cloning and molecular analysis of the poly(3-Hydroxyalkanoic Acid) gene locus of *Pseudomonas aeruginosa* pao1. Eur J Biochem 1992;209:15–30.
53. Langenbach S, Rehm BHA, Steinbüchel A. Functional expression of the Pha synthase gene Phac1 from *Pseudomonas aeruginosa* in *Escherichia Coli* results in poly(3-hydroxyalkanoate) synthesis. FEMS Microbiol Lett 1997;150:303–309.
54. Qi Q, Rehm BHA, Steinbüchel A. Synthesis of poly(3-Hydroxyalkanoates) in *Escherichia Coli* expressing the pha synthase gene Phac2 from Pseudomonas aeruginosa: comparison of Phac1 and Phac2. FEMS Microbiol Lett 1997;157:155–162.
55. Qi Q, Steinbüchel A, Rehm BHA. Metabolic routing towards polyhydroxyalkanoic acid synthesis in recombinant *Escherichia Coli* (Fadr): inhibition of fatty acid ß-oxidation by acrylic acid. FEMS Microbiol Lett 1998;167:89–94.
56. Tsuge T, Fukui T, Matsusaki H, Taguchi S, Kobayashi G, Ishizaki A, Doi Y. Molecular cloning of two (R)-specific enoyl-CoA hydratase genes from *Pseudomonas aeruginosa* and their use for polyhydroxyalkanoate synthesis. FEMS Microbiol Lett 2000;184:193–198.
57. Fiedler S, Steinbüchel A, Rehm B. The role of the fatty acid ß-oxidation multienzyme complex from *Pseudomonas oleovorans* in polyhydroxyalkanoate biosynthesis: molecular characterization of the fadba operon from *P. oleovorans* and of the enoyl-CoA hydratase genes phaj from *P. oleovorans* and *Pseudomonas putida*. Arch Microbiol 2002;178:149–160.
58. Park SJ, Lee SY. Identification and characterization of a new enoyl coenzyme a hydratase involved in biosynthesis of medium-chain-length polyhydroxyalkanoates in recombinant *Escherichia Coli*. J Bacteriol 2003;185:5391–5397.
59. Tsuge T, Taguchi K, Seiichi Taguchi, Doi Y. Molecular characterization and properties of (R)-specific enoyl-CoA hydratases from *Pseudomonas aeruginosa*: metabolic tools for synthesis of polyhydroxyalkanoates via fatty acid ß-oxidation. Int J Biol Macromol 2003;31:195–205.
60. Rehm BHA, Mitsky TA, Steinbüchel A. Role of fatty acid de novo biosynthesis in polyhydroxyalkanoic acid (Pha) and rhamnolipid synthesis by pseudomonads: establishment of the transacylase (Phag)-mediated pathway for Pha biosynthesis in *Escherichia Coli*. Appl Environ Microbiol 2001;67:3102–3109.
61. Rehm BHA, Steinbüchel A. Heterologous expression of the Acyl-Acyl carrier protein thioesterase gene from the plant Umbellularia californica mediates polyhydroxyalkanoate biosynthesis in recombinant *Escherichia coli*. Appl Microbiol Biotechnol 2001;55:205–209.
62. Fukui T, Shiomi N, Doi Y. Expression and characterization of (R)-specific enoyl coenzyme a hydratase involved in polyhydroxyalkanoate biosynthesis by Aeromonas caviae. J Bacteriol 1998;180:667–673.
63. Lu X-Y, Wu Q, Zhang W-J, Zhang G, Chen G-Q. Molecular cloning of polyhydroxyalkanoate synthesis operon from *Aeromonas hydrophila* and its expression in *Escherichia Coli*. Biotechnol Prog 2004;20:1332–1336.
64. Takase K, Matsumoto Ki, Taguchi S, Doi Y. Alteration of substrate chain-length specificity of type Ii synthase for polyhydroxyalkanoate biosynthesis by *in vitro* evolution: *in vivo* and *in vitro* enzyme assays. Biomacromolecules 2004;5:480–485.

65. Nomura CT, Taguchi K, Taguchi S, Doi Y. Coexpression of genetically engineered 3-ketoacyl-acp synthase Iii (Fabh) and polyhydroxyalkanoate synthase (Phac) genes leads to short-chain-length-medium-chain-length polyhydroxyalkanoate copolymer production from glucose in *Escherichia Coli* Jm109. Appl Environ Microbiol 2004;70:999–1007.

66. Nomura CT, Taguchi K, Gan Z, Kuwabara K, Tanaka T, Takase K, Doi Y. Expression of 3-ketoacyl-acyl carrier protein reductase (Fabg) genes enhances production of polyhydroxyalkanoate copolymer from glucose in recombinant *Escherichia Coli* Jm109. Appl Environ Microbiol 2005;71:4297–4306.

67. Nomura CT, Tanaka T, Eguen TE, Appah AS, Matsumoto Ki, Taguchi S, Ortiz CL, Doi Y. Fabg mediates polyhydroxyalkanoate production from both related and nonrelated carbon sources in recombinant *Escherichia Coli* Ls5218. Biotechnol Prog 2008;24:342–351.

68. Park S, Lee S. Biosynthesis of poly(3-Hydroxybutyrate- Co -3-Hydroxyalkanoates) by metabolically engineered *Escherichia Coli* strains. Appl Biochem Biotechnol 2004;114:335–346.

69. Leaf TA, Peterson MS, Stoup SK, Somers D, Srienc F. *Saccharomyces cerevisiae* expressing bacterial polyhydroxybutyrate synthase produces poly-3-hydroxybutyrate. Microbiology 1996;142:1169–1180.

70. Carlson R, Srienc F. Effects of recombinant precursor pathway variations on poly[(R)-3-hydroxybutyrate] synthesis in *Saccharomyces cerevisiae*. J Biotechnol 2006;124:561–573.

71. Zhang B, Carlson R, Srienc F. Engineering the monomer composition of polyhydroxyalkanoates synthesized in *Saccharomyces cerevisiae*. Appl Environ Microbiol 2006;72:536–543.

72. Poirier Y, Dennis DE, Klomparens K, Somerville C. Polyhydroxybutyrate, a biodegradable thermoplastic, produced in transgenic plants. Science 1992;256: 520–523.

73. Bohmert K, Balbo I, Kopka J, Mittendorf V, Nawrath C, Poirier Y, Tischendorf G, Trethewey RN, Willmitzer L. Transgenic arabidopsis plants can accumulate polyhydroxybutyrate to up to 4% of their fresh weight. Planta 2000;211:841–845.

74. Hahn JJ, Eschenlauer AC, Sleytr UB, Somers DA, Srienc F. Peroxisomes as sites for synthesis of polyhydroxyalkanoates in transgenic plants. Biotechnol Prog 1999;15:1053–1057.

75. Houmiel KL, Slater S, Broyles D, Casagrande L, Colburn S, Gonzalez K, Mitsky TA, Reiser SE, Shah D, Taylor NB, Tran M, Valentin HE, Gruys KJ. Poly(ß-Hydroxybutyrate) production in oilseed leukoplasts of brassica napus. Planta 1999;209:547–550.

76. Purnell MP, Petrasovits LA, Nielsen LK, Brumbley SM. Spatio-temporal characterization of polyhydroxybutyrate accumulation in sugarcane. Plant Biotechnol J 2007;5:173–184.

77. Somleva MN, Snell KD, Beaulieu JJ, Peoples OP, Garrison BR, Patterson NA. Production of polyhydroxybutyrate in switchgrass, a value-added Co-product in an important lignocellulosic biomass crop. Plant Biotechnol J 2008;6:663–678.

78. Mooney BP. The second green revolution? Production of plant-based biodegradable plastics. Biochem J 2009;418:219–232.

79. Choi J, Lee SY. Factors affecting the economics of polyhydroxyalkanoate production by bacterial fermentation. Appl Microbiol Biotechnol 1999;51:13–21.
80. DiGregorio BE. Biobased performance bioplastic: mirel. Chem Biol 2009;16:1–2.
81. Kim S, Dale BE. Energy and greenhouse gas profiles of polyhydroxybutyrates derived from corn grain: a life cycle perspective. Environ Sci Technol 2008;42: 7690–7695.
82. Castilho LR, Mitchell DA, Freire DMG. Production of polyhydroxyalkanoates (Phas) from waste materials and by-products by submerged and solid-state fermentation. Bioresour Technol 2009;100:5996–6009.
83. Ishizaki A, Tanaka K, Taga N. Microbial production of poly-D-3-hydroxybutyrate from Co_2. Appl Microbiol Biotechnol 2001;57:6–12.
84. Kadouri D, Jurkevitch E, Okon Y, Castro-Sowinski S. Ecological and agricultural significance of bacterial polyhydroxyalkanoates. Crit Rev Microbiol 2005;31:55–67.
85. Wallen LL, Rohwedder WK. Poly-ß-hydroxyalkanoate from activated sludge. Environ Sci Technol 1974;8:576–579.
86. Liu W-T, Mino T, Matsuo T, Nakamura K. Isolation, characterization and identification of polyhydroxyalkanoate-accumulating bacteria from activated sludge. J Biosci Bioeng 2000;90:494–500.
87. Bengtsson S, Werker A, Christensson M, Welander T. Production of polyhydroxyalkanoates by activated sludge treating a paper mill wastewater. Bioresour Technol 2008;99:509–516.
88. Satoh H, Iwamoto I, Mino T, Matsuo T. Activated sludge as a possible source of biodegradable plastic. Water Sci Technol 1998;32:103–109.
89. Dionisi D, Beccari M, Gregorio SD, Majone M, Papini MP, Vallini G. Storage of biodegradable polymers by an enriched microbial community in a sequencing batch reactor operated at high organic load rate. J Chem Technol Biotechnol 2005;80:1306–1318.
90. Dias JML, Lemos PC, Serafim LS, Oliveira C, Eiroa M, Albuquerque MGE, Ramos AM, Oliveira R, Reis MAM. Recent advances in polyhydroxyalkanoate production by mixed aerobic cultures: from the substrate to the final product. Macromol Biosci 2006;6:885–906.
91. Johnson K, Jiang Y, Kleerebezem R, Muyzer G, van Loosdrecht MCM. Enrichment of a mixed bacterial culture with a high polyhydroxyalkanoate storage capacity. Biomacromolecules 2009;10:670–676.
92. Drumright RE, Gruber PR, Henton DE. Polylactic acid technology. Adv Mater 2000;12:1841–1846.
93. Tsai S, Coleman R, Moon S, Schneider K, Millard C. Strain screening and development for industrial lactic acid fermentation. Appl Biochem Biotechnol 1993;39–40:323–335.
94. Yoo DK, Lee DS, Kim D. Synthesis of lactide from oligomeric Pla: effects of temperature, pressure, and catalyst. Macromol Res 2006;14:510–516.
95. Mehta R, Kumar V, Bhunia H, Upadhyay SN. Synthesis of poly(Lactic Acid): a review. J Macromol Sci Polym Rev 2005;45:325–349.
96. Williams CK, Hillmyer MA. Polymers from renewable resources: a perspective for a special issue of polymer reviews. Polymer Rev 2008;48:1–10.

97. Spassky N, Wisniewski M, Pluta C, Borgne AL. Highly stereoselective polymerization of Rac-(D,L)-lactide with a chiral schiff's Base/Aluminium alkoxide initiator. Macromol Chem Phys 1996;197:2627–2637.
98. Hormnirun P, Marshall EL, Gibson VC, White AJP, Williams DJ. Remarkable stereocontrol in the polymerization of racemic lactide using aluminum initiators supported by tetradentate aminophenoxide ligands. J Am Chem Soc 2004;126:2688–2689.
99. Tsuji H. Poly(Lactide) stereocomplexes: formation, structure, properties, degradation, and applications. Macromol Biosci 2005;5:569–597.
100. Zhang D, Hillmyer MA, Tolman WB. Catalytic polymerization of a cyclic ester derived from a "Cool" natural precursor. Biomacromolecules 2005;6:2091–2095.
101. Wanamaker CL, Bluemle MJ, Pitet LM, Leary LE, Tolman WB, Hillmyer MA. Consequences of polylactide stereochemistry on the properties of polylactide-polymenthide-polylactide thermoplastic elastomers. Biomacromolecules 2009;10:2904–2911.
102. Taguchi S, Yamada M, Matsumoto Ki, Tajima K, Satoh Y, Munekata M, Ohno K, Kohda K, Shimamura T, Kambe H, Obata S. A microbial factory for lactate-based polyesters using a lactate-polymerizing enzyme. Proc Natl Acad Sci USA 2008;105:17323–17327.
103. Yamada M, Matsumoto Ki, Nakai T, Taguchi S. Microbial production of lactate-enriched poly[(R)-lactate-Co-(R)-3-Hydroxybutyrate] with novel thermal properties. Biomacromolecules 2009;10:677–681.
104. Jung YK, Kim TY, Park SJ, Lee SY. Metabolic engineering of *Escherichia Coli* for the production of polylactic acid and its copolymers. Biotechnol Bioeng 2010;105:161–171.
105. Yang TH, Kim TW, Kang HO, Lee S-H, Lee EJ, Lim S-C, Oh SO, Song A-J, Park SJ, Lee SY. Biosynthesis of polylactic acid and its copolymers using evolved propionate CoA transferase and pha synthase. Biotechnol Bioeng 2010;105:150–160.
106. Gerngross TU, Martin DP. Enzyme-catalyzed synthesis of poly[(R)-(-)-3-Hydroxybutyrate]: formation of macroscopic granules *in Vitro*. Proc Natl Acad Sci USA 1995;92:6279–6283.
107. Satoh Y, Murakami F, Tajima K, Munekata M. Enzymatic synthesis of poly(3-hydroxybutyrate-co-4-hydroxybutyrate) with CoA recycling using polyhydroxyalkanoate synthase and Acyl-CoA synthetase. J Biosci Bioeng 2005;99:508–511.
108. Jaipuri FA, Bower BD, Pohl NL. Protic acid-catalyzed polymerization of ß-lactones for the synthesis of chiral polyesters. Tetrahedron: Asymmetry 2003;14:3249–3252.
109. Tokiwa Y, Ugwu CU. Biotechnological production of (R)-3-hydroxybutyric acid monomer. J Biotechnol 2007;132:264–272.
110. Matsumura S, Suzuki Y, Tsukada K, Toshima K, Doi Y, Kasuya K-I. Lipase-catalyzed ring-opening polymerization of ß-butyrolactone to the cyclic and linear poly(3-Hydroxybutyrate). Macromolecules 1998;31:6444–6449.
111. Gorke JT, Okrasa K, Louwagie A, Kazlauskas RJ, Srienc F. Enzymatic synthesis of poly(Hydroxyalkanoates) in ionic liquids. J Biotechnol 2007;132:306–313.
112. Auras R, Harte B, Selke S. An overview of polylactides as packaging materials. Macromol Biosci 2004;4:835–864.
113. Seebach DB, Albert K, Breitschuh Richard, Job Kurt. Direct degradation of the biopolymer poly[(R)-3-Hydroxybutyric Acid] to (R)-3-hydroxybutanoic acid and its methyl ester. Org Synth 1993;71:39–47.

114. Lee SY, Lee Y, Wang F. Chiral compounds from bacterial polyesters: sugars to plastics to fine chemicals. Biotechnol Bioeng 1999;65:363–368.
115. Ren Q, Grubelnik A, Hoerler M, Ruth K, Hartmann R, Felber H, Zinn M. Bacterial poly(Hydroxyalkanoates) as a source of chiral hydroxyalkanoic acids. Biomacromolecules 2005;6:2290–2298.
116. Misra SK, Valappil SP, Roy I, Boccaccini AR. Polyhydroxyalkanoate (Pha)/Inorganic phase composites for tissue engineering applications. Biomacromolecules 2006;7:2249–2258.
117. Dennis D, Liebig C, Holley T, Thomas KS, Khosla A, Wilson D, Augustine B. Preliminary analysis of polyhydroxyalkanoate inclusions using atomic force microscopy. FEMS Microbiol Lett 2003;226:113–119.
118. Russell RA, Holden PJ, Garvey CJ, Wilde KL, Hammerton KM, Foster LJ. Investigation of the phase morphology of bacterial Pha inclusion bodies by contrast variation sans. Phys Rev B Condens Matter 2006;385–386:859–861.
119. Jendrossek D. Polyhydroxyalkanoate granules are complex subcellular organelles (Carbonosomes). J Bacteriol 2009;191:3195–3202.
120. Banki MR, Gerngross TU, Wood DW. Novel and economical purification of recombinant proteins: intein-mediated protein purification using *in vivo* polyhydroxybutyrate (Phb) matrix association. Protein Sci 2005;14:1387–1395.
121. Barnard GC, McCool JD, Wood DW, Gerngross TU. Integrated recombinant protein expression and purification platform based on *Ralstonia eutropha*. Appl Environ Microbiol 2005;71:5735–5742.
122. Wang Z, Wu H, Chen J, Zhang J, Yao Y, Chen G-Q. A novel self-cleaving phasin tag for purification of recombinant proteins based on hydrophobic polyhydroxyalkanoate nanoparticles. Lab Chip 2008;8:1957–1962.
123. Atwood J, Rehm B. Protein engineering towards biotechnological production of bifunctional polyester beads. Biotechnol Lett 2009;31:131–137.
124. Yao Y-C, Zhan X-Y, Zhang J, Zou X-H, Wang Z-H, Xiong Y-C, Chen J, Chen G-Q. A specific drug targeting system based on polyhydroxyalkanoate granule binding protein phap fused with targeted cell ligands. Biomaterials 2008;29:4823–4830.
125. de Koning GJM, Lemstra PJ. The amorphous state of bacterial poly[(R)-3-Hydroxyalkanoate] *in Vivo*. Polymer 1992;33:3292–3294.
126. Noel BJ. Process for preparing poly-beta-hydroxybutyric acid. US 3044942. 1962.
127. Lafferty RM, Heinzle E. Cyclic carbonic acid esters as solvents for poly-(ß-hydroxybutyric acid. US 4101533. 1978.
128. McChalicher CWJ, Srienc F, Rouse DP. Solubility and degradation of polyhydroxyalkanoate biopolymers in propylene carbonate. AIChE J 2010;56:1616–1625.
129. Traussnig H, Kloimstein E, Kroath H, Estermann R. Extracting agents for poly-D(-)-3-hydroxybutyric acid. US 4968611. 1990.
130. Hanggi UJ. Pilot scale production of P(3hb) with Alcaligenes latus. In: Dawes EA, editor. Novel biodegradable microbial polymers. Dordrecht, Netherlands: Kluwer; 1990. pp. 65–70.
131. Zinn M, Weilenmann H-U, Hany R, Schmid M, Egli T. Tailored synthesis of poly([R]-3-Hydroxybutyrate-Co-3-Hydroxyvalerate) (Phb/Hv) in *Ralstonia eutropha* Dsm 428. Acta Biotechnol 2003;23:309–316.
132. Jacquel N, Lo C-W, Wei Y-H, Wu H-S, Wang SS. Isolation and purification of bacterial poly(3-hydroxyalkanoates). Biochem Eng J 2008;39:15–27.

133. Lee KM, Chang HN, Chang YK, Kim BS, Hahn SK. The lysis of gram-negative *Alcaligenes eutrophus* and *Alcaligenes latus* by palmitoyl carnitine. Biotechnol Tech 1993;7:295–300.

134. Kim M, Cho K-S, Ryu HW, Lee EG, Chang YK. Recovery of poly(3-Hydroxybutyrate) from high cell density culture of *Ralstonia eutropha* by direct addition of sodium dodecyl sulfate. Biotechnol Lett 2003;25:55–59.

135. Berger E, Ramsay BA, Ramsay JA, Chavarie C, Braunegg G. Phb recovery by hypochlorite digestion of non-Phb biomass. Biotechnol Tech 1989;3:227–232.

136. Holmes PA, Lim GB. Separation process. US 4910145. 1990.

137. de Koning GJM, Witholt B. A process for the recovery of poly(Hydroxyalkanoates) from pseudomonads Part 1: solubilization. Bioprocess Biosyst Eng 1997;17:7–13.

138. Kapritchkoff FM, Viotti AP, Alli RCP, Zuccolo M, Pradella JGC, Maiorano AE, Miranda EA, Bonomi A. Enzymatic recovery and purification of polyhydroxybutyrate produced by *Ralstonia eutropha*. J Biotechnol 2006;122:453–462.

139. Yasotha K, Aroua MK, Ramachandran KB, Tan IKP. Recovery of medium-chain-length polyhydroxyalkanoates (Phas) through enzymatic digestion treatments and ultrafiltration. Biochem Eng J 2006;30:260–268.

140. Tamer IM, Moo-Young M, Chisti Y. Disruption of Alcaligenes latus for recovery of poly(ß-Hydroxybutyric Acid): comparison of high-pressure homogenization, bead milling, and chemically induced lysis. Ind Eng Chem Res 1998;37:1807–1814.

141. Ghatnekar MS, Pai JS, Ganesh M. Production and recovery of poly-3-hydroxybutyrate from methylobacterium Sp V49. J Chem Technol Biotechnol 2002;77:444–448.

142. Hwang K-J, You SF, Don TM. Disruption kinetics of bacterial cells during purification of Poly-.Beta.-hydroxyalkanoate using ultrasonication. J Chin Inst Chem Eng 2006;37:209–216.

143. Resch S, Gruber K, Wanner G, Slater S, Dennis D, Lubitz W. Aqueous release and purification of poly(ß-Hydroxybutyrate) from *Escherichia Coli*. J Biotechnol 1998;65:173–182.

CHAPTER 14

ENZYMATIC DEGRADATION OF LIGNOCELLULOSIC BIOMASS

FENG XU
Novozymes Inc., Davis, California

14.1 INTRODUCTION

Entering the twenty-first century, sustainability of our natural resources is recognized as vital for the future of humanity. Biologically derivable (thus renewable) materials are being aggressively pursued as potential replacers of fossil-based chemicals. Among various potential bio-feedstocks, lignocellulosic materials, which account for about a half of the photosynthesized matters in the world, are of particular interest because of their rich energy content, diverse chemical architecture, decisive involvement in natural carbon cycling, and vast availability scale. This is exemplified by the development of the so-called cellulosic bioethanol as transportation fuel. Although tens of billions of liters are now produced annually from sugary/starchy biomaterials, often, more are needed to be produced from lignocellulosic biomaterials to meet the demand for the coming decades [1–4]. Lignocellulosic feedstocks may come from dedicated forest or crops, agricultural residues, forestry by-products, low-value plants, or urban wastes. The industrial processes relying on enzymatic conversion of lignocellulosic feedstocks for the production of valuable chemicals, are not only attractive in terms of consumption of energy and impact on the environment but also feasible because many microorganisms can effectively degrade lignocellulosic substances with their enzymes [5,6].

In this chapter, the basic aspects of enzymatic lignocellulose degradation are introduced. Because of space limit, only some of the recent reviews and representative reports are cited. These and other uncited references, as well as the *c*arbohydrate-*a*ctive en*z*ymes (CAZy) (http://www.cazy.org/), *f*ungal

Biocatalysis for Green Chemistry and Chemical Process Development, First Edition.
Edited by Junhua (Alex) Tao and Romas Kazlauskas.

*o*xidative *l*ignin enzymes (FOLy) (http://foly.esil.univ-mrs.fr/), and BRENDA (http://www.brenda-enzymes.info/) databases, may be consulted for more authoritative and comprehensive information on biomass-active enzymes.

14.2 COMPOSITION AND STRUCTURE OF LIGNOCELLULOSIC MATERIALS

Natural lignocelluloses are found mostly in plant cell wall substances or biomass. Being very complex and heterogeneous, biomass consists of three major components: cellulose, hemicelluloses, and lignin, plus minor components, including proteins, lipids, and other substances. These components interact at intra-, inter-, and supramolecular levels, making them physically entangled or chemically linked [7,8]. To support the structure and shape of plants as well as to defend against pathogens, lignocellulose is not only physically sturdy but also chemically recalcitrant, made possible by its high compositional and structural heterogeneity as well as low porosity and accessibility for water and exogenous biomolecules (e.g., enzymes).

Cellulose, which may account for ~40–55% of a plant's dry weight, is a linear polysaccharide simple in composition and primary structure but complex in conformation and morphology. Individual cellulose chains comprise repeating cellobiosyl (CB) units linked by $\beta(1 \rightarrow 4)$ glycosidic bond (Figure 14.1). With four or more CB units, cellulose chains become insoluble in water. Via H-bonding, cellulose chains can bundle together to form either amorphous or crystalline (monoclinic I_β- or triclinic I_α-type) microfibrils. In biomass, cellulose microfibrils are often embedded in or entangled with complex hemicelluloses and lignin. Glucose (Glc), the sole basic component of cellulose, can serve as not only a feed of fermentative microorganisms but also a platform molecule of chemoenzymatic syntheses, for making valuable chemicals.

Hemicellulose, which may account for ~25–40% of a plant's dry weight, is a group of often branched polysaccharides complex in composition and linkage. Major hemicelluloses include arabino- or glucuronoxylan ($\beta(1 \rightarrow 4)$ linked D-xylopyranosyl (Xyl) polysaccharides with different O-substitutions by acetyl, D-glucuronoyl (GlcU), or α-L-arabinofuranosyl (Ara) groups which may

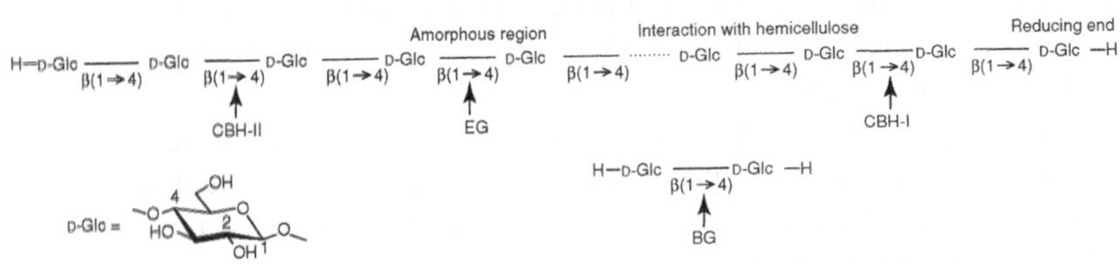

Figure 14.1 Cellulose and action of cellulase. The simplified scheme depicts the main features of cellulose, including composition (D-Glc), backbone glycosidic bonds ($\beta(1 \rightarrow 4)$), and morphology. Arrows show representative glycosidic bond targeting by major cellulases.

be esterified by hydroxycinnamoyl), galacto- or glucomannan (β(1 → 4)-D-mannosyl or co-present Glc polysaccharides with variable α(1 → 6) D-galactosyl (Gal) side chains as well as O2 and/or O3 acetylations), xyloglucan (β(1 → 4) glucan substituted at O6 by α(1 → 6)-linked Xyl modified (at O2) by either α(1 → 2) Ara or β(1 → 2) Gal (partially acetylated or substituted by α(1 → 2) L-fucopyranosyl (Fuc) at O2)), noncellulose β-glucan (β(1 → 3)-, (1 → 4)-, or (1 → 6)-linked Glc polymer), polygalacturonan (α(1 → 4)-linked galacturonoyl (GalU), often embedded with α(1 → 2)-linked rhamnosyl (Rha), polysaccharide acids, with variable methylation/acetylation on backbone GalU as well as carbohydrate side chains (containing mainly Ara and Gal) linked to backbone Rha), arabinan, and galactan (Figure 14.2). In softwood, hardwood, and grass,

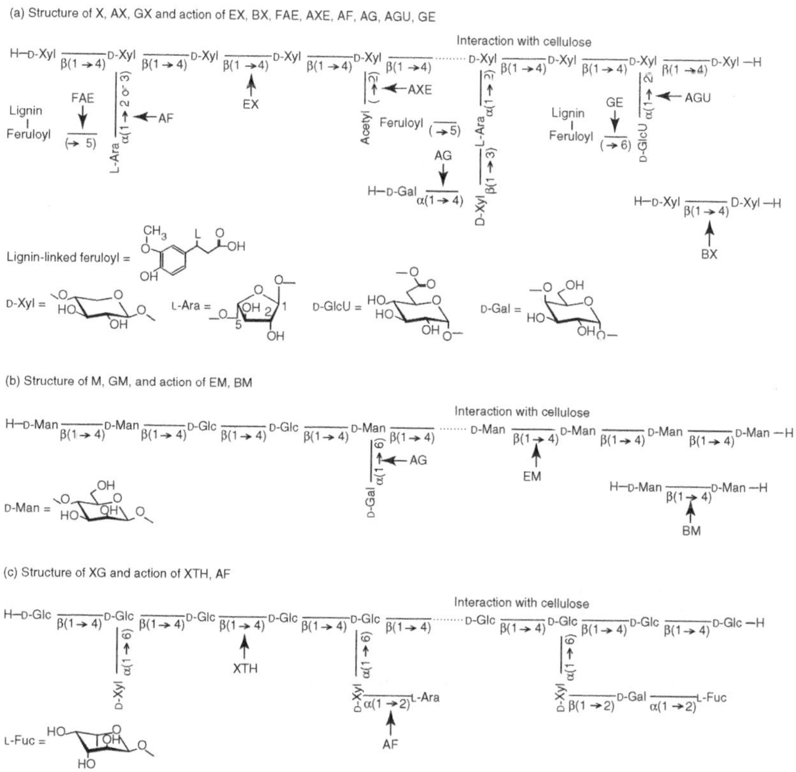

Figure 14.2 Hemicelluloses and action of hemicellulases. The simplified schemes depict the main features of hemicelluloses, including compositions (D-Xyl, L-Ara, etc.), backbone glycosidic bonds (β(1 → 4), α(1 → 2), etc.), and side chains. Arrows show representative glycosidic bond targeting by major hemicellulases. (a) (Arabino or glucurono)xylan and xylan-active enzymes. (b) (Galacto(gluco))mannan and mannan-active enzymes. (c) Xyloglucan and xyloglucan-active enzymes. (d) (Noncellulose) β-glucan and β-glucan-active enzymes. (e) Poly(rhamno)galacturonan and poly(rhamno)galacturonan-active enzymes.

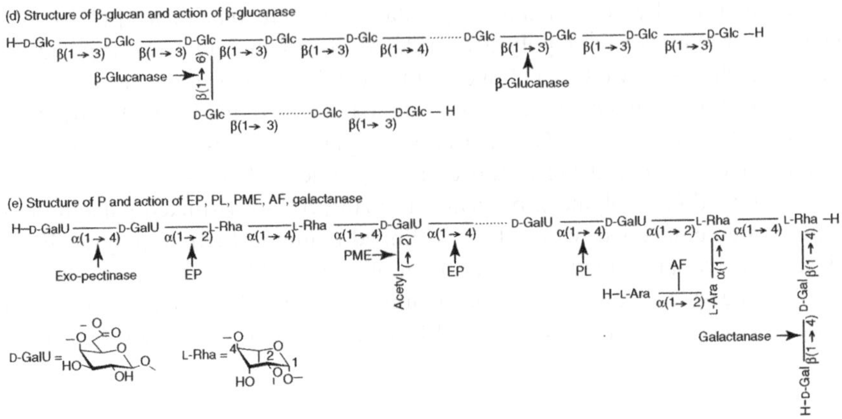

Figure 14.2 (*Continued*)

acetylated (galacto)glucomannan, glucuronoxylan, and arabinoxylan constitute the principal hemicelluloses, respectively. Hemicellulose contains many pentoses, hexoses, and derivatives useful for industrial microbial fermentation or synthesis.

Lignin, which may account for ~20–30% of a plant's dry weight, is a group of branched aromatic polymers highly complex and heterogeneous in configuration. Synthesized by radical-driven polymerization reactions in plant, different phenylpropanoids, including coumaryl (mainly in softwood), coniferyl, and sinapyl alcohols (mainly in hardwood), are fused into lignin with diverse C–C and C–O ether bonds (Figure 14.3). Lignin may be covalently cross-linked (via hydroxycinnamoyls) to hemicellulose and cover cellulose (leading to limited exposure of the latter). Many aromatic compounds may be derived from lignin and replace their petrochemical counterparts for various industrial uses.

14.3 DEGRADATION, MODIFICATION, OR CONVERSION OF LIGNOCELLULOSE

14.3.1 Natural Lignocellulose Degradation

Many archaea, bacteria, and fungi, as well as some protozoa, nematodes, insects, crustaceans, mollusks, and herbivores (aided by symbiotic microbes), can break down biomass with elaborate enzymatic machineries. Cellulose and hemicellulose are degraded directly by (hydro)lytic enzymes (cellulase, hemicellulase, esterase, lyase), which may be extracellular and freely dissociated, as in the case of cellulolytic aerobic fungi, or cell-bound and supramolecularly assembled (into cellulosomes), as in the case of many cellulolytic anaerobic bacteria, or a combination of intracellular and extracellular ones, as in the case of cellulolytic aerobic bacteria. The glycoside hydrolase (GHs) and lyases cleave various backbone or side chain glycosidic bonds, while esterases cleave various ester substituents in

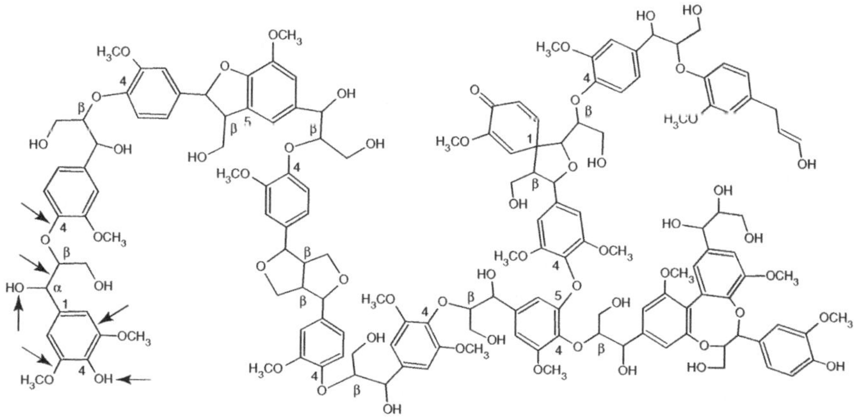

Figure 14.3 Lignin and action of delignifying enzymes. Depicting the main acknowledged features of hardwood and softwood lignins [9], the simplified and hypothetic structure contains the following compositions and linkages: (from left) syringyl –(β-*O*-4)– guaiacyl –(β-*O*-4)– guaiacyl –(β-5)– phenyl coumaran unit –(β-*O*-4)– resinol unit (β–β internal link) –(4-*O*-β)– syringyl –(4-*O*-β)–biphenyl ether unit (5-*O*-4 internal link), branch point; right branch:–(4-*O*-β)– dibenzodioxocin (5–5 internal link) with a glycerol end; top branch: –(β-*O*-1)– spirodienone –(β-*O*-4)– guaiacyl with a cinnamyl alcohol end. Arrows show representative functional group and linkage targeting by major delignifying enzymes (at the left region only).

hemicellulose (Figures 14.1 and 14.2). Cellulose may also be degraded by nonspecific attacks from radicals, exemplified by the action of brown rot fungi which use H_2O_2-generating enzymes [10,11].

Lignin is degraded either directly or indirectly by an array of oxidative enzymes produced in saprophytic microbes, particularly fungi. White rots produce peroxidase and laccase capable of directly and indirectly oxidizing lignin, respectively, as well as various oxidases capable of generating H_2O_2 to support the peroxidase. Indirect enzymatic delignification is accomplished with certain metabolites or lignin-degradation products serving as redox-active mediators. During lignin modification/degradation, C_α aliphatic side chains, C_α–C_β bonds, aryl methoxy substituents, and phenoxyl groups are preferentially cleaved or modified (Figure 14.3). Brown rots apparently lack delignifying peroxidase and laccase, but may demethoxylate lignin to yield methanol for H_2O_2 production. Some soft-rots, aquatic, and symbiotic microbiota in insects and ruminants can also have delignification capability [10].

14.3.2 Artificial Lignocellulose Degradation

For many lignocellulose degradation applications, enzymes are preferred as catalysts over either live microbes or chemicals. Advantages of an enzyme-based technology include high specificity, ambient conditions, inexpensive solvent and reactor, low pollution, and competitive cost.

Owing to the complexity and heterogeneity of biomass feedstocks, it is desirable to first separate cellulose, hemicellulose, lignin, or other constituents, before subjecting them to different enzymes. This could enhance the efficiency of biomass degradation, by maximizing the specificity/activity of different enzymes (which may require different pH, temperature, and other conditions) and minimizing cross-inhibition of the enzymes by nonsubstrate biomass components or degradation products. Various combinations of different physicochemical pretreatments, enzymatic hydrolyses, and microbial fermentations are being intensively pursued at present, aimed at removing hemicellulose and/or lignin from or weakening their holds on cellulose, so that cellulose could be effectively converted to Glc by cellulases [3,4,12].

Industrial lignocellulose degradations (such as those already practiced in commercial pulping and being developed for cellulosic bioethanol production) are, in general, very different from their natural counterparts, in terms of substrate's form, substrate-to-enzyme stoichiometry, water content, reaction time, pH, temperature, presence of deactivators, and so on. Some cellulolytic and ligninolytic enzymes may be applied without major structural changes, while others may need protein engineering to make them (more) active under selected, nonnative conditions. Being either wild type or engineered, enzymes active in degrading lignocellulose need to be thoroughly investigated with regard to their structure, activity, specificity, and stability, as well as their interdependence on or cooperation with other enzymes, so that their potential as industrial biocatalysts for biomass conversion can be assessed, explored, and developed.

14.4 LIGNOCELLULOSE-DEGRADING ENZYMES: AN OVERVIEW

14.4.1 Classifications

In general, biomass-active enzymes are hydrolytic, lytic, or oxidoreductive (Table 14.1). On the basis of the main reactions that they catalyze, these enzymes belong to the EC 3.2.1 glycosidases (including traditionally called cellulases and *hemicellulases*), EC 4.2.2 lyases, EC 3.1.1 esterases, EC 1.11.1 peroxidases, EC 1.1.3, and EC 1.10.3 oxidases, and EC 1.1.99 dehydrogenases sub-subclasses from the enzyme nomenclature and classification (http://www.chem.qmul.ac.uk/iubmb/enzyme/). Detailed enzymological information of these enzymes can be found in the BRENDA database (http://www.brenda-enzymes.info/). Although it may allow preliminary selection of lignocellulose-degrading enzymes based on targeted biomass feedstocks, the EC classification overlooks structural information, which is vital for the present and future enzyme technology development.

The ever-expanding genetic and protein structural knowledge has enabled sequence-/structure-based enzyme classification. For the (hydro)lytic enzymes active in lignocellulose degradation, the CAZy classification, based on sequence and three-dimensional structure homology, groups GHs, polysaccharide lyase (PL), carbohydrate esterase (CEs), and carbohydrate-binding module (CBMs) into different families. Members of the same family have the same catalytic

TABLE 14.1 Representative Major Lignocellulose-Degrading Enzymes

	Classification			Typical Substrate		
Enzyme	EC	CAZy or FOLy	Mechanistic Characteristics	Main	Side	Typical Source
CBH-I	3.2.1.-	GH7	Retaining, processive, reducing end-specific	C	Chitin	*Hypocrea jecorina*
CBH-II	3.2.1.91	GH6	Inverting, processive, nonreducing end-specific	C	Chitin	*H. jecorina*
CBH	3.2.1.91	GH48	Inverting, processive, nonreducing end-specific	C		*Cellulomonas fimi*
EG-I	3.2.1.4	GH7	Retaining	C(a)	X, XG, AX	*H. jecorina*
EG-II	3.2.1.4	GH5	Retaining	C(a)	M, GM	*H. jecorina*
EG-III	3.2.1.4	GH12	Retaining	C(a)	XG	*H. jecorina*
EG-V	3.2.1.4	GH45	Inverting	C(a)		*Humicola insolens*
EG	3.2.1.4	GH9	Inverting, processive, nonreducing end-specific	C	Chitosan	*Clostridium thermocellum*
EG	3.2.1.-	GH48	Inverting, processive, reducing end-specific	C		*Clostridium cellulolyticum*
BG	3.2.1.21	GH3	Retaining, nonreducing end-specific	Cellodextrin		*Aspergillus niger*
EX-III	3.2.1.8	GH10	Retaining	X, AX, GX	C(a)	*H. jecorina*
EX-I	3.2.1.8	GH11	Retaining	X, AX, GX	C(a)	*H. jecorina*
BX-I	3.2.1.37	GH3	Retaining	Xylooligo	Ara_f-side	*H. jecorina*
EM	3.2.1.78	GH5	Retaining	M, GM	C(a)	*H. jecorina*
AF	3.2.1.55	GH51	Retaining	Ara_f-side, AX		*C. thermocellum*
AGU	3.2.1.139	GH67	Inverting	GX		*Aspergillus nidulans*
AXE	3.1.1.72	CE2	Ser-His-Asp catalytic triad	X, AX, GX		*C. thermocellum*

(*continued overleaf*)

TABLE 14.1 (*Continued*)

Enzyme	Classification: EC	Classification: CAZy or FOLy	Mechanistic Characteristics	Typical Substrate: Main	Typical Substrate: Side	Typical Source
FAE	3.1.1.73	CE1	Ser-His-Asp catalytic triad	Fer ester, AX		*A. niger*
XTH	3.2.1.151	GH74	Inverting	XG	C	*H. jecorina*
AG	3.2.1.22	GH4	Retaining, redox, NAD^+-dependent	GM, P		*Bacillus subtilis*
EP	3.2.1.15	GH28	Inverting	P		*A. niger*
PL	4.2.2.2	PL1	β-Elimination	P		*A. nidulans*
PME	3.1.1.11	CE28	Asp-Asp-Arg catalytic triad	Pectin		*Erwinia chrysanthemi*
LiP	1.11.1.14	LO2	Heme-containing, surface cocatalytic Trp, high E^o	L	Polycyclic aromatic	*Phanerochaete chrysosporium*
MnP	1.11.1.13	LO2	Heme-containing, Mn(II)-dependent	L		*P. chrysosporium*
VP	1.11.1.16	LO2	Heme-containing, surface catalytic Trp and Mn(II)	L		*Pleurotus eryngii*
Lac	1.10.3.2	LO1	Cu-containing, mediator-dependent, variable E^o	Phenol, L		*Trametes versicolor*
CDH	1.11.99.18	LO3	Flavin/heme-containing	Cellodextrin		*H. insolens*
GO	1.1.3.4	LDA6	Flavin-containing	Glucose		*A. niger*
AAO	1.1.3.7	LDA1	Flavin-containing	Aryl alcohol		*P. eryngii*

Abbreviation: CBH, cellobiohydrolase; EG, endo-1,4-β-glucanase; BG, β-glucosidase; EX, endo-β-1,4-xylanase; BX, β-xylosidase; EM, endo-1,4-β-mananase; AF, α-L-arabinofuranosidase; AGU, α-glucuronidase; AXE, acetyl xylan esterase; FAE, feruloyl esterase; XTH, xyloglucan transferase/hydrolase; AG, α-galactosidase; EP, endo-polygalacturonan hydrolase; PL, polygalacturonan lyase; PME, pectin methyl esterase; LiP, lignin peroxidase; MnP, Mn peroxidase; VP, versatile peroxidase; Lac, laccase; CDH, cellobiose dehydrogenase; GO, glucose oxidase; AAO, aryl-alcohol oxidase; C, cellulose; C(a), amorphous cellulose; X, xylan; XG, xyloglucan; AX, arabinoxylan; GX, glucuronoxylan; M, mannan; GM, galacto(gluco)mannan; Ara_f-side, α-L-arabinofuranoside; Fer, feruloyl; P, polygalacturonan; L, lignin or (non)phenolic models.

mechanism for their function. Different families, with different structures, may carry out the same catalysis. The classification can allow bioinformatic mining for potential biomass-active enzymes. For the oxidoreductive (or ligninolytic) enzymes active in lignocellulose degradation, the FOLy classification groups lignin oxidase (LOs), lignin degradation auxiliary enzyme (LDAs), and lignin degradation auxiliary esterase (ESTs) into different families.

14.4.2 Architecture and Enzymology

Many glycosidases are modular, comprising one or more catalytic modules, CBMs, dockerins, and other functionally known or unknown domains, whose individual functions contribute to the overall enzymatic activity [2,4,11]. Many anaerobic bacteria assemble their biomass-active enzymes into supramolecular cellulosomes, in which scaffolding molecules not only anchor an array of different enzymes (via cohesin–dockerin interaction) but also bind to targeted lignocellulose (via CBM) as well as host cells (via surface-layer-homolog proteins). It is postulated that in a cellulosome, the stoichiometry and vicinity of different but synergistic/interdependent enzymes, as well as their mobility to map topological features of a substrate, may render the enzyme assembly highly efficient in degrading biomass substances [13,14].

Biomass-active GHs generally employ two acidic amino acid residues (Glu or Asp) for their catalysis, one as proton donor (or general acid) and the other as nucleophile (or base). Two glycosidic bond-hydrolyzing mechanisms have been elucidated on the basis of the stereochemistry [15]. One mechanism retains the configuration of the targeted anomeric C after a double-displacement reaction, while the other inverts the configuration after a single-displacement reaction. The "retaining" feature enables many GHs acting as transglycosylases, a property that might affect the extent of lignocellulose degradation when the product concentration is high.

Compared to many "conventional" enzymes, biomass-active enzymes often possess four special properties. The first one is the existence of multiple subsites in these enzymes' active site, which can result in requirement of optimal enzyme–substrate interaction at the "binding" subsites for an effective catalysis at the catalytic subsite, as well as vulnerability toward substrate or product inhibition (caused by undesirable enzyme–substrate or product binding). The second one is the heterogeneous and even fractal nature of these enzymes' catalysis, imposed by their substrate's insolubility and reduced spatial degrees of freedom when they act on the surface or inside the pores of an insoluble/polymeric substrate. The third one is the synergism among many of these enzymes, because they may act on different regions of a polymeric substrate or on lignocellulosic substances inhibitory to the cooperating enzymes. The fourth one is the indispensability of the targeting, binding, or mediating functions, provided by CBM (for (hemi)cellulolytic hydrolases) or endogenous redox-active amino acid residues (for delignifying peroxidases), to these enzymes' optimal lignocellulose-degrading activity.

14.5 CELLULASES AND CELLULOSE DEGRADATION

Cellulose-hydrolyzing cellulases (Cels) possess catalytic modules belonging to ~14 GH families and CBMs to ~15 CBM families. Functionally, cellulases include cellobiohydrolase (CBH), endo-1,4-β-D-glucanase (EG), and β-glucosidase (BG) (Table 14.1), whose different substrate specificity allow them to act synergistically on cellulose. Most cellulases are active at pH ~4–8 and temperature ~20–60°C, but some cellulases from extremophiles can extend their activity/stability far beyond this pH and temperature range.

Besides CBH, EG, and BG, other lesser known cellulolytic enzymes include cellobiose phosphorylase (EC 2.4.1.20, 2.4.1.49) and glucuronan lyase (EC 4.2.2.14). Cellobiose phosphorylase (belonging to the GH94 family) splits cellobiose by phosphorylating one Glc unit [16]. Glucuronan lyase (belonging to the PL20 family) degrades polyglucuronan by β-elimination, splitting a GlcU-β(1 → 4)-GlcU unit into a GlcU and a Δ4: 5 unsaturated GlcU units [17]. The potential of these enzymes as biocatalysts for industrial biomass conversion has yet to be fully explored.

14.5.1 Cellobiohydrolases and Hydrolysis of both Crystalline and Amorphous Cellulose

Among cellulases, CBHs are special because of their ability to hydrolyze crystalline cellulose, which is recalcitrant to other cellulases. Although also found in the GH5, 9, and 74 families, GH6, 7 (fungal only), and 48 (mostly bacterial) CBHs are the most encountered. CBHs can account up to ~70% by weight of all secreted proteins by cellulolytic fungi. CBH may possess one or more catalytic module, CBM (critical for CBH's action on crystalline cellulose), or other domains. Having its active site (with ~5–10 subsites) inside a tunnel, CBH hydrolyzes a cellulose chain in a "processive" way, by which the CBH first threads in the chain into its active site, then cuts off a CB, followed by gliding down the chain for the next CB [15]. Such "processive" dynamics (which may lead to premature detachment from or unproductive lock onto cellulose chain) limits CBH's degree of freedom, resulting in non-Michaelis–Menten or even fractal kinetics [18,19]. Like other GHs, CBHs bind and activate cellulose through H-bonding (exerted by polar amino acid residues) and π-stacking (exerted by aromatic amino acid residues such as Trp).

On the basis of their specificity, there are two kinds of CBH. CBH-I (EC 3.2.1.-, e.g., *Hypocrea jecorina* or *Trichoderma reesei* Cel7A) is specific to the "reducing (hemi-acetal) end" of a cellulose chain, and CBH-II (EC 3.2.1.91, e.g., *H. jecorina* Cel6A) is specific to the opposing "nonreducing end" (Figure 14.1). This difference in specificity makes CBH-I and CBH-II synergistic in hydrolyzing cellulose, and is likely the reason why many cellulolytic microbes produce both CBHs [4,11].

Kinetically, CBHs are slow-acting enzymes on insoluble cellulose. In general, CBH has low, if any, activity on modified/substituted celluloses and noncellulose polysaccharides, likely due to unfavorable steric hindrance and/or H-bond disruption at its active site. Nevertheless, it has been reported that Cel6, Cel7, and Cel48 CBHs may act on chitin [11], and Cel74 CBH may act on oligoxyloglucan [20].

Among various CBH deactivations, cellobiose (product) inhibition and unproductive adsorption unto lignocellulose (substrate or by-product) may be the most relevant in industrial application [21,22]. They could be mitigated by supplementing BG and lignin removal or CBH engineering (to reduce its lignin affinity), respectively.

14.5.2 Endo-1,4-β-D-Glucanases and Hydrolysis of Amorphous Cellulose

Unlike CBHs, EGs (EC 3.2.1.4), in general, hydrolyze internal $\beta(1 \rightarrow 4)$ glycosidic bonds, using their groove-shaped active sites (with up to six subsites). Although also found in GH8, 44, 48, 51, and 74, EGs whose catalytic modules belong to the GH5–7, 9 (mostly plants), 12, and 45 families are the most encountered. EGs can account up to 20% of all secreted proteins of cellulolytic fungi. These cellulases may possess one or more catalytic module, CBM, and other domains. Although most EGs act on amorphous cellulose in an "on-off" manner (Figure 14.1), several Cel5, 9, and 48 EGs apparently hydrolyze crystalline cellulose "processively," a property attributed to their specific CBM [11,23]. For CBH-lacking brown rots, having processive EG is considered key for their cellulolytic activity [24,25].

The difference between EG and CBH's specificity makes these cellulases highly synergistic, because EG may generate new reducing and nonreducing ends for CBHs to attack, and CBH may degrade the outer layer of cellulose microfibril so that inner cellulose parts can be exposed to EG [4,11]. Many EGs have side activity on noncellulose polysaccharides, such as the case of Cel5, 7, 8, 9, 12, and 74 EGs on mannan, xylan, chitosan, chitin, β-glucan, and xyloglucan, respectively [26]. Cellulolytic microbes secrete more than one EG, probably taking advantage of their differential specificity and side activity, so that more effective hydrolysis of a heterogeneous biomass substance may be achieved.

14.5.3 β-D-Glucosidases and Mitigation of CBH and EG's Product Inhibition

Unlike CBH and EG, BG (EC 3.2.1.21) carries out only a homogeneous catalysis of hydrolyzing cellobiose (or longer but soluble cellodextrins) into Glc. Although also found in GH9, BGs from GH1 and 3 families are the most encountered. While aerobic fungi produce extracellular BGs (account for ~1% of secreted proteins), the BGs of anaerobic bacteria, in general, are cytoplasmic.

Having a pocket-shaped active site, BG binds the nonreducing Glc unit and tolerates steric and electronic variations at the "aglycon" part in clipping it off from cellobiose [27]. BG's activity on cellobiose is crucial in a "complete" cellulolytic enzyme system, because it mitigates CBH and EG's product inhibition.

Of various deactivations, BG's inhibition by Glc (product), transglycosylation side activity at high Glc concentration, and adsorption toward lignocellulose may be the most important in industrial enzymatic biomass conversion. The product inhibition and transglycosylation might be mitigated by Glc removal, such as that in a simultaneous saccharification and fermentation process.

14.6 HEMICELLULASES AND HEMICELLULOSE DEGRADATION

Hemicellulases possess catalytic modules belonging to ~29 GH, 5 PL, and 9 CE families as well as CBMs to ~37 CBM families. Functionally, hemicellulases include hydrolases, esterases, and lyases, whose different substrate specificity allows them to act synergistically among them or with cellulase when degrading biomass feedstock, as reflected by the coexpression of multiple hemicellulases and cellulases by biomass-active microbes [28,29]. Such synergism has been explored for the improvement of enzymatic biomass conversion [4,30,31]. Most hemicellulases are active at pH ~4–8 and temperature ~40–70°C, but some from extremophiles can significantly extend their activity/stability beyond this pH and temperature range.

14.6.1 Endo-β-xylanases, β-Xylosidase, and (Glucurono or Arabino)xylan Hydrolysis

Common xylanase (Xyns) may be divided into three kinds: endo-1,4-β-xylanase (EXs, EC 3.2.1.8) that cleave internal β(1 → 4) backbone bonds in xylan, endo-1,3-β-xylanases (EC 3.2.1.32) that cleave internal β(1 → 3) branch bonds in xylan, and β-xylosidase (BXs, EC 3.2.1.37) that degrade xylobiose or longer xylooligosaccharides into Xyl (Figure 14.2a). Although also found in the GH5, 8 (bacterial only), 26 (bacterial 1,3-EX only), and 43 families, GH10 and GH11 EXs are the most encountered (often accounting for $<1\%$ of secreted proteins by cellulolytic fungi). EX may possess one or more catalytic modules, with or without CBM and other domains. The active sites are cleft shaped with 4–7 subsites, and EXs bind and clip a segment of a xylan chain in an on-off manner. Xyn3 BX apparently occurs more than the Xyn39, 43, 52, and 54 counterparts. Although xylan is considered as ubiquitous as cellulose, natural occurrence of Xyn is much less diverse than cellulase in terms of structure. BXs have a pocket-like active site, and cleave Xyl off from the nonreducing end of xylooligosaccharides.

Xyn10 and Xyn11 enzymes may have different specificity. Compared to Xyn11s, Xyn10s often degrade xylan into shorter oligosaccharides, may act on β(1 → 3) branch, and have broader side activity on other polysaccharides (e.g., cellulose) [32,33]. This difference may enable Xyns to act in concert to better assist cellulases in degrading the cellulose in biomass. BXs can mitigate EX's product inhibition by splitting xylobiose, and may remove xylan's Ara side chain by using their side activity on α-L-arabinofuranoside. Many multifunctional enzymes contain catalytic modules of Xyn and other biomass-active enzymes, implying synergism between Xyn and these enzymes. Because of the co-presence of xylan with cellulose in many biomass feedstocks (particularly hardwood or grass derived), Xyns are often indispensable for an effective cellulolytic enzyme system [30,31,34].

Among various deactivations, the inhibition of EX by specific plant proteins (e.g., *Triticum aestivum* Xyn inhibitor-I and II [35]) might affect the enzyme's biomass conversion application.

14.6.2 Endo-β-Mannanase, β-Mannosidase, and (Galacto or Gluco)mannan Hydrolysis

Endo-mannanase (EMs, EC 3.2.1.78), able to cleave internal β(1 → 4) bonds in mannan (Figure 14.2b) [36], have catalytic modules belonging to the GH5, 26 (mostly bacterial), and 113 (mostly bacterial) families. EMs may possess one or more catalytic modules (whose active site can have up to six subsites), with or without CBM (often with dual specificity to mannan and cellulose) and other domains. β-Mannosidase (BMs, EC 3.2.1.25), able to cleave mannosyl units from nonreducing ends of mannooligosaccharides, belong to the GH1, 2, and 5 families.

Many biomass-degrading microbes produce EM and BM together with various other biomass-active enzymes [28]. Synergism between EM and EX, EG, BG, α-galactosidase (AG), and acetyl mannan esterase has been reported, in which EM's action on mannan and other enzymes' action on mannan's side chains (hindrance of EM's action) are believed to be beneficial for the overall hydrolysis of the targeted biomass feedstocks [36,37]. Considering the existence of (galacto)glucomannan in biomass stocks (particularly softwood derived), mannanases may play an important role for an effective biomass-converting enzymatic system.

14.6.3 α-L-Arabinofuranosidase, Endo-1,5-α-arabinanase, α-Glucuronidase, as well as Ara and GlcU Side Chain Removal in Hemicellulose

α-L-Arabinofuranosidase (AF, EC 3.2.1.55), whose catalytic modules belong to the GH3, 43, 51, 54, and 62 families, can cleave many hemicellulose side chains anchored by Ara α-glycosidic bonds (Figure 14.2a,c and e). Endo-1,5-α-arabinanase (EC 3.2.1.99), whose catalytic modules belong to the GH43 family, may also cleave off such side chains, in addition to internal α(1 → 5) glycosidic bonds in arabinan. The CBM found in many AFs may guide or anchor them onto targeted regions in a biomass substrate.

For arabinoxylan, different AFs may exert different specificity, depending on whether the Ara is linked to the O2 or 3 site of the Xyl unit [29,38]. Unlike their GH51 counterparts, GH54 AFs act on polymeric arabinoxylan, in addition to oligomeric arabinoxylan [39]. Being an exo–endo pair, AF and arabinase may synergistically degrade Ara side chains [40], thus facilitating other enzymes' (e.g., Xyn, pectinase) action on Ara-substituted hemicelluloses (e.g., xylan, pectin) [41]. α-Glucuronidase (AGUs, EC 3.2.1.139, 3.2.1.131), belonging to the GH4 and 67 (exclusively) families, can cleave off unsubstituted or 4-*O*-methylated GlcU side chains, linked by $\alpha(1 \rightarrow 2)$ bonds, in xylan or other hemicelluloses. Different AGUs may prefer glucuronated xylooligosaccharides, polymeric, or branched glucuronoxylan [33,42].

Being able to transform the Ara and GlcU side chains in biomass feedstocks (particularly hardwood- and grass derived), AF and AGU may contribute to enzymatic systems for industrial biomass conversion.

14.6.4 Acetyl Xylan Esterase, Feruloyl Esterase, Glucuronoyl Esterase, as well as Deacetylation, Deesterification, and Lignin Detaching in Hemicellulose

Acetyl xylan esterase (AXE, EC 3.1.1.72) can hydrolyze acetyl ester bonds (at the O2 or 3 site of backbone Xyl units), feruloyl esterase (FAE, EC 3.1.1.73) can hydrolyze the hydroxycinnamoyl (including feruloyl) ester bonds (at the side chain α-L-Ara's O2 or O5 or β-Gal's O6 site)s, and glucuronoyl esterase (GE, EC 3.1.1.-) can hydrolyze the methyl ester bonds (at the side chain methylated $\alpha(1 \rightarrow 2)$ GlcU's O6 site), in xylan or other hemicelluloses (Figure 14.2a). As the catalytic module is concerned, FAEs belong to the CE1 family, GEs to the CE15 family, and AXEs belong to the CE1, 2, 3, 4, 5, 6, 7, and 12 families. Many FAEs are modular and equipped with CBM. Similar to other esterases and serine proteases, many AXEs and FAEs in general use a Ser-His-Asp catalytic triad and "oxy anion hole" for their catalysis, although some AXEs and FAEs may also rely on Zn^{2+} or other metal ions. Different FAEs may exhibit different specificities depending how the target cinnamoyls are substituted (e.g., hydroxylation vs methoxylation) and/or linked to hemicellulose (e.g., Ara vs Gal ester bond onto xylan vs pectin) [43].

AXE, FAE, and GE may assist hemicellulases' action on complex hemicellulose, by removing various hemicellulose substituents (e.g., acetyl, methyl, feruloyl) that hinder the activity of the hemicellulases [31,34]. FAE and GE are postulated to be capable of cleaving the bond linking hemicellulose's hydroxycinnamoyl sites to lignin. For hardwood- and grass-derived biomass feedstocks, AXE and FAE may be indispensable to any enzymatic conversion systems, considering the abundance of xylan acetylation and arabinoxylan ferulation in such feedstocks, respectively.

Including AXE, FAE, or GE in an enzymatic biomass-converting system can acidify the overall reaction, because of the esterase's deesterification action. Proper pH control may be in need to avoid the inactivation of pH-sensitive cellulase and/or hemicellulase. The action of FAE can result in free cinnamoyl or other phenolic/aromatic substances that may be inhibitory to cellulase [44].

14.6.5 Xyloglucan Hydrolase, Xyloglucan Endo-transglycosylase, and Xyloglucan Degradation

Xyloglucan hydrolases (EC 3.2.1.151, 3.2.1.155, 3.2.1.150), which hydrolytically fragment xyloglucan into oligoxyloglucans (Figure 14.2c), and xyloglucan endo-transglycosylases (EC 2.4.1.207), which attach one oligoxyloglucan to another (oligo)xyloglucan, are part of the xyloglucan transferase/hydrolase (XTH) superfamily (http://labs.plantbio.cornell.edu/XTH/). Typical cellulolytic fungi produce xyloglucan hydrolases at levels of $<1\%$ w/w of their secreted proteins. Plants have multiple xyloglucan transglycosylases for regulating the growth of plant cell wall. The catalytic modules of XTH belong to the GH5, 12, 16, 44, and 74 families, and are in general cleft shaped with 4–6 subsites. Some XTH possess CBM, whose main function is targeting the enzyme onto lignocellulose [45]. Many XTHs, particularly XTH74, can have significant side activity on cellulose or other polysaccharides.

XTHs are of interest as potential biocatalysts for biomass conversion. Xyloglucan hydrolase can assist cellulases' activity by degrading xyloglucan that often shield cellulose [46]. The ability of xyloglucan transglycosylase to loosen rigid lignocellulosic cell wall might be exploited to weaken recalcitrant polysaccharides.

14.6.6 α-Galactosidase, Endo-galactanase, Exo-β-glucanase (Non-BG), Endo-β-glucanase (Non-EG), and Hydrolysis of Galactosyl Side Chain, Galactan, and β-Glucan

The α-glycosidic bond-linked Gal substituents in hemicellulose can be cleaved off by AG (EC 3.2.1.22; Figure 14.2a), whose catalytic modules are found in the GH4 (bacterial only), 27, 36, 57, and 110 (bacterial only) families. The GH4 AGs stand out in the GH world: The enzymes use an Zn/NAD^+-based redox mechanism to hydrolyze their substrates [47]. The Gal side chains in hemicellulose, as well as the backbone $\beta(1 \rightarrow 4)$ bonds in galactan, can be hydrolyzed by endo-galactanase (EC 3.2.1.89; Figure 14.2e), whose catalytic modules belong to the GH53 family. Because of the abundance of Gal $\alpha(1 \rightarrow 6)$ substitution and Gal side chain in galactomannan, pectin, and other hemicelluloses, AG and endo-galactanase could contribute to enzymatic conversion of such hemicelluloses.

Non-EG endo-β-glucanases (EC 3.2.1.6, 3.2.1.39, 3.2.1.73) fragment noncellulose β-glucans, while non-BG exo-β-glucanases (EC 3.2.1.58, 3.2.1.75) cut off the nonreducing ends of β-glucans (Figure 14.2d). Different β-glucanases are found in the GH3, 5, 8, 12, 16 ((1,3;1,4)-β-D-glucanases only), 17, 55, 64 ((1,3)-β-D-glucanases only), and 81 families, and different exo-β-glucanases are found in the GH3, 5, 17, and 30 families. Many microbes produce multiple β-glucanases, which probably act collaboratively on complex β-glucan with complementary specificities (toward β(1 → 3), (1 → 4), or (1 → 6) backbone or branching bonds) [48].

14.6.7 Polygalacturonan Hydrolases, Polygalacturonan Lyases, and Polygalacturonan Degradation

The complex poly(rhamno)galacturonyl (or pectic) polysaccharides in plant cell wall can be degraded by various specific hydrolases or lyases (Figure 14.2e) [49,50]. Polygalacturonases (belonging to the GH28 family), including endo-polygalacturonase (EP, EC 3.2.1.15), exo-polygalacturonidase (EC 3.2.1.67), and exo-polygalacturonosidase (EC 3.2.1.82), hydrolyze α(1 → 4) backbone bonds. Like CBH, exo-polygalacturonases act "processively." Pectin/pectate lyases (PL, belonging to the PL1, 2, 3, 9, and 10 families), including endo-pectate lyase (EC 4.2.2.2), endo-oligogalacturonide lyase (EC 4.2.2.6), endo-pectin lyase (EC 4.2.2.10), and exo-pectate disaccharide lyase (EC 4.2.2.9), cleave the backbone bonds by β-elimination (yielding Δ4: 5 C=C bonds at the nonreducing side GalU). Pectate lyases, but not pectin lyases, require Ca^{2+} for activity. Pectin methyl esterase (PME, EC 3.1.1.11), belonging to the CE8 family, demethylates pectin (or polymethylgalacturonic ester) into pectate.

Like endo-(hemi)cellulases, exo-(hemi)cellulases, and other CEs, different pectinolytic enzymes can synergize with each other. Pectinolytic enzymes may assist (hemi)cellulases by degrading pectic substances in co-presence with (hemi)cellulose, or may be assisted by debranching hemicellulases (e.g., AF, AG) that are capable of removing the side chains in poly(rhamno)galacturonan. Because of the abundance of pectic polysaccharides in sugar beet and fruits, pectinolytic enzymes could be important for enzymatic conversion of sugar beet pulp, fruit peels, and other related biomass feedstocks.

14.7 OXIDOREDUCTASES AND LIGNIN DEGRADATION

Organized in a three-dimensional, branched, and nonrepetitive architecture, lignin is far more recalcitrant than (hemi)cellulose toward biological degradation. Compared with the vast number of GHs, PLs, and CEs, lignin-active enzymes have significantly less variety not only in structure but also in mechanism. Almost all known delignifying enzymes are oxidoreductases, particularly peroxidase, phenol oxidase, or H_2O_2-generating carbohydrate/alcohol oxidoreductase [10].

It is noticed that the oxidative enzymes and/or their immediate oxidized products can attack not only lignin but also (hemi)cellulose. For enzymatic biomass conversion, adding oxidative enzymes may be beneficial under certain conditions [51], but detrimental under others [34], because certain oxidations may inactivate (hemi)cellulases or render cellulose more recalcitrant [52,53].

14.7.1 Peroxidases and Direct Lignin Degradation

Four delignifying peroxidases (belonging to FOLy LO2 family), all extracellular fungal heme proteins and structural homologs of other heme peroxidases, are well studied in terms of structure, function, and mechanism. They include lignin peroxidase (LiP, EC 1.11.1.14), Mn peroxidase (MnP, EC 1.11.1.13), versatile peroxidase (VP, EC 1.11.1.16), and "generic" peroxidase (EC 1.11.1.7) [10,54]. Relying on the canonical heme peroxidase activation mechanism, the active intermediates (Fe(IV)-O^+ ("compound II") and Fe(IV)-O^+-porphyrin or protein radical ("compound I")) of the peroxidases can have oxidation potential (E^o) of ~1 V or above. When the enzymes extract electrons from lignin, they may initiate radical-driven lignin degradation (delignification) in lignocellulose. However, the peroxidases are also prone to autooxidation, a key limiting factor in their development as industrial biocatalysts.

Unlike most peroxidases, LiP can oxidize nonphenolic benzene derivatives (e.g., aryl ethers, polycyclic aromatics) lacking "electron-donating" substituents (whose presence may lower the E^{o} of a benzene molecule, thus making it more susceptible to oxidation). Produced mostly by white rots, LiP uses a surface Trp or Tyr to mediate its catalysis. First, the catalytic Trp or Tyr is oxidized into a radical by surrendering an electron to the active heme intermediate (either compound I or II, kept deep inside the active pocket and not accessible to the reducing substrate), then the activated Trp or Tyr radical extracts an electron from a reducing substrate that is in contact with LiP. LiP's action on lignin is believed to involve C4 ether bond scission, C_α–C_β cleavage, aromatic ring opening, demethoxylation, or radical generation/propagation in lignin (Figure 14.3). In laboratory studies, LiP is directly active on mono and dimeric lignin models, indirectly active on oligomeric compounds (up to ~20 cyclic aromatic units) with a redox mediator (e.g., veratryl alcohol, which may also protect LiP from autooxidation), but inactive on polymeric lignin [53]. The involvement of natural mediators in *in vivo* LiP function is suspected, but their identities remain to be conclusively elucidated. There is a multiplicity of LiP genes in many white rots (e.g., 10 in *Phanerochaete chrysosporium*), whose relation with natural delignification is currently unknown.

Widespread among white rots, MnP does not have the cocatalytic surface Trp or Tyr found in LiP, but has a cocatalytic Mn that also serves as a redox mediator. Bound to a specific site in which one of the ligands is a propionate of the (heme) porphyrin, Mn(II) can be oxidized into Mn(III) by surrendering an electron to the active heme intermediate. Two main schemes for the resulting Mn(III)'s delignifying action have been proposed. In one scheme, the Mn(III) dissociates from MnP, binds to a small chelator (e.g., oxalate), and becomes a diffusible oxidant that attacks lignin, even that hidden behind small pores (which limit access for an enzyme) of a biomass substrate. In laboratory studies, however, MnP is unable to directly oxidize nonphenolic lignin models, partly attributed to the significant decrease in Mn(III)'s oxidative potency when it is chelated. In another more plausible scheme, the Mn(III) oxidizes an unsaturated lipid into lipid peroxide, which then oxidizes lignin. The *in vivo* mechanism of MnP's delignifying reaction has yet to be fully elucidated [10,54].

Found so far in a few white rots, VP combines the key, catalysis-relevant structural features of LiP and MnP, with both the cocatalytic surface Trp and the Mn site, making it of particular interest in industrial delignification. Like LiP, VP can directly oxidize lignin, although in terms of catalytic proficiency (k_{cat}/K_m), VP in general is inferior to LiP.

14.7.2 Laccase, Catechol Oxidase, and Indirect Lignin Degradation

The phenolic units in lignin may be oxidized by Cu-containing microbial phenol oxidases (belonging to FOLy LO1 family), including catechol oxidase (EC 1.10.3.1) and, particularly, laccase (Lac, EC 1.10.3.2). Laccase is a group of fungal (mostly extracellular), bacterial, and plant multi-Cu oxidases [55,56]. In general, laccases have four catalytic Cu atoms at a "type I" site (responsible for oxidizing reducing substrate via electron extraction) and a "type 2–type 3"-coupled trinuclear site (responsible for taking electrons from the type 1 Cu and reducing O_2 to H_2O). Laccases have E^o of ~0.4–0.7 V, making them oxidatively potent toward phenols, but not nonphenolics in lignin. Compared with peroxidases (whose activity depend on H_2O_2), laccase is more attractive as an industrial biocatalyst, partly because of its oxidizing substrate O_2 is inexpensive and readily available (from air).

When a lignin substance has exposed phenolic groups, a laccase may directly oxidize them into phenoxy radicals, and subsequently initiate a radical propagation in the lignin. When such phenolic groups are not available (which is the case for many lignocellulosic feedstocks), a laccase may still oxidize a lignin substrate with the help of small redox-active mediators [57]. Natural mediators in *in vivo* laccase delignification have yet to be conclusively identified, and the cost and/or environmental effects of synthetic mediators are currently preventing the commercialization of laccase as biocatalyst for industrial delignification.

14.7.3 Cellobiose Dehydrogenase, Carbohydrate Oxidase, Alcohol Oxidase, and Delignification-Related Peroxide Generation

Production and use of H_2O_2 (or other peroxide and related reactive oxygen species (ROS)) are important for microbial degradation of lignocellulose. Delignifying peroxidases rely on H_2O_2 as their oxidizing cosubstrate and the ultimate electron acceptor of their catalysis. Fenton-type chemistry, based on OH radical or other ROS generated from the interaction between H_2O_2 and transition redox metal ions (e.g., Fe), is believed to be responsible for many apparently nonenzymatic delignification and cellulolysis (e.g., that by brown rots) [58]. Many oxidoreductases can produce H_2O_2 *in situ* and play an accessory role in lignin degradation.

Cellobiose dehydrogenase (CDH, EC 1.11.99.18), constituting the FOLy LO3 family, is probably the only known extracellular dehydrogenase. Produced by numerous fungi, the dehydrogenase uses cellobiose (or cellodextrins and some other carbohydrates) as an electron donor, instead of reduced pyridine nucleotides used by many other dehydrogenases, thus making it attractive as an industrial enzyme. Being a flavoheme enzyme, the dehydrogenase can reduce phenol into (semi)quinone or O_2 into H_2O_2, with a concomitant oxidation of cellobiose into cellobionolactone [59]. The (semi)quinone products of the enzyme may reduce Fe(III) or other oxidized transition metal ions, thus sustaining the Fenton-type ROS generation catalyzed by Fe. CDHs in general have high affinity to cellulose, either due to a CBM or due to another distinct cellulose-binding protein region [51]. The exact *in vivo* role of CDH in microbial biomass degradation has yet to be fully elucidated, although supporting OH radical formation and delignification, causing radical-based cellulolysis, or providing H_2O_2 for peroxidase action is postulated.

Many ligninolytic/cellulolytic microbes secrete various carbohydrate oxidases that generate H_2O_2 from O_2 with concomitant oxidation of mono- or oligomeric carbohydrates. Such oxidases include glucose oxidase (GO, EC 1.1.3.4), pyranose oxidase (EC 1.1.3.10), and galactose oxidase (EC 1.1.3.9). The flavin-containing GO (FOLy LDA6 family) oxidizes many pyranoses at the C1 site, converting a reducing hemi-acetal into a lactone (or aldonic acid on spontaneous hydrolysis). The flavin-containing pyranose oxidase (LDA4 family) oxidizes many pyranoses at the C2 site, converting a secondary alcohol into a ketone. The Cu-containing galactose oxidase (LDA5), which uses a cocatalytic Tyr radical paired with Cu, oxidizes a D-Gal (or some other primary alcohols) at the C6 site, converting a primary alcohol into an aldehyde. The carbohydrate substrates of these oxidases can be provided by various (hemi)cellulases (as the products of their catalyses). Besides carbohydrate oxidases, alcohol oxidases can also supply H_2O_2 (derived from O_2) to microbial lignocellulose degradation. Such flavin-containing oxidases include aryl-alcohol oxidase (AAO, EC 1.1.3.7, LDA1 family), vanillyl alcohol oxidase (EC 1.1.3.38, LAD2), and alcohol oxidase (EC 1.1.3.13, LAD8), which oxidize many aromatic ring-anchored, vanillyl (hydroxymethoxybenzyl), and aliphatic primary alcohols into corresponding aldehydes, respectively. The alcohol substrates of these oxidases may be provided by peroxidases during their lignin demethoxylation.

Glyoxal oxidase (LAD3), a structural homolog of galactose oxidase, can generate H_2O_2 from O_2 with concomitant oxidation of glyoxal or other small, microbial metabolite aldehydes into corresponding carboxylic acid.

14.8 BIOLOGICAL AND SYNTHETIC ENHANCERS FOR ENZYMATIC DEGRADATION OF LIGNOCELLULOSE

Natural lignocellulose degradations often involve contributors other than the (hemi)cellulolytic/ligninolytic enzymes described above. For instance, several kinds of noncatalytic "accessory proteins," including CBM, swollenin, and expansin, may assist biomass-active enzymes by disrupting H-bonds in cellulose or modifying lignin surface (to prevent "nonproductive" adsorption of cellulase). Certain artificial substances can also enhance enzymatic lignocellulose degradation by similar mechanisms. Such enhancers have the potential in reducing the overall cost of industrial lignocellulose degradation.

14.8.1 CBM, Expansin, Swollenin, and Cellulose Modification

To enable/enhance (modular) biomass-active enzymes, nature has a repertoire of effective CBMs with versatile composition, assemblage, and specificity. Belonging to ~53 families based on sequence homology, CBMs can bind to the surface of bundled polysaccharides (mainly via π-stacking), individual oligosaccharide chains (similar to the catalytic modules of many GH enzymes), or small saccharides (similar to lectins) [39,60]. A few studies suggest that certain free CBMs and noncatalytic, CBM-containing proteins might disrupt the H-bonds and packing in cellulose microfibrils to make them more susceptible toward enzymatic cellulolysis. Such observations have potential ramification for industrial enzymatic cellulolysis, but need further investigation [2,11].

Along with EG, XTH, and OH radical, expansin has attracted significant attention for its role in the loosening of rigid primary cell walls during a plant's growth, which requires disentanglement of lignocellulose [61]. Having an N- and C-terminal domain homologous to Cel45 EG and grass pollen allergen, respectively, expansin may disrupt noncovalent bonds among cell wall polysaccharides. Microbial proteins homologous to plant expansions may also loosen plant cell wall [62,63]. Swollenin, secreted by several cellulolytic fungi and consisting of an expansin-like domain linked to a CBM and/or other modules, is reported to be capable of dispersing cellulosic substances [64]. Both expansin and swollenin may enhance enzymatic lignocellulose hydrolysis at laboratory scale, and the applicability of such enhancement should be further investigated.

14.8.2 Apparent Accessory (Noncatalytic) Proteins in Enzymatic Hydrolysis of Lignocellulose

The proteins currently classified in the GH61 family are found in numerous cellulolytic fungi, often with multiple alleles. Although early studies annotated the proteins as EG (e.g., *H. jecorina* EG-IV), later studies indicated no credible EG or other typical GH activities attributable to GH61 [65,66]. Having a fold similar to that of a chitin-binding protein, a metal ion-binding site, and sometimes a CBM [66], GH61 has recently been shown to be capable of enhancing enzymatic hydrolysis of lignocellulose [67]. This intriguing effect from GH61 remains to be mechanistically elucidated.

Besides GH61, there are a few other apparently noncatalytic proteins that may be involved in *in vivo* enzymatic cellulolysis [2]. They include *H. jecorina* Cip1 protein, which has a CBM and a "core" homologous to a putative *Streptomyces coelicolor* hydrolase [68], and *Clostridium cellulovorans* HbpA, which has a surface layer module and a cohesin [69].

Such accessory proteins are in general coregulated with GHs in cellulolytic microbes, implying their active role in natural lignocellulose degradation. Understanding these accessory proteins' mechanism and developing them along with cellulolytic/ligninolytic enzymes are of significant interest for industrial biomass conversion.

14.8.3 Microbial Cells, Other Proteins, and Enhanced Enzymatic Hydrolysis of Lignocellulose

In addition to the noncatalytic, lignocellulose-interacting proteins described above, other biological entities may also positively impact enzymatic hydrolysis of lignocellulose. For cellulase-producing *H. jecorina*, co-presence of the fungal cells with secreted enzymes can lead to enhanced biomass hydrolysis, probably due to an interfacial interaction between the cells and insoluble lignocellulose [70]. For cellulosome-producing *Clostridium thermocellum*, cell-adhered cellulosomal cellulases can hydrolyze cellulose more efficiently than cell-free cellulases, an effect attributed to close proximity of the cell–cellulosome complex to cellulose and subsequent low local concentration of cellulase inhibitors [71]. Such an effect may be of interest to commercial enzymatic systems for biomass conversion, because it may reduce not only enzyme dosing for hydrolysis but also enzyme cost from production (no need to separate enzymes from cells).

Adding exogenous proteins may also lead to enhanced enzymatic hydrolysis. It is known that albumin and gelatin, both widely used as inert bulk proteins in various applications, may significantly reduce cellulase dosage while maintaining the overall hydrolysis of biomass substance [21,72]. The proteins have little affinity toward cellulose, but bind to lignin. It is believed that the nonspecific binding of these proteins to lignin may prevent nonproductive lignin adsorption of cellulases, leading to more active enzymes and higher overall hydrolysis. Nevertheless, the applicability of this effect to commercial biomass hydrolysis waits to be further evaluated, because currently known dosing levels of such proteins seem prohibitively high, and much cheaper and widely available bulk protein sources need to be explored.

14.8.4 Alkoxylate Polymers and Enhanced Enzymatic Hydrolysis of Lignocellulose

Poly(ethylene glycol) (PEG) and certain types of nonionic surfactants (e.g., Tween 80) are known to improve cellulase action on lignocellulose [21]. Like bulk proteins, these polymers may adsorb onto lignin to force lignin-adsorbed cellulases back into solution. Because the effective nonionic surfactants also contain alkoxylate chains similar to those in PEG, most likely it is the alkoxylate composition that renders the nonionic surfactants the ability to enhance cellulases. Such polymer effect seems to depend more on the composition and structure, than on the content, of lignin in biomass substrates [73]. Like bulk proteins, the applicability of this polymer effect needs further evaluation, in terms of both net cost (saving in cellulases against expense in polymers) and source of the polymers (currently derived from fossil feedstocks).

14.9 RECENT AND FUTURE DEVELOPMENTS OF ENZYMATIC LIGNOCELLULOSE HYDROLYSIS

Developing more active, more robust, and less costly enzyme catalysts is vital for the success of biomass-related- and biotech industries. To do this, intensive research is being carried out to discover better biomass-active enzymes, explore their synergism, and optimize their integration with up- or downstream biomass processings. Expanding natural and artificial diversity, particularly the parts from animals, protists, and metagenomics, provides more biomass-active enzymes of desired properties [74]. New insights into natural lignocellulose-degrading processes continue to emerge [11]. Recombinant gene technology allows combination or coproduction of selected enzymes to achieve collective activity and specificity suited for a given lignocellulose degradation.

14.9.1 Discovery of Better Biomass-Active Enzymes

Recently, the emphasis of research on biomass-active enzymes has gradually shifted from basic mechanistic analysis to performance-oriented protein engineering. In-depth understanding of an enzyme's structure–function relationship, large natural and artificial enzyme diversity, and insightful phylogenetic analysis assists us in creating enzyme variants by rational design, gene shuffling, or other recombinatory techniques. Various cellulases, hemicellulases, and delignifying oxidoreductases have been constructed for improved kinetic or thermodynamic properties. Engineering well-expressed mesophilic enzymes to render them with high-thermal activity/stability is of particular interest, because of the potential benefits of having an enzymatic lignocellulose hydrolysis at high temperature (increased rate, better compatibility with high temperature upstream pretreatment, or decreased microbial contamination during prolonged hydrolysis). Such enzyme engineering may complement the effort of finding wild-type thermophilic biomass-active enzymes [75].

The modular structure of biomass-active enzymes and the quaternary assemblage of cellulosomal enzymes are believed to be vital for these enzymes' amazing specificity, versatility, efficiency, and synergism. Many of these enzymes with multiple/different catalytic modules, CBMs, and other functional domains are known in nature. Designer enzymes and "mini-cellulosomes" have been made to possess selected catalytic modules, CBMs, dockerins, or other desired functionalities. Biomass-active enzymes may be surface anchored on selected microbes (cellulolytic or fermentative), to supplement deficient lignocellulose-degrading functions or consolidate different lignocellulose conversion reactions. Such protein engineerings could lead to simplified biomass-active enzyme systems with improved specificity, activity, or synergism.

14.9.2 Multienzyme Synergism, "One-Pot" Biomass Conversion, and Novel Lignocellulose Degradation Media

Secretomes of biomass-degrading organisms contain tens or even hundreds of distinct proteins or isoforms [28]. Among the interactions of these proteins, many are synergistic and have ramification for the development of industrial biomass-degrading biocatalysts. Although a few synergisms (e.g., that between CBH-I and CBH-II, CBH and EG, or cellulase and Xyn) are known at present, more are suspected because there likely are enzymes whose importance in *in vivo* biomass degradation is underestimated, due to the difference between the *in vitro* secretomes obtained under culture conditions (controlled medium, pH, temperature, aeration, solid–liquid ratio, agitation, etc.) and the *in vivo* ones. The dissimilarity between artificially pretreated and natural biomass substances (in terms of composition and morphology) may require enzyme mixtures dissimilar to those produced *in vivo* by biomass-active microbes. Thus, it is imperative to search for optimal enzyme mixture and explore enzymatic synergism pertaining to specific lignocellulose degradations at laboratory and industrial scales [30,31,34,41,46,67].

Natural biomass degradation can take place "on spot," from native lignocellulose all the way to CO_2, with a consortium of microbes of differential specificity and interdependence, over relatively long time periods (years). "One-pot" enzymatic conversion of native lignocellulose at industrial scale is attractive (in terms of savings in energy, material, and infrastructure), but currently not viable, mainly because of the ineffectiveness of enzymatic degradation of native biomass feedstocks within short time (days). Nevertheless, developing a common biocatalyst for hydrolyzing pretreated lignocellulose (into simple sugars) and fermenting simple sugars (into ethanol or other chemicals) is currently being pursued. For instance, ethanogenic *Saccharomyces cerevisiae* has been equipped with surface-attached cellulases for a consolidated bioprocessing of lignocellulose into fuel ethanol [2,76,77].

As a novel type of industrial solvent, ionic liquids are tested for potential application in lignocellulose degradation. Being polar, ionic liquids can dissolve cellulose and separate lignin [78], or allow homogeneous enzymatic cellulose degradation [79]. Being essentially nonvolatile, ionic liquids may be recycled and reused for multiple rounds of biomass hydrolysis. However, the enzymology of biomass-active enzymes in ionic liquids and the economy of using ionic liquid need to be thoroughly studied, before ionic liquid-based enzymatic biomass hydrolysis can be applied industrially.

14.9.3 Specific Assays and High-Throughput Screening for Biomass-Active Enzymes

Suitable assays are important for the research and development of biomass-active enzymes, from preliminary screening of candidates to upscaled, industry-relevant evaluation of selected enzymes. Many biomass-active enzymes have complex interactions (at multiple sites) with substrates, invoke heterogeneous catalysis, are promiscuous (with significant side activity), or co-depend on synergistic partners. Such properties make an assay that is specific (to individual enzyme), robotizable (for high-throughput screening), and predictive (for industrial scale performance) difficult to find [80].

To identify individual enzymes or analyze an enzyme mixture's composition, immunochemical or mass spectrometry assays coupled with electrophoretic or chromatographic protein separation can be applied. To screen large libraries of engineered variants (with the same host background) or conduct on-line monitoring, activity-based photogenic assays are desirable. However, conventional activity assays for biomass-active enzymes generally lack specificity (particularly toward enzymes of the same class), which often leads to severe overestimation of a targeted enzyme when impurity enzymes also respond to the assay and synergize with the target enzyme. Conventional laboratory assays often do not correlate with each other, depending on the type and state (insoluble or soluble) of the substrates used, or with upscaled (pilot or production) assays, depending on the consistency of the hydrolysis mix [80–82]. To overcome these difficulties, new assays based on microplate-hosted hydrolysis systems with more application-relevant substrates and conditions are being developed [83].

14.9.4 Enzyme Cost Reduction, Integration of Enzymatic and Nonenzymatic Steps, and Commercialization

To compete with conventionally produced (fossil) chemicals, biomass-based chemicals need to be cost effective. This requires inexpensive production of biomass-active enzymes, in addition to low-enzyme dosing/cost as biocatalysts. At present and in the near future, fermentative production of secreted microbial enzymes is and will remain the most economical means on an industrial scale. Maximizing the expression of targeted biomass-active enzymes, ideally in one microbial host with desired stoichiometry, is one of the most important tasks for developing viable biomass conversion industries. This needs advanced recombinant gene technology aimed at gene transcription, gene translation, protein secretion, or pathway engineering (especially, for the biosynthesis of enzyme cofactors such as heme). Transgenic plants (particularly the dedicated energy crops) might become viable sources of the enzymes, but the enabling and complex biotechnology related with transgenetic plants need further development.

Current models for industrial lignocellulose conversion envision processes whose major steps include physicochemical pretreatment of feedstock, enzymatic (hemi)cellulolysis, and biological or chemical conversion of simple carbohydrates. Viable enzymatic biomass hydrolyses have to take into account the up- and downstream steps from the point of view of the overall efficiency and economy. For instance, different pretreatments may generate different substrates or inhibitors that would require enzymes of different specificity or resilience. Hence, the developments of prospective pretreatment, enzymatic, and other technologies for biomass conversion should be in sync.

Microbe- or enzyme-based lignocellulose modifications/degradations have been intensively investigated for a few industrial applications, especially the biopulping (using Xyn or lignocellulolytic microbes) and biobleaching (using laccase) for paper manufacturing, waste paper recycling (using cellulase), and, more recently, cellulosic bioethanol production (using (hemi)cellulase). The enzyme technology being developed for lignocellulose degradation holds great promise for the future biorefinery industry, in which the abundant and renewable biomass (or other sources such as algae and municipal waste) would be converted into platform carbohydrate or aromatic compounds capable of supporting the production of a range of valuable chemicals.

ACKNOWLEDGMENTS

Drs. Robert L. Starnes and Claus C. Fuglsang from Novozymes provided critical readings of the manuscript.

REFERENCES

1. Margeot A, Hahn-Hagerdal B, Edlund M, Slade R, Monot F. New improvements for lignocellulosic ethanol. Curr Opin Biotechnol 2009;20:372–380.
2. Wilson DB. Cellulases and biofuels. Curr Opin Biotechnol 2009;20:295–299.
3. Lynd LR, Laser MS, Bransby D, Dale BE, Davison B, Hamilton R, Himmel M, Keller M, McMillan JD, Sheehan J, Wyman CE. How biotech can transform biofuels. Nat Biotechnol 2008;26:169–172.
4. Kumar R, Singh S, Singh OV. Bioconversion of lignocellulosic biomass: biochemical and molecular perspectives. J Ind Microbiol Biotechnol 2008;35:377–391.
5. Himmel ME, Ding SY, Johnson DK, Adney WS, Nimlos MR, Brady JW, Foust TD. Biomass recalcitrance: engineering plants and enzymes for biofuels production. Science 2007;315:804–807.
6. Pérez J, Muñoz-Dorado J, de la Rubia T, Martínez J. Biodegradation and biological treatments of cellulose, hemicellulose and lignin: an overview. Int Microbiol 2002;5:53–63.
7. Wei H, Xu Q, Taylor LE, Baker JO, Tucker MP, Ding SY. Natural paradigms of plant cell wall degradation. Curr Opin Biotechnol 2009;20:330–338.
8. Pauly M, Keegstra K. Cell-wall carbohydrates and their modification as a resource for biofuels. Plant J 2008;54:559–568.
9. Ralph J, Brunow G, Harris PJ, Dixon RA, Schatz PF, Boerjan W. Lignification: are lignins biosynthesized via simple combinatorial chemistry or via proteinaceous control and template replication? In: Daayf F, Lattanzio V, editors. Recent advances in polyphenol research. Blackwell; 2008. pp. 36–66.
10. Martínez AT, Ruiz-Dueñas FJ, Martínez MJ, del Río JC, Gutiérrez A. Enzymatic delignification of plant cell wall: from nature to mill. Curr Opin Biotechnol 2009;20:348–357.
11. Wilson DB. Three microbial strategies for plant cell wall degradation. Ann N Y Acad Sci 2008;1125:289–297.

12. da Costa Sousa L, Chundawat SP, Balan V, Dale BE. "Cradle-to-grave" assessment of existing lignocellulose pretreatment technologies. Curr Opin Biotechnol 2009;20:339–347.
13. Ding SY, Xu Q, Crowley M, Zeng Y, Nimlos M, Lamed R, Bayer EA, Himmel ME. A biophysical perspective on the cellulosome: new opportunities for biomass conversion. Curr Opin Biotechnol 2008;19:218–227.
14. Gilbert HJ. Cellulosomes: microbial nanomachines that display plasticity in quaternary structure. Mol Microbiol 2007;63:1568–1576.
15. Vocadlo DJ, Davies GJ. Mechanistic insights into glycosidase chemistry. Curr Opin Chem Biol 2008;12:539–555.
16. Ye X, Wang Y, Hopkins RC, Adams MWW, Evans BR, Mielenz JR, Zhang YHP. Spontaneous high-yield~production of hydrogen from cellulosic materials and water catalyzed by enzyme cocktails. ChemSusChem 2009;2:149–152.
17. Konno N, Ishida T, Igarashi K, Fushinobu S, Habu N, Samejima M, Isogai A. Crystal structure of polysaccharide lyase family 20 endo-β-1,4-glucuronan lyase from the filamentous fungus *Trichoderma reesei*. FEBS Lett 2009;583:1323–1326.
18. Xu F, Ding H. A new kinetic model for heterogeneous (or spatially confined) enzymatic catalysis: contributions from the fractal and jamming (overcrowding) effects. Appl Catal A Gen 2007;317:70–81.
19. Kipper K, Valjamae P, Johansson G. Processive action of cellobiohydrolase Cel7A from *Trichoderma reesei* is revealed as 'burst' kinetics on fluorescent polymeric model substrates. Biochem J 2005;385:527–535.
20. Yaoi K, Kondo H, Hiyoshi A, Noro N, Sugimoto H, Tsuda S, Mitsuishi Y, Miyazaki K. The structural basis for the exo-mode of action in GH74 oligoxyloglucan reducing end-specific cellobiohydrolase. J Mol Biol 2007;370:53–62.
21. Xu F, Ding H, Osborn D, Tejirian A, Brown K, Albano W, Sheehy N, Langston J. Partition of enzymes between the solvent and insoluble substrate during the hydrolysis of lignocellulose by cellulases. J Mol Catal B Enzym 2007;51:42–48.
22. Berlin A, Balakshin M, Gilkes N, Kadla J, Maximenko V, Kubo S, Saddler J. Inhibition of cellulase, xylanase and beta-glucosidase activities by softwood lignin preparations. J Biotechnol 2006;125:198–209.
23. Mejia-Castillo T, Hidalgo-Lara ME, Brieba LG, Ortega-Lopez J. Purification, characterization and modular organization of a cellulose-binding protein, CBP105, a processive beta-1,4-endoglucanase from *Cellulomonas flavigena*. Biotechnol Lett 2008;30:681–687.
24. Yoon JJ, Cha CJ, Kim YS, Kim W. Degradation of cellulose by the major endoglucanase produced from the brown-rot fungus *Fomitopsis pinicola*. Biotechnol Lett 2008;30:1373–1378.
25. Cohen R, Suzuki MR, Hammel KE. Processive endoglucanase active in crystalline cellulose hydrolysis by the brown rot basidiomycete *Gloeophyllum trabeum*. Appl Environ Microbiol 2005;71:2412–2417.
26. Vlasenko E, Cherry J, Xu F. Substrate specificity of family 5, 6, 7, 9, 12, and 45 endoglucanases. Bioresour Technol 2010;101:2405–2411.
27. Langston J, Sheehy N, Xu F. Substrate specificity of *Aspergillus oryzae* family 3β-glucosidase. Biochim Biophys Acta 2006;1764:972–978.
28. Bouws H, Wattenberg A, Zorn H. Fungal secretomes–nature's toolbox for white biotechnology. Appl Microbiol Biotechnol 2008;80:381–388.

29. Shallom D, Shoham Y. Microbial hemicellulases. Curr Opin Microbiol 2003;6: 219–228.

30. Gupta R, Kim TH, Lee YY. Substrate dependency and effect of xylanase supplementation on enzymatic hydrolysis of ammonia-treated biomass. Appl Biochem Biotechnol 2008,148:59–70.

31. Selig MJ, Knoshaug EP, Adney WS, Himmel ME, Decker SR. Synergistic enhancement of cellobiohydrolase performance on pretreated corn stover by addition of xylanase and esterase activities. Bioresour Technol 2008;99:4997–5005.

32. Ustinov BB, Gusakov AV, Antonov AI, Sinitsyn AP. Comparison of properties and mode of action of six secreted xylanases from *Chrysosporium lucknowense*. Enzyme Microb Technol 2008;43:56–65.

33. Pastor FIJ, Gallardo O, Sanz-Aparicio J, Diaz P. Xylanases: molecular properties and applications. In: Polaina J, MacCabe AP, editors. Industrial enzymes. Dordrecht, The Netherlands: Springer; 2007. pp. 65–82.

34. Tabka MG, Herpoel-Gimbert I, Monod F, Asther M, Sigoillot JC. Enzymatic saccharification of wheat straw for bioethanol production by a combined cellulase xylanase and feruloyl esterase treatment. Enzyme Microb Technol 2006;39:897–902.

35. Berrin JG, Juge N. Factors affecting xylanase functionality in the degradation of arabinoxylans. Biotechnol Lett 2008;30:1139–1150.

36. Moreira L, Filho E. An overview of mannan structure and mannan-degrading enzyme systems. Appl Microbiol Biotechnol 2008;79:165–178.

37. Beukes N, Chan H, Doi RH, Pletschke BI. Synergistic associations between *Clostridium cellulovorans* enzymes XynA, ManA and EngE against sugarcane bagasse. Enzyme Microb Technol 2008;42:492–498.

38. Numan MT, Bhosle NB. α-L-arabinofuranosidases: the potential applications in biotechnology. J Ind Microbiol Biotechnol 2006;33:247–260.

39. Hashimoto H. Recent structural studies of carbohydrate-binding modules. Cell Mol Life Sci 2006;63:2954–2967.

40. Yang H, Ichinose H, Nakajima M, Kobayashi H, Kaneko S. Synergy between an alpha-L-arabinofuranosidase from *Aspergillus oryzae* and an endo-arabinanase from *Streptomyces coelicolor* for degradation of arabinan. Food Sci Technol Res 2006;12:43–49.

41. Sorensen HR, Pedersen S, Jorgensen CT, Meyer AS. Enzymatic hydrolysis of wheat arabinoxylan by a recombinant "minimal" enzyme cocktail containing beta-xylosidase and novel endo-1,4-beta-xylanase and alpha-(L)-arabinofuranosidase activities. Biotechnol Prog 2007;23:100–107.

42. Ryabova O, Vrsanská M, Kaneko S, van Zyl WH, Biely PA. A novel family of hemicellulolytic alpha-glucuronidase. FEBS Lett 2009;583:1457–1462.

43. Benoit I, Danchin EGJ, Bleichrodt RJ, de Vries RP. Biotechnological applications and potential of fungal feruloyl esterases based on prevalence, classification and biochemical diversity. Biotechnol Lett 2008;30:387–396.

44. Tejirian A, Xu F. Mechanistic study of enzymatic cellulolysis inhibitions. 31st Symposium on Biotechnology for Fuels and Chemicals; 2009; San Francisco (CA), poster 5–29.

45. Ishida T, Yaoi K, Hiyoshi A, Igarashi K, Samejima M. Substrate recognition by glycoside hydrolase family 74 xyloglucanase from the basidiomycete *Phanerochaete chrysosporium*. FEBS J 2007;274:5727–5736.

46. Benko Z, Siika-aho M, Viikari L, Reczey K. Evaluation of the role of xyloglucanase in the enzymatic hydrolysis of lignocellulosic substrates. Enzyme Microb Technol 2008;43:109–114.

47. Yip VLY, Withers SG. Family 4 glycoside hydrolases are special: The first-elimination mechanism amongst glycoside hydrolases. Biocatal Biotransformation 2006;24:167–176.

48. Martin K, McDougall BM, McIlroy S, Chen JZ, Seviour RJ. Biochemistry and molecular biology of exocellular fungal beta-(1,3)- and beta-(1,6)-glucanases. FEMS Microbiol Rev 2007;31:168–192.

49. Payasi A, Sanwal R, Sanwal GG. Microbial pectate lyases: characterization and enzymological properties. World J Microbiol Biotechnol 2009;25:1–4.

50. Niture SK. Comparative biochemical and structural characterizations of fungal polygalacturonases. Biologia (Bratislava) 2008;63:1–19.

51. Krusa M, Lennholm H, Henriksson G. Pre-treatment of cellulose by cellobiose dehydrogenase increases the degradation rate by hydrolytic cellulases. Cellul Chem Technol 2008;41:105–111.

52. Bendl RF, Kandel JM, Amodeo KD, Ryder AM, Woolridge EM. Characterization of the oxidative inactivation of xylanase by laccase and a redox mediator. Enzyme Microb Technol 2008;43:149–156.

53. Xu F, Ding H, Tejirian A. Detrimental effect of cellulose oxidation on cellulose hydrolysis by cellulase. Enzyme Microb Technol 2009;45:203–209.

54. Hammel KE, Cullen D. Role of fungal peroxidases in biological ligninolysis. Curr Opin Plant Biol 2008;11:349–355.

55. Xu F, Damhus T, Danielsen S, Østergaard LH. Catalytic applications of laccase. In: Schmid RD, Urlacher V, editors. Modern biooxidations - enzymes, reactions and applications. Weinheim: Wiley-VCH; 2007. pp. 43–76.

56. Zhukhlistova NE, Zhukova YN, Lyashenko AV, Zaitsev VN, Mikhailov AM. Three-dimensional organization of threedomain copper oxidases: a review. Crystallogr Rep 2008;53:92–109.

57. Morozova OV, Shumakovich GP, Shleev SV, Yaropolov YI. Laccase-mediator systems and their applications: a review. Appl Biochem Microbiol 2007;43:523–535.

58. Baldrian P, Valaskova V. Degradation of cellulose by basidiomycetous fungi. FEMS Microbiol Rev 2008;32:501–521.

59. Cameron MD, Aust SD. Cellobiose dehydrogenase - an extracellular fungal flavocytochrome. Enzyme Microb Technol 2001;28:129–138.

60. Shoseyov O, Shani Z, Levy I. Carbohydrate binding modules: Biochemical properties and novel applications. Microbiol Mol Biol Rev 2006;70:283–295.

61. Choi D, Cho HT, Lee Y. Expansins: expanding importance in plant growth and development. Physiol Plant 2006;126:511–518.

62. Kim E, Lee H, Bang WG, Choi IG, Kim KH. Functional characterization of a bacterial expansin from Bacillus subtilis for enhanced enzymatic hydrolysis of cellulose. Biotechnol Bioeng 2009;102:1342–1353.

63. Kerff F, Amoros A, Herman R, Sauvage E, Petrella S, Filee P, Charlier P, Joris B, Tabuchi A, Nikolaidis N, Cosgrove DJ. Crystal structure and activity of *Bacillus subtilis* YoaJ (EXLX1), a bacterial expansin that promotes root colonization. Proc Natl Acad Sci USA 2008;105:16876–16881.

64. Yao Q, Sun TT, Liu WF, Chen GJ. Gene cloning and heterologous expression of a novel endoglucanase, swollenin, from *Trichoderma pseudokoningii* S38. Biosci Biotechnol Biochem 2008;72:2799–2805.
65. Koseki T, Mese Y, Fushinobu S, Masaki K, Fujii T, Ito K, Shiono Y, Murayama T, Iefuji H. Biochemical characterization of a glycoside hydrolase family 61 endoglucanase from *Aspergillus kawachii*. Appl Microbiol Biotechnol 2008;77:1279–1285.
66. Karkehabadi S, Hansson H, Kim S, Piens K, Mitchinson C, Sandgren M. The first structure of a glycoside hydrolase family 61 member, Cel61B from *Hypocrea jecorina*, at 1.6 angstrom resolution. J Mol Biol 2008;383:144–154.
67. Merino S, Cherry J. Progress and challenges in enzyme development for biomass utilization. In: Olssen L, Ahring BK, editors. Volume 108, Advanced biochemical engineering & biotechnology. Berlin: Springer; 2007. pp. 95–120.
68. Scott BR, Hill C, Tomashek J, Liu C. Enzymatic hydrolysis of lignocellulosic feedstocks using accessory enzymes. US patent US20090061484. 2009
69. Matsuoka S, Yukawa H, Inui M, Doi RH. Synergistic interaction of *Clostridium cellulovorans* cellulosomal cellulases and HbpA. J Bacteriol 2007;189:7190–7194.
70. Berka R, Cherry J. Methods for degrading or converting plant cell wall polysaccharides. US patent US7413882-B2. 2008.
71. Lu YP, Zhang YHP, Lynd LR. Enzyme-microbe synergy during cellulose hydrolysis by *Clostridium thermocellum*. Proc Natl Acad Sci USA 2006;103:16165–16169.
72. Yang B, Wyman CE. BSA treatment to enhance enzymatic hydrolysis of cellulose in lignin containing substrates. Biotechnol Bioeng 2006;94:611–617.
73. Kristensen JB, Borjesson J, Bruun MH, Tjerneld F, Jorgensen H. Use of surface active additives in enzymatic hydrolysis of wheat straw lignocellulose. Enzyme Microb Technol 2007;40:888–895.
74. Rubin EM. Genomics of cellulosic biofuels. Nature 2008;454:841–845.
75. Blumer-Schuette SE, Kataeva I, Westpheling J, Adams MW, Kelly RM. Extremely thermophilic microorganisms for biomass conversion: status and prospects. Curr Opin Biotechnol 2008;19:210–217.
76. Xu Q, Singh A, Himmel ME. Perspectives and new directions for the production of bioethanol using consolidated bioprocessing of lignocellulose. Curr Opin Biotechnol 2009;20:364–371.
77. van Zyl WH, Lynd LR, den Haan R, McBride JE. Consolidated bioprocessing for bioethanol production using *Saccharomyces cerevisiae*. Adv Biochem Eng Biotechnol 2007;108:205–235.
78. Lee SH, Doherty TV, Linhardt RJ, Dordick JS. Ionic liquid-mediated selective extraction of lignin from wood leading to enhanced enzymatic cellulose hydrolysis. Biotechnol Bioeng 2009;102:1368–1376.
79. Kamiya N, Matsushita Y, Hanaki M, Nakashima K, Narita M, Goto M, Takahashi H. Enzymatic *in situ* saccharification of cellulose in aqueous-ionic liquid media. Biotechnol Lett 2008;30:1037–1040.
80. Zhang YHP, Himmel ME, Mielenz JR. Outlook for cellulase improvement: Screening and selection strategies. Biotechnol Adv 2006;24:452–481.
81. Kabel MA, van der Maarel MJEC, Klip G, Voragen AGJ, Schols HA. Standard assays do not predict the efficiency of commercial cellulase preparations towards plant materials. Biotechnol Bioeng 2006;93:56–63.

82. Andersen N, Johansen KS, Michelsen M, Stenby EH, Krogh KBRM, Olsson L. Hydrolysis of cellulose using mono-component enzymes shows synergy during hydrolysis of phosphoric acid swollen cellulose (PASC), but competition on Avicel. Enzyme Microb Technol 2008;42:362–370.
83. Chundawat SPS, Balan V, Dale BE. High-throughput microplate technique for enzymatic hydrolysis of lignocellulosic Biomass. Biotechnol Bioeng 2008;99:1281–1294.

CHAPTER 15

BIOCONVERSION OF RENEWABLES—PLANT OILS

MANFRED P. SCHNEIDER
FB C, Bergische Universität Wuppertal, Wuppertal, Germany

15.1 INTRODUCTION

About 120 billion tons of carbon in biomass, equivalent to >80 billion tons of oil equivalents (toe), are produced annually by photosynthesis. Only about 5% of these is presently used by humans. From a chemical point of view, about 75% is carbohydrates, 20% lignin, and about 5% fats, oils, and others. Thus, next to carbohydrates (starch, cellulose, hemicelluloses) and lignin, plant oils constitute the most important plant-derived class of renewable resources. They are derived from major agricultural crops such as palm, soy, canola (rape seed), sunflower, palm kernel, coconut, and others. The annual global production of these major vegetable oils has reached 137 Mt in 2009 [1] (Table 15.1). Another 31 Mt of other oils (and nonedible oils such as *Jatropha* species) and animal fats were also produced.

About 80% of these oils is used for food and feed applications, while about 20% is used industrially. They constitute sustainable and ecologically important alternatives to petroleum products for the production of fatty acid methyl esters (FAME), biodiesel (energy), lubricants and hydraulic fluids [2], oleochemicals, and surface-active compounds such as detergents and emulsifiers. With the large amounts of biodiesel being produced, 1.1 Mt of glycerol became globally available in 2008, providing a novel platform chemical for the production of, for example, 1,2-propanediol, 1,3-propanediol, epichlorohydrin, acrolein, and others [3].

Plant oils have much lower gross energy requirements (GERs) than petroleum-based products. GER indicates the total primary energy consumption of an entire

Biocatalysis for Green Chemistry and Chemical Process Development, First Edition.
Edited by Junhua (Alex) Tao and Romas Kazlauskas.

TABLE 15.1 Global Annual Production of Oilseeds [1]

Plant Oil	Production in Million Metric Tons (2009)
Coconut	3.63
Cottonseed	4.75
Olive	2.99
Palm	45.13
Palm kernel	5.31
Peanut	4.45
Rapeseed	21.93
Soybean	37.69
Sunflower seed	11.45
Total	137.33

process chain. Thus, the GER of rapeseed oil (20 GJ/t) is much less than that of even bulk petroleum-based chemicals such as ethene (60 GJ/t) [3]. In an attempt to replace classical chemical technologies, which are energy intensive, produce large amounts of waste, and use toxic chemicals, green technologies are increasingly studied and employed, also on an industrial scale. Thus, the use of lipases for the bioconversion of native fats and oils into value-added products is increasingly studied and applied both in industrial and university laboratories worldwide. Work in this area ranges from the production of food-related products such as cocoa butter substitutes, diglycerides as dietary cooking oils (Enova™), structured lipids of nutritional value, personal care products, emollient esters such as cetyl ricinoleate or myristyl myristate, esters of polyglycerol, emulsifiers such as mono- and diglycerides and sugar esters; the oxidative functionalization of fatty acids (FAs); all the way to the enzymatic production of biodiesel. In the sections below, selected examples for such applications are discussed on reviewing both the literature and the studies from the author's laboratory. The discussion is limited to plant-derived oils and does not include similar commodities such as animal fats (beef tallow) and phospholipids. After a brief description of the important tools for the bioconversion of lipids, that is, lipases and their most important catalytic transformations, examples for the above product portfolio are discussed in the sections to follow.

15.2 LIPASES—TOOLS FOR THE BIOCONVERSION OF PLANT OILS

Clearly, the most important tools for the bioconversion of plant oils (lipids) are (microbial) lipases of different origin (bacteria, fungi). Their role in nature is the hydrolysis of triglycerides into FAs and glycerol.

15.2.1 Lipases—Structure and Function

Lipases (triacylglycerol (TAG) acylhydrolases) belong to the group of ester hydrolases (EC 3.1.1.3). Their molecular weights are typically between 18 and 60 kDa. The major difference between esterases and lipases is a hydrophobic helix in lipases creating a so-called "lid" which—in an aqueous environment—covers the active (catalytic) site. As a consequence and in contrast to esterases and proteases, most lipases are inactive in water until in contact with a hydrophobic substrate such as a lipid. *Via* a process called *interfacial activation*—the interaction of the lipid with the hydrophobic lid—the cover is opened and the active site is exposed for the entering substrate.

15.2.2 Lipases—Mechanism of Action

The generally accepted mechanism for lipase-catalyzed transformations is closely related to that of esterases and proteases in having a catalytic triad usually consisting of serine, histidine, and aspartic (glutamic) acid. *Via* concomitant increase of the nucleophilicity of the primary hydroxy group of serine and at the same time stabilization of the sp^3-transition state by hydrogen bonds in the so-called *oxyanion hole* the sp^2 ester functions are attacked, ultimately resulting in the formation of the so-called acyl-enzyme, the central intermediate of the catalytic cycle (Figure 15.1). It constitutes a covalently bound derivative that could be termed an *activated ester*, which can now be attacked by various nucleophiles (H_2O, H_2O_2, ROH,RNH_2, including amino acids and carbohydrates), resulting in a nearly unlimited number of products such as free FAs, various alcohols, esters, amides, peracids, acylated amino acids, and carbohydrates, examples of which are discussed throughout this chapter.

15.2.3 Reactions Catalyzed by Lipases

On the basis of this generally accepted mechanism, lipases catalyze various reactions as applied to triglycerides, such as hydrolysis and alcoholysis (Figure 15.2a), interesterification (Figure 15.2b), and acyl transfer reactions, leading to a wide variety of products as is described in the sections below (Figure 15.2c).

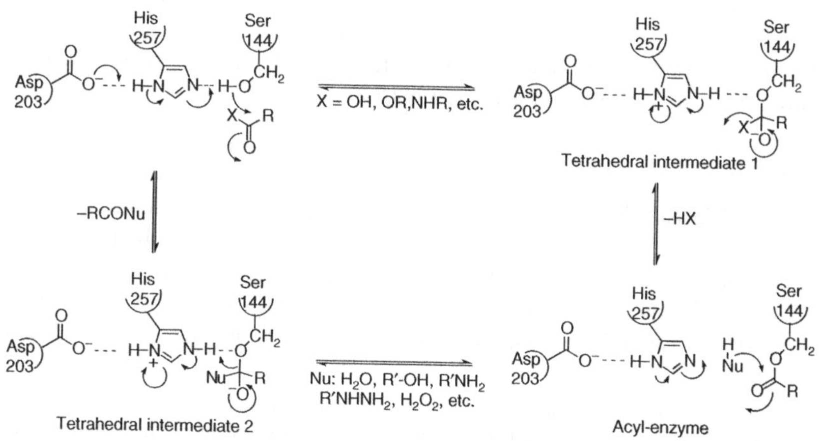

Figure 15.1 Lipases—mechanism of action (numbering of the amino acids refers to the lipase from *Rhizomucor mihei*).

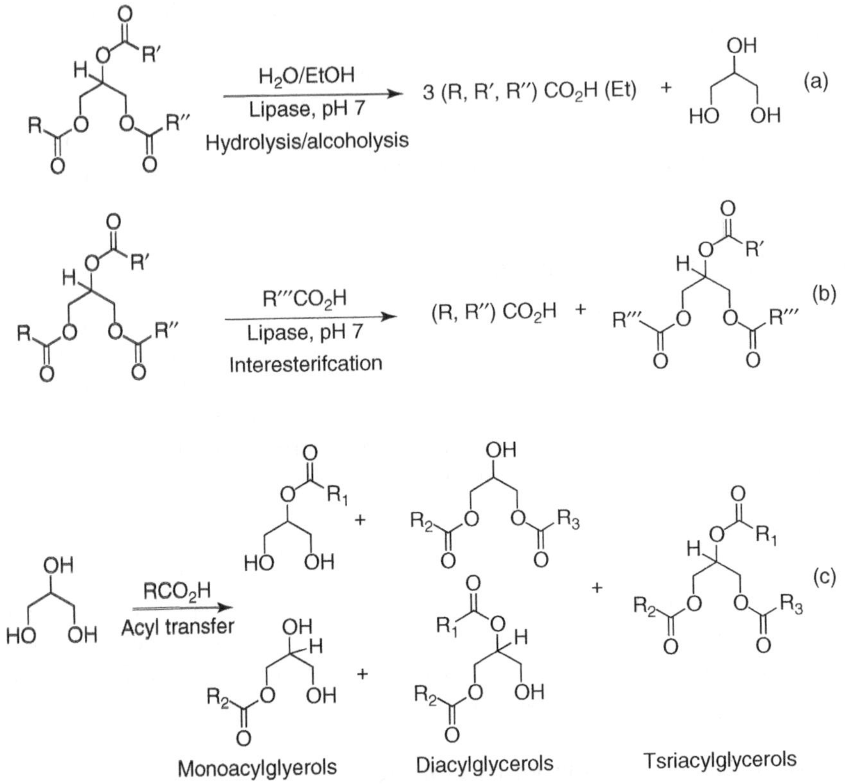

Figure 15.2 (a–c) Schematic representation of lipase-catalyzed transformations as applied to triglycerides.

In a purely aqueous environment, hydrolysis is clearly dominating, leading in the simplest cases to free FAs and glycerol. However, the situation is not quite as simple as that. The particular lipase may show positional (even enantio-) selectivity and/or selectivity regarding the degree of unsaturation in the FA, so that a wide variety of products can be obtained both in acyl transfer reactions (Figure 15.2c) and in hydrolyses (Figure 15.3).

15.2.4 Selectivities of Lipases

15.2.4.1 Regioselectivity of Lipases. Triglycerides are potentially chiral molecules and contain stereochemically different positions termed *sn*-1, *sn*-2, and *sn*-3 (*sn*) stands for *s*tereospecific *n*umbering (Figure 15.4).[1]

While positions *sn*-1 and *sn*-3 carry acylated primary hydroxy groups, the esterified secondary hydroxy group occupies position *sn*-2. Lipases

[1] http://www.chem.qmul.ac.uk/iupac/lipid/. This nomenclature—although different from the CIP-nomenclature (*R, S*), commonly used in organic chemistry—has become well established in the area of lipids.

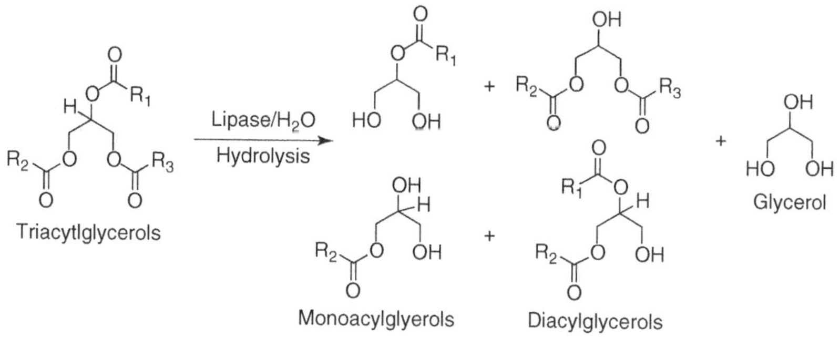

Figure 15.3 Lipase-catalyzed hydrolysis of triglycerides—possible products.

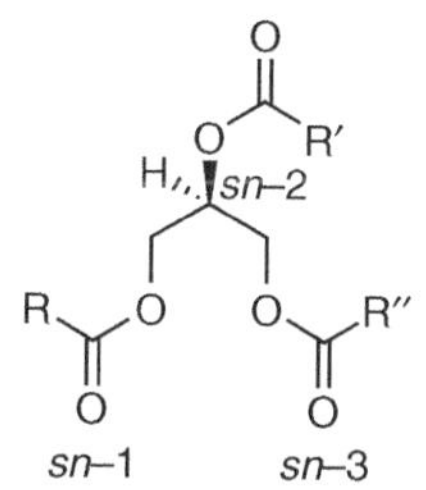

Figure 15.4 *S*tereospecific *n*umbering of triglycerides.

are able to distinguish between these positions and are thus classified as *sn*-1,3-selective, *sn*-2-selective, and nonselective. Owing to rapid acyl group migrations in substrates with glycerol substructures, when stored in protic media or in the presence of traces of acids or bases, information that is obtained from studies in aqueous or biphasic media regarding the regioselectivity of microbial lipases is inherently unreliable. Therefore, most data in the literature describing regioselectivities of lipases are at best of qualitative nature. In contrast, mono- and diglycerides are quite stable in aprotic organic solvents of low water contents (<2%). On this basis, we developed a system for the quantitative and accurate determination of regioselectivities some time ago. For this, we defined regioselectivity (RE) as follows:

$$\mathrm{RE} = \%r.ee. = \%(1,3\text{-diglyceride}) - \%(1,2\text{-diglyceride})$$

Thus, by incubating *sn*-1,3-dilaurin with various lipases in organic solvents, the ratio of 1,3- and 1,2-diglycerides asymptotically approaches a constant value and thereby allows the calculation of the Regiosomeric Excess (RE) values. As a result, it was found that the lipases from *Pseudomonas cepacia* and *Rhizomucor mihei* are only 1,3-selective with RE = 79 and 84, respectively, while the lipase from *Rhizopus delemar* (RE = 98) can be termed *1,3-specific* [4]. A selection of representative, frequently used lipases that display such differences is summarized in Table 15.2 [5].

TABLE 15.2 Commercially Available and Commonly Used Lipases [5]

Lipase	Abbreviation	*M* (kDa)	Specificity
Porcine pancreatic lipase	PPL	50	*sn*-1,3
Candida rugosa	CRL	60	Nonspecific
Candida antarctica B	CAL	60	*sn*-1,3
Geotrichum candidum	GCL	60	*Cis*-9,10-
Humicula lanuginosa	HLL	30	Nonspecific
Rhizomucor mihei	RML	30	*sn*-1,3
Aspergillus oryzae	AOL		
Penicillium camemberti	PEL	30	*sn*-1,3
Rhizopus delemar	RDL	41	*sn*-1,3
Rhizopus oryzae	ROL	41	*sn*-1,3
Rhizopus arrhizus	RAL	41	*sn*-1,3
Pseudomonas glumae	PGL	33	Nonspecific
Burkholderia cepacia	PCL/BCL	33	Nonspecific
Chromobacterium viscosum	CVL	33	*sn*-1,3-

M = molecular weight.

15.2.4.2 Lipases—Fatty Acid Profile. The FAs of triglycerides can vary widely regarding (i) chain length and (ii) degree of unsaturation, and lipases display distinct preferences in this respect. Several methods for the determination of these properties have been described in the literature, mainly based on lipase-catalyzed hydrolyses of triglycerides in aqueous emulsions. Unfortunately, for every determination a separate experiment is required. We, therefore, developed a simple and rapid procedure by which the complete FA profile of a given lipase can be determined in one single experiment. Thus, in a competitive experiment the glycerolysis of equimolar mixtures of triglycerides in t-BuOCH$_3$ as solvent was studied, leading to competition factors α as a measure of the FA selectivity. As a result it became clear that *R. mihei* (Lipozyme™) shows a strong activity maximum for tricapryline and very low activity toward unsaturated FAs. In contrast, the lipase from *Geotrichum candidum* shows a particularly high activity toward unsaturated FAs. The short time required for the analysis makes the method a good practical tool for the screening of lipase preparations [7]. The results can be used in practice, for example, in the synthesis of structured triglycerides (Section 15.4.2).

15.3 SURFACE-ACTIVE COMPOUNDS

15.3.1 Partial Glycerides

The in-part hydrolysis of triglycerides usually leads to mixtures of mono- and diglycerides (compare Figure 15.3). Such "partial glycerides" are classically synthesized *via* base-catalyzed glycerolysis of native plant oils (Figure 15.5). The resulting mixtures consist of free FAs, practically all positional isomers of mono- and diglycerides and even triglycerides. They can be used as mixtures for a

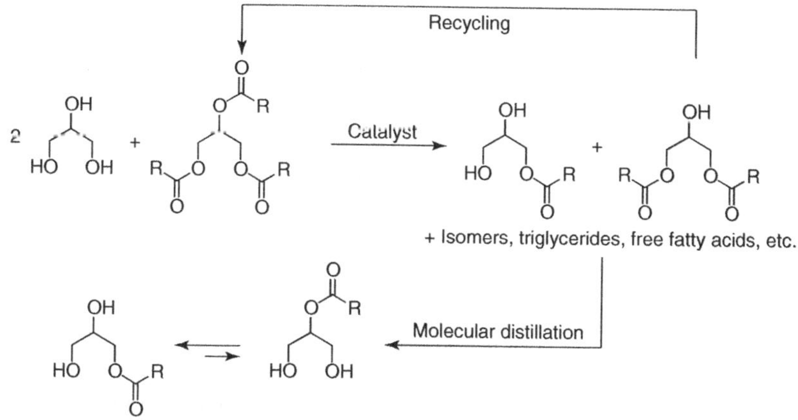

Figure 15.5 Industrial synthesis of *partial glycerides* (*mono acylglycerols*).

number of applications; however, the so-called "monoglycerides" (mono acylglycerols) (MAGs) must be isolated (purified) by short path molecular distillation, a highly energy-intensive process leading to liquids that are not color free because of the high temperatures involved. The resulting emulsifiers are used in various food and cosmetic applications.

Using such chemical approaches, pure positional isomers of MAGs cannot be obtained. In contrast, lipase-catalyzed acylations of glycerol have been demonstrated to achieve this goal fairly easily on the basis of the selectivity of the biocatalyst and/or the methodology employed. Besides having products of high quality and being colorless and odorless, these materials are also suitable for high value-added applications such as synthetic building blocks, the lipid modification of drugs and proteins, or the effective synthesis of other surfactants such as "monoglyceride sulfates" (compare Section 15.3.5).

15.3.2 Diacylglycerols

Diacylglycerols (DAGs) can exist in three isomeric forms, a symmetrical (*sn*-1,3-) derivative and two enantiomers (*sn*-1,2 and *sn*-2,3). These are thermodynamically less stable than the symmetrical molecules and can undergo acyl group migrations, resulting in equilibrium mixtures in favor of the symmetrical, achiral *sn*-1,3-isomer (Figure 15.6).

Owing to their claimed high nutritional value, *sn*-1,3-DAGs have become of interest as constituents of cooking oils (Section 15.4.1), they are interesting as building blocks for structured lipids (Section 15.4.2) and also as building blocks for organic syntheses (Section 15.3.3 ff).

15.3.3 Isomerically Pure *sn*-1,3-Diacylglycerols (DAG) as Synthetic Building Blocks

To use *sn*-1,3-DAG as synthetic building blocks, these molecules must be of high isomeric purity, which cannot be achieved by purely chemical methods.

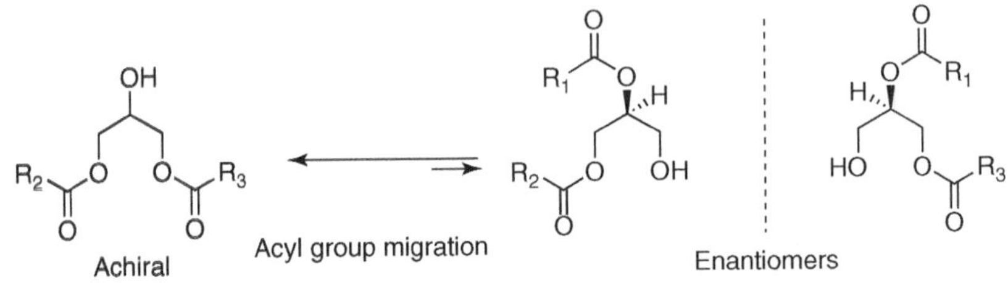

Figure 15.6 Stereoisomers of diacyl glycerols: acyl group migrations leading to racemization.

These lack positional selectivity and the isolation of the pure isomers requires additional purifications. In contrast, enzymatic transformations in the presence of 1,3-selective lipases have been shown to achieve this goal relatively easily, and thus numerous reports of such transformations can be found in the literature.

15.3.3.1 Isomerically Pure sn-1,3-Diacylglycerols (DAG)—Strategy. Clearly, the best way for the synthesis of *sn*-1,3-DAG is the lipase-catalyzed esterification of glycerol, employing 1,3-selective lipases and using a variety of acyl donors such as free FAs, their esters, and also the corresponding vinyl esters for irreversible acyl transfer (Figure 15.8).

Lipases display their highest specific activity in aprotic organic solvents of low polarity such as *n*-hexane or *t*-$BuOCH_3$. Unfortunately, hydrophilic glycerol is immiscible with such solvents and thus many earlier attempts for direct esterifications in such media have been unsuccessful. This problem can easily be overcome by prior adsorption of glycerol onto a solid support [6,8,9].

Typically, glycerol and the corresponding inert support (e.g., silica gel) are mechanically mixed until free flowing "dry" powders are obtained, which contain up to 50% (w/w) of immobilized glycerol. Suspensions of the thus immobilized glycerol in nonpolar organic solvents react as if these materials are homogeneously dissolved. Using a variety of acyl donors and various lipases, acyl transfer reactions proceed smoothly under these conditions; the reaction progress can be conveniently monitored by TLC. On completion of the reactions, both the immobilized enzyme and the solid support are removed by simple filtration or centrifugation, allowing the convenient isolation of the resulting glycerides. The method can be employed both for the syntheses of *sn*-1,3-DAG and *sn*-1(3)-MAGs (Section 15.3.4; Figure 15.7).

This solution to the solubility problem—the immobilization of glycerol on a solid support—followed by subsequent esterifications in the presence of lipases has also been successfully used by other groups [10–12]. Mechanistic studies [13] indicate that the success of the method is not based on the creation of an artificial interface for improved solubility of glycerol, as was previously assumed, but by preventing the hydrophilic glycerol from covering the immobilized lipase and thus not allowing access of the acyl donor to the active site of the biocatalyst; SiO_2 simply serves as a depot for the hydrophilic glycerol.

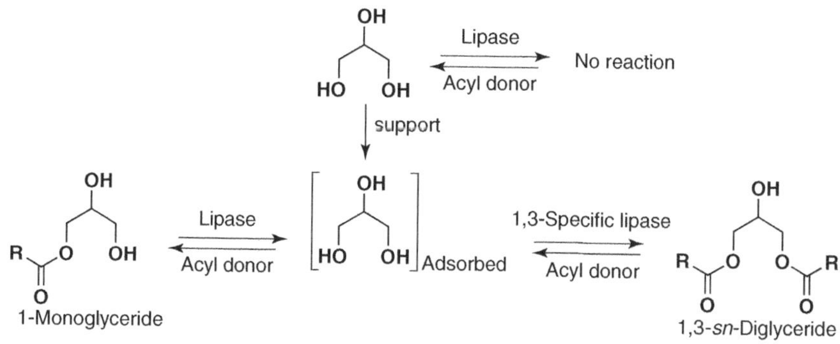

Figure 15.7 Facile lipase-catalyzed acylation of adsorbed glycerol in organic media.

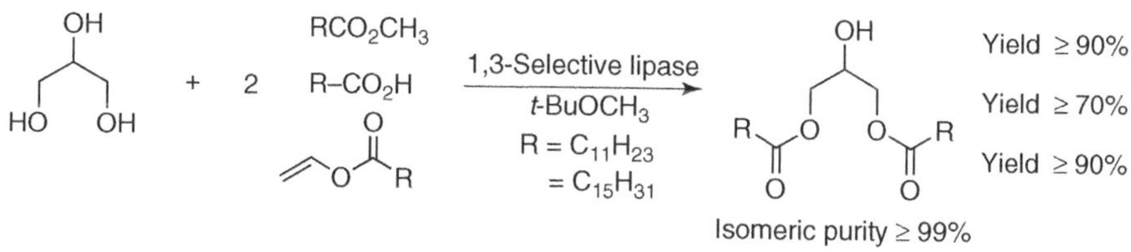

Figure 15.8 Isomerically pure *sn*-1,3-diacyl glycerols by lipase-catalyzed acylations.

15.3.3.2 Syntheses of Regioisomerically Pure sn-1,3-DAG. Using a molar ratio of glycerol and acyl donor of 1:2, numerous isomerically pure 1,3-*sn*-DAGs were prepared in high yields and purities. Examples for such preparations, using free FAs, FAME and FA vinyl esters, are shown in Figure 15.8 [14]. Yields are excellent (>90% in the case of vinyl esters and free FAs) to satisfactory (>70%) in the case of FAME [8].

The method is suitable for the preparation of such molecules with FAs of widely variable chain length and also various degrees of unsaturation. Owing to the simplicity of the purification process by recrystallization, the method is best suited for the preparation of saturated *sn*-1,3-DAG of moderate (C_8) to long (C_{18}) chain length. There are, of course, numerous alternative methodologies published in the literature, such as employing esterifications in solvent-free systems [15a], fluorous solvents or supercritical solvents such as $scCO_2$ [15b], or ionic liquids [15c]. From an environmental point of view, solvent-free transformations are certainly highly desirable and should be used whenever possible. Although solvents used for lipase-catalyzed transformations are usually nonpolar and aprotic, the need for (energy-intensive) recycling remains; also, yields of recovered solvents are rarely quantitative. From a practical, preparative point of view, solvent-free transformations are only of advantage if the reactants get mixed well and the reaction rates and conversions are high. Frequently, one of the reactants (FA ester, low-boiling alcohols such as CH_3OH and EtOH, etc). is used in excess, thus acting both as solvent and as reaction partner. Generally speaking, the reactants should be miscible. For improved solubility, the addition of surfactants has been employed or supercritical solvents have been used as alternatives to classical solvents. Frequently, by-products such as water or low-boiling alcohols have to be removed in order to drive the conversions to completion, requiring processes which may be difficult to achieve in the case of highly viscous reaction mixtures. In such cases, a special reactor design may be the answer [15d]. For the removal of reaction products such as water or methanol, applied vacuum (sometimes in tandem with a stream of nitrogen) seems to be the most widely applied methodology. The use of solvent-free systems in biocatalysis has recently been reviewed [15b], including the use of small amounts of solvents in the case of solid–liquid reaction mixtures (compare Section 15.4.2). Although it is doubtful whether ionic liquids can qualify as green solvents, numerous examples for their use can be found in the literature, which has also been reviewed recently [15c].

The *sn*-1,3-DAGs resulting from the above transformations (Figure 15.8) are stable and isomerically pure and possess a free hydroxy group in the *sn*-2-position, which can be modified both chemically and enzymatically. Thus, the introduction of long-chain, unsaturated FAs at this position leads to structured TAG of nutritional value (Section 15.4.2). They can also serve as synthetic building blocks in organic synthesis [16–19]. For example, various lipid-modified pro-drugs can be obtained by esterification with pharmaceuticals such as aspirin, (*S*)-ibuprofen, niflumic acid [20], bupranol [21] or phenytoin [22]. Also amino acids and various other functionalities such as phosphocholine residues can be introduced into this position, leading to nearly unlimited synthetic possibilities. Representative examples are shown in Figure 15.9.

15.3.4 Monoglycerides

Mixtures of long-chain MAGs and other partial glycerides of varying composition and constitution are widely employed as nonionic surfactants (emulsifiers) in the processing of food, in cosmetics, as excipients for pharmaceuticals, and

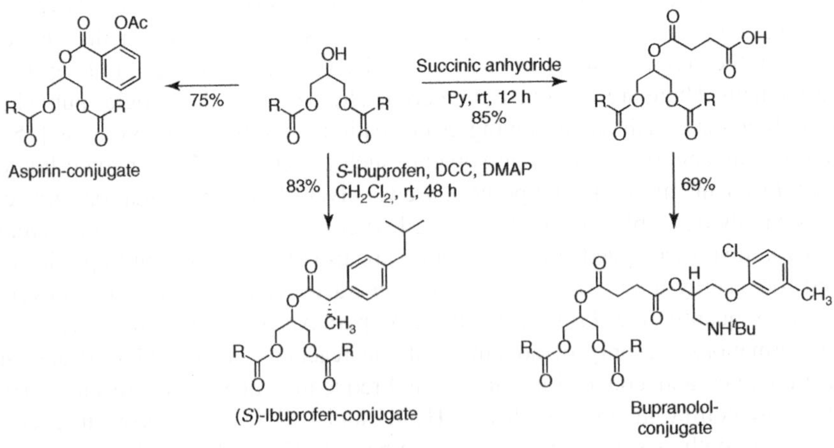

Figure 15.9 Isomerically pure 1,3-*sn*-DAG for the lipid modification of drugs.

related applications. They also serve for the synthesis of numerous conjugates, for example, with citric, lactic, tartaric, and acetic acid. Isomeric mixtures of such "monoglycerides" are produced in kiloton quantities. They are classically made by glycerolysis of native oils with 2 equivalents of glycerol in the presence of inorganic catalysts at 210–240°C and under a nitrogen atmosphere (compare Figure 15.5).

The resulting isomeric mixtures contain only about 50–60% of the desired MAGs together with isomeric diglycerides, triglycerides, and free FAs. Owing to the high temperatures employed, these materials are usually colored and not free of odor [23]. "Monoglycerides" of higher chemical (but not isomeric!) purity (>90%) can be obtained only by cost- and energy-intensive molecular distillation of these crude mixtures. From a stereochemical point of view, monoglycerides can exist as three positional isomers, two being enantiomers and one an achiral molecule (Figure 15.10).

Thus, we must differentiate between the synthesis of monoglycerides as mixtures (*sn*-1, *sn*-3, and *sn*-2), pure isomers (i.e., *sn*-2), and single enantiomers (*sn*-1 or *sn*-3). Examples for all three possibilities are given in this section.

15.3.4.1 Enantiomerically Enriched Monoglycerides. To the best of the author's knowledge, none of the existing enzymatic methods allows the *direct* synthesis of enantiomerically *pure* monoglycerides such as *sn*-1 (*S*) or *sn*-3 (*R*). Moderate enantioselectivities were obtained by the lipase-catalyzed desymmetrization of glycerol using vinyl benzoate or benzoic anhydride as acyl donors (54% *ee*; 98% *ee* after recrystallization) [24a] or by the kinetic resolution of racemic isopropylidene glycerol (up to 65% *ee*) [24b]. Very high enantioselectivities were achieved in the lipase-catalyzed desymmetrization of 2-*O*-benzyl glycerol by both acyl transfer and hydrolysis (Figure 15.11). Interestingly, and typical of desymmetrizations, depending on the process (hydrolysis vs esterification) opposite enantiomers were obtained [25].

15.3.4.2 Isomerically Pure 2-Monoglycerides. The synthesis of achiral *sn*-2-MAGs can be achieved by alcoholysis (ethanolysis) of triglycerides in the presence of 1,3-selective lipases [26,27] (Figure 15.12).

Practically, only the 1- and 3-positions are attacked, leaving the 2-position largely untouched. The most important prerequisites for the use of these

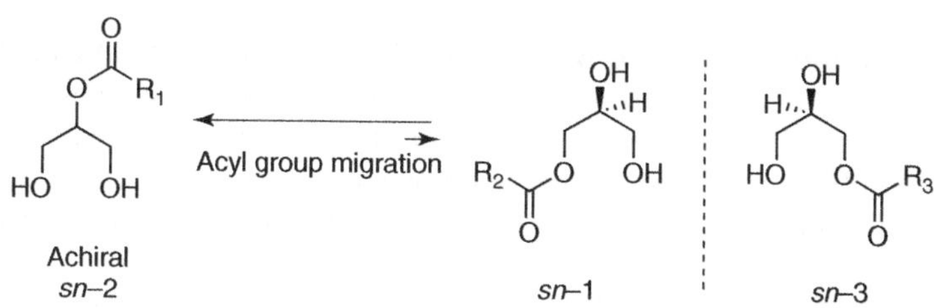

Figure 15.10 Stereoisomers of mono acylglycerols: acyl group migrations lead to racemization.

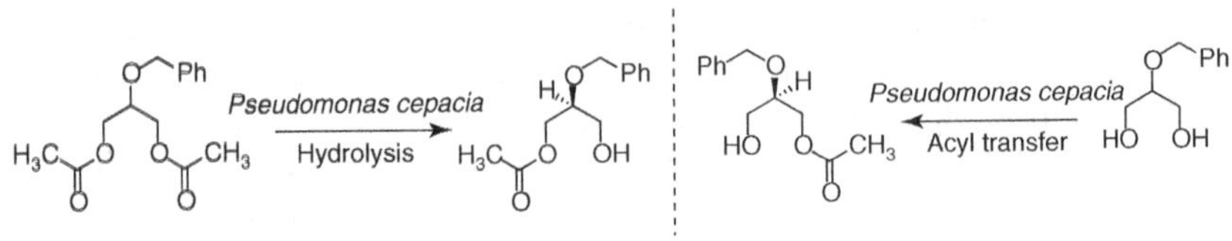

Figure 15.11 Lipase-catalyzed desymmetrizations of 2-*O*-benzyl-1,3-diacetyl glycerol and 2-*O*-benzyl glycerol.

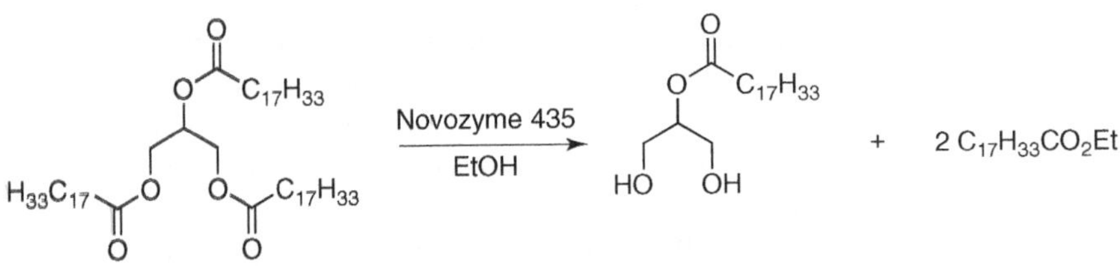

Figure 15.12 Synthesis of isomerically pure 2-*O*-acyl glycerols using 1,3-selective lipases.

molecules as synthetic building blocks are (i) their synthesis as single isomers and (ii) the stability of these isomers under the employed reaction conditions. The 2-monoglycerides obtained this way are thermodynamically instable (compare Figure 15.10) and under a variety of conditions undergo the so-called acyl group migrations, leading to a thermodynamic equilibrium mixture with the corresponding *sn*-1(3)-MAGs. If, however, care is taken to avoid such migrations [28,29], which usually take place in protic solvents, under acid/base catalysis or elevated temperatures, they can be used as building blocks for various applications, such as the synthesis of structured lipids [26,27,30] (see also Section 15.4.2).

15.3.4.3 Isomeric Mixtures of Mono Acylglycerols. As already pointed out, these so-called "monoglycerides" are important emulsifiers for the processing of food, for cosmetical applications, as excipients for pharmaceuticals, and many more. The lipase-catalyzed syntheses of these molecules have been reviewed extensively [31–33], and thus this section only addresses (i) some leading principles and (ii) the preparation of single isomers.

15.3.4.4 Isomerically Pure Monoglycerides—Principles and Methods. Two different strategies can be employed for lipase-catalyzed syntheses of these target molecules: (i) the use of temporary protection groups such as phenylboronic acid [34,35] (Figure 15.13) or the isopropylidene group [36] (Figure 15.14), followed by removal of the protection group; and (ii) direct lipase-catalyzed acylations of glycerol with the options shown in Figure 15.15.

Clearly, the most economical way to prepare the target molecules is the lipase-catalyzed glycerolysis of native fats and oils (Figure 15.15a). In a stoichiometric reaction, 3 mol of monoglycerides can be obtained from 1 mol of TAG and 2 mol

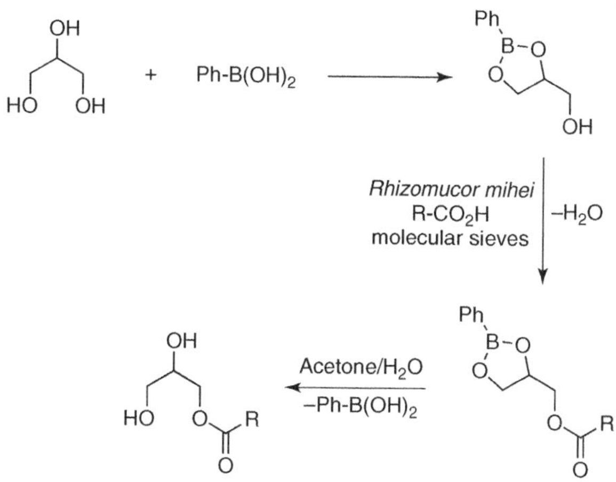

Figure 15.13 Synthesis of mono acylglycerols using temporary protection groups—phenylboronic acid.

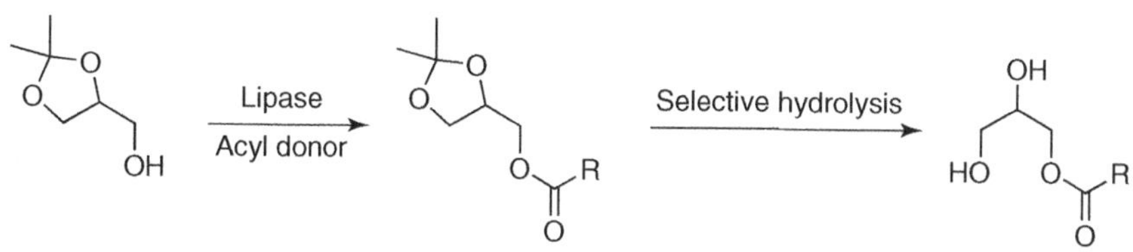

Figure 15.14 Synthesis of mono acylglycerols using temporary protection groups—acetonides.

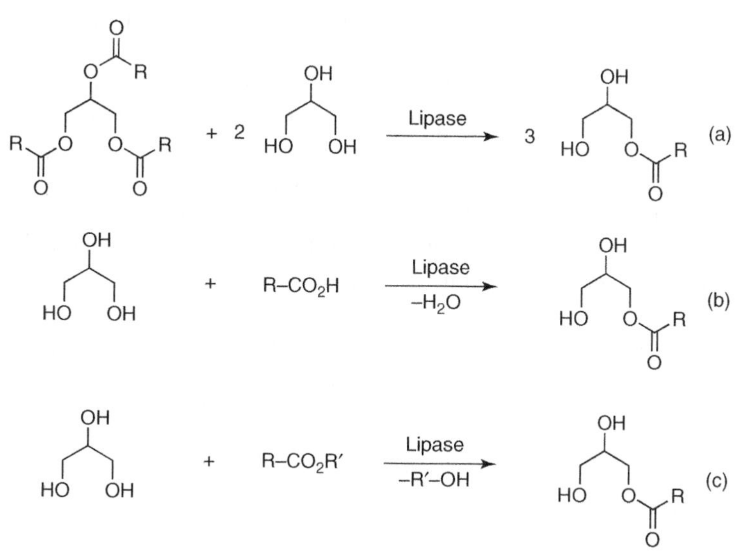

Figure 15.15 Mono acylglycerols *via* acyl transfer reactions.

of glycerol. Under ambient temperatures usually mixtures of isomers are obtained with only low yields of the desired monoglycerides. By lowering the temperature to a value specific for every substrate, the so-called *critical temperature* [37], yields are increased to 70% or higher. Thus, in carrying out the enzymatic glycerolysis at ambient temperatures or above (42°C) for 8–16 h, followed by incubation at 5°C for four days, 90% of monoglycerides was obtained from beef tallow, palm oil, and palm stearin [38]. Triolein was converted into "monoolein" with a yield (purity) of 99% in 120-h reaction time at 8°C [39]. Clearly, the monoglycerides thus prepared crystallize out preferentially at the lower temperature, thus shifting the equilibrium in favor of the monoglyceride. The critical temperature for the production of MAG correlates with the melting point of the MAG. The same principle has been used for the syntheses of MAG derived from physiologically interesting polyunsaturated fatty acids (PUFAs) such as eicosapentaenoic acid (EPA) and docosahexaenoic acid (DHA), where the chemical synthesis cannot be applied owing to the heat sensitivity of these FAs or in case of synthetic conjugated linoleic acids (*c*9,*t*11). Yields up to 93% were achieved in carrying out the glycerolysis at 5°C [40,41] (Figure 15.16).

In addition, direct esterifications using free FA or their esters can be employed (Figure 15.15b and 15.15c). To achieve reasonable yields the produced water (alcohol) must be removed, which can be achieved by (i) addition of molecular sieves, (ii) applying vacuum (sometimes in tandem with a nitrogen flow), (iii) azeotropic removal by distillation, and (iv) reactions on solid phases. Further applications involve the use of inverse micelles [32,33].

15.3.5 Isomerically Pure 1(3)-*rac*-mono acylglycerols

Frequently, the "monoglycerides" obtained by the methods described above are not chemically and isomerically pure, but also contain free FAs, other partial glycerides, and triglycerides. They may be useful for practical and technical applications as outlined above, but certainly not as synthetic building blocks. The same is true for technically produced "monoglycerides." For the synthesis of isomerically pure 1(3)-*rac*-MAGs, a special strategy was developed on the basis of the synthetic procedure described above for the synthesis of isomerically pure 1,3-*sn*-DAG (Section 15.3.4; Figures 15.7 and 15.8).

In principle, it should be straightforward to extend the method described above also to the production of regioisomerically pure 1(3)-*sn*-MAGs.

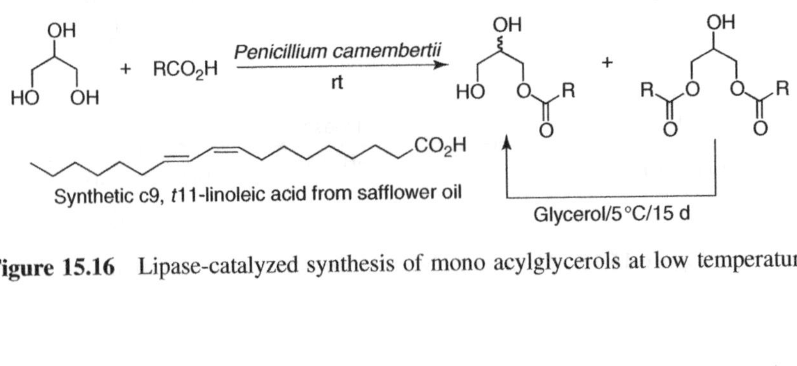

Figure 15.16 Lipase-catalyzed synthesis of mono acylglycerols at low temperature.

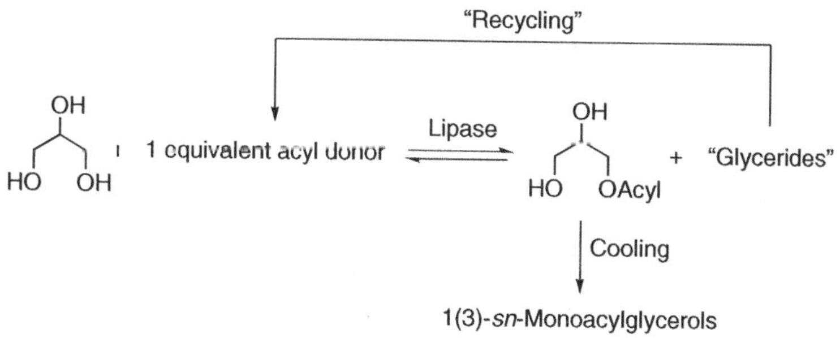

Figure 15.17 Regioisomerically pure 1(3)-*sn*-mono acylglycerols: synthetic process.

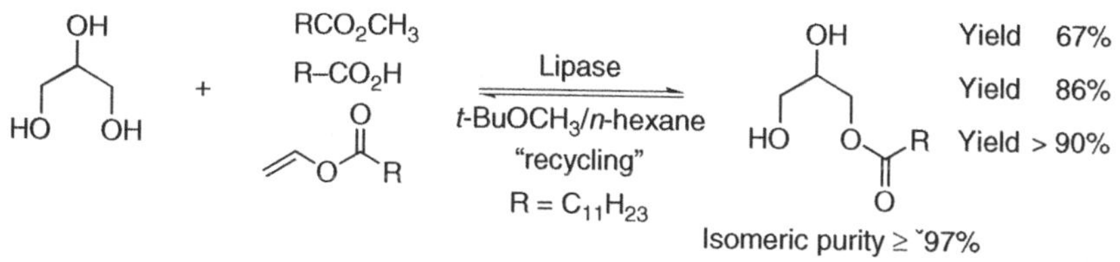

Figure 15.18 Isomerically pure mono acylglycerols *via* lipase-catalyzed acylations [6,9].

Unfortunately, however, these products were produced under the above conditions only in low to moderate yields. At first glance, and in view of the high regioselectivities displayed by many lipases regarding the primary positions of glycerides, it seemed possible to increase their yield by simply employing an excess of glycerol. Nonetheless, even with a 10-fold excess of glycerol the content of 1(3)-*rac*-MAGs in the reaction mixture did not exceed 70%. Obviously, the MAGs primarily produced are excellent substrates for the lipase catalysts and are rapidly converted into the corresponding 1,3-*sn*-DAGs.

Clearly, a practical and useful method requires the following:

1. stoichiometric reactions with quantitative conversions of all employed reactants and without the need of excess glycerol;
2. a simple method for the continuous separation of the desired products from the reaction mixtures; and
3. an effective recycling method for all undesirable materials, including 2-*sn*-MAGs, DAGs, and TAGs.

All three goals were achieved simultaneously as outlined in Figure 15.17 by a compartmental separation of the two steps of the process—synthesis and isolation. The enzymatic esterification is carried out in the reactor vessel with stoichiometric amounts of both glycerol and the corresponding acyl donor under the desired reaction conditions (solvent, temperature). The solution of the reaction mixture thus obtained is circulated into a second vessel—termed *separator*—in which the desired MAGs are separated out at lower temperatures, the so-called *critical temperature* (see above [40,41]).

Crucial for the success of this method is the choice of the solvent or solvent mixture in which the desired 1(3)-*rac*-MAGs are less soluble than all of the other products, at least at the low temperatures employed. While the desired products are thus precipitating out in the cooled separator, all other products, as well as unreacted acyl donors, remain in solution and are fed back to the reactor, which contains both the enzyme and the glycerol on the solid support. High yields of monoglycerides are obtained this way using a variety of acyl donors. The success of the method was demonstrated with numerous acyl donors, such as free FAs, FA methyl- and vinylesters (Figure 15.18), synthetic triglycerides (Figure 15.21), as well as natural fats and oils (Figure 15.22).

The method is not limited to natural FAs; unnatural FAs can also be employed. In Figure 15.19, the synthesis of mono pentadecanoin, a material used in cosmetic applications, is shown.

Important from a practical point of view is the fact that unpurified technical FA mixtures can be used directly, leading to the corresponding monoglycerides in high yields and excellent chemical purities (Figure 15.20).

Unspecific lipases are best suited for the conversion of triglycerides (Figure 15.21) and natural fats and oils into regioisomerically pure *sn*-1(3)-monoglycerides. All acyl groups should be used by the employed biocatalysts, including those in the *sn*-2-positions of triglycerides.

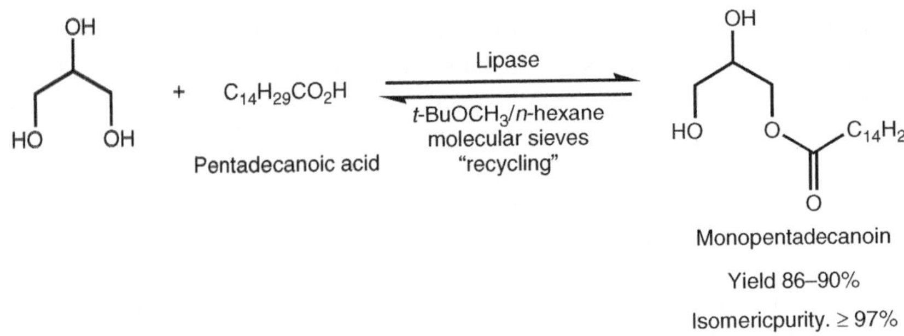

Figure 15.19 Lipase-catalyzed synthesis of 1(3)-*rac*-mono acylglycerols using nonnatural fatty acids.

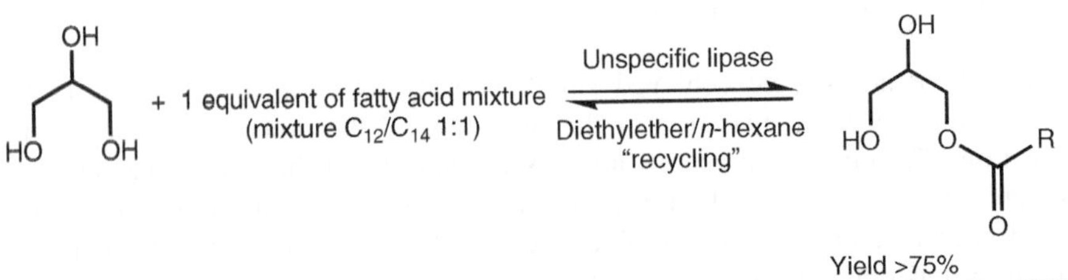

Figure 15.20 Mono acylglycerols from technical mixtures of free fatty acids.

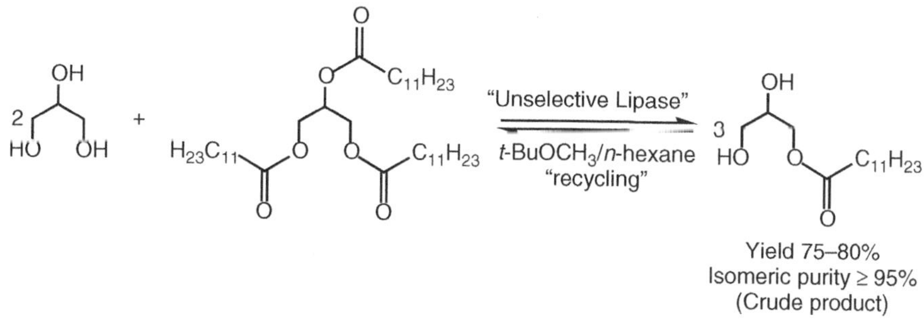

Figure 15.21 Mono acylglyerols by glycerolysis of triglycerides.

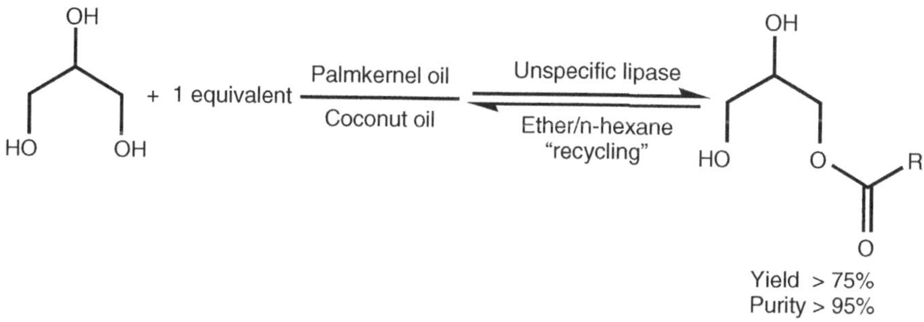

Figure 15.22 Mono acylglycerols by lipase-catalyzed glycerolysis of native oils.

The conversion of natural coconut oil or palm kernel oil leads to mixtures of the corresponding monoglycerides in high yields and purities without the need for further purification (Figure 15.22). The natural distribution of FAs, shown as triglycerides in the GC—diagram of the starting oil (Figure 15.23a)—is converted exclusively into the corresponding mixture of isomerically pure 1(3)-*rac*-mono acylglycerols (Figure 15.23b) with only traces of the corresponding isomers or diglycerides being present.

A selection of the 1(3)-*rac*-MAGs thus prepared can be found in Refs [6,9]. This method (the prior immobilization of glycerol onto silica gel) for the syntheses of MAG from native fats and oils was also employed successfully by other groups [42]. Isomerically pure *sn*-1(3)-monoglycerides can be potentially used for the synthesis of numerous interesting and useful compounds such as structured triglycerides (Section 15.4.2), phospho- and glycolipids, liquid crystals, drug carriers, surfactants, pharmaceuticals reagents for the lipid modification of drugs and proteins, and many more. From the numerous possibilities, just two examples were selected.

15.3.6 Isomerically Pure Monoglyceride Sulfates

Regioisomerically pure monoglycerides are ideally suited for the environmentally benign synthesis of monoglyceride sulfates, a class of skin-friendly surfactants. Treatment of the isomerically pure mixture of monoglycerides derived from

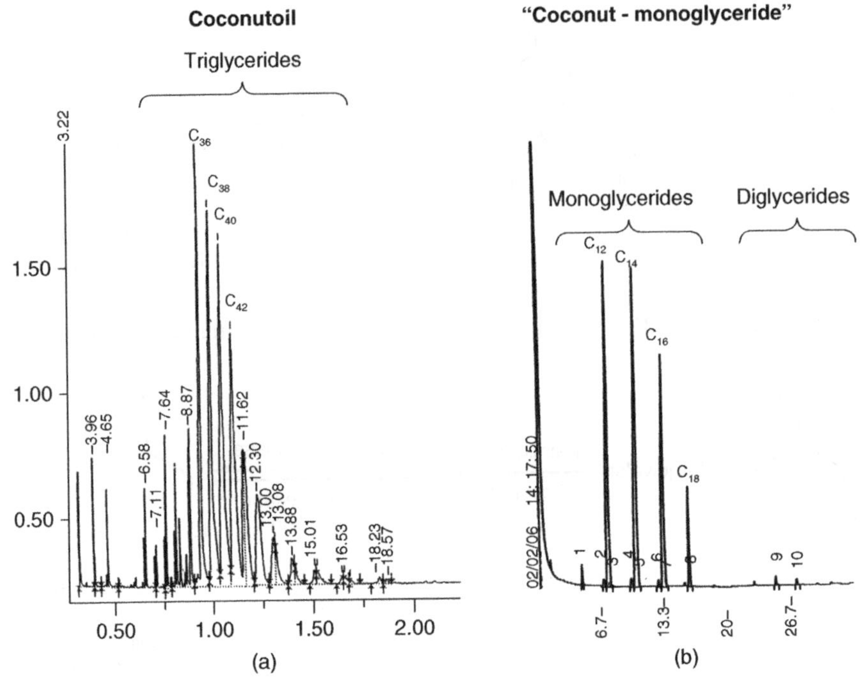

Figure 15.23 Isomerically pure mono acylglycerols from technical grade coconut oil: (a) starting material and (b) products.

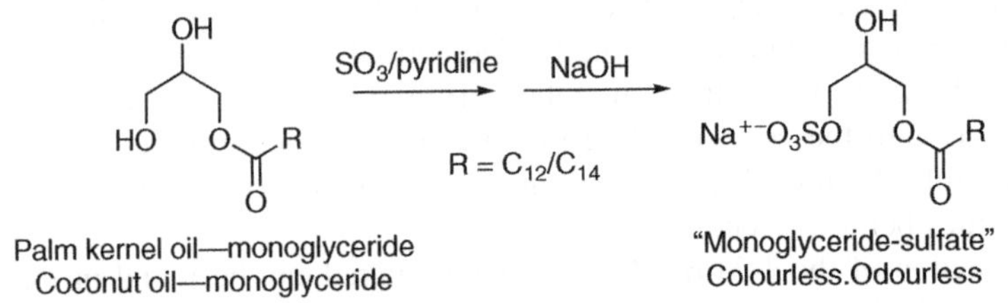

Figure 15.24 Monoglyceride sulfates from mono acylglycerols.

coconut oil with SO_3·pyridine followed by neutralization with NaOH yields these compounds as colorless and odorless powders [43] (Figure 15.24). However, for this conversion, monoglycerides of high purity are required. Monoglyceride sulfates are usually prepared on an industrial scale by reacting a mixture of 1 mol of coconut oil (palm kernel oil) and 2 mol of glycerol with 18 mol of fuming sulfuric acid, the large excess being required for operational (stirring) reasons. Neutralization produces an unacceptable 9.5 kg of Na_2SO_4 for every kilogram of product, which is also colored and not free of odor. In contrast to this classical approach, when using the above enzyme-assisted method, all transformations are stoichiometric and no waste products are being produced. The high-quality

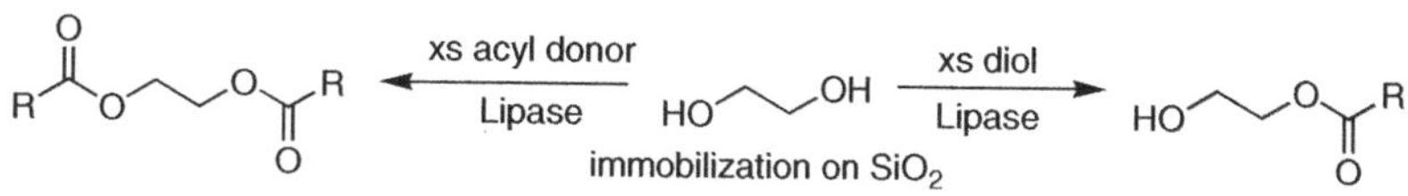

Figure 15.25 Lipase-catalyzed acylation of hydrophilic diols (solvent *n*-hexane), R = C_{11} H_{23}.

products thus obtained form lyotropic liquid crystalline phases in water and are also thermotropic [16]. They could thus also be used as valuable surfactants for the formulation of pharmaceuticals and cosmetics.

As shown above (Figure 15.19), regioisomerically pure *sn*-1(3)-monoglycerides can also be synthesized from nonnatural FAs. This approach can be used to prepare highly pure synthetic building blocks. Thus, 1(3)-*rac*-6-bromo monocaproin was synthesized using this method and converted in just two steps into the corresponding Platelet-Activating Factor (PAF)—antagonists BN52111 and BN52116 [16].

The synthesis of mono- and diglycerides based on the immobilization of glycerol onto SiO_2 was further extended toward various hydrophilic diols, exemplified here for ethylene glycol [44] (Figure 15.25).

15.4 LIPASE-CATALYZED SYNTHESES OF LIPIDS FOR DIETARY APPLICATIONS

Lipase-catalyzed transformations for the modification of lipids, especially for dietary applications, have a long tradition, and numerous products have been commercialized. Most prominent examples are cocoa butter equivalents (CBEs) and Betapol™ for infant formulations, which closely resemble mother's milk. CBE carries mainly stearic acid in the *sn*-1,3-positions and oleic acid in the *sn*-2 position; Betapol has palmitic acid in the *sn*-2 and unsaturated FAs in the *sn*-1,3-positions. Both are manufactured by lipase-catalyzed interesterifications (compare Figure 15.2b)—CBE from the interesterification of high oleic sunflower oil and stearic acid and Betapol from interesterification of palm stearin with the FAs from high oleic sunflower oil [45].

In the section below, the lipase-catalyzed preparation of diglycerides as novel cooking oils and the synthesis of structured, functional lipids are discussed.

15.4.1 Diglycerides as Cooking Oils

DAGs (mainly *sn*-1,3-DAG) have attracted considerable attention as healthy components in cooking oils [46]. Studies on the nutritional properties of DAG have revealed their ability to reduce serum lipids (TAG) [47] and as a result to decrease body weight and visceral fat mass [48]. Using classical chemical methods that employ basic catalysts, products with a high content of the desired *sn*-1,3-DAG cannot be obtained [49]. In contrast, lipase-catalyzed esterifications of glycerol in the presence of 1,3-selective lipases have been shown to produce these materials with high yield and acceptable purity (Sections 15.3.2 and 15.3.3). The most recent method for the commercial production of *sn*-1,3-DAGs as constituents of cooking oils involves (i) partial hydrolysis of native plant oils containing mainly unsaturated FAs followed by (ii) direct esterification of glycerol in a solvent-free process employing 1,3-selective lipases [50]. With the use of a commercially available lipase of *R. mihei* (Lipozyme RM IM™), a maximum yield of 84% with a purity of 90% was claimed for this process [51] (Figure 15.26). These DAG cooking oils have been approved by the FDA (termed *GRAS*) and are commercially available in the United States under the trade name Enova. A review describing the nutritional characteristics of DAG oils has been published [52].

As an alternative to the direct esterification shown in Figure 15.26, the glycerolysis of TAG in the presence of a nonspecific lipase (*Candida antarctica* B) in a solvent-free system allowed the synthesis of these molecules in reasonable yields (60–65%) [53].

15.4.2 Structured and Functional Lipids

The nutritional value of triglycerides is not only determined by the FA composition but also by their positional distribution within the glycerol substructure. Structured lipids of the MLM (*m*edium, *l*ong, *m*edium)-type, carrying FAs with medium chain length in the *sn*-1 and *sn*-3 positions and a long chain, preferably unsaturated FA in the *sn*-2 position, possess high nutritional value in providing rapid delivery of energy *via* oxidation of the medium-chain FAs, while at the same time providing an adequate supply of essential FAs [54]. This is of particular importance for patients with malabsorption of fats and oils due to pancreatic insufficiency. Medium-chain triglycerides are more easily absorbed in the human intestine than long-chain triglycerides. Unfortunately, lipids of the MMM (medium, medium, medium)-type do not contain essential FAs such as linoleic and linolenic acids. This drawback can be overcome by using structured lipids of the MLM type. Pancreatic lipase hydrolyzes MLM faster than LLL (long, long, long)-type lipids and the resulting 2-monoglycerides are absorbed more efficiently. Mixtures of MMM and LLL types will not serve this purpose owing to these adsorptive problems.

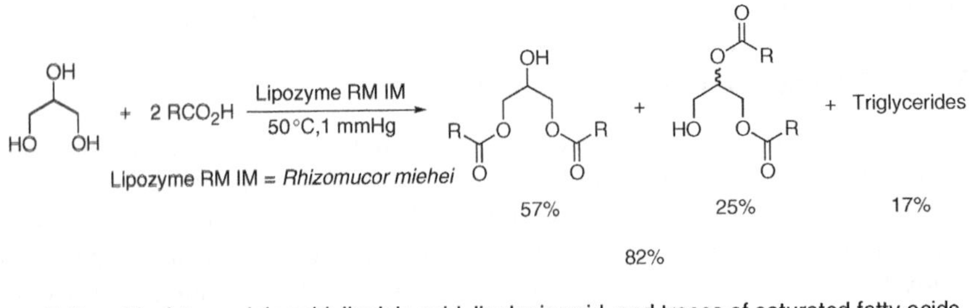

Figure 15.26 Lipase-catalyzed synthesis of diacylglycerols: dietary cooking oils (Enova).

A range of methods have been published for the preparation of these materials, using both purely chemical and enzymatic methods [55,56].

Two examples for the synthesis of such structured lipids were chosen to demonstrate the general principle. If, for example, triolein is used as substrate, the regioselective hydrolysis of the ester functions using a 1,3-specific lipase such as *R. delemar* leads to the corresponding 2-olein (compare Section 15.3.4.2), which can be obtained isomerically pure by recrystallization. In a second step and using caprylic acid as acyl donor and a 1,3-specific lipase such as *R. delemar*, the free 1- and 3-positions can be esterified, leading to the corresponding structured lipid in high isomeric purity [26,57] (Figure 15.27).

Next to triolein, trilinolein and native peanut oil were used for similar transformations. In a complementary approach, *sn*-1,3-DAG, prepared by lipase catalysis, can be used as the starting material. As shown above, these compounds can be obtained with high isomeric purity by direct lipase-catalyzed acylation of glycerol, followed by recrystallization. Using an FA-specific lipase, which does not act on the FAs at the 1- and 3-positions and at the same time displays high specificity toward oleic acid, leads to structured lipids in which up to 90% of oleic acid is in the desired 2-position [58] (Figure 15.28).

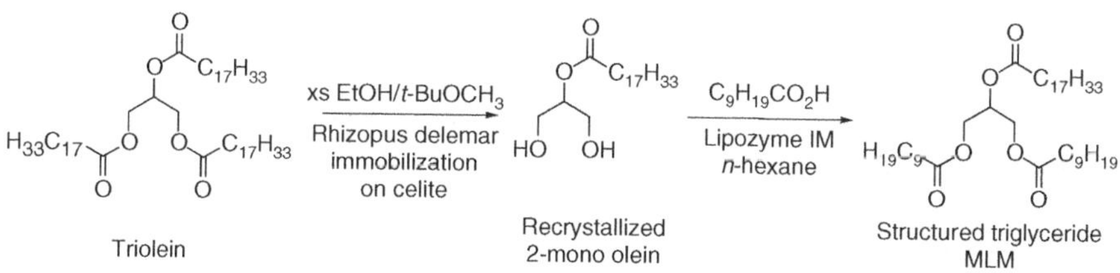

Figure 15.27 Lipase-catalyzed synthesis of structured lipids *via sn*-2-mono acylglycerols.

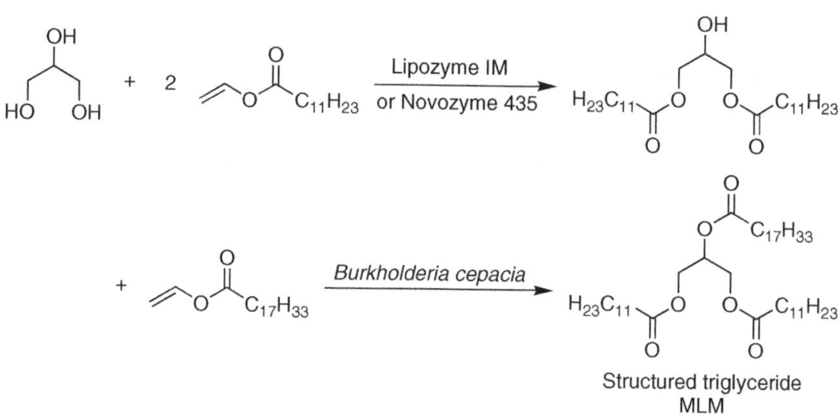

Figure 15.28 Lipase-catalyzed synthesis of structured lipids *via sn*-1,3-diacylglycerols.

Particularly beneficial is the introduction of PUFAs such as DHA, EPA, and similar FAs into the *sn*-2 position while saturated medium-chain FAs occupy the *sn*-1,3-positions [59–62]. PUFAs have received considerable attention in recent years because of health benefits such as reduction of coronary diseases, prevention of certain cancers, and improvement of the immune system. The enzymatic methods for the syntheses of lipids containing PUFAs have recently been reviewed [63].

15.5 SURFACE-ACTIVE COMPOUNDS II

15.5.1 Sugar Esters

Alkylpolyglucosides (APGs) as surfactants and industrial emulsifiers based on anhydro-sugars (sorbitans) are manufactured on a scale of many thousand tons per year by acid-catalyzed esterification with FAs, followed, for example, in the case of Tween, by ethoxylation. Mixtures of sucrose esters are commercialized as food grade emulsifiers (E 473). Sugar esters based on mono- and disaccharides are environmentally benign, biodegradable, and nontoxic, and can be manufactured from cheap and renewable starting materials. In view of their excellent surface-active properties, they are of increasing importance in the food, cosmetic, and pharmaceutical industries.

In view of the numerous hydroxy groups in sugars, all purely chemical methods for the syntheses of sugar esters suffer from low selectivity, leading to product mixtures. High energy needs and the formation of colored products are further disadvantages.

In contrast, lipase-catalyzed reactions are characterized by high regioselectivities and mild reaction conditions [64]. Lipases are known to display their highest specific activities in nonpolar organic solvents in which sugars are insoluble or of very low solubility. Unfortunately, solvents in which sugars are soluble, such as pyridine and dimethyl formamide (DMF) or dimethyl sulfoxide (DMSO), are problematic in view of their (i) environmental safety (toxicity) and (ii) the low specific activity of lipases in such solvents. Thus, next to the screening for suitable lipases, most scientific efforts have addressed the problem of solubility. Various approaches along these lines, such as the use of detergents, temporary protection groups, selected solvents, the use of heterogeneous reaction systems (suspension of substrates), ionic liquids, and microwave/ultrasound-assisted syntheses are published in the literature. A few examples are described below.

Thus, by using a detergent-modified lipase and with continuous removal of water by molecular sieves the synthesis of 6-*O*-palmitoylglucose from D-glucose and palmitic acid in 90% yield was reported [65].

Phenylboronic acid solubilizes sugars by forming complexes that can be cleaved again by the simple addition of water. Thus, employing phenylboronic acid, the monoacylation of fructose by esterification with stearic acid in *n*-hexane was achieved in the presence of the immobilized lipase of *R. mihei* (Lipozyme) [66].

Using a highly heterogeneous, nearly solid phase system in which most of the substrates (sugar and FA) are present in the form of suspended particles together with a small amount of ethyl methyl ketone (EMK) (as catalytic phase) and lipase B from *C. antarctica* (Novozyme 435) as catalyst, in combination with a water removal and solvent regeneration system, 6-*O*-stearoyl-D-glucose was prepared in a yield of about 80% on a preparative (169 g) scale [67] (Figure 15.29).

Using a similar approach, it was found that regioselectively acylated monosaccharides such as 6-*O*-acyl-D-glucose, 6-*O*-acyl-D-mannose, 6-*O*-acyl-D-galactose, acylated D-fructose, 6-*O*-acylated L-ascorbic acid (vitamin C), and also sugar alcohols can be conveniently obtained by lipase-catalyzed acylations in various solvents using the corresponding vinyl esters in the presence of Novozyme 435. For this, the unprotected, native sugars are suspended in the solvent of choice, for example, THF, both the enzyme and the acyl donor are added, and the reaction mixture stirred at 50°C. Although only milligram quantities of the sugars are soluble in THF, the reaction proceeds smoothly with high conversions [68]. Although THF turned out to be the solvent of choice, dioxane, monoglyme, and diglyme can also be employed. The best yields are obtained with vinyl esters; however, other esters and even free FAs can also be used (Figure 15.30).

L-Ascorbic acid is a natural antioxidant. It is, however, highly hydrophilic and insoluble in fats and oils where it could be used as an antioxidant for increased stability; the same holds true for cosmetic formulations. The regioselective acylation of L-ascorbic acid leads to oil soluble esters that can be prepared using this very method [68–70] (Figure 15.31).

The method can also be extended to lipase-catalyzed acylations of disaccharides. Unfortunately, the lipase-catalyzed acylation of sucrose seems to remain a

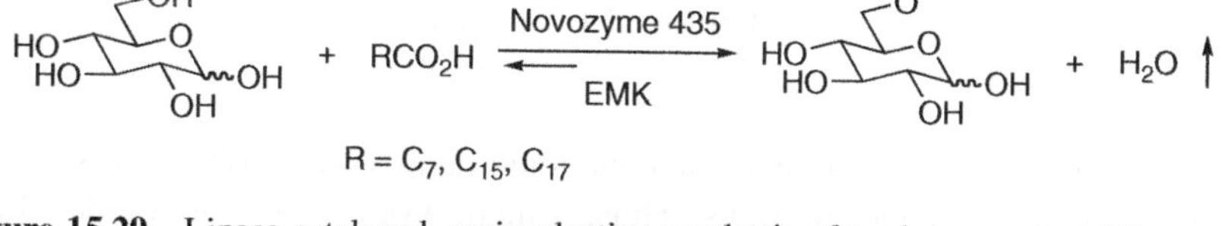

Figure 15.29 Lipase-catalyzed, regioselective synthesis of D-glucose esters (Novozyme 435 = *Candida antarctica* B; EMK = ethyl methyl ketone).

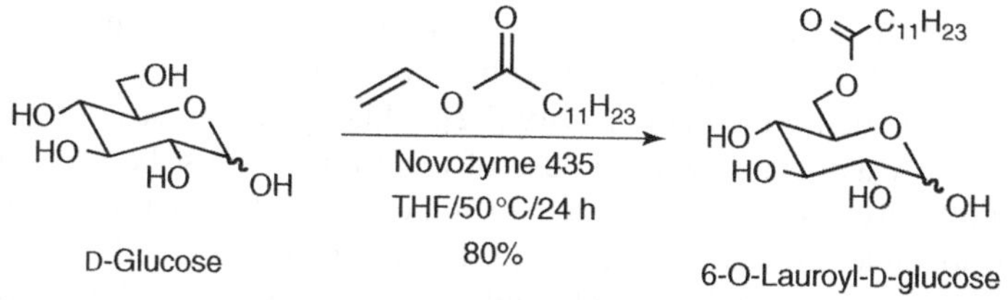

Figure 15.30 Lipase-catalyzed, regioselective synthesis of sugar esters [68].

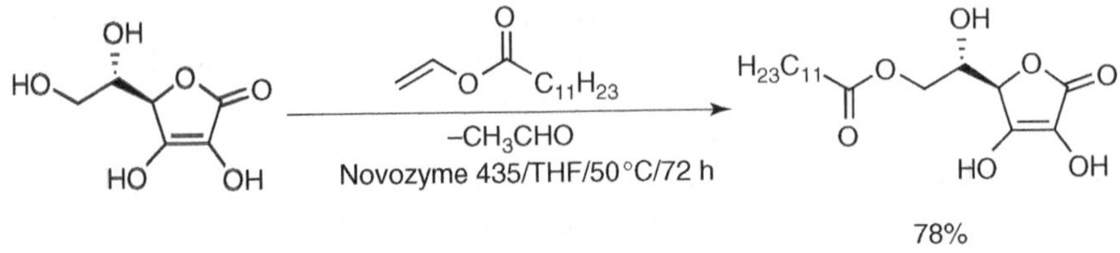

Figure 15.31 Lipase-catalyzed acylation of L-ascorbic acid [68].

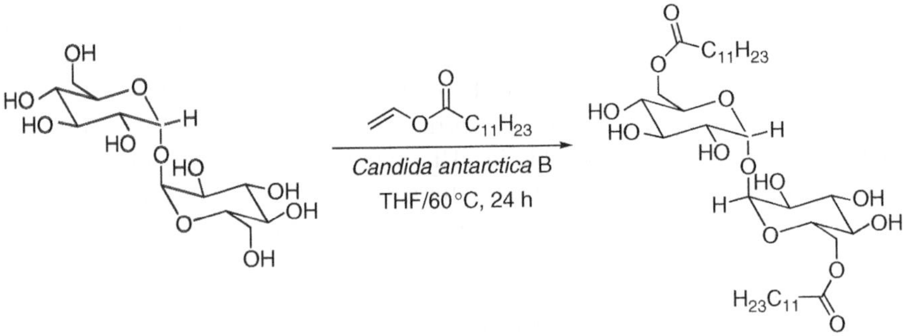

Figure 15.32 Lipase-catalyzed synthesis of 6,6′-dilauroyl-α, α′-trehalose.

problem. From the existing literature, it seems that either unacceptable solvents such as pyridine or Dimethyl Acetamide (DMA) must be employed; otherwise, the yields are only low to moderate. An exception seems to be α, α′-trehalose, which can be acylated in good yield (Figure 15.32). The resulting 6,6′dilauroyl-α, α′-trehalose has been shown to be of considerable interest owing to its antiviral, antitumor, and antibiotic activities [64] (Figure 15.32). The molecule proved to be an interesting gelator; by addition of even small quantities, various organic solvents and oils turn into gels, the properties of which have recently been reported [68d].

In contrast to classical organic solvents, ionic liquids are capable of dissolving sugars in very high concentrations. Thus, 1-methoxyethyl-3-methylimidazolium tetrafluoroborate ([MOEMIm][BF4]) dissolves 100 times more D-glucose than acetone. Some of these materials even dissolve up to 10% of cellulose. Thus, by using such a system the lipase-catalyzed esterification of sugars was successfully achieved [71]. In a very recent example, ascorbyl oleate was obtained in high yield using [BMIM][BF4] as solvent with *C. antarctica* lipase B as catalyst and through careful control of water activity [72]. Although the issues of toxicity and biodegradability associated with the use of ionic liquids must be clarified [15c], one can faithfully assume that the use of ionic liquids for such transformations holds considerable promise for the future. It should also be noted that the use of microwave- and ultrasound-assisted syntheses is an attractive alternative, especially if such acylations are to be carried out under solvent-free conditions [64].

15.5.2 Ionic Surfactants—*N*-Acylated Amino Acids

The *N*-acylation of amino acids and protein hydrolysates with medium- or long-chain FAs leads to another group of surface-active materials, "biosurfactants," which are useful for applications mainly in the cosmetic sector and also for certain medical applications. Numerous such derivatives of amino acids such as sarcosin, alanine, aspartic and glutamic acid, leucine, and valine are described in the literature. Protein hydrolysates of wheat gluten, gelatin, and other proteins have been acylated with mixtures of FA chlorides under the conditions of the well-established Schotten–Baumann or Einhorn reactions. Also silylated amino acids can be acylated [73]. A series of such materials have been commercialized. All of these classical chemical methods involve multistep sequences and toxic solvents, and require additional purification steps and the use of toxic acid chlorides. From an ecological, and possibly also economical, point of view, biocatalysts could provide an attractive alternative. Again, hydrolases are potential biocatalysts for this purpose. While lipases have been successfully employed for the *N*-acylation of chiral amines (including the *N*-acylation of the *o*-amino function of L-lysine), the direct *N*-acylation of the amino function in the α-position of amino acids is problematic since (i) amino acids exist as zwitter ions and the protonated amino function is thus of low nucleophilicity, and (ii) they are insoluble in aprotic organic solvents in which lipases display their highest specific activities. One method to circumvent these problems is the use of amino acid esters, in particular, *t*-butyl esters. This simultaneously improves the solubility in organic solvents and increases the nucleophilicity of the amino function thus liberated.

Thus, under optimized conditions (column reactor) and in a stoichiometric reaction, using glycine *t*-butyl ester and methyl caproate, the *N*-caproyl-glycine-*t*-butyl ester was obtained in yields up to 92% [74] (Figure 15.33). The same method was employed for the synthesis of the corresponding arachidonic acid derivative [75] (Figure 15.34).

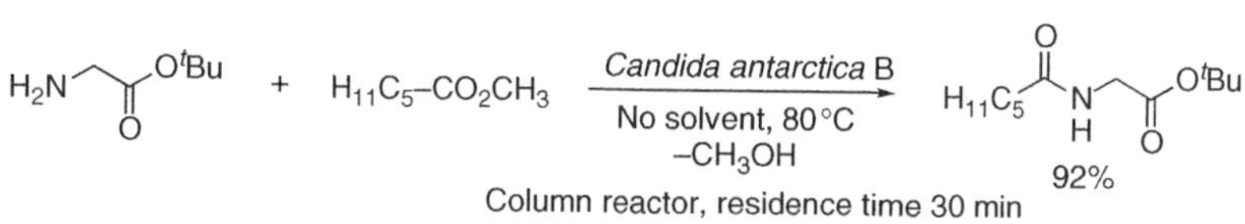

Figure 15.33 Stoichiometric, solvent-free *N*-acylation of glycine *t*-butyl ester.

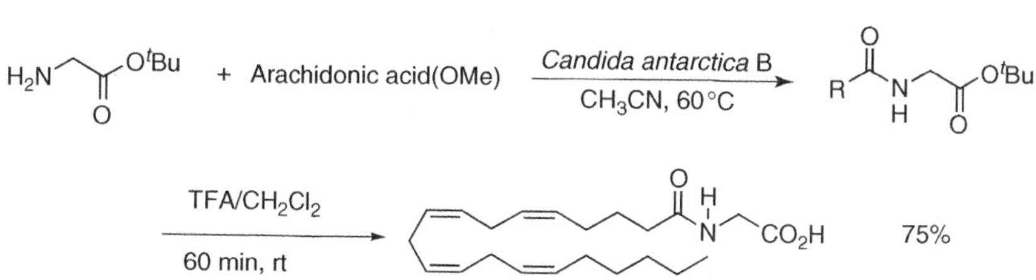

Figure 15.34 Lipase-catalyzed *N*-acylation of glycine *t*-butyl ester with arachidonic acid.

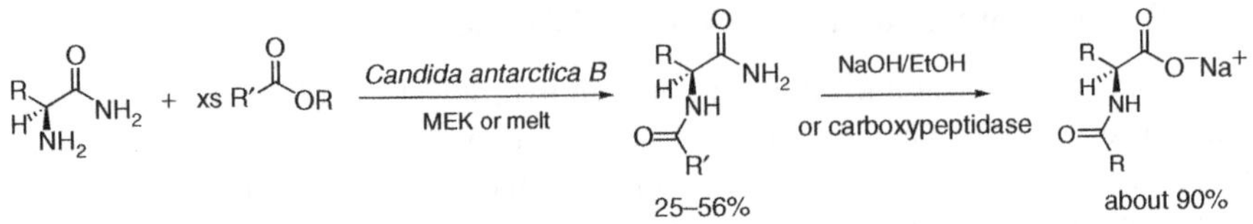

Figure 15.35 Lipase-catalyzed N-acylation of amino acid amides.

As an alternative to amino acid esters amino acid amides have also been employed [76] (Figure 15.35).

In all cases, the desired products, salts of *N*-acylated amino acid are only obtained in multistep sequences. Clearly, the best approach would be the direct N-acylation of free (i.e., unprotected) amino acids in aqueous media and in the presence of amino acylases. Although there are several reports in the literature along these lines, the obtained yields are generally low. In a purely aqueous environment, the hydrolysis competes effectively with the synthesis, resulting in equilibrium mixtures, which contain only about 6% of the desired product. Since the reactions are thermodynamically controlled, an effective method for the removal of the desired products would provide a solution for an effective synthetic route to such molecules.

15.6 OXIDATIVE FUNCTIONALIZATIONS OF UNSATURATED FATTY ACIDS

Oxidative transformations of unsaturated FAs are usually carried out with *in situ* prepared peracidic acids, hydrogen peroxide, ozone, singlet oxygen, and other classical oxidative reagents [77]. As an alternative to the usually harsh conditions of such transformations, more green approaches can be visualized. In view of the space limitations of this chapter, only two examples have been chosen to exemplify such approaches: (i) lipase-catalyzed epoxidations employing the lipase-catalyzed *in situ* formation of peracidic acids and (ii) the use of lipoxygenases.

15.6.1 Lipase-Catalyzed Epoxidations of Unsaturated Fatty Acids and Lipids

Chemo-enzymatic epoxidations of unsaturated FAs and the corresponding triglycerides are of considerable interest, since this method completely suppresses ring opening of the epoxides formed, which can be isolated and used for further transformations (e.g., formation of diols and their esters or the corresponding aziridines and thiiranes). About two decades ago, scientists at Novo Nordisk A/S discovered that medium-chain FAs can be converted into the corresponding peracids by reaction with H_2O_2 in the presence of Novozyme 435 (lipase B from *C. antarctica*) [78,79] (Figure 15.36). These can be employed for the *in situ* epoxidation of simple alkenes. The only by-product is water.

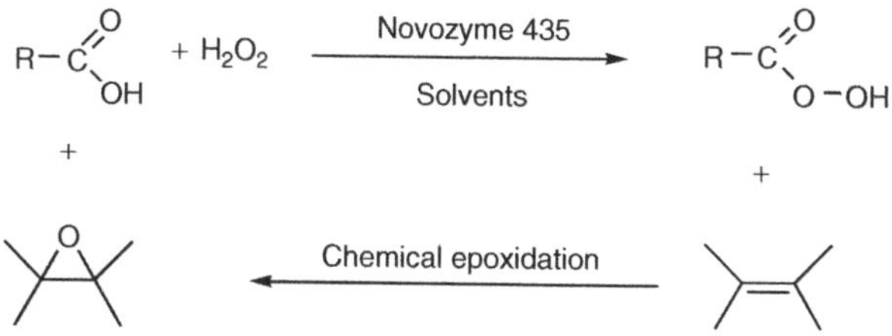

Figure 15.36 Chemo-enzymatic epoxidation of alkenes.

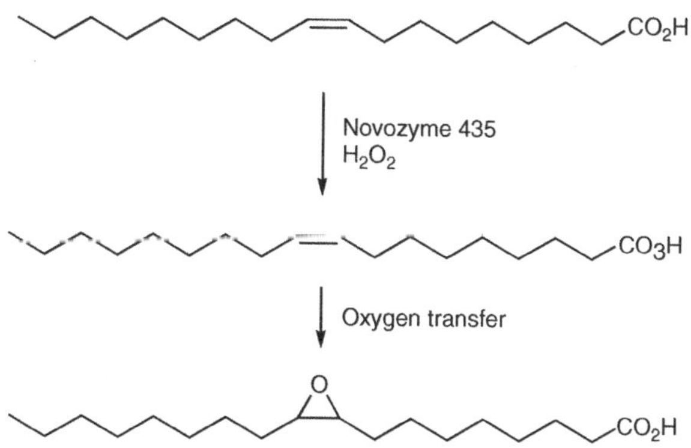

Figure 15.37 "Self" epoxidation of unsaturated fatty acids.

In a continuation of this work, this observation was used for the "self" epoxidation of unsaturated FAs [80]. Thus, if oleic acid is treated with H_2O_2 in the presence of Novozyme 435, the primarily formed peroxy oleic acid is transferring its oxygen unto the double bond of oleic acid (Figure 15.37). Yields are usually between 70% and 95%. The biocatalyst can be used many times, with calculated turnover numbers (TONs) of 200,000.

If a small amount of free FA (about 5%) is added to the reaction mixture, intact triglycerides can also be directly epoxidized. Several plant oils such as canola, sunflower, soybean, linseed, castor oils, and also FAME (biodiesel) were epoxidized with yields of 88–95%. The method was further extended to the epoxidation of the corresponding unsaturated fatty alcohols and their derivatives [81].

15.6.2 Hydroxylation of Unsaturated Fatty Acids Using Lipoxygenases

Lipoxygenases are nonheme iron dioxygenases that catalyze the stereospecific perhydroxylation of unsaturated FAs carrying a 1*Z*,4*Z*-pentadiene substructure such as linoleic or arachidonic acid, resulting in enantiomerically pure (or enriched) conjugated dienic hydroperoxides. High regio-(positional) and stereospecificities are usually observed. Thus, depending on the lipoxygenase employed (soybean or potato), linoleic acid can be converted into either the 13(*S*)- or 9(*S*)-hydroperoxide (Figure 15.38).

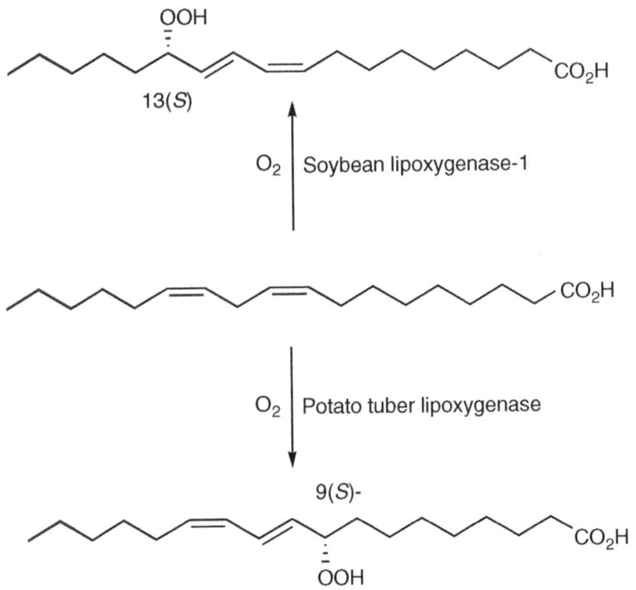

Figure 15.38 Enantiomeric hydroperoxides derived from linoleic acid using plant lipoxygenases.

The resulting hydroperoxides can be easily reduced chemically, for example, with $NaBH_4$, leading to the corresponding enantiomeric hydroxy carboxylic acids. The most widely employed enzyme for synthetic application is soybean lipoxygenase (SBLOX-1). It is commercially available and has been studied extensively. One of the drawbacks in employing these enzymes is the low solubility of the FA substrates in the aqueous reaction media. However, optimization of the synthetic procedure now allows the lipoxygenation of linoleic acid on a 0.1 M scale using ethanol as a cosolvent in an oxygen pressurized reactor. Here, the substrate is not dissolved but dispersed; the oxygen pressure maintains sufficient oxygen dissolved in the medium for an optimal catalytic process. The high pH (9–10) further assists the solubility, whereby the formed hydroperoxides act as surfactants, further increasing the solubility. The reaction is complete (95% conversion) within 30 min. After purification of the hydroperoxide and its reduction with $NaBH_4$, (+)-coriolic acid was obtained in an overall yield of 54% and with 99% *ee*. Numerous natural and nonnatural, PUFAs have been converted this way and numerous synthetic applications for the preparation of chiral synthons have been described [82,83].

15.7 BIODIESEL—ENZYMATIC

As an alternative and supplement to conventional, petroleum-based fuels, biodiesel (FAME) is produced from plant oils such as canola, soybean, and palm oils. In 2006, the production reached 4.9 Mt in Europe (5.7 million tons in 2007) and 0.9 Mt in the United States. In 2008, biodiesel production and capacity amounted globally to 11.1 and 32.6 Mt, respectively [84]. A recent EU directive requires the addition of 5.75% biodiesel to conventional diesel with an increase to 10% anticipated by 2020. Clearly, the demand for biodiesel will increase considerably in the years to come. Although beneficial from an ecological point of view in (i) using renewables and thus helping to balance the carbon dioxide cycle, (ii) being biodegradable, (iii) having a higher positive energy balance (lower GER) than petroleum products (Section 15.1), and (iv) reducing emissions of particles, sulfur, carbon monoxide, and hydrocarbons, the increased use of biodiesel (and biofuels in general) is not without problems in view of a number of opposing arguments. The use of native oils as feedstock frequently competes directly with their use as food and feed. They are also needed as raw materials for the chemical industry and here for the production of materials such as detergents, various oleochemicals, lubricants, and hydraulic fluids, and thus should not be wasted (burned) as fuel. For the production of biodiesel only oil seeds are used, whereas other techniques use the whole plant such as in *biomass to liquid*. It is also highly disputable whether biodiesel (or other biofuels for that matter) based on presently used raw materials is the answer to the world's need for sustainable fuel or can contribute significantly toward the global need for energy. In order to achieve this goal, other bulk materials such as cellulose, hemicelluloses, and lignin will have to be employed. At present, the amounts of biofuels producible will replace only a minor fraction of the presently dominating mineral-oil-based fuels. A complete replacement of petroleum-based fuels by biofuels using present feedstocks seems to be out of the question in view of the acreage required for that purpose. It is essential, therefore, to employ increasingly nonedible plant oils (i.e., from *Jathropha* species or microalgae), oils of low quality (used cooking oil, palm FA distillates), and the so-called side-stream refining products (soap-stocks, acid oil, deodorizer distillates, and spent bleaching earth) as feedstock for biodiesel [85].

Presently, nearly all biodiesel is produced by methanolysis of *refined* plant oils in the presence of alkaline chemical catalysts. For this, oils of high and constant quality are required. Impurities with partial glycerides and free FAs are not tolerated very well and cause complications within the production process. In contrast, lipase-catalyzed methods tolerate low-quality feedstocks, free FAs, and impurities, and are thus attractive, alternative catalysts. They have been shown to be stable under a variety of process conditions. Additional benefits are fewer process steps, high quality of the produced glycerol, improved phase separation (no emulsification from soaps), reduced energy consumption, and wastewater volumes (Figure 15.39).

Unfortunately, at present, most of the published enzymatic methodologies cannot yet compete with the purely chemical process, at least not on the required large scale. There seems to be a unanimous consensus that, in order to compete effectively, the cost for the immobilized enzymes used must come down [86].

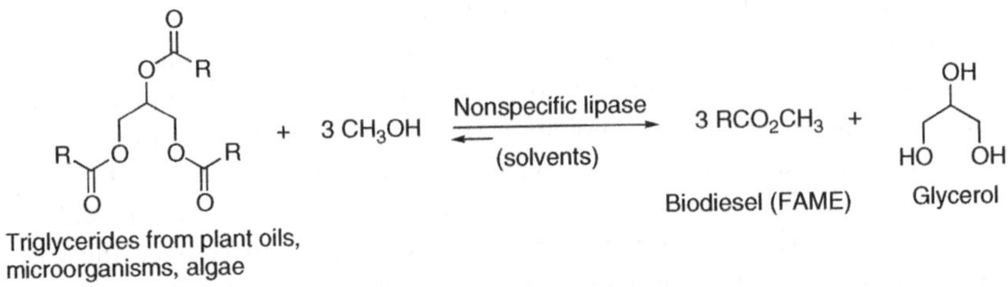

Figure 15.39 Schematic representation of lipase-catalyzed preparation of FAME (biodiesel).

The application of biotechnological (green) methods for the synthesis of biodiesel has been reviewed by several authors in recent years [87,88], and thus in this section only some selected aspects are discussed.

The use of enzymes for the alcoholysis of plant oils dates back to a publication by Mittelbach [89]. Such alcoholyses can be carried out under very mild conditions and almost ambient temperatures (30–50°C). In order to avoid competing hydrolyses, the reactions are usually carried out under low water conditions and/or in aprotic solvents (THF) or in sterically hindered alcohols (*t*-BuOH). Nonspecific microbial lipases (i.e., *R. mihei, Candida rugosa, P. cepacia, Thermomyces lanuginosa*) are usually employed with the lipase of *C. antarctica* B (Novozyme 435) being one of the most commonly used. Since CH_3OH inactivates this lipase (requiring to keep the concentration of CH_3OH low by slow addition) [90], other acyl donors such as CH_3OAc, EtOAc, and dimethylcarbonate have also been used for this purpose. Instead of CH_3OH other alcohols can also be used, with (bio-) ethanol (EtOH) being the most popular; branched alcohols can also be employed. There are indications that the use of *t*-BuOH as solvent is beneficial in protecting and stabilizing the lipase. There are first hints that the process can be made economical on a large scale. In China, the first pilot plant with 20,000 Mt per year is operating using *t*-BuOH as solvent and Lipozyme TL IM as biocatalyst [91]. Also, recent initiatives of enzyme producers indicate high interest in the development of such processes on a commercial scale [92].

It can be hoped that new immobilization techniques presently developed for commercialization by a major enzyme producer can reduce the cost for the biocatalysts further and—together with process optimizations—lower the cost for lipase-catalyzed biodiesel production to an economically acceptable level.

15.7.1 Looking into the Future

Other raw materials for the production of biodiesel can be envisaged [87]. Thus, microalgae produce oils with FA compositions attractive for biodiesel production, and their oil content can exceed up to 80% of algae dry weight. Heterotrophic algae such as *Clorella* or *Rhodosporidium toruloides* Y4 seem to be attractive producers of triglycerides for this purpose. In view of the high oil yield and small acreage required for their growth, this could be preferential to the use of

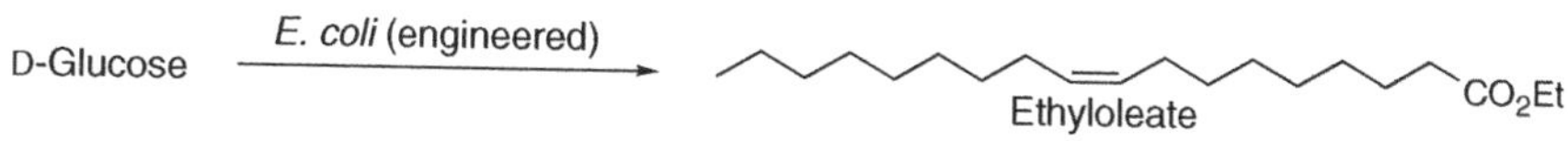

Figure 15.40 Schematic representation of the production of *microdiesel* by engineered microorganisms.

traditional agricultural cultivation of oil seeds. Experimental algae farms already exist in the United States and Israel. It should be noted, however, that in the manufacture of large fuel quantities, considerable amounts of water are needed for algae growth and efficient recycling methods for water must be applied.

Last but not least, numerous microorganisms produce directly esters of FAs, termed *microdiesel*. Thus, *Escherichia coli* was recently engineered for fuel production [93] (Figure 15.40). In a recent paper, it was successfully demonstrated that engineered *E. coli* is able to convert not only simple sugars into biodiesel but possibly also hemicelluloses and thus directly plant biomass [94].

Lastly, there are also strains of microorganisms, which produce directly hydrocarbons such as *Gliocladium roseum* NRRL 50072, the products being named *mycodiesel* [87]. In times of dwindling supplies of mineral oil, one of the most urgent problems to be solved in the future by mankind is to provide sustainable sources of energy and raw materials for the chemical industry. In other words, we need to make use of alternative carbon sources, that is, photosynthesis. Therefore, in addition to the current approaches discussed in the sections above, we must make increasing use of the abundantly available plant materials, in particular, cellulose, hemicelluloses, and lignin. Clearly, numerous developments along these lines are to be expected in the very near future.

15.8 SUMMARY AND CONCLUSIONS

The bioconversion of native fats and oils using green technologies allows the production of a large number of valuable and useful products ranging from numerous surface-active materials *via* functional and structured lipids for dietary applications, food supplements, personal care products, epoxidized/hydroxylated FAs all the way to biodiesel. In view of the space limitations, only the most important aspects in using enzymes (lipases) for such bioconversions could be discussed here, focusing mainly on the most important issues and principles. Although at present only a selection of processes made it into the industrial production, the author hopes that he was able to convince the interested reader that both biocatalysts and renewable resources have a lot to offer for the needs of future generations.

Lastly an apology: within the scope of this chapter it was virtually impossible to review even a fraction of the literature in this vast and rapidly growing field. It was the main intention of the author to inform the reader regarding general principles, trends, and developments, and I have to apologize to all scientists whose work was not or only partly cited in this chapter.

REFERENCES

1. USDA. http://www.fas.usda.gov/oilseeds/circular/Current.asp. Accessed year: 13/12/2009
2. Schneider M. Review: plant-oil-based lubricants and hydraulic fluids. J Sci Food Agric 2006;86:1769–1780.
3. Metzger JO. Fats and oils as renewable feedstock for chemistry. Eur J Lipid Technol 2009;111:865–876.
4. Berger M, Schneider MP. Regioselectivity of Lipases in organic Solvents. Biotechnol Lett 1991;13:333–338.
5. Schmid RD, Verger R. Lipases: interfacial enzymes with attractive applications. Angew Chem Int Ed 1998;37:1608–1633.
6. Berger M, Schneider MP. Enzymatic esterification of glycerol II. Lipase-catalyzed synthesis of regioisomerically pure 1(3) - *rac*-monoacylglycerols. J Am Oil Chem Soc 1992;69:961–965.
7. Berger M, Schneider MP. Lipases in organic solvents: the fatty acid chain length profile. Biotechnol Lett 1991;13:641–645.
8. Berger M, Laumen K, Schneider MP. Enzymatic esterification of glycerol I. Lipase-catalyzed synthesis of regioisomerically pure 1,3-*sn*-diacylglycerols. J Am Oil Chem Soc 1992;69:955–960.
9. Waldinger C, Schneider MP. Enzymatic esterification of glycerol III. Lipase catalyzed synthesis of regioisomerically pure 1,3-*sn*-diacylglycerols and 1(3)-*rac*-monoacylglycerols derived from unsaturated fatty acids. J Am Oil Chem Soc 1996;73:1513–1519.
10. Wongsakul S, H-Kittikun A, Bornscheuer UT. Lipase-catalyzed synthesis of structured triacylglycerides from 1,3-diacylglycerides. J Am Oil Chem Soc 2004;81:151–155.
11. Halldorsson A, Magnusson CD, Haraldsson GG. Chemoenzymatic synthesis of structured triacylglycerols. Tetrahedron Lett 2001;42:7675–7677.
12. Kristensen JB, Xu X, Mu H. Diacylglycerol synthesis by enzymatic glycerolysis: screening of commercially available lipases. J Am Oil Chem Soc 2005;82:329–334.
13. Castillo E, Dossat V, Marty A, Condoret JS, Combes D. The role of silica gel in lipase-catalyzed esterifcation reactions of high polar substrates. J Am Oil Chem Soc 1997;74:77–85.
14. Lobell M, Schneider M. Synthesis of hydroxycarboxylic acid vinyl esters. Synthesis 1994:375–377.
15. (a) Rosu R, Yasui M, Iwasaki Y, Yamane T. Enzymatic synthesis of symmetrical 1,3-diacylglycerols by direct esterification of glycerol in solvent-free system. J Am Oil Chem Soc 1999;76:839–843; (b) Hobbs HR, Thomas NR. Biocatalysis in supercritical fluids, in fluorous solvents, and under solvent-free conditions. Chem Rev 2007;107:2786–2820; (c) van Rantwijk F, Sheldon RA. Biocatalysis in ionic liquids. Chem Rev 2007;107:2757–2785; (d) Hilterhaus L, Thum O, Liese A. Reactor concept for lipase-catalyzed solvent-free conversion of highly viscous reactants forming two-phase systems. Org Process Res Dev 2008;12:618–625.
16. Berger M, Schneider MP. Regioisomerically pure mono- and diacylglycerols as synthetic building blocks. Fat Sci Technol 1993;95:169–175.
17. Haftendorn R, Ulbrich-Hofmann R. Synthesis of 2-modified diacylglycerols. Tetrahedron 1995;51:1177–1186.

18. Haftendorn R, Schwarze G, Ulbrich-Hofmann R. 1,3-Diacylglycero-2-phosphocholines—synthesis, aggregation behavior and properties as inhibitors of phospholipase D. Chem Phys Lipids 2000;104:57–66.

19. Ulbrich-Hofmann R. Phospholipases used in lipid transformations. In: Bornscheuer UT, editor. Enzymes in lipid modification. Weinheim, Germany: Wiley-VCH; 2000. pp. 219–262.

20. El Kihel L, Bourass J, Richomme P, Petit JY, Letourneux P. Synthesis and evaluation of the anti-inflammatory effects of niflumic acid lipophilic prodrugs in brain edema. Drug Res 1996;46:1040–1044.

21. Mantelli S, Speiser P, Hauser H. Phase behavior of diglyceride prodrugs: spontaneous formation of unilamellar vesicles. Chem Phys Lipids 1985;37:329–343.

22. Scriba GKE. Phenytoin-lipid conjugates as potential prodrugs of phenytoin. Arch Pharm 1993;326:477–481.

23. Baumann H, Bühler M, Fochem H, Hirsinger F, Zoebelein H, Falbe J. Natural fats and oils—renewable raw materials for the chemical industry. Angew Chem Int Ed Engl 1988;27:41–62.

24. (a) Kato Y, Fujiwara I Asano Y. Synthesis of optically active α-monobenzoyl glycerol by asymmetric transesterification of glycerol. J Molec Catal B Enzym 2000;9:193–200; (b) Heidt M, Bornscheuer U, Schmid RD. Studies on the enantioselectivity in the lipase-catalyzed synthesis of monoacyl glycerols from isopropylidene glycerol. Biotechnol Tech 1996;10:25–50.

25. (a) Breitgoff D, Laumen K, Schneider MP. Enzymatic differentiation of the enantiotopic hydroxymethyl groups of glycerol; synthesis of chiral building blocks. J Chem Soc, Chem Commun 1986:1523–1524; (b) Breitgoff D, dissertation Wuppertal. 1989; (c) Schneider M, Laumen K, Breitgoff D, Wullbrandt D, Schlingmann M, Keller R. Chiral synthetic building blocks from prochiral glycerol. 3624703 A1. 1986.

26. Soumanou MM, Bornscheuer UT, Schmid RD. Two-step enzymatic reaction for the synthesis of pure structured triacylglycerols. J Am Oil Chem Soc 1998;75:703–710.

27. Irimescu R, Iwasaki Y, Hou CT. Study of TAG ethanolysis to 2-MAG by immobilized *Candida antarctica* lipase and synthesis of symmetrically structured TAG. J Am Oil Chem Soc 2002;79:879–883.

28. Millquist A, Adlercreutz P, Mattiason B. Lipase catalyzed alcoholysis of triglycerides for the preparation of 2-monoglycerides. Enzyme Microb Technol 1994;16:1042–1047.

29. Millquist A, Virto C, Adlercreutz P, Mattiason B. Acyl group migration in 2-monoolein. Biocatalysis 1996;14:89–111.

30. Wongsakul S, Prasertsan P, Bornscheuer UT, H-Kittikun A. Synthesis of 2-monoglycerides by alcoholysis of palm oil and tuna oil using immobilized lipases. Eur J Lipid Sci Technol 2003;105:68–73.

31. Diks RMM, Bosley JA. The exploitation of lipase selectivities for the production of acyl glycerols. In: Bornscheuer UT, editor. Enzymes in Lipid Modifications. Wiley-VCH; 2000. pp 3–22 and references therein.

32. Bornscheuer U. Lipase-catalyzed syntheses of mono acylglycerols (in German). Fat Sci Technol 1995;97:241–249.

33. Bornscheuer UT. Lipase-catalyzed syntheses of mono acylglycerols. Enzyme Microb Technol 1995;17:578–586.

34. Steffen B, Ziemann A, Lang S, Wagner F. Enzymatic monoacylation of trihydroxy compounds. Biotechnol Lett 1992;14:773–778.
35. Steffen B, Lang S, Hamann D, Schneider P, Cammenga HK, Wagner F. Enzymatic formation and physico-chemical behaviour of uncommon monoglycerides. Fat Sci Technol 1995;97:132–136.
36. Akoh CC. Lipase-catalyzed synthesis of partial glyceride. Biotechnol Lett 1993;15:949–954.
37. McNeill GP, Shimizu S, Yamane T. High-yield enzymatic glycerolysis of fats and oils. J Am Oil Chem Soc 1991;68:1–5.
38. McNeill GP, Yamane T. Further improvement in the yield of monoglycerides during enzymatic glycerolysis of fats and oils. J Am Oil Chem Soc 1991;68:6–10.
39. Bornscheuer UT, Yamane T. Activity and stability of lipase in the solid-phase glycerolysis of triolein. Enzyme Microb Technol 1994;16:864–869.
40. Watanabe Y, Yamauchi-Sato Y, Nagao T, Yamamoto T, Tsutsumi K, Sugihara A, Shimada Y. Production of MAG of CLA in a solvent-free system at low temperature with *Candida rugosa* lipase. J Am Oil Chem Soc 2003;80:909–914.
41. Pinsirodom P, Watanabe Y, Nagao T, Sugihara A, Koayashi T, Shimada Y. Critical temperature for production of MAG by esterification of different FA with glycerol using *Penicillium camembertii* lipase. J Am Oil Chem Soc 2004;81:543–547.
42. (a) Elfman-Börjesson I, Härröd M. Synthesis of monoglycerides by glycerolysis of rapeseed oil using immobilized lipase. J Am Oil Chem Soc 1999;76:701–707; (b) see also: Selmi B, Gontier E, Ergan F, Thomas D. Enzymatic synthesis of tricaprylin in a solvent-free system: lipase regiospecificity as controlled by glycerol absorption on silica gel. Biotechnol Tech 1997;11:543–547, and many others.
43. Waldinger C. dissertation Wuppertal; 1995.
44. Berger M, Laumen K, Schneider MP. Lipase-catalyzed esterification of hydrophilic diols in organic solvents. Biotechnol Lett 1992;14:553–558.
45. For detailed discussions see: Xu X. Modification of oils and fats by lipase-catalyzed interesterification: aspects of process engineering. In: Bornscheuer UT, editor. Enzymes in Lipid Modifications. Weinheim, Germany: Wiley-VCH; 2000. pp. 190–215.
46. Matsuo N, Tokimitsu I. Metabolic characteristics of diacylglycerol. Inform 2001;12:1098–1102.
47. Hara K, Onizawa K, Honda T, Ide T, Otsuji K, Murata M. Dietary diacylglycerol - dependent reduction in serum triacylglycerol concentrations in rats. Ann Nutr Metab 1993;37:185–191.
48. Nagao T, Watanabe H, Goto N, Onizawa K, Taguchi H, Matsuao N, Yasukawa T, Tsushima R, Shimazaki H, Itakura H. Dietary diacylglycerol suppresses accumulation of body fat compared to triacylglycerol in men in a double blind trial. J Nutr 2000;130:792–797.
49. Sonntag NOV. New developments in the fatty acid industry in America. J Am Oil Chem Soc 1984;61:229–232.
50. Yokomichi H, Okauchi Y, Oka T, Okisaka M, Kokumai H, Koike S, Mori K, Tabeta M. Frying oil or fat composition. 5,514,405. 1996.
51. Watanabe T, Shimizu M, Sugiura M, Sato M, Kohori J, Yamada N, Nakanishi K. Optimization of reaction conditions for the production of DAG using immobilized, 1,3-regiospecific lipase Lipozyme RM IM. J Am Oil Chem Soc 2003;80:1201–1207.

52. Flickinger BD, Matsuo N. Nutritional characteristics of DAG oil. Lipids 2003;38:129–132.
53. Kristensen JB, Xu X, Hu M. Diacylglycerol synthesis by enzymatic glycerolysis: screening of commercially available lipases. J Am Oil Chem Soc 2005;82:329–334.
54. Jandacek RJ, Whitesides JA, Holcombe BN, Volpenheim RA, Taulbee JD. The rapid hydrolysis and efficient absorption of triglycerides with octanoic acid in the 1- and 3-positions and long chain fatty acids in the 2-position. Am J Clin Nutr 1997;45:940–945.
55. (a) Bornscheuer UT, Adamczak M, Soumanou MM. In: Gunstone FD, Barnes PJ, associates, editors. Lipase-catalyzed synthesis of modified lipids. UK: Bridgewater; 2002. pp. 149–182; (b) Akoh CC. Structured lipids—enzymatic approach. Inform 1995;6:1055–1061.
56. Xu X. Review: Production of structured triacylglycerols by lipase—catalyzed reactions. Eur J Lipid Sci Technol 2000;102:287–303.
57. Soumanou MM, Bornscheuer UT, Schmid U, Schmid RD. Synthesis of structured triglycerides by lipase catalysis. Fett/Lipid 1998;100:156–160.
58. Wongsakul S, H-Kiikun A, Bornscheuer UT. Lipase-catalyzed synthesis of structured triacylglycerides from 1,3—diacylglycerides. J Am Oil Chem Soc 2004;81:151–155.
59. Christensen MS, Höy C-E, Becker CC, Redgrave TG. Intestinal absorption and lymphatic transport of eicosapentaenoic (EPA), docosahexaenoic (DHA) and decanoic acids: dependence on intramolecular triacylglycerol structure. Am J Clin Nutr 1995;61:56–61.
60. Haraldsson GG, Halldorsson A, Kulas E. Chemoenzymatic synthesis of structured triacylglycerols containing eicosapentaenoic and docosahexaenoic acids. J Am Oil Chem Soc 2000;77:1139–1145.
61. Haraldsson GG. Enrichment of lipids with EPA and DHA by lipase. In: Bornscheuer UT, editor. Enzymes in Lipid Modification. Weinheim, Germany: Wiley-VCH; 2000. pp. 170–189.
62. Yamane T. Lipase-catalyzed synthesis of structured triacylglycerols containing polyunsaturated fatty acids: monitoring the reaction and increasing the yield. In: Bornscheuer UT, editor. Enzymes in Lipid Modification. Weinheim, Germany: Wiley-VCH; 2000. pp. 148–169.
63. Adamczak M, Bornscheuer UT, Bednarski W. Review: properties and biotechnological methods to produce lipids containing conjugated linoleic acid. Eur J Lipid Sci Technol 2008;110:491–504.
64. Kennedy JF, Kumar H, Panesar PS, Marwaha SS, Goyal R, Parmar A, Kaur S. Review: Enzyme-catalyzed regioselective synthesis of sugar esters and related compounds. J Chem Technol Biotechnol 2006;81:866–876.
65. Tsuzuki W, Kitamura Y, Suzuki T, Kobayashi S. Synthesis of sugar fatty acid esters by modified lipase. Biotechnol Bioeng 1999;64:267–271.
66. Schlottterbeck A, Lang S, Wray V, Wagner F. Lipase-catalyzed monoacylation of fructose. Biotechnol Lett 1993;15:61–64.
67. Yan Y, Bornscheuer UT, Stadler G, Lutz-Wahl S, Otto RT, Reuss M, Schmid RD. Regioselective glucose ester synthesis on a preparative scale. Eur J Lipid Sci Technol 2001;103:583–587, and literature cited therein.

68. (a) Schneider MP, Haase B, Machmüller G. Enzyme-catalyzed procedure for the preparation of mono esters of mono-, di- and oligosaccharides. DE 19626943.1. 1996; (b) Schneider M, Haase B, Machmueller G, Chem Abstr 1998;128:127145t; (c) Machmüller G. dissertation Wuppertal; 2004; (d) for studies of the gel properties see also: John G, Zhu G, Li J, Dordick JS. Enzymatically derived sugar-containing self-assembled organogels with nanostructured morphologies. Angew Chem Int Ed 2006;45:4772–4775.
69. Yan Y, Bornscheuer UT, Schmid RD. Lipase-catalyzed synthesis of vitamin C fatty acid esters. Biotechnol Lett 1999;21:1051–1054.
70. Adamczak M, Bornscheuer UT, Bednarski W. Synthesis of ascorbyloleate by immobilized *Candida Antarctica* lipases. Process Biochem 2005;40:3177–3180.
71. Park S, Kazlauskas RJ. Improved preparation and use of room-temperature ionic liquids in lipase-catalyzed enantio- and regioselective acylations. J Org Chem 2001;66:8395–8401.
72. Adamczak M, Bornscheuer UT. Improving ascorbyl oleate synthesis catalyzed by *Candida antarctica* lipase B in ionic liquids and water activity control by salt hydrates. Process Biochem 2009;44:257–261.
73. Infante MR, Perez L, Pinazo A, Clapes P, Moran MC, Angelet M, Garcia MT, Vinardell MP. Amino acid-based surfactants. C R Chimie 2004;7:583–592.
74. Müller S. dissertation Wuppertal. 2000.
75. Goujard L, Figuera MC, Villeneuve P. Chemo-enzymatic synthesis of *N*-arachidonyl glycine. Biotechnol Lett 2004;26:1211–1216.
76. Godtfredsen SE, Björkling F. An enzyme-catalyzed process for preparing *N*-acyl amino acids and *N*-acyl amino acid amides. WO 90/14429. 1990.
77. Biermann U, Friedt W, Lang S, Lühs W, Machmüller G, Metzger JO, Rüsch gen Klass M, Schäfer HJ, Schneider MP. New syntheses with oils and fats as renewable raw materials for the chemical industry. Angew Chem Int Ed 2000;39:2206–2224.
78. Björkling F, Godtfredsen SE, Kirk O. Lipase - mediated formation of peroxycarboxylic acids used in catalytic epoxidations of alkenes. J Chem Soc, Chem Commun 1990:1301–1303.
79. Björkling F, Frykman SE, Godtfredsen SE, Kirk O. Lipase-catalyzed synthesis of peroxy carboxylic acids and lipase-mediated epoxidations. Tetrahedron 1992;48:4587–4592.
80. Rüsch gen Klaas M, Warwel S. Lipase-catalyzed peroxy fatty acid generation and lipid oxidation. In: Bornscheuer UT, editor. Enzymes in Lipid Modification. Weinheim, Germany: Wiley-VCH; 2000. pp. 116–127.
81. Warwel S, Rüsch gen Klaas M. New peroxycarbonic acids by enzyme catalysis—syntheses and oxidation properties. In: Adam W, editor. Peroxide chemistry, mechanistic and preparative aspects of oxygen transfer. Wiley-VCH: 2001.
82. Iacazio G, Martini-Iacazio D. Properties and applications of lipoxygenases. In: Bornscheuer U, editor. Enzymes in Lipid Modification. Weinheim, Germany: Wiley-VCH; 2000. pp. 337–359.
83. Fussner I, Kühn H. Application of lipoxygenases and related enzymes for the preparation of oxygenated lipids. In: Bornscheuer U, editor. Enzymes in Lipid Modification. Weinheim, Germany: 2000; Wiley-VCH. pp. 309–336.
84. http://www.emerging-markets.com/biodiesel. Access year: 10/12/2009

85. Echim C, Verhe R, De Greyt W, Stevens C. Production of biodiesel from side-stream refining products. Energy Environ Sci 2009;2:1131–1141.
86. Nielsen PM, Brask J, Fjerbaeck L. Enzymatic biodiesel production: technical and economical considerations. Eur J Lipid Sci Technol 2008;110:692–700.
87. Adamczak M, Bornscheuer UT, Bednarski W. Review: the application of biotechnological methods for the synthesis of biodiesel. Eur J Lipid Sci Technol 2009;111:808–813, and literature cited therein.
88. Fjerbaeck L, Christensen KK, Norddahl B. A Review of current state of biodiesel production using enzymatic transesterification. Biotechnol Bioeng 2009;102:1298–1315.
89. Mittelbach M. Lipase catalyzed alcoholysis of sunflower oil. J Am Oil Chem Soc 1990;67:168–170.
90. Shimada Y, Watanabe Y, Sugihara A, Tominaga Y. Enzymatic alcoholysis for biodiesel production and application of the reaction to oil processing. J Mol Catal B Enzym 2002;17:133–142.
91. Du W, Li W, Sun T, Chen X, Liu D. Perspectives for biotechnological production of biodiesel and impacts. Appl Microbiol Biotechnol 2008;79:331–337.
92. http://www.biodieselmagazine.com/article.jsp?articleid_2416 and 2537. Accessed year: 9/12/2009
93. Kalscheuer R, Stolting T, Steinbüchel A. Microdiesel:*Escherichia coli* engineered for fuel production. Microbiology 2006;152:2529–2536.
94. Steen EJ, Kang Y, Bokinsky G, Hu Z, Schirmer A, McClure A, del Cardayre SB, Keasling JD. Microbial production of fatty acid derived fuels and chemicals from plant biomass. Nature 2010;463:559–563.

CHAPTER 16

MICROBIAL BIOPROCESSES FOR INDUSTRIAL-SCALE CHEMICAL PRODUCTION

TODD BANNER, ARLENE FOSMER, HOLLY JESSEN, ERIN MARASCO, BRIAN RUSH and JON VELDHOUSE
Cargill Biotechnology Development Center (BioTDC), Minneapolis, Minnesota

MERVYN DE SOUZA
Cargill Specialty Canola Oils, Fort Collins, Colorado

16.1 GENERAL INTRODUCTION

Concerns about the long-term environmental effects and the supply of petrochemicals have spurred interest in using biotechnology in the sustainable production of chemicals from renewable resources. Using microbial and enzymatic biocatalysts to produce industrially relevant products is termed *white biotechnology*. Recent advances in genomics, systems biology, metabolic engineering, and catalysis have contributed to making white biotechnology industrially viable [1]. White biotechnology integrates the idea of green chemistry in that it also has a positive effect on our environment and society. Both green chemistry and white biotechnology share the challenge that products made using sustainable processes must continue to have an economical advantage to displace currently used petrochemical processes and gain widespread industrial acceptance [2].

In implementing biotechnology for the production of industrial chemicals, several considerations must be taken into account including market growth rate and size potential, platform potential of the chemical, and degree of technical development required. Platform potential refers to the ability of the chemical

Biocatalysis for Green Chemistry and Chemical Process Development, First Edition.
Edited by Junhua (Alex) Tao and Romas Kazlauskas.

to be derivatized or serve as a building block for multiple compounds [3]. A 2004 Department of Energy (DOE) report detailed several of these platform chemicals, all of which possess multiple functional groups that can be utilized to form new families of molecules. The chemicals discussed in this chapter are examples of successful and potential industrial biotechnology applications, with some fulfilling the role of platform chemicals as described in the 2004 DOE report. We review the different factors for consideration during the sustainable production of selected chemical building blocks that are at different stages of technology and market maturity. Chemicals discussed have been divided into three categories based on maturity of both production technology and market. Mature production technology is defined as a process currently operating on an industrial scale with limited process innovation potential. A mature market is characterized as having limited projected growth.

Citric acid represents a chemical with both a mature market and a well-established commercial microbial production process. In the case of a high-volume commodity-scale chemical such as citric acid, cost is the most significant driver to maintain economical viability. Although the potential for lowering production costs through fermentation improvement is low in mature processes, value can be gained through utilization of by-product streams, traditionally considered to be of minimal value. These streams are now more commonly referred to as coproduct streams. We describe some examples of utilization of citric acid coproducts to generate value-added streams that make the overall economics of a mature product such as citric acid more attractive.

Our second category comprises chemicals that are currently being produced on a commercial scale using green chemistry or white biotechnology processes, but are entering developing product markets. Select examples of chemicals in this category include 1,3-propanediol (1,3-PDO), lactic acid, and itaconic acid. All chemicals discussed in this category are primarily applied in the production of polymers and share the challenge of competing directly with petroleum-based products not only in cost but also in end-product properties. All three face different challenges to grow their markets. Lactic acid for polylactic acid (PLA) and 1,3-PDO face cost constraints versus their respective competing petrochemical product. PLA competes with polyethylene terephthalate (PET), while 1,3-PDO competes with petrochemically derived ethylene glycol and propylene glycol. In the case of itaconic acid, the production capacity exceeds the current demand [4].

A third category includes organic acids for which there have been many recent advances in microbial production options. Although the microbial production routes to make these chemicals are novel, either the chemical itself or its derivatives have well-established product markets. In this category, we include the four-carbon diacid family consisting of succinic acid, fumaric acid, and malic acid, as well as the three-carbon chemical, 3-hydroxypropionic acid (3-HP). 3-HP is unique in that its basic chemistry is not represented by a current petrochemically derived technology and thus direct uses of 3-HP require the development of efficient catalysts [3]. However, some 3-HP derivatives such as acrylic acid, acrylamide, and ethyl ethoxypropionate currently have established markets.

A common challenge for green chemicals that compete with the current petrochemically derived products is the reduction of overall costs associated with fermentation and downstream processing. An additional challenge is that relatively new biobased chemical production facilities must compete against the cost advantage of depreciated and optimized petrochemical production facilities. Established, as well as future biotechnology applications rely on integration of the microbial or enzymatic process with a feedstock source. This idea is commonly referred to as *biorefining*. Although future biorefining may use cellulosics as base feedstocks, current biorefining operation is exemplified in corn wet milling [5]. Corn wet milling is the initial step of processing to produce carbohydrate as well as protein feedstocks. Corn kernels are sequentially steeped, milled, and separated into their main constituent fractions, germ, fiber, starch, protein, and solubles during wet milling. Corn oil can be extracted from the germ and the fiber and protein streams have animal feed applications. The starch slurry from wet milling is subjected to liquefaction and saccharification for dextrose production. Among other applications, dextrose can be used as an industrial fermentation substrate, and nitrogen sources produced from milling, such as corn steep liquor, can be used as nitrogen sources for fermentation.

16.2 MATURE MARKET AND ESTABLISHED TECHNOLOGY (CITRIC ACID)

Citric acid exemplifies the biorefining process by using carbohydrate feedstocks from corn refining and creating valued coproducts from by-product streams once considered waste. Citric acid is an intermediate in the Krebs (TCA (tricarboxylic acid)) cycle (Figure 16.1) and the ability of fungi to produce it was discovered in 1893 [6]. Citric acid production by the filamentous fungus *Aspergillus niger* is one of the world's most prominent industrial-scale fermentation processes [7,8] with a global production capacity of about 1.6 million tons per year [9]. However, citric acid was extracted from lemon before it was discovered that it can be accumulated by *Aspergillus* under specific conditions [7]. The large volume demand of citric acid is due to its broad use in a variety of applications, including industrial, food and beverage, household, personal care, and pharmaceutical [9]. The variety of applications of citric acid has increased the demand for this organic acid. Thus, citric acid is a classic example of an organic acid with a well-established fermentation and recovery process along with a mature market.

A. niger has been the production organism of choice for several decades, but in recent years the production of citric acid by yeast has attracted attention. A variety of yeasts such as *Yarrowia* and *Hansenula* produce citric acid from carbohydrates and hydrocarbons that are not used as carbon sources by *A. niger* [10]. Strains of *Yarrowia lipolytica* appear promising, and good yields of citric acid from various raw materials have been reported [11]. Any new production technology for citric acid will have to compete with the well-established fungal process.

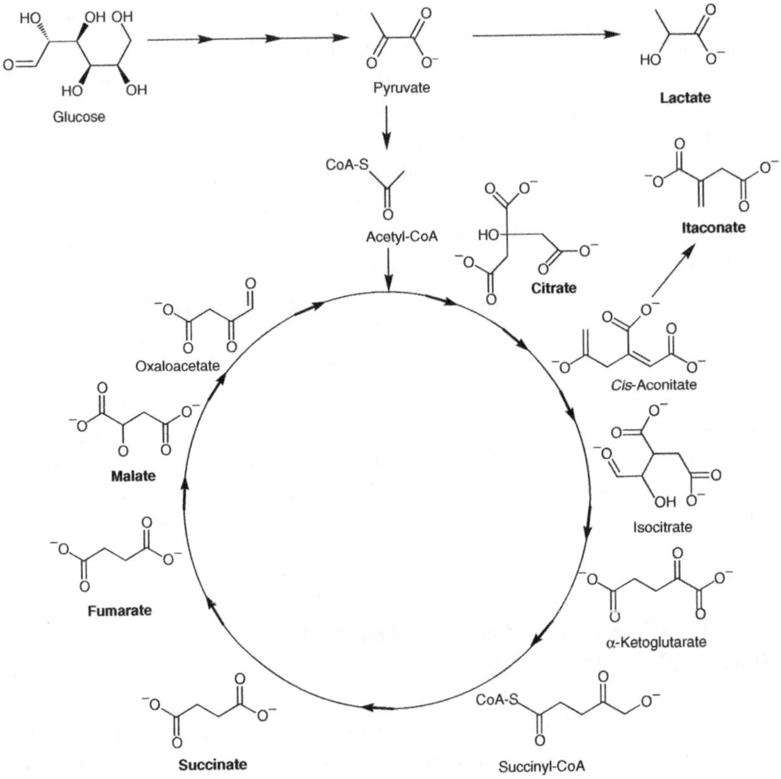

Figure 16.1 Schematic of the tricarboxylic acid (TCA) cycle highlighting key acids discussed in this chapter, as well as the formation of lactic acid from the key TCA cycle precursor pyruvate.

16.3 CHEMICALS OBTAINED FROM CITRIC ACID COPRODUCT STREAM CONVERSION

The traditional approach to improve economics of microbial processes, such as the production of citric acid, is to increase fermentation or conversion rates, titers, or yields. However, a whole process approach viewing fermentation as a unit operation in an entire process creates opportunities to understand the potential for coproducts or waste reduction, and can result in improved process economics and reduced waste. Spent fungal biomass from citric acid, which was previously viewed as a waste stream, is now being utilized to create value-added products. The global annual production capacity of mycelium (fungal biomass) is estimated to be about 300,000Mt [12] and is closely related to citric acid production. This coproduct was traditionally used as an ingredient in animal feeds or disposed of by composting. Recent efforts, however, have focused to convert this large quantity of industrially produced biomass into a platform for several higher value biobased products by thermochemical and/or biocatalytic processing.

A. niger mycelium is composed of substantial amounts of chitin and β-glucans [13,14], which are synthesized in the fungal cell walls [15]. In 2003, Cargill launched Regenasure™ Glucosamine, the first commercial non-shellfish glucosamine produced by acid hydrolysis of the chitin portion of the mycelium [16,17].

Chitosan/glucan [18] and chitosan biopolymers can also be produced from the mycelium by alkaline hydrolysis [19]. There are several promising applications of chitosan in cosmetics, pharmaceutical, and food applications [20]. A process for producing water-soluble β-glucans from the mycelium has been developed by Cargill, Inc. [21]. β-Glucans are known *immunostimulants* [22,23], and recent research has provided evidence that fungal β-glucans from citric acid mycelia also have stimulating properties [24].

Conversion of the citric acid coproducts to value-added products maintains the economical viability in a mature market. Citric acid markets are well developed and mature, but for white biotechnology to be truly transforming, microbial processes must penetrate new markets, including those currently occupied by petroleum-derived chemicals. This, however, brings new challenges in terms of price competitiveness, while retaining or providing added functionality, which are required to gain market traction.

16.4 CHEMICALS CURRENTLY PRODUCED ON A COMMERCIAL SCALE WITH MARKETS THAT STILL NEED DEVELOPMENT

Three examples of commercial-scale microbial processes with growing markets are itaconic acid, 1,3-PDO, and lactic acid production. In the case of itaconic acid, production currently exceeds the worldwide demand. Production cost is an important factor that affects market penetration of green chemicals. Chemicals such as 1,3-PDO and lactic acid are recent examples of how white biotechnology processes can generate economically viable production processes. Both chemicals have been known for over a century. 1,3-PDO production at a competitive price has changed its specialty chemical status. Lactic acid, long known for its use in food applications, now has an increasing market for PLA production. With both lactic acid and 1,3-PDO, existing microbial fermentation technologies were known, but metabolic engineering has created bioprocesses with increased cost competitiveness and sustainability. For the commercial success of itaconic, lactic, and 1,3-PDO to match the technological success of their bioproduction processes, there is urgent need to develop the existing market or create new markets for these chemicals and their derivatives.

16.4.1 Itaconic Acid

Itaconic acid production (Figure 16.1) is a mature microbial technology; however, unlike citric acid, the market demand has lagged behind the global production capacity. Itaconic acid was originally discovered as a pyrolysis product of citric acid [25], and additional chemical synthetic routes have been reported [26]. Today, fermentation is the dominant, low-cost industrial production method of itaconic acid [4,26]. The industrial-scale production of itaconic acid has not been economically favorable to many producers in recent years, as global production capacity has exceeded the demand. Global production capacity is about 80,000 t, but production demand is only about 15,000 t [4].

Aspergillus terreus was used in the original industrial-scale fermentation developed by Pfizer [27]. The fermentation technology was adapted to air-lift reactors from stirred tank to increase energy efficiency [28]. Inexpensive sources of carbon such as those derived from biorefined corn starch [5] in the case of citric acid, or beet or sugarcane molasses are required for economical production. The recovery process can include filtration of the fermentation broth, followed by crude crystallization, resolubilization and decolorization, and a second crystallization [4].

The predominant use for itaconic acid is as a carboxylating agent for styrene/butadiene latexes [29,30] where it competes with petrochemical acrylates and methacrylates [31]. In the current situation where itaconic supply exceeds the demand, itaconic acid market growth requires new, large volume applications. One area that is gaining momentum is the positioning of itaconic acid as a sustainably produced green chemical to replace petroleum-derived chemicals (Figure 16.2). Itaconic acid has chemistry that is similar to petrochemicals derived from maleic anhydride (MA) [3] and fumaric acid, and may be substituted in some applications. Unsaturated polyester resins (UPRs), with a global production capacity of about 4 million metric tons annually [32], could incorporate itaconic acid [33] and acquire biodegradable functionality [33–35].

16.4.2 1,3-Propanediol (1,3-PDO)

1,3-PDO has attracted interest as an important production chemical due to its utility in the manufacture of polyester fibers and polyurethanes, as well as its use as a solvent. Until recently, the predominant methods for producing 1,3-PDO involved chemical catalysis of petrochemical feedstocks [36]. The two common chemical routes to 1,3-PDO formation are the three-step Degussa method and the two-step Shell method [37]. The former method oxidizes propylene to acrolein, followed by a selective hydration to 3-hydroxypropionaldehyde (3-HPA) and a catalytic hydrogenation to 1,3-PDO [38]. The Shell process involves the hydroformylation of ethylene oxide to 3-HPA and then a catalytic hydrogenation to 1,3-PDO [39]. The economical viability of the Shell process allowed 1,3-PDO market development.

More recently, biological routes to 1,3-PDO were investigated to allow sustainable production of this monomer. Many native bacterial strains such as *Klebsiella pneumonia, Citrobacter freundii*, and *Clostridium pasteurianum* have long been known to produce 1,3-PDO under anaerobic conditions with glycerol as the sole carbon source [40,41]. The genes involved in 1,3-PDO formation are known and well characterized [42]. In the absence of exogenous reducing equivalent acceptors, the formation of 1,3-PDO from glycerol is a two-step enzymatic conversion. Glycerol is first converted to 3-HPA and water by a dehydratase. The

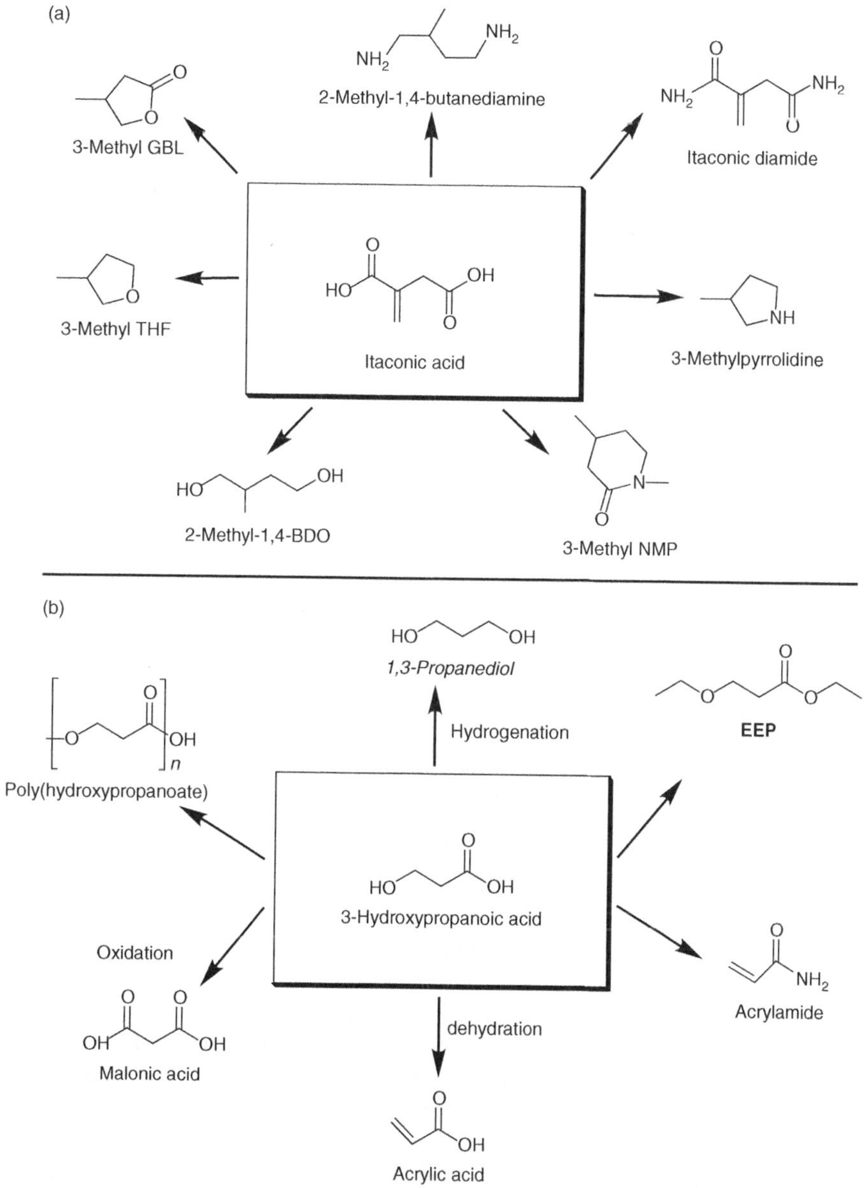

Figure 16.2 Itaconic acid and 3-HP as examples of platform chemicals.

3-HPA is then reduced to 1,3-PDO via an oxidoreductase, consuming a reducing equivalent. Once produced, 1,3-PDO is not metabolized by the cells and it accumulates in the media. Raw materials, especially the carbon source, are the most significant costs of traditional biosynthetic 1,3-PDO production [37]. Anaerobic production of 1,3-PDO from glycerol has been studied extensively due to the availability of low-cost crude glycerol as a by-product of biodiesel production [43]. Issues involved with the use of glycerol as a substrate are similar to those discussed for 3-HP (Section 16.5.1). Biosynthetic production of 1,3-PDO from glycerol has also been limited due to low titer. A biosynthetic route to 1,3-PDO from carbohydrates (i.e., glucose or fructose) would provide a consistent and cost-effective substrate supply. However, the growth of 1,3-PDO producers (*C. freundii, Clostridium autobutylicum, Clostridium butylicum*, and *K. pneumonia*) is limited in the presence of hydrogen donors such as glucose [44,45]. Other strains (i.e., various lactic acid bacteria) can grow in the presence of glucose and glycerol, but do not produce 1,3-PDO from glucose nor grow on glycerol alone [46]. Even *Escherichia coli* transformed with the *dha* regulon encoding glycerol dehydratase could not produce 1,3-PDO from glucose or fructose without glycerol [47].

An alternative to the production of 1,3-PDO from glycerol is a two-step conversion, in which glycerol produced by one strain is then fed to bacterial cells for conversion into 1,3-PDO [48]. Examples include the combinations of the green algae *Dunaliella tertiolecta* with *C. pasteurianum* (yielding 5 g/L 1,3-PDO), and the yeast *Pichia farinose* with *K. pneumoniae* [49,50]. In addition to low product titers, production processes using more than one fermentation organism can be challenging to commercialize.

A third option, pursued by Dupont deals with conversion of carbohydrates such as glucose or hydrolyzed cellulosic biomass into 1,3-PDO using a single microorganism [51–55]. Dupont and Genencor (Danisco) collaborated to develop a whole cell biocatalyst that could biologically produce 1,3-PDO at high titer. The recombinant *E. coli* K12 strain expressed genes encoding dehydratase activity (*dhaB*1–3), dehydratase reactivation factor (*dhaB*X, orfX), 1,3-PDO oxidoreductase (*yqhD*), glycerol-3-phosphate dehydrogenase (DAR1), and a glycerol-3-phosphatase (GPP2). Other metabolic engineering targets included down-regulation of glyceraldehyde 3-phosphate dehydrogenase, replacement of the Phosphoenol Pyruvate (PEP)-dependent system with an ATP-dependent phosphorylation system, and elimination of genes allowing glycerol to reenter central metabolism [56]. Additional genes were modified to increase the flux of carbon away from glycolysis and the TCA cycle toward 1,3-PDO. Unlike fermentations by natural organisms using glycerol, the DuPont process is aerobic and produces 50% more 1,3-PDO. The genetic modifications described above improved the strain for the energetic and redox demands associated with 1,3-PDO production. Aerobic 1,3-PDO production rates with the Dupont biocatalyst are reported to be 3.5 g/L/h with a titer of 135 g/L and w/w yield of 51%, compared to anaerobic fermentations from glycerol of 3.0 g/L/h with a titer of 78 g/L and 55% w/w yield [56].

The creation of a single biocatalyst capable of 1,3-PDO production from glucose and other carbon sources was realized through the efforts of three major contributors. The genes were isolated in 1994 and a Genencor (Danisco)–DuPont collaboration started in 1995. The first-generation biocatalyst was patented in 1997 and the second-generation biocatalyst followed shortly thereafter in 1999. In 2000, a collaboration with Tate & Lyle for fermentation development began and resulted in piloting of polymer grade material in 2001 and a commercial performance biocatalyst in 2002 [57]. The commercial plant start-up occurred in 2006 with a 45,000 ton capacity. The commercialization of the 1,3-PDO bioprocess took approximately seven years and significant financial investment [58].

Similar to petrochemically produced 1,3-PDO, Bio-PDO™ is clear, colorless, odorless, and has low toxicity. 1,3-PDO can be formulated into a variety of industrial products, including composites, adhesives, laminates, coatings, moldings, aliphatic polyesters, and copolyesters. Bio-PDO can replace petrochemically derived 1,3-PDO as a monomer for polytrimethylene terephthalate (PTT) production. Biosourced 1,3-PDO is copolymerized with terephthalic acid to PTT, requiring a purity of the 1,3-PDO to be between 95% and 99%. The biosynthetic 1,3-PDO must be free of reducing sugars, other organic acids, amino acids and proteins, complex nitrogen, and solvents [59]. Reacting 1,3-PDO with terephthalic acid resulted in a new fiber Sorona® Polymer with enhanced functionality for applications such as carpets, autos, fiber, and apparel [60]. The demand for PTT polymers is expected to be 1–2 billion pounds per year in the next 10 years.

Besides having a bioprocess that is competitive with the original petrochemically derived production process, an advantage of Bio-PDO over chemical 1,3-PDO is improved public perception of environmental friendliness. Life cycle assessment, a standardized method for comparing the environmental footprint of biobased materials versus petroleum-based material throughout the supply chain, was performed on the 1,3-PDO process (for additional information on life cycle assessment [61]). Dupont reported that the biological production of 1,3-PDO requires 40% less energy and produces 56% lower greenhouse gas emissions than petrochemically derived 1,3-PDO. A direct comparison between the manufacturing process for nylon 6 and renewably sourced Sorona showed that the latter used 30% less energy [57].

16.4.3 Lactic Acid

Lactic acid (2-hydroxypropionic acid) has been produced commercially by fermentation for more than a century. During this time, very few changes in production organisms or fermentation techniques have transpired; however, a significant knowledge of the underlying physiology of the organisms and process has been gained [62].

Historically, lactic acid has been used in food, tanning, cosmetics, and pharmaceuticals [62–64]. Although these markets are still present, over the past decade world lactic acid production has expanded 10-fold, due, in large part, to increased demand for green products derived from lactic acid. On a worldwide basis, Galactic, PURAC Corporation, Cargill, Inc., Archer Daniels Midland Company (ADM), and joint ventures derived from these companies account for approximately 75% of world lactic acid capacity [65].

The primary industrial applications of lactic acid are green solvents (derived from ethyl lactate) and PLA [63]. Although the low human toxicity of ethyl lactate compared to petrochemically derived solvents is attractive, price is cited as the primary reason for limited market use [65]. PLA, a polymer, is a green alternative to petroleum-derived plastics due to its biodegradability and reduced carbon footprint [66]. PLA products are on the market in a wide range of applications, including packaging, fibers, and foams [65–68]. A major producer of PLA is Natureworks LLC, which is currently wholly owned by Cargill Inc. [69]. Other companies produce smaller quantities of PLA in Europe and Japan, including Mitsui Chemical, Galactic, Treofan, and Shimadzu Corporation [65]. Like ethyl lactate, PLA is also limited in market penetration. Several key challenges currently prevent PLA from gaining a greater share of the 19 MM metric ton PET market [70]. The primary challenges for PLA market penetration are polymer attributes, disposal/life cycle, and cost [65,71].

The primary cost in the production of PLA and ethyl lactate is the cost of raw material, for example, lactic acid [65,70]. Lactic acid production from dextrose also accounts for >50% of the energy and greenhouse gas emissions in the PLA process; thus, reducing raw material and energy use associated with lactic acid improves carbon footprint and cost competitiveness of PLA versus petroleum polymers [61].

Cost estimates suggest that to be viable, lactic acid fermentation titers must exceed 100 g/L and overall production costs be at or below \$1.0 kg^{-1} of lactate [63]. As of 2006, the published price for food grade lactic acid was \$1.35 kg^{-1} of lactate [65]. This cost should be taken as an estimate and is expected to be highly variable trending primarily with raw material (e.g., sugar) costs [65]. As a result, a production strain for industrial lactic acid must fit the following criterion: production of >100 g/L lactic acid at yields near theoretical (>90% w/w lactic per dextrose) with rates, media, and recovery costs able to meet the above cost targets [63]. Lactic acid intended for use in PLA production must meet the additional requirements of high enantiomeric purity (>99% in L or D form); low residual sugars; low organic acids other than lactic; and minimal amounts of amino acids and proteins, complex nitrogen compounds, solvents, and color compounds. Presence of these compounds can result in the termination of PLA chain or negative effects on polymer properties [72].

Lactic acid fermentation is traditionally carried out by lactic acid bacteria [62]. Lactic acid is produced directly from glucose or other carbohydrate feedstocks (Figure 16.1). Pyruvate produced from glycolysis is converted by lactate dehydrogenase (LDH) to lactic acid, regenerating NAD+ in the process. Optimal production conditions vary with strain, but typically the temperature is from 40 to 45°C, the process is not aerated, and pH is maintained between 5 and 7 due to the strong inhibitory effect of free lactic acid on metabolic activity in bacteria (e.g., pH <5.0 typically halts the bacterial fermentation) [62]. Production rates, titers, and yields typically meet the criteria discussed above. Known limitations are present with the bacterial processes, especially when lactic acid is to be used for PLA manufacture. Requirements for complex, nutritionally rich media add raw material and purification costs [73]. The nutritionally rich media and pH range of these fermentations require that process sterility must be maintained so that growth of contaminating organisms do not impact the quality of the final product [63]. Lactic acid bacteria, like all bacteria, are subject to attack by bacteriophages. This translates into a risk of significant production downtime [62].

Finally, one of the major limitations of the bacterial lactic acid fermentation are the costs associated with purification, which can account for up to 50% of the production cost of lactic acid [65]. The pK_a of lactic acid is 3.86; therefore, at a pH of 6.0, <5% of lactic acid is in the free acid form, meaning >95% must be neutralized [63]. Typical neutralizing agents used at the industrial scale are lime ($Ca(OH)_2$) or calcium carbonate [65], both of which result in calcium lactate as the end product of fermentation. In this form, lactic acid has very little usability or value as an industrial chemical. Thus, the lactic acid must be freed from the calcium lactate [72]. This is typically done via acidulation of the broth with sulfuric acid. The stoichiometry of the neutralization and acidulation is as follows:

Neutralization: 1 mol lactic acid$^-$ + 1 mol H^+ + 0.5 mol $Ca(OH)_2$ = 0.5 mol calcium lactate + 1 mol H_2O

Acidulation: 0.5 mol calcium lactate + 0.5 mol H_2SO_4 = 1 mol lactic acid + 0.5 mol $CaSO_4$

Acidulation decreases the pH of the broth to < 2.0, freeing the lactic acid and resulting in the formation of calcium sulfate. As a by-product in the production of lactic acid, gypsum is either land filled as waste or land applied in agricultural applications [65,68,72,73]. The crude lactic acid is subsequently purified via filtration, ion exchange, evaporation, and/or other unit operations to meet product specifications; for PLA, these requirements consist of low residual nitrogen and sugars, and removal of other organic acids [72]. Typically, lactic acid is sold at a concentration of 50–80%, so the crude lactic acid is evaporated to make it concentrated (which is present in fermentation broth at 10–20%) [71].

Purification costs associated with lactic acid, primarily the acidulation requirement, can be addressed through either modifying the downstream operations or, more preferably, adjusting the fermentation process [70–72]. Although the calcium lactate process is the most common, alternate neutralization and recovery processes have been examined for lactic acid, including extractive solvents, ultrafiltration, reverse osmosis, and electrodialysis, but none have proven economical on an industrial scale [71,72].

To minimize lactic acid production costs and reduce the environmental footprint of the process, an ideal lactic acid production strain should fit the previously discussed criteria (titer >100 g/L, with high yields and chiral purity), and additionally have the following attributes to reduce purification costs: use a low-cost chemically defined media, not be susceptible to bacteriophages, and, most significantly, operate at a pH value well below the pK_a of lactic acid (3.86, e.g., high concentrations of free lactic acid). This "low-pH" operation would reduce or eliminate production and LCA (life cycle analysis) costs associated with lime, sulfuric acid, and gypsum [68]. Although conceptually a good idea, low-pH operation for the production of lactic acid is not straightforward. Lactic acid is a potent antimicrobial, especially as the amount of free lactic acid increases. One approach to identify, genetically engineer, mutagenize, or adapt lactic acid bacteria is to increase their relatively low tolerance to free lactic acid [63,74–79]. But to this point, no acid-tolerant bacteria has been genetically engineered or isolated that is able to produce lactic acid at industrially relevant rates or yields at pH values <4.5, meaning that the key solution for low-cost, low-carbon lactic acid has yet to be found in bacterial research [63,73].

Although filamentous fungi are known for tolerance to acidic environments, and are used in the production of organic acids such as citric and fumaric, a fungal organism that is able to produce lactic acid efficiently at pH values <4.5 has yet to be developed [80]. So, although *Rhizopus oryzae* can produce lactate at the titers necessary for a commercial lactic acid process, it does provide significant benefit over bacterial lactic acid production as the majority of lactic acid produced is still present in the salt form and not in the free (anion) form [80]. Further, issues with by-product flux, aeration requirements, and challenges associated with filamentous cell morphology limit industrial use of this organism [81,82].

The final approach that has received significant attention is the production of lactic acid in yeasts. Many yeast grow and ferment sugar to ethanol at pH <3, utilizing a chemically defined media. However, no yeast natively produces the concentrations or yields of lactic acid required for commercial operation [83]. Therefore, metabolic engineering approaches have focused on transferring flux from the native yeast fermentation product, ethanol, to lactic acid, which is stoichiometrically a redox and energy neutral fermentation product substitution [84]. The engineered pathway requires the introduction of a lactate dehydrogenase gene (*LDH*), which can convert natively produced pyruvate into lactic acid. Lactic acid production efficiency varies both with host strain and *LDH* used [85]. Multiple yeast species have been engineered to produce both stereoisomers (D and L) using genes from bacteria, fungi, bovine, frog, and humans [86–89].

As the most widely studied yeast in the world, strains of *Saccharomyces cerevisiae* have received considerable attention in this area, and in 1994 it was the first yeast to be metabolically engineered for the production of lactic acid [90]. Initial engineered strains produced only 10–20 g/L lactate [91,92]. Depending on host strain, PDC (pyruvate decarboxylase) and/or ADH (alcohol dehydrogenase) genes have been deleted additionally to effectively reduce ethanol production and increase lactic acid yields [87]. However, deletion of PDC reduces strain performance in *S. cerevisiae* [86,93]. When PDC is deleted, a growth requirement for exogenous two carbon compounds (C2s: ethanol, acetate, or acetaldehyde) is introduced, although this has been eliminated with evolution or mutagenesis [87,94,95]. The C2 requirement has not been reported in non-*Saccharomyces* PDC deletion yeast strains [96,97].

Changes in the aeration requirements occur in certain metabolically engineered lactic acid producing yeast strains. For example, in a strain of *Kluyveromyces lactis*, engineered by deletion of *PDC* and expression of *LDH*, reducing aeration rates from 0.5 to 0.05 L/min resulted in arrested lactic acid production while this same aeration change in the ethanol producing parent stimulates ethanol production [98]. Deletion of *PDC* and expression of *LDH* in *S. cerevisiae* strains eliminate the ability to grow anaerobically, even in the presence of anaerobic growth factors [94,99]. This inability to grow anaerobically may stem from increased ATP demand due to lactic acid transport from the cell [71]. Export of lactic acid out of the cell is not well understood in yeasts [100]. Some have suggested that the lactate anion, a polar molecule, is not likely to exit the cell through passive diffusion [98,101]. If lactic acid leaves the cell through a transport mechanism similar to other organic acids, such as benzoic as studied in *S. cerevisiae*, ATP would be required for transport [102]. Coupling this transport hypothesis with the inability to grow anaerobically suggests that lactic acid export requires energy [100]. As such, energy from the production of lactic acid may not be sufficient to sustain cellular maintenance and transport of lactic acid, thus additional energy, in the form of respiration is required when the lactic acid anion is present in significant quantities [84].

To improve lactic acid production in yeasts, several groups have gone beyond the basic engineering of overexpressing LDH and reducing ethanol production, and have examined the ability to improve lactic acid tolerance and production (Table 16.1). Evolutionary engineering approaches, and classical mutagenesis and selection have resulted in increased tolerance to lactic acid and enhanced strain performance [101,103–106]. Porro *et al*. reported increased tolerance to lactic acid in *S. cerevisiae* by increasing the activity of the plasma membrane ATPase (*PMA1*) [103]. Toyota has reported improved yields to lactic acid and improved quality of lactic acid for PLA production by deleting or reducing flux through the glycerol pathway [87,107]. Porro *et al*. have suggested that PDH (pyruvate dehydrogenase) deletion redirects pyruvate flux to lactic acid [93]. At this time, only Cargill has publically stated that their low-pH yeast process is commercially competitive with the established bacterial process for the production of lactic acid [106].

To address the primary challenges to PLA market growth, lactic acid producers will continue to strive for lower production cost and environmental effects. To achieve this, next-generation low-pH yeasts or bacteria need to be further developed to efficiently ferment low-cost substrates to free lactic acid (lactic acid anion). By using cellulosic feedstocks and wind power, the LCA of PLA could be a net negative greenhouse gas emission product. Vink *et al*.[61], Dornburg *et al*. [70], and Ilmen *et al*. [108] have reported the production of lactic acid from xylose in yeast [108]. Finally, Park *et al*. have demonstrated PLA and PLA copolymer production in *E. coli* by engineering in propionyl-CoA transferase and polyhydroxylalkanoate (PHA) synthase genes. Synthesis of PLA *in vivo* has the potential to eliminate the issues associated with the inhibition of lactic acid on microbial performance, in addition to reduced costs by potentially eliminating several downstream unit operations for PLA production [109].

Lactic acid, 1,3-PDO, and itaconic acid microbial production processes are well developed and have been commercialized. Although cost optimization of

TABLE 16.1 Summary of Published Reports on Bench-Scale Production of Lactic Acid in Yeast

Titer Produced (g/L)	Yield (%w/w)	Rate of Production (g/L/h)	Host Organism	pH of Production	Scale	Group	Citation
35	44	0.4	*K. lactis*	3.0	0.25 L flask	Daniel Porro (Tate and Lyle funded)	[96]
70	93	1.0	*S. cerevisiae*	< 3.0	0.25 L flask	Daniel Porro (Tate and Lyle funded)	[101]
120	90	9.0	*S. cerevisiae*	3.5	1 L fermentor (repeated batch)—13 g/L biomass pitch	Toyota	[87]
86	83	6.1		2.6	100 mL flask		
86	101	1.5	*Candida boidinii*	6.1	5 L fermentor	Toyota/Kirin	[98]
61	82	0.75	*S. cerevisiae*	2.6	Flask	Tate and Lyle (AE Staley)	[104]
107	Not reported	5.7	Not disclosed	3.0	2 L fermentor	Cargill	[106]

these processes will aid market expansion by making them more cost competitive with petrochemical alternatives, all three products are currently limited in the market, more so because of low market penetration and limited use of their end products, rather than because of the efficiency of the fermentation process. Replacing a route to a commodity chemical from petroleum with a white biotechnology route avoids problems with market acceptance, but requires the cost to be competitive.

16.5 EMERGING TECHNOLOGY FOR SUSTAINABLE PRODUCTION OF CHEMICALS THAT CAN FEED INTO EXISTING MARKETS

We now turn our attention from chemicals with technologies that are already commercialized to chemicals that for which sustainable production technologies are currently being developed, for example, the family of C4 diacids, 3-HP, and biological production of isoprene. The development of biological routes to produce acrylic acid (via 3-HP), isoprene, and substitute MA (via malic acid, fumaric acid, or succinic acid) is used as examples of the current state of biotechnology potential to replace petrochemically derived platform chemicals.

Organic acid toxicity issues discussed in detail regarding lactic acid are common to any discussion of organic acids and are a key challenge in the microbial production of organics acids such as 3-HP, fumaric, succinic, and malic acids. The C4 building block diacids are in many ways more similar to lactic acid than 3-HP. Unlike 3-HP, which requires significant metabolic engineering to achieve commercially relevant titers, all three C4 diacids can be produced by native organic acid producers.

In the area of emerging bioprocess technologies, strategic partnerships and collaborations are essential for commercial success. The impact of effective collaborations is discussed for the ongoing development of succinic acid, 3-HP, and bioisoprene.

16.5.1 3-Hydroxypropionic Acid

3-HP itself does not have an established market but it can be chemically converted to acrylic acid that has a very large and mature market. Although 3-HP is an isomer of lactic acid, it is not produced in significant amounts in nature, and therefore microbial production requires a metabolic engineering approach. A major technical consideration in the production of 3-HP was the development of an organism with the appropriate biosynthetic pathways. This has been accomplished as is illustrated below.

3-HP is an isomer of lactic (2-hydroxypropionic) acid that has a low toxicity, and is an attractive renewable platform chemical. Its theoretical yield from glucose is 100%, and it has two functional groups that allow it to participate in a variety of chemical reactions. 3-HP can be used as a substrate to form several commodity chemicals such as 1,3-PDO (Section 16.4.2) malonic acid, acrylamide, and acrylic acid. Acrylic acid is a large-volume chemical (>3 million metric tons/year) used to make acrylate esters and superabsorbent polymers, and is currently derived from the catalytic oxidation of propylene. Fermentative production of 3-HP would provide a sustainable alternative to petrochemicals as the feedstock for these commercially significant chemicals, thus reducing nonrenewable energy consumption, US dependence on foreign oil, and the production of greenhouse gases.

There are currently no established markets for 3-HP; however, besides its potential as a building block for commodity chemicals (Figure 16.2), 3-HP also has useful properties as an organic acid and as a polymer. The 3-HP acid has low metal corrosivity, high mineral salt solubility, and low toxicity attributes that could be applied in the cosmetics, oil well servicing, semiconductor, metal processing, fertilizer, and food industries [110]. Poly(3-HP) has been synthesized and is very heat stable, tough, and ductile, with a high glass transition temperature. Its barrier properties are similar to PET and thus poly(3-HP) could perform well in single layer films or stretch wraps [111]. In addition, copolymers using 3-HP, such as dimethyl terephthalate-3-HP-ethylene glycol and malic acid-glycolic acid-3-HP, have useful properties [112,113].

The potential of 3-HP as a renewable platform chemical prompted the DOE, and Cargill Incorporated to cofund a $12 million project in 2001 to develop a two-stage biobased process to produce a portfolio of chemicals based on 3-HP. The first stage consists of a conventional fermentation of renewable carbohydrates to 3-HP by a metabolically engineered organism. In the second stage, 3-HP is processed to high-volume products such as acrylic acid, its salts and esters, and acrylamide using chemical catalysis.

Unlike lactic acid, 3-HP is not a major end product of any pathway known in nature, being found in only trace amounts in some bacteria and yeast [114–116]. Before Cargill's research, only one fermentative route to 3-HP had been described [117], which used glycerol as the feedstock. Before the start of the DOE grant, Cargill had developed several novel pathways for the fermentation of 3-HP from glucose, passing through the intermediates of lactic acid, malonyl CoA, or β-alanine (Figure 16.3) [118]. Organism design criteria considered included maximum theoretical yield, pathway energetics, availability of enzymes/genes, cofactor requirements of the pathway enzymes, and the thermodynamic driving force to 3-HP production. Two pathways (through lactic acid and through β-alanine via alanine) have theoretical yields of 100% and sufficient ATP production to maintain the biocatalyst under anaerobic conditions, whereas the alternate pathways produce net zero or negative ATP and thus require some oxygenation for biocatalyst maintenance. Aeration requirements increase capital and operational costs compared to an anaerobic process.

Although 3-HP production was demonstrated using the pathway through lactate, this route is inherently inefficient since the conversion of lactate to 3-HP is in equilibrium, resulting in a mixture. Therefore, work was focused on a pathway through β-alanine that uses an alanine 2,3-aminomutase to convert alanine to β-alanine. Although this enzyme has not been discovered in nature, enzymes with

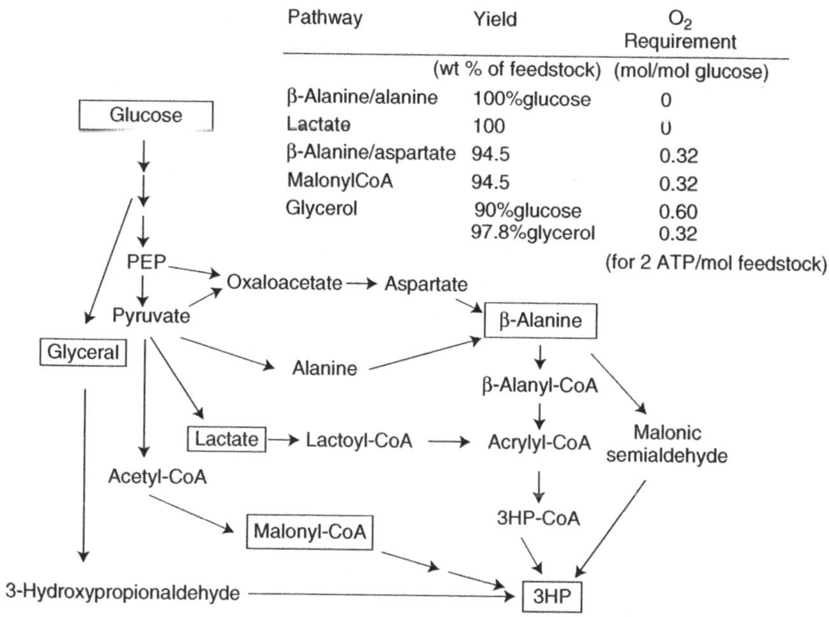

Figure 16.3 Various possible pathways for 3-hydroxypropionic acid biosynthesis.

this activity were developed by mutation of lysine 2,3-aminomutases from four different bacterial species. The novel enzymes were selected using a β-alanine auxotrophic strain, exploiting the fact the β-alanine is required for coenzyme A synthesis in *E. coli*. The alanine aminomutases were then improved by Codexis, using their directed evolution technology. Publication of the crystal structure of lysine 2,3-aminomutase in 2005 allowed the use of structure-based evolution techniques. Several rounds of evolution yielded a 300-fold improved alanine aminomutase over the starting enzyme. Alternative oxygen-requiring pathways were also engineered and tested.

In December 2007, Cargill entered into a Joint Development Agreement with Novozymes Inc. for continuation of the project. Building on the work done at Cargill, Novozymes is using a combination of deletions, gene overexpressions, and enzyme improvement to maximize carbon flow to 3-HP and increase pathway yield and titer.

In addition to the pathways for 3-HP from glucose outlined by Cargill, routes for its production from glycerol have also been proposed. Glycerol is gaining an increasing notice as a substrate for chemical production because of the availability of low-cost crude glycerol as a by-product of biodiesel production. The biodiesel process globally produced 1.2 million metric tons of crude glycerol in 2008, which could potentially fulfill only a fraction of the needs of a fully developed 3-HP platform. Crude glycerol prices have also been very volatile depending on supply and demand. An additional concern is that the impurities in crude glycerol (80% purity) could lead to increased costs for purification of fermentation broth or cause problems such as catalyst fouling, low quality product, or reduced product yield in the production of derivatives. Refined glycerol is 99.5% pure after undergoing an energy-intensive refining process, but has not achieved large supplies due to a refining bottleneck. The pathways for the production of 3-HP from glycerol utilize a dehydratase to produce 3-HPA from glycerol, as described for 1,3-PDO (Section 16.4.2). However, the subsequent reaction utilizes an aldehyde dehydrogenase to produce 3-HP from 3-HPA in place of the oxidoreductase that is used to produce 1,3-PDO. Raj *et al*. [119] reported an *E. coli* strain engineered expressing a *K. pneumonia* glycerol dehydratase and overexpressing a native aldehyde dehydrogenase, *aldH*. By using a fed batch procedure and controlling pH, a titer of 31 g/L could be achieved in 72 h [120]. Further pathway engineering led to titers of nearly 39 g/L and an average molar yield of 35% [121]. Pathways for 3-HP production from glucose with a glycerol intermediate have also been envisioned. The primary pathway is thus similar to the Dupont pathway for 1,3-PDO biosynthesis from glucose, except the last reaction from 3-HPA to 3-HP (Section 16.4.2).

Twelve biosynthetic pathways for 3-HP production were recently evaluated for technical viability, including eight pathways from glucose, three pathways from glycerol, and the 3-HP CO_2 fixation pathway [122]. Pathways were appraised on the basis of mass and redox balances, pathway energetics, cofactor requirements, and thermodynamics. The authors proposed the Cargill 2,3-aminomutase pathway, described above, and an ATP-generating pathway from glycerol [123] to be attractive based on these appraisal criteria.

E. coli was chosen as a host organism for the Cargill research owing to its well-developed genetic tools and availability of an annotated sequence to facilitate metabolic engineering. Furthermore, *E. coli* strains have been used for industrial-scale fermentation to make products such as 1,3-PDO, ethanol, succinic acid, and lactic acid. These strains can accumulate high levels of product and can grow on minimal media. They can also use five-carbon sugars from crude lignocellulosic hydrolysates as a carbon source. Because *E. coli* strains vary significantly in growth requirements, fermentative capacity, and growth rates, several *E. coli* strains were screened for these traits. In addition, tolerance to 3-HP and other organic acids was tested under different growth conditions, and in bench-scale fermentations the selected host exhibited moderate to good tolerance to organic acids, including 3-HP, fast growth rate, minimal nutrient requirements, growth up to 42°C, and a high fermentative capacity. Challenges with using *E. coli* as the production host include risk of contamination or bacteriophage infection and the requirement for a neutral pH fermentation as discussed in detail above for lactic acid production.

Toxicity issues resulting from the production of organic acids have also been described above for lactic acid. As is the case with lactic acid, current knowledge about 3-HP transport or anion effects is limited. Studies with 3-HP have demonstrated higher toxicity under anaerobic than aerobic growth conditions, and a greater effect on growth than on carbon metabolism.

Toxicity of 3-HP has been a main research focus of OPX Biotechnologies. Using their SCALES technology, OPX has identified genes and pathways that are up-regulated in strains selected for tolerance to 3-HP, including transporters and pathways for the synthesis of the amino acids tyrosine, phenylalanine, and threonine [124,125]. Toxicity of intermediates in the pathways to 3-HP must also be considered. OPX reported that reuterin, a mixture of compounds that includes 3-HPA along with its dimer and hydrate, was toxic at 0.03–0.05 g/L [124,125]. The toxic effects may be attributable to aldehyde-induced DNA damage, similar to what occurs during formaldehyde exposure. Intermediate toxicity can be addressed through pathway engineering or selection, or engineering of a tolerant host. OPX reported the production of approximately equal amounts of 3-HP and 3-HPA, requiring several gene knockouts to eliminate the unwanted aldehyde production. Toxicity of malonate semialdehyde in a Cargill pathway was used to identify a host enzyme that reduced this intermediate to 3-HP, increasing cell growth when overexpressed [126].

In addition to using engineering strategies to address toxicity issues, evolution, mutagenesis, and selection techniques have also been successfully utilized to improve host tolerance to organic acids. Mutagenesis and selection specific for 3-HP tolerance were conducted at Cargill under the DOE grant.

Similar to lactic acid production, impurities in the recovered 3-HP product could affect downstream processing and process economics by reducing catalyst lifetime, and product quality or yield. One possible recovery process for 3-HP is the lime/sulfuric acid addition, as described for lactic acid, with the requisite production of gypsum. Alternatively, an extractive salt-splitting purification process that would eliminate the need for the acidification process, produce recycled by-products, and could be run in either a batch or continuous mode has been described [127]. *In situ* product removal can be useful to decrease the effects of product toxicity [128]. As part of the development program at Novozymes, three potential processes for recovery of 3-HP from fermentation broth, in addition to the gypsum process, have been evaluated. One of these processes was successfully demonstrated on a kilogram scale for concentrating and semipurifying 3-HP. This process, along with the gypsum process, has been confirmed to be technically viable for industrial application. Cost analyses of these processes are underway.

The above-mentioned DOE 3-HP grant included a Cooperative Research and Development Agreement (CRADA) with Pacific Northwest National Laboratory (PNNL), which focused on catalyst development for derivatives of 3-HP. At the conclusion of the CRADA, the formation of acrylic acid and from 3-HP and sodium acrylate from sodium 3-HP was achieved with over 98% and 86% conversion and 99% and 90% selectivity, respectively. In addition, the formation of ammonium acrylate from ammonium 3-HP with high conversion rate and good selectivity was achieved, and a catalyst that could produce high yields of polyacrylates from sodium 3-HP was also identified. Optimum operating conditions for a commercial Titania-catalyzed reaction for the conversion to acrylic acid were identified; variables tested included 3-HP concentration, catalyst particle size, space velocity, operating temperature, and polymerization inhibitor concentration. Work also included evaluation of the lifetime potential of catalysts for 3-HP dehydration to acrylic acid in the presence and absence of simulated fermentation impurities and the effects of impurities on selectivity and conversion [129].

The 3-HP production pathways that are currently under development use an *E. coli* host. Future production pathways might use more acid-tolerant hosts, as for lactic acid. Other second-generation advances could include host organisms that allow production from cellulosic biomass. Such development programs are the focus of the EuroBioRef, a European program for the development of over 10 chemicals from biomass, scheduled to run from 2010 to 2014 with a funding of €23 million. Lastly, although the markets for many 3-HP derivatives have been fully developed, a significant amount of process optimization is required for conversion of 3-HP to these chemicals. This is partly due to these chemicals being currently produced using petrochemical routes that do not use 3-HP as an intermediate. The development of a market for 3-HP as an organic acid could provide lower volume but higher margin applications.

16.5.2 C4-Diacid Family: Succinic, Fumaric, and Malic Acids

The four-carbon 1,4-diacids (malic, fumaric, and succinic acids) are listed in the DOE top 12 value-added chemicals from sugars as potential biologically based building blocks with good reason [3]. The four-carbon diacids (succinic, fumaric, and malic acids) share not only common biosynthetic pathways but also a common petrochemical precursor, MA (Figure 16.4).

When considering petrochemical production processes to displace by biological fermentation, the redox state of MA makes it a particularly attractive target.

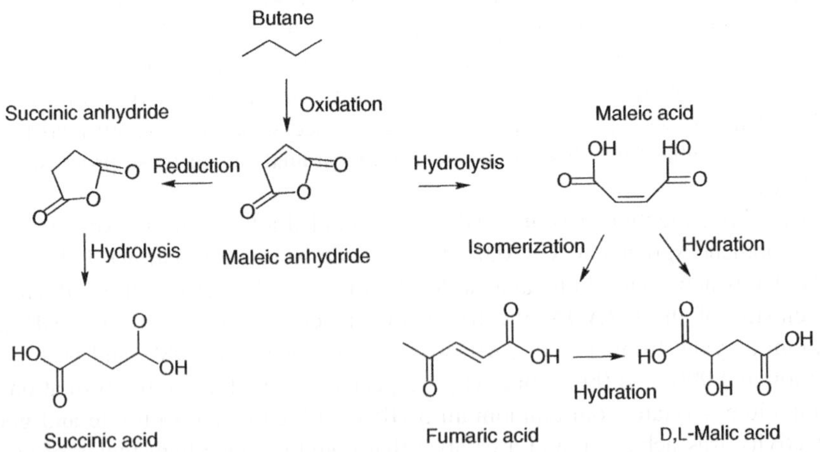

Figure 16.4 Schematic showing that the four-carbon diacid family have a common petrochemical origin: maleic anhydride is derived from butane with subsequent conversion to succinate, fumarate, and malate [130–132].

The carbohydrate feedstocks typical of most biological transformations contain carbon at a higher oxidation state than petroleum-derived carbon. When considering the production of one of the highly reduced platform chemicals such as ethylene or propylene, petrochemical feedstocks have the advantage of greater chemical similarity, fewer transformation steps, less energy input, and higher carbon yields [133]. All these factors translate to advantages in production cost. However, when considering the proposition of addressing the MA market, the situation is reversed. MA is highly oxidized relative to most petrochemical feedstocks, and from an oxidation state perspective it is more similar to carbohydrate feedstocks. In fact, the incumbent route to MA involves oxidation from butane, which is costly:

$$CH_3CH_2CH_2CH_3(n\text{-butane}) + 3\tfrac{1}{2}O_2 \rightarrow C_4H_2O_3\ (\text{maleic anhydride}) + 4H_2O$$

On the basis of the rationale above, a renewable route to succinic acid could be more cost effective than the current petrochemical production process [134]. Biologically derived succinic acid has an added advantage with a 50% reduction in green house gas emissions per ton of succinic acid produced [135].With both environmental and economical benefits, it is clear why bioprocesses for succinic acid production have begun to attract commercial interest. At least four ventures have announced their intent to produce fermentation-derived succinic acid on commercial scales between 2010 and 2012 [136]. Fumaric acid is commercially produced via the acid-catalyzed isomerization of maleic acid that originates as MA [130,137,138]. Conversion costs of MA to fumaric acid ($\sim$\$0.15 lb^{-1}) limit its substitution for MA in UPR applications, except where the economical benefit of improved resin hardness properties meets or exceeds the price spread. Thus, the use of fumaric acid could grow significantly in the UPR market if biobased fumaric acid could be produced cost competitively with MA. Fumaric and maleic acids are used in the commercial production of malic acid. Thus, these four-carbon dicarboxylic acids have the potential to be platform chemicals for the sustainable production of both commodity and specialty chemicals.

16.5.2.1 Succinate. The current succinate market, estimated at 25–30 tons in 2007 [139], is relatively modest, and is derived primarily from petrochemically produced MA as mentioned above [140]. Succinic acid is structurally similar to maleic acid, differing only by the single, rather than the double, bond between the second and third carbon atoms. MA and succinic acid have a wide variety of potential applications, including liquid antigels, heat transfer fluids, the solvents GBL (γ-butyrolactone) and dimethyl succinate, pigments, the polyesters PBS (polybutylene succinate) and PEIT (poly(ethylene-*co*-isosorbide)terephthalate), synthesis intermediates, and plasticizers [141]. While it has been estimated that the potential market for succinic acid could be as large as €2.5 billion per year [141], current production costs have limited the market size for potential applications [140].

There are three predominant metabolic routes available for succinate production:

1. *Reductive TCA* [$1C_6H_{12}O_6$ (glucose) + $2CO_2 + 2(NADH + H^+) \rightarrow$ $2C_4H_6O_4$ (succinic acid) + $2H_2O + 2NAD^+$].
2. *Oxidative TCA* [$1C_6H_{12}O_6$ (glucose) + $2H_2O + 5NAD^+ \rightarrow 1C_4H_6O_4$ (succinic acid) + $2CO_2 + 5\,(NADH + H^+)$].
3. *Glyoxylate shunt* [$1C_6H_{12}O_6$ (glucose) + $\frac{2}{3}H_2O + 2\,\frac{2}{3}NAD^+ \rightarrow 1\frac{1}{3}C_4H_6O_4$ (succinic acid) + $\frac{2}{3}CO_2 + 2\frac{2}{3}(NADH + H^+)$].

None of these pathways, in isolation, are capable of satisfying the requirement for an oxidation/reduction balance as seen from the net pathway stoichiometries.

In considering which pathway or pathways to employ, there is a strong economical driver to maximize the yield of fermentation products, through homofermentative routes. However, the formation of homofermentative succinate can be achieved only with pathways whose net stoichiometry can satisfy all elemental and redox balances. As shown in Table 16.2, none of the three metabolic routes, in isolation, can satisfy the redox balance, which is shown as net NADH production. This means that a succinic acid–producing organism using only one of these pathways will be heterofermentative. Alternatively, one can see that a proper NADH balance could be achieved with some combination of the reductive TCA with one or both of the other two pathways. As such, true homofermentative succinic acid fermentation could be achieved if multiple pathways are employed.

The technologies that are likely to be commercially viable in the near term are based on bacterial hosts and fall roughly into two categories: (i) native succinic acid producers, some of which have been metabolically engineered to improve yields, and (ii) nonproducers metabolically engineered for succinate production. Next-generation technologies with a longer time to commercialization include fungal hosts such as filamentous fungi or yeast. Advantages of these hosts include being able to operate at lower pH and subsequently avoid the costs associated with lime and gypsum that are a part of all near-neutral pH organic acid fermentations, as described for lactic acid production.

The natural succinate producers isolated from the bovine ruminate, *Actinobacillus succinogenes* and *Mannheimia succiniciproducens*, primarily use the reductive TCA route to produce succinic acid, and the production of coproducts is required to satisfy the need for NADH production as discussed above. These natural producers commonly carry out mixed acid fermentations and produce succinic acid in combination with a mixture of other organic acids. These

TABLE 16.2 Summary of Succinate Yields and Net NADH Production from Three Metabolic Pathways for Succinate Biosynthesis

Pathway	Molar Succinate Yield on Glucose (mol/mol)	Net NADH Production (mol NADH/mol Succinate)
Reductive TCA	2	−1
Oxidative TCA	1	5
Glyoxylate shunt	1.33	2

organic acid by-products represent not just a yield loss, but also a significant complication in downstream separations. Despite the mixed acid fermentation, these natural producers are able to achieve high titers (80 g/L succinate) [142], and high production rates (3.9 g/L/h) [140]. A major limitation in the further development of *A. succinogenes* is the paucity of genetic tools [143]. For *M. succiniciproducens*, there currently exists no system for expression of exogenous genes, but there have been some successes in knocking out genes by insertional inactivations [140]. Elimination of by-products remains a prime consideration for both of these strains.

E. coli naturally produces trace levels of succinic acid under fermentative conditions. Substantial metabolic engineering, required to make succinic acid a primary fermentation product in *E. coli*, is aided by the availability of a well-developed genetic toolbox. While there has been substantial success, the specific productivities of the best *E. coli* strains still lag those of the natural producers [140,143]. Natural succinic acid–producing organisms both grow and ferment under anaerobic conditions. In contrast, *E. coli* strains produce succinic acid at the highest rates and yields under anaerobic conditions, but do not grow without oxygen. This situation requires a two-phase fermentation with air sparged aerobic growth, followed by CO_2 sparged anaerobic fermentation. Despite the low productivities and two-phase fermentation, engineered *E. coli* strains are the first-generation fermentation hosts for three of the four ventures that have announced near-term plans for commercialization. This likely indicates the perceived economical advantages in downstream separations associated with an *E. coli* fermentation that generates lower levels of organic acid by-products.

Four companies have recently announced plans to produce succinic acid by fermentation. Bioamber is a collaboration between Diversified Natural Products, Inc. (DNP) and the French agricultural consortium Agro-industries Recherches et Développements (ARD) [141]. The Bioamber fermentation technology is based on an *E. coli* strain that produces succinic acid anaerobically with a CO_2 feed [136,144]. Succinium is a joint venture between the French agricultural conglomerate Roquette and the Netherlands-based chemical company DSM. Succinium purchased the *E. coli*-based succinic acid fermentation intellectual property from the University of Georgia and Rice University [145,146]. The University of Georgia technology reduces the formation of pyruvate-derived by-products by overexpression of a pyruvate carboxylase [147]. The Rice technology achieves higher yields of succinic acid by utilizing two pathways with complementary NADH balances: reductive TCA and the glyoxylate shunt. Additionally, DSM has several patent applications concerning succinic acid production in yeast that would potentially enable a second generation, low-pH fermentation technology [148–151]. Myriant, formerly known as BioEnergy International, has technologies for *E. coli*-based succinic acid fermentation, separation, and purification processes. The most recent succinic acid venture to be announced is a collaboration between the chemical company BASF and the lactic acid producer PURAC.

16.5.2.2 Malic Acid. Malic acid (hydroxybutanedioc acid, hydroxyl-succinic acid) (Figure 16.1) was isolated in 1785 [152]. Commercially, malic acid is produced by the hydration of maleic and fumaric acids, both derived from the MA at high temperature and pressure [131,153]. For this reason, currently malic acid is often produced at the same facility as fumaric acid [131]. This chiral molecule is commercially available as a racemate and is considered GRAS by the US FDA [131]. In addition to the racemate, biological processes are able to yield the L-enantiomer from carbohydrates by fermentation, as well as enzymatically starting from fumarate [154–159].

There are a wide range of applications for malic acid as a food acidulant, feed additive, for skin care, and polymers [131,153]. Malic acid competes directly with citric acid as a food acidulant. Malic acid is reported to mask some aftertaste associated with high-intensity sweeteners and has a synergistic impact in combination with some high-intensity sweeteners [131,153]. In 2002, BASF launched a water treatment application based on a malic acid polymer to prevent limescale build-up in desalination plants [160].

From a stoichiometric standpoint, both malic and fumaric acids present much simpler metabolic engineering targets compared with succinic acid. Succinic acid requires either the use of multiple pathways or the production of coproducts to satisfy a redox balance. In contrast, when fixing CO_2 by the reductive TCA route, both malic and fumaric acids directly redox balance with glucose, enabling high yield, homofermentative production using a single pathway. This reductive TCA route to malic acid has been genetically engineered in both *S. cerevisiae* and *E. coli*.

As with similar commercially available organic acids, filamentous fungi establish the baseline knowledge for the production of malic acid from carbohydrates. Of those reported in the literature, *Aspergillus flavus* has demonstrated the best results reaching commercially viable concentrations (113 g/L) with good yield (1.28 M yield), however, at relatively low rates (0.59 g/L/h) [161]. Also of concern with this approach is the production of aflatoxin by this organism, which makes this process especially undesirable for food applications [162]. The production of malic acid from microbes isolated from nature is not only limited to filamentous fungi but has also been reported in a number of yeast strains from the genera *Saccharomyces, Saccharomycodes, Zygosaccharomyces*, and *Schizosaccharomyce* [163]. Later investigations with *Zygosaccharomyces rouxii* showed that this sugar-tolerant strain would accumulate 74.9 g/L malic acid at a yield of 32.8% (w/w), which is stimulated, in part, with small amounts of glutamic acid added to the culture [164]. As with other fungal-produced organic acids such as citric acid, nutrient limitation, neutralizing agents, and trace metals have been focused on during process characterization and improvements for malic acid production. For example, high carbon-to-nitrogen ratios are preferred when fermenting *Rhizopus arrhizus* to produce malic acid [165,166]. This is again reiterated in *A. flavus* ATCC 13697 when Goldberg points out that under nitrogen limitation, increased iron concentration, and the inclusion of biotin and $CaCO_3$, malic acid production is stimulated markedly (from below detection to 48.6 g/L) as compared to a medium that is more suited for citric acid production (relatively high nitrogen and low iron concentration) [154].

The metabolic route to malic acid in filamentous fungi has been discussed recently by Goldberg *et al.* in detail [154]. The initial presumption was that malic acid was the product of the oxidative route of the TCA cycle or the glyoxylate cycle [165–167]. However, subsequent investigations have provided enzyme and ^{13}C labeling data to support the hypothesis of a cytosolic reductive route [154,168,169].

To that point, and as outlined by Zelle [170], a preferred pathway for malic acid production in *S. cerevisiae* starts with the carboxylation of pyruvate to oxaloacetate, followed by reduction to malic acid [170]. In this instance, the researchers successfully demonstrated enhanced malic acid production by the overexpression of pyruvate carboxylase (*PYC2*), increased expression of a peroxisomal malate dehydrogenase (*MDH3*) that was redirected to the cytosol, and the expression of a heterologous malate transporter from *Schizosaccharomyces pombe* (Sp*MAE1*) in a PDC-negative strain. The researchers demonstrated in shake flask cultures an approximately fivefold improvement in malic acid accumulation over the previous best genetically altered *S. cerevisiae* in which the cytosolic malate dehydrogenase (*MDH2*) was overexpressed [171]. More improvements are needed for this to become a viable commercial process (rate, titer, and yield); however, this sets a good basis for malic acid production in *S. cerevisiae*.

In *E. coli*, another interesting approach using stoichiometric pathway analysis is reported for the production of malic acid in a genetically modified strain. Here, researchers at the Korean Advanced Institute of Science and Technology applied this approach to identify an optimal metabolic network for the production of malic acid in *E. coli* lacking phosphate acetyltransferase (*pta*). The modeling analysis indicated a preferred route of engineering that led the researchers to introduce the phosphoenolpyruvate carboxykinase (*pck*) gene from *M. succiniciproducens* into the *pta*-deficient mutant [172]. These successes suggest that future improvement efforts in malic acid productions will likely benefit from the use of numerous tools, including fermentation process optimization, pathway engineering, and *in silico* pathway modeling.

16.5.2.3 Fumaric Acid. Fumaric acid (*trans*-butenedioic acid; *trans*-1,2-ethylenedicarboxylic acid) (Figure 16.1) is the geometric isomer of maleic acid (*cis*-butenedioic acid) and is found in plants and mushrooms [152]. As an isomer of maleic acid, fumaric acid has potentially broad ranging applications similar to that of maleic acid or MA, especially in polyester resins [3,138].

Fumaric acid is used as an acidulant in the food and beverage industry, a pharmaceutical in the polymer industry, and as discussed above as the precursor to malic acid. One of the unique properties of fumaric acid is its low water solubility (4.28 g/kg H_2O at 15.5°C), which makes it well suited for nonhygroscopic applications [130]. As an alternative to MA, fumaric acid is used in the production of UPRs and alkyd resins, especially when MA is in short supply, or for improving resin hardness properties [130,137]. Fumaric acid is also used as a feed additive [173].

Commercialization of a fungal fermentation for the production of fumaric acid in the 1940s by Pfizer was soon displaced by the lower cost petrochemical process that was revived 40 years later by Dupont for the production of Nylon [154,174–177].

Native fumaric acid production is reported in a wide variety of fungal genera, including *Aspergillus, Rhizopus, Cunninghamella*, and *Circinella* [176,178,179]. A key component for fumaric acid production is CO_2 for the carboxylation of pyruvate to oxaloacetate by pyruvate carboxylase. Most researchers have suggested that $CaCO_3$ should be used as a neutralizing agent to control pH and provide CO_2 [180–182]. In the absence of this CO_2 source, fermentation yield to fumaric acid will likely suffer as glucose, for example, becomes the source of CO_2 through the oxidative TCA cycle.

As similarly discussed for malic acid, Goldberg *et al*. suggest that in filamentous fungi fumaric acid is produced primarily through the reductive pathway of the TCA cycle, with some amount diverted through the oxidative route to provide energy for cellular growth and maintenance [154]. However, unlike malic and succinic acids, currently there is no reported pursuit of a genetically modified microorganism for the production of fumaric acid, although in the future a process that reduces or eliminates the need for neutralizing agent, such as an acid-tolerant yeast, would be a favorable approach [3,138].

Alternatively, an enzyme-catalyzed process for the isomerization of maleic acid to fumarate has been described in the literature as well. A number of enzyme isomerization processes utilizing maleate isomerase (EC 5.2.1.1) from different species of the genera *Pseudomonas, Alcaligens*, and *Bacillus* have been studied over the years [183–186]. A process utilizing *Pseudomonas alcaligenes* XD-1, for instance, has demonstrated conversion rates of 6.98 g/L/h, and later 14.3 g/L/h, with as high as 95% conversion of fumaric acid from maleic acid [187,188].

16.5.3 Bioisoprene

Isoprene (2-methyl-1,3-butadiene; C_5H_8), is the five-carbon building block of a class of compounds known as *terpenoids* and is predominantly produced by plants, although identification of microbial isoprene producers led to the development of biological production from inexpensive starting materials to compete with petroleum-based isoprene [189]. Compared to isoprene produced by the cracking of C5 hydrocarbons from petroleum, isoprene of a microbial origin can be isolated as a pure hydrocarbon for the manufacture of synthetic isoprene-based rubber and elastomers.

The commercialization of biosynthetic isoprene is the result of a joint venture between the Genencor division of Danisco and Goodyear Tire and Rubber Company with an exclusive license to the Kuzma patent [189], to serve a potential world market estimated at USD 1–2 billion [22,179–187189–191]. About 800,000 tons of *cis*-1,4-polyisoprene are produced per year from the polymerization of isoprene, most serving the tire and rubber industry. Goodyear, one of the world's largest users of isoprene, has a strategic supply arrangement with Genencor/Danisco to receive the manufactured BioIsoprene™.

Synthetic isoprene rubber is a component of tires (comprising *cis*-1,4-polyisoprene) [190]. Isoprene is obtained from petroleum, but the purification of this material is expensive and time consuming, resulting in only about 15% isoprene. The target of the Danisco/Goodyear Tire collaboration is the development of sustainable methods for isoprene production at high rates, titers, and purity that are sufficient to meet the demands of a robust commercial process from inexpensive starting materials. The 2009 Danisco patent [191] describes engineered cells that have high isoprene yield per cell, carbon yield, isoprene purity and productivity combined with reduced energy usage, production costs and investment, and minimal side reactions. BioIsoprene is made from corn- and sugarcane-based carbohydrates, but Genencor would like to switch from using such feedstock materials to using cheaper biomass sources such as corn cobs, corn leaves, bagasse, and switchgrass. According to the press release, BioIsoprene could be commercially available by 2013 [192].

16.6 CONCLUSIONS

Among the various chemicals that can be produced sustainably, organic acids represent a class of chemicals with a viable present and future. There is significant academic and industrial attention focused on the microbial production of organic acids as platform chemicals from renewable carbon sources, while keeping in mind the factors that will ensure commercial success once the technology is developed. The development of both the production technology and the product market is equally important for successful industrial microbial production of chemicals.

During the production of mature chemicals using microbial processes, the spent biomass after product recovery is generated as a waste. Traditionally, such spent biomass is inactivated, and then incinerated or used in animal feed applications. Using an integrated process approach that views fermentation as a unit operation in chemical production, as described in the case of citric acid, results in more attention being paid to value-added products that may be derived from microbial biomass or other waste streams following fermentations.

With chemicals that have a developed commercial-scale microbial production technology, there is a strong focus on developing the markets for these chemicals and their derivatives as seen with itaconic acid, lactic acid, and 1,3-PDO. The lactic acid and 1,3-PDO bioprocesses also illustrate how metabolic engineering can be successfully employed to make white biotechnology competitive and relevant by improving production costs and reduction of the carbon footprint of these processes. Emerging technology processes for green chemicals such as lactic acid and 1,3-PDO will test the power of metabolic engineering to provide commercial solutions and the risks associated with the introduction of a new product.

In the case of emerging technologies for sustainable production of chemicals, there is a strong emphasis on industrial and academic partnerships for the rapid development and commercialization of technology. This is exemplified in the creation of multiple collaborative ventures to explore the development of succinate using biological processes. These ventures approach an established market with a novel green process. The key to success for these ventures will not be building the market, but creating product at a cost low enough to compete with the petroleum-derived material in the bulk chemicals market.

There are many future considerations for the economic production of green bulk chemicals through microbial process. The primary drivers will continue to be costs associated with starting substrate and downstream processing. Feedstock prices can contribute up to 30–60% of the final production costs. The choice and purity of the carbon source can have a profound effect on the production process, product purification, and applications. In the quest for cheaper feedstocks, a number of novel strategies are being evaluated to use lignocellulosic biomass as a carbohydrate feedstock. While no clear winning technology has been established yet, we should expect to see that future green chemical production move from using corn-, sugarcane-, and sugar beet-derived feedstocks to cellulosic derived feedstocks once biorefineries are able to generate inexpensive clean feedstocks. Algal production of biofuels and chemicals using photosynthesis is theoretically promising but is far from technical and economical feasibility. Successful cellulosic or algal technologies have the potential to transforming by making biobased chemicals truly cost competitive with petrochemically derived production processes.

For industrial microbial processes, the production organism selected needs to be robust, and able to produce high rates, titers, and yields of product with a fermentation medium and downstream product separation regime that is economically viable. For organic acid production to be truly green and competitive, as demonstrated with citric, itaconic, and lactic acids, product tolerance under conditions of high free acid (respective anion) is vital for reducing costs and carbon footprint. The microbe chosen as the production host and the biosynthetic pathway selected for a specific chemical production also influence the extent of production strain development and bioprocess engineering that might be required. Lactic acid and 1,3-PDO illustrate the viability of integrating metabolic engineering into industrial processes; they also illustrate the complexity and the time required to develop a microbial process that can be competitive on an industrial scale.

To succeed, the microbial process must be considered as a key unit operation and not treated in isolation, as discussed above, and both downstream and upstream operations are critical to success and reducing production costs. Furthermore, coproduct streams must be evaluated as added cost drivers. Citric acid provides an example of using the spent mycelial biomass for secondary value-added products, while lactic acid provides an example of reducing waste streams through careful consideration of the fermentation and production host strain.

Finally, the development of green chemical production processes requires highly interdisciplinary teams with collaboration between engineers, biologists, and chemists. Consideration of both process engineering requirements and metabolic engineering approaches during the design phase of a green production process increases the chances of commercial success. Collaborations and partnerships between companies, and between industry and academic institutions, seek to leverage strengths from technology areas such as metabolic engineering, microbial physiology, systems biology, and fermentation, to inexpensive raw material supply and market access for products derived from green chemistry. Examples of partnerships such as the relationships between DuPont, Genencor (Danisco), and Tate & Lyle for 1,3-PDO production, Genencor (Danisco) and Goodyear Tires for Bioisoprene production, Cargill and Novozymes for 3-HP development illustrate industrial commitment to developing sustainable production and the potential for success using biotechnology in the green production of chemicals.

While the "green" features of biobased chemical production are important from an environmental and societal standpoint, it is the functionality and production cost of green chemicals that will dictate long-term success. Government mandates, subsidies, and funding of research can also have a significant influence on the production of green chemicals. Current industrial efforts are not directed toward complete petrochemical replacement; however, the gap between the economical viability of green chemicals compared with their corresponding petrochemical derivatives is closing.

REFERENCES

1. Frazzetto G. White biotechnology. EMBO Rep 2003;4:835–837.
2. Anastas PT, Warner JC. Green chemistry: theory and practice. New York: Oxford University Press; 1998. p. 30.
3. Werpy T, Petersen G, Aden A, Bozell J, Holladay J, White J, Manheim A. Top value added chemicals from biomass. Pacific Northwest National Laboratory and the National Renewable Energy Laboratory Publication; 2004. http://www.fischer-tropsch.org/DOE/DOE_reports/35523/PNN-NREL-35523_toc.htm
4. Okabe M, Lies D, Kanamasa S, Park E. Biotechnological production of itaconic acid and its biosynthesis in *Aspergillus terreus*. Appl Microbiol Biotechnol 2009;84:597–606.
5. Johnson L, May J. Wet milling: the basis for corn biorefineries. In: White P, Johnson L, editors. Corn: chemistry and technology. St. Paul.: American Association of Cereal Chemists, Inc.; 2003. pp. 449–494.
6. Wehmer C. Note sur la fermentation citrique. Bull Soc Chem Fr 1893;9:728.
7. Papagianni M. Advances in citric acid fermentation by *Aspergillus niger*: Biochemical aspects, membrane transport and modeling. Biotechnol Adv 2007;25:244–263.
8. Soccol C, Vandenberghe L, Rodrigues C, Pandey A. New perspectives for citric acid production and application. Food Technol Biotechnol 2006;44:141–149.
9. Malveda M, Janshekar H, Inoguchi Y. Citric acid. Chemical Economics Handbook Marketing Research Report. Menlo Park (CA): SRI Consulting; 2006.
10. Mattey M. Biochemistry of citric acid production by yeasts. Citric Acid Biotechnology. London: Taylor and Francis; 1999. pp. 33–54.
11. Forster A, Aurich A, Mauersberger S, Barth G. Citric acid production from sucrose using a recombinant strain of the yeast *Yarrowia lipolytica*. Appl Microbiol Biotechnol 2007;75:1409–1417.
12. Saad M. Chelating ability of chitosan-glucan complex from *Aspergillus niger* nrrl595 biomass recycling in citric acid production. Int J Biol Sci 2006;2:132–136.

13. Nud'ga LA, Petrova VA, Kever EE, Makarova TV. Hydrolysis of chitin-glucan complex of *Aspergillus niger* fungus with phosphoric acid. Russ J Appl Chem 2002;75:1864–1867.
14. Stagg C, Feather M. The characterization of a chitin-associated d-glucan from the cell walls of *Aspergillus niger*. Biochim Biophys Acta 1973;320:64–72.
15. Ruiz-Herrera J. Biosynthesis of β-glucans in fungi. Antonie van Leeuwenhoek 1991;60:73–81.
16. Fan W, Bohlmann J, Trinkle J, Steinke J, Hwang K, Henning J. Glucosamine and method of making glucosamine from microbial biomass. US patent 7,049,433. 2006.
17. Fosdick L, Bohlmann J, Trinkle J, Ray B. Glucosamine and method of making glucosamine from microbial biomass. 2004/0077055. 2004.
18. Teslenco A, Nikolaewna W. Process for producing chitosan-glucan complexes, compounds producible therefrom and their use. 6,333,399. 1997.
19. Cai J, Yang J, Du Y, Fan L, Qiu Y, Li J, Kennedy JF. Enzymatic preparation of chitosan from the waste *Aspergillus niger* mycelium of citric acid production plant. Carbohydr Polym 2006;64:151–157.
20. Pillai CKS, Paul W, Sharma CP. Chitin and chitosan polymers: Chemistry, solubility and fiber formation. Prog Polym Sci 2009;34:641–678.
21. Zhou S, Han T, Fosdick L, Vasina J. Water soluble beta-glucan, glucosamine, and *n*-acetylglucosamine compositions and methods for making the same. 2006/0172392. 2006.
22. Brown GD, Gordon S. Immune recognition of fungal β-glucans. Cell Microbiol 2005;7:471–479.
23. Soltanian SS, Stuyven EE, Cox EE, Sorgeloos P, Bossier PP. Beta-glucans as immunostimulant in vertebrates and invertebrates. Crit Rev Microbiol 2009;35:109–138.
24. Dergunova MA, Alexeenko TV, Zhanaeva SY, Filyushina EE, Buzueva II, Kolesnikova OP, Kogan G, Korolenko TA. Characterization of the novel chemically modified fungal polysaccharides as the macrophage stimulators. Int Immunopharmacol 2009;9:729–733.
25. Carlsson M, Habenicht C, Kam LC, Antal MJ Jr, Bian N, Cunningham RJ, Jones M Jr. Study of the sequential conversion of citric to itaconic to methacrylic acid in near-critical and supercritical water. Ind Eng Chem Res 2002;33:1989–1996.
26. Willke T, Vorlop KD. Biotechnological production of itaconic acid. Appl Microbiol Biotechnol 2001;56:289–295.
27. Pfeifer V, Vojnovich C, Heger E. Itaconic acid by fermentation with *Aspergillus terreus*. Ind Eng Chem 1952;44:2975–2980.
28. Yahiro K, Takahama T, Jai S-R, Park Y, Okabe M. Comparison of air-lift and stirred tank reactors for itaconic acid production by *Aspergillus terreus*. Biotechnol Lett 1997;19:619–621.
29. Lock MR, El-Aasser MS, Klein A, Vanderhoff JW. Investigation of the persulfate/itaconic acid interaction and implications for emulsion polymerization. J Appl Polym Sci 1990;39:2129–2140.
30. Lock MR, El-Aasser MS, Klein A, Vanderhoff JW. Role of itaconic acid in latex particle nucleation. J Appl Polym Sci 1991;42:1065–1072.
31. Ceska GW. The effect of carboxylic monomers on surfactant-free emulsion copolymerization. J Appl Polym Sci 1974;18:427–437.

32. Chinn H, Lochner U, Ishikawa Y. CEH marketing research report: unsaturated polyester resins, SRI Consulting, CEH; 2009.

33. Teramoto N, Ozeki M, Fujiwara I, Shibata M. Crosslinking and biodegradation of poly(butylene succinate) prepolymers containing itaconic or maleic acid units in the main chain. J Appl Polym Sci 2005;95:1473–1480.

34. Seppälä JV, Helminen AO, Korhonen H. Degradable polyesters through chain linking for packaging and biomedical applications. Macromol Biosci 2004;4:208–217.

35. Wallach J, Huang S. Copolymers of itaconic anhydride and methacrylate-terminated poly(lactic acid) macromonomers. Biomacromol 2000;1:174–179.

36. Zeng A-P, Biebl H. Bulk chemicals from biotechnology: The case of 1,3-propanediol production and the new trends. Tools Appl Biochem Eng Sci 2002;74:239–259.

37. Kraus GA. Synthetic methods for the preparation of 1,3-propanediol. CLEAN Soil Air Water 2008;36:648–651.

38. Haas T, Neher A, Arntz D, Klenk H, Girke W. Process for the simultaneous production of 1,2- and 1,3-propanediol. Degussa Aktiengesellschaft (Frankfurt, DE): United States. US patent 5426219. 1995.

39. Che TM. Production of propanediols. Celanese Corporation US patent 4642394. 1987.

40. Cameron DC, Altaras NE, Hoffman ML, Shaw AJ. Metabolic engineering of propanediol pathways. Biotechnol Progr 1998;14:116–125.

41. Wolf-Dieter D. Microbial conversion of glycerol to 1,3-propanediol. FEMS Microbiol Rev 1995;16:143–149.

42. Biebl H, Menzel K, Zeng AP, Deckwer WD. Microbial production of 1,3-propanediol. Appl Microbiol Biotechnol 1999;52:289–297.

43. Johnson DT, Taconi KA. The glycerin glut: Options for the value-added conversion of crude glycerol resulting from biodiesel production. Environ Prog 2007;26:338–348.

44. Gottschalk G. Prof. Dr. Averhoff Beate Dipl-Biol. Process for the microbiological preparation of 1,3-propane-diol from glycerol. European patent Office (EPO) EP0373230. Unilever Nv; Unilever Plc. 1990.

45. Daniel R, Boenigk R, Gottschalk G. Purification of 1,3-propanediol dehydrogenase from *Citrobacter freundii* and cloning, sequencing, and overexpression of the corresponding gene in *Escherichia coli*. J Bacteriol 1995;177:2151–2156.

46. Veiga-Da-Cunha M, Foster MA. 1,3-propanediol:Nad+ oxidoreductases of *Lactobacillus brevis* and *Lactobacillus buchneri*. Appl Environ Microbiol 1992;58:2005–2010.

47. Tong IT, Liao HH, Cameron DC. 1,3-propanediol production by *Escherichia coli* expressing genes from the *Klebsiella pneumoniae* dha regulon. Appl Environ Microbiol 1991;57:3541–3546.

48. Selembo PA, Perez JM, Lloyd WA, Logan BE. Enhanced hydrogen and 1,3-propanediol production from glycerol by fermentation using mixed cultures. Biotechnol Bioeng 2009;104:1098–1106.

49. Nakas JP, Schaedle M, Parkinson CM, Coonley CE, Tanenbaum SW. System development for linked-fermentation production of solvents from algal biomass. Appl Environ Microbiol 1983;46:1017–1023.

50. Hartlep M, Hussmann W, Prayitno N, Meynial-Salles I, Zeng A-P. Study of two-stage processes for the microbial production of 1,3-propanediol from glucose. Appl Microbial Biotechnol 2002;60:60–66.
51. Laffend LA, Nagarajan V, Nakamura CE. Bioconversion of a fermentable carbon source to 1,3-propanediol by a single microorganism. E. I. Du Pont de Nemours and Company, Wilmington, DE. US patent 6025184. 1997.
52. Gatenby AA, Haynie SL, Nagarajan V, Nair RV, Nakamura CE, Payne MS, Picataggio SK, Dias-Torres M, Hsu AK-H, Lareau RD, Trimbur DE, Whited GM. Method for the production of 1,3-propanediol by recombinant organisms. WO US9720292. 1998.
53. Nair RV, Payne MS, Trimbur DE, Valle F. Method for the production of glycerol by recombinant organisms. WO. US9825551. 1999.
54. Dunn-Coleman NS, Diaz-Torres M, Chase MW, Trimbur D. In: Dunn-Coleman NS, Diaz-Torres M, Chase MW, Trimbur D, editors. Method for the recombinant production of 1,3-propanediol. I.N.C. Genencor International. WO. US9720873. 1998.
55. Nakamura CE, Gatenby AA, Hsu AK, La Reau RD, Haynie SL, Diaz-Torres R, Trimbur DE, Whited GM, Nagarajan V, Payne MS, Picataggio SK, Nair RV. Method for the production of 1,3-propanediol by recombinant microorganisms. E. I. du Pont de Nemours and Company (Wilmington, DE),Genencor International (Palo Alto, CA). US patent 6013494. 2000.
56. Willke T, Klaus V. Biotransformation of glycerol into 1,3-propanediol. Eur J Lipid Sci Technol 2008;110:831–840.
57. Dupont. Clothing from cornfields: DuPont develops a fermentation-based process to create sorona® polymer from renewable resources. 2009. Available at www.dupont.com/sorona/literature.html. Accessed 2009 Jan 10.
58. Sauer M, Marx H, Mattanovich D. Microbial production of 1,3-propanediol. Recent Pat Biotechnol 2008;2:191–197.
59. Xiu Z-L, Zeng A-P. Present state and perspective of downstream processing of biologically produced 1,3-propanediol and 2,3-butanediol. Appl Microbiol Biotechnol 2008;78:917–926.
60. Kurian JV. A new polymer platform for the future—sorona® from corn derived 1,3-propanediol. J Polym Environ 2005;13:159–167.
61. Vink E, Rabago K, Glassner D, Gruber P. Applications of life cycle assessment to natureworks polylactide (pla) production. Polym Degrad Stab 2003;80:403–419.
62. Benninga H. A history of lactic acid making, a chapter in the history of biotechnology. Dordrecht, Boston and London: Kluwer Academic Publishers; 1990.
63. Litchfield J. Lactic acid, microbially produced. Encyclopedia of Microbiology. Oxford: Academic Press; 2009. pp. 362–372.
64. Hassan A, Frank J. Starter cultures and their use. In: Marth EH, Steele J, editors. Applied dairy microbiology. New York: Marcel Dekkar, Inc.; 2001. pp. 151–206.
65. Malveda M, Blagoev M, Kishi A. Lactic acid, its salts and esters. CEH Marketing Research Report; 2006.
66. Gupta B, Revagade N, Hilborn J. Poly(lactic acid) fiber: An overview. Prog Polym Sci 2007;32:455–482.
67. Ingeo. Ingeo plastic made from plants reaches a new plateau, reducing co2 emissions another 60%, energy consumption 30%. CSR Press Release; 2009.

68. Vink E, Glassner D, Kolstad J, Wooley R, O'connor R. The eco-profiles for current and near-future natureworks polylactide (pla) production. Ind Biotechnol 2007;3:58–81.

69. Defosse M. Teijin exits jv, leaving Cargill as sole owner of natureworks bioplastic. Plast. Today; 2009.

70. Dornburg V, Faaij A, Patel M, Turkenburg WC. Economics and ghg emission reduction of a pla bio-refinery system–combining bottom-up analysis with price elasticity effects. Resour Conserv Recycl 2006;46:377–409.

71. Averous L. Polylactic acid: synthesis in properties and applications. In: Belgacem N, et Gandini A, editors. Monomers, polymers and composites from renewable resources. Elsevier; 2008. pp. 433–450. http://science.cjb.net/science/book/9780080453163

72. Joglekar HG, Rahman I, Babu S, Kulkarni BD, Joshi A. Comparative assessment of downstream processing options for lactic acid. Sep Purif Technol 2006;52:1–17.

73. Singh SK, Ahmed SU, Pandey A. Metabolic engineering approaches for lactic acid production. Process Biochem 2006;41:991–1000.

74. Wang Y, Li Y, Pei X, Yu L, Feng Y, Biotechnol J. Genome-shuffling improved acid tolerance and l-lactic acid volumetric productivity in *Lactobacillus rhamnosus*. J Biotechnol 2007;129:510–515.

75. Kyla-Nikkila K, Hujanen M, Leisola M, Palva A. Metabolic engineering of *Lactobacillus helveticus* cnrz32 for production of pure l-(+)-lactic acid. Appl Environ Microbiol 2000;66:3835–3841.

76. Dien BS, Nichols NN, Bothast RJ. Fermentation of sugar mixtures using *Escherichia coli* catabolite repression mutants engineered for production of l -lactic acid. J Ind Microbiol Biotechnol 2002;29:221–227.

77. Zhou S, Causey TB, Hasona A, Shanmugam KT, Ingram LO. Production of optically pure d-lactic acid in mineral salts medium by metabolically engineered *Escherichia coli* w3110. Appl Environ Microbiol 2003;69:399–407.

78. Zhou S, Shanmugam KT, Ingram LO. Functional replacement of the *Escherichia coli* d-(-)-lactate dehydrogenase gene (ldha) with the l-(+)-lactate dehydrogenase gene (ldhl) from *Pediococcus acidilactici*. Appl Environ Microbiol 2003;69:2237–2244.

79. Budhavaram NK, Fan Z. Production of lactic acid from paper sludge using acid-tolerant, thermophilic *Bacillus coagulan* strains. Bioresour Technol 2009;100:5966–5972.

80. Zhang ZY, Jin B, Kelly JM. Production of lactic acid from renewable materials by *Rhizopus* fungi. Biochem Eng J 2007;35:251–263.

81. Liu T, Miura S, Yaguchi M, Arimura T, Park EY, Okabe M. Scale-up of l-lactic acid production by mutant strain *Rhizopus sp*. Mk-96-1196 from 0.003 m^3 to 5 m^3 in airlift bioreactors. J Biosci Bioeng 2006;101:9–12.

82. Baia D-M, Jia M-Z, Zhao X-M, Ban R, Shen F, Li X-G, Xu S-M. -lactic acid production by pellet-form *Rhizopus oryzae* r1021 in a stirred tank fermentor. Chem Eng Sci 2003;58:785–791.

83. Robbins EI, Gross S. Acidophilic and acid-tolerant fungi and yeasts. Hydrobiologia 2000;433:91–109.

84. Van Maris AJA, Winkler AA, Porro D, Van Dijken JP, Pronk JT. Homofermentative lactate production cannot sustain anaerobic growth of engineered *Saccharomyces cerevisiae*: possible consequence of energy-dependent lactate export. Appl Environ Microbiol 2004;70:2898–2905.

85. Branduardi P, Sauer M, De Gioia L, Zampella G, Valli M, Mattanovich D, Porro D. Lactate production yield from engineered yeasts is dependent from the host background, the lactate dehydrogenase source and the lactate export. Microb Cell Fact 2006;5:4.

86. Ishida N, Saitoh S, Tokuhiro K, Nagamori E, Matsuyama T, Kitamoto K, Takahashi H. Efficient production of l-lactic acid by metabolically engineered *Saccharomyces cerevisiae* with a genome-integrated l-lactate dehydrogenase gene. Appl Environ Microbiol 2005;71:1964–1970.

87. Onishi T, Matsushita H, Tada N, Shimamura T, Ishida N. Conference kf efficient production of lactic acid by metabolically engineered. Saccharomyces cerevisiae; 2009.

88. Rajgarhia V. Methods for the synthesis of lactic acid using crabtree-negative yeast; 2002.

89. Sawai H, Sawai K, Sonoki T, Hatahira S. Yeast and method of producing l-lactic acid. Editor: WO.I.N.C.N.-M.C.C.-K.T. Toray Industries; 2007.

90. Dequin S, Barre P. Yeast strains expressing lactic dehydrogenase gene and vectors useful in the producing said strains. 1998.

91. Dequin S, Barre P. Mixed lactic-alcoholic fermentation by *Saccharomyces cerevisiae* expressing the *Lactobacillus casei l*() -ldh. Biotechnol (NY) 1994;12:173–177.

92. Porro D, Brambilla L, Ranzi BM, Martegani E, Alberghina L. Development of metabolically engineered *Saccharomyces cerevisiae* cells for the production of lactic acid. Biotechnol Progr 1995;11:294–298.

93. Porro D, Bianchi M, Ranzi B, Frontali L, Vai M, Winkler A, Alberghina L. Yeast strains for the production of lactic acid transformed with a gene coding for lactate dehydrogenase. 1998.

94. Van Maris AJA, Geertman J-MA, Vermeulen A, Groothuizen MK, Winkler AA, Piper MDW, Van Dijken JP, Pronk JT. Directed evolution of pyruvate decarboxylase-negative *Saccharomyces cerevisiae*, yielding a c2-independent, glucose-tolerant, and pyruvate-hyperproducing yeast. Appl Environ Microbiol 2004;70:159–166.

95. Van Maris A, Pronk J, Van Djiken J. Pyruvate producing yeast strain. 2005.

96. Porro D, Bianchi MM, Brambilla L, Menghini R, Bolzani D, Carrera V, Lievense J, Liu C-L, Ranzi BM, Frontali L, Alberghina L. Replacement of a metabolic pathway for large-scale production of lactic acid from engineered yeasts. Appl Environ Microbiol 1999;65:4211–4215.

97. Rajgarhia V. Methods and materials for the production of d-lactic acid in yeast. 2003.

98. Osawa F, Fujii T, Nishida T, Tada N, Ohnishi T, Kobayashi O, Komeda T, Yoshida S. Efficient production of lactic acid by crabtree-negative yeast *Candida boidinii*. Yeast 2009;26:485–496.

99. Ikada Y, Jamshidi K, Tsuji H, Hyon SH. Stereocomplex formation between enantiomeric poly(lactides). Macromol 2002;20:904–906.

100. Van Maris AJA, Konings WN, Dijken JPV, Pronk JT. Microbial export of lactic and 3-hydroxypropanoic acid: Implications for industrial fermentation processes. Metab Eng 2004;6:245–255.

101. Valli M, Sauer M, Branduardi P, Borth N, Porro D, Mattanovich D. Improvement of lactic acid production in *Saccharomyces cerevisiae* by cell sorting for high intracellular ph. Appl Environ Microbiol 2006;72:5492–5499.
102. Mollapour M, Piper P. Weak organic acid resistance of spoilage yeasts. Br Mycol Soc Symp Ser 2008;27:143–155.
103. Porro D, Dato L, Brandaurdi P. Method for improving acid and low ph tolerance in yeast. 2008.
104. Liu C, Lievense J. Lactic acid producing yeast. US patent 20050112737. 2003.
105. Minoska V, Sauer M, Porro D, Paoka B, Diethard M. Improved strains for the production of organic acids. WO00738130. 2007.
106. Mcmullin T. Biotechnology to the bottom line-low ph lactic acid production at commercial scale. in SIM. 2009.
107. Yamaguchi I, Saotome O, Kuromiya S, Ohnishi T, Yasutani N, Saitoh S, Mouri M, Nakano M, Usuki A, Ishida N, Tokuhiro K, Nagamori E, Takahashi H. Method for producing lactic acid. US patent 20070161098. 2004.
108. Ilmen M, Koivuranta K, Ruohonen L, Suominen P, Penttila M. Efficient production of l-lactic acid from xylose by *Pichia stipitis*. Appl Environ Microbiol 2007;73:117–123.
109. Park S, Yang T, Lee S, Lee E, Kang H, Kim T. Recombinant microorganism having a producing ability of polylactate or it copolymers and methods for preparing polylactate or its copolymers using the same. 2009.
110. Cameron D. Engineering of a new metabolic system: the conversion of sugars to 3-hydroxypropionic acid. Meriden (NH); 2004
111. Liao H, Jessen H. Control of chromosomal gene expression by promoter replacement. J Am Chem Soc 2006.
112. Kudo S, Yamane H, Kitao T, Kimura Y, Shirotani K. Manufacture of hydroxycarboxylic acid copolymers for medical goods. 1990.
113. Hirose MT, Kuratsuji TM. Polyester copolymers for packaging materials. 1986.
114. Holo H. *Chloroflexus aurantiacus* secretes 3-hydroxypropionate, a possible intermediate in the assimilation of co2 and acetate. Arch Microbiol 1989;151:252–256.
115. Loh KD, Gyaneshwar P, Markenscoff Papadimitriou E, Fong R, Kim K-S, Parales R, Zhou Z, Inwood W, Kustu S. A previously undescribed pathway for pyrimidine catabolism. Proc Natl Acad Sci USA 2007;103:5114–5119.
116. Andersen G, Björnberg O, Polakova S, Pynyaha Y, Rasmussen A, Møller K, Hofer A, Moritz T, Sandrini MPB, Merico A-M, Compagno C, Âkerlund H-E, Gojkovic Z, Piskur J. A second pathway to degrade pyrimidine nucleic acid precursors in eukaryotes. J Mol Biol 2008;380:656–666.
117. Suthers P, Cameron D. Production of 3-hydroxypropionic acid in recombinant organisms. 2001.
118. Gokarn R, Selifonova OV, Jessen H, Gort S, Gselmer T, Buchel W. 3-hydroxypropionic acid and other organic compounds. 2007.
119. Raj SM, Rathnasingh C, Jo J-E, Park S. Production of 3-hydroxypropionic acid from glycerol by a novel recombinant *Escherichia coli* bl21 strain. Process Biochem 2008;43:1440–1446.
120. Raj SM, Rathnasingh C, Jung W-C, Park S. Effect of process parameters on 3-hydroxypropionic acid production from glycerol using a recombinant *Escherichia coli*. Appl Microbiol Biotechnol 2009;84:649–657.

121. Rathnasingh C, Raj SM, Jo J-E, Park S. Development and evaluation of efficient recombinant *Escherichia coli* strains for the production of 3-hydroxypropionic acid from glycerol. Biotechnol Bioeng Symp 2009;104:729–739.
122. Jiang X, Meng X, Xian M. Biosynthetic pathways for 3-hydroxypropionic acid production. Appl Microbiol Biotechnol 2009;82:995–1003.
123. Yasuda S, Mukoyama M, Horikawa H, Toraya T, Morita H. Process for producing 1,3-propanediol and/or 3-hydroxypropionic acid. US2007148749 A1. 2007.
124. Gill R, Lynch MD. Compositions and methods for enhancing tolerance for the production of organic chemicals produced by microorganisms. 2008.
125. Warnecke T, Gill RT. Organic acid toxicity, tolerance, and production in *Escherichia coli* biorefining applications. Microb Cell Fact 2005;4:25.
126. Jessen H, Liao H, Gort S, Selifonova O. Beta-alanine/alpha-ketoglutarate amintransferase for 3-hydroxypropionic acid production. US20090291480. 2008.
127. Meng X, Tsobanakis P, Malsom J, Abraham T. Process for preparing 3-hydroxycarboxylic acids and derivatives thereof. US7186856. 2007.
128. Schügerl K, Hubbuch J. Integrated bioprocesses. Curr Opin Microbiol 2005;8: 294–300.
129. Hallen R, Cooper A. Conference—High throughput Catalyst Screening for Bioproducts Process Development. Regional ACS Meeting; Salt Lake City, Utah.
130. Bizzari S, Blagoev M. CEH marketing research report-fumaric acid. 2007.
131. Bizzari S. CEH marketing research report, DL-malic acid. C EH-SRI Consulting. 2007.
132. Delhomme C, Weuster-Botz D, Kuhn F. Succinic acid from renewable resources as a c4 building-clock chemical- a review of the catalytic possibilities in aqueous media. Green Chem 2009;11:13–26.
133. Banholzer WF, Watson KJ, Jones ME. How might biofuels impact the chemical industry? Chem Eng Prog 2008;104:S7–S14.
134. Willke T, Vorlop KD. Industrial bioconversion of renewable resources as an alternative to conventional chemistry. Appl Microbiol Biotechnol 2004;66:131–142.
135. Hermann BG, Blok K, Patel MK. Producing bio-based bulk chemicals using industrial biotechnology saves energy and combats climate change. Environ Sci Technol 2007;41:7915–7921.
136. Challener C. Sugar never tasted so sweet. Spec Chem Mag 2009;29:42–44.
137. Lohbeck K. Ullmann's chemical encyclopedia–maleic and fumaric acids. Weinheim: Wiley-VCH; 2005.
138. Roa Engel CA, Straatholf JJA, Zijilmans TW, Van Gulik WM, Van Der Wielen LAM. Fumaric acid production by fermentation. Appl Microbiol Biotechnol 2008;78:379–389.
139. Badina J. Green chemistry. Intermediates. Roquette and DSM companies will produce "Bio" Succinic acid. New isosorbide industrial production for roquette. Chimie verte. Interme'diaires. Roquette et DSM vont produire de l'acide succinique "bio" Nouvelle production industrielle d'isosorbide pour Roquette. 2008;45:28.
140. Song H, Lee SY. Production of succinic acid by bacterial fermentation. Enzyme Microb Technol 2006;39:352–361.
141. Cottineau J. Green chemistry. Intermediates. The bioamber company builds a succinic acid bio-refinery in the marine. Chimie verte. Intermédiaires. Bioamber contruit une bioraffinerie d'acide succinique dans la Marne. 2009;46:23.

142. Guettler M, Mahendra J, Bhupendra S. Process for making succinic acid, microorganisms for use in the process and methods of obtaining the microorganisms. Lansing (MI): Michigan, Biotechnology Institute; 1996.

143. Mckinlay JB, Vieille C, Zeikus JG. Prospects for a bio-based succinate industry. Appl Microbiol Biotechnol 2007;76:727–740.

144. Timizar N. File on the environment. Green chemistry. The bioamber company bio-produces succinic acid. Dossier environnement. Chimie verte. Bioamber bioproduit l'acide succinique. 2008;45:50–51.

145. Ruth D. 'Green' chemicals closer to market. Rice university, Roquette freres sign deal for bio-based succinic acid. Rice University News Release; 2008. http://www.eurekalert.org/pub_releases/2008-12/ru_cc121908.php

146. Hastings TM. University of Georgia signs licensing deal with European company for a building block of hundreds of industrial and household products. The University of Georgia news Release; 2008.

147. Gokarn RR, Eitman MA, Altman E. Metabolically engineered *E. coli* for enhanced production of oxaloacetate-derived biochemcials. 2002.

148. Verwaal RM, Nootdorp NL, Wu LS, Delft VK, Damveld RA, Sagt CM, Jacobus KM, Utrecht HV. Succinic acid production in a eukaryotic cell. H.O.N.L.T.E.H. Dsm Ip Assets B.V., Editor: WO; 2009.

149. Verwaal RM, Nootdorp NL, Wu LS, Delft VK, Damveld RA, Slufterhof NK, Rodenrijs BE, Sagt CM, Millockerhof JK, Utrecht HV. Dicarboxylic acid production in a recombinant yeast. H.O.N.L.T.E.H. Dsm Ip Assets B.V., Editor: WO; 2009.

150. Verwaal RM, Nootdorp NL, Wu LS, Delft VK, Damveld RA, Slufterhof NK, Rodenrijs BE, Sagt CM, Jacobus KM, Utrecht HV. Dicarboxylic acid production in eukaryotes. H.O.N.L.T.E.H. Dsm Ip Assets B.V., Editor: WO; 2009.

151. Verwaal RM, Nootdorp NL, Wu LS, Delft VK, Damveld RA, Slufterhof NK, Rodenrijs BE, Sagt CM, Jacobus KM, Utrecht HV. Process for the production of a dicarboxylic acid. H.O.N.L.T.E.H. Dsm Ip Assets B.V., Editor: WO; 2009.

152. Jensen W. The origin of the names malic, maleic, and malonic acids. J Chem Educ 2007;84:924.

153. Miltenberger K. Ullmann's chemical encyclopedia–hydroxycarboxylic acids, aliphatic. Weinheim: Wiley; 2005.

154. Goldberg I, Rokem J, Pines O. Review organic acids: old metabolites, new themes. J Chem Technol Biotechnol 2006;81:1601–1611.

155. Neufeld R. Malic acid formation by immobilized *Saccharomyces cerevisiae* amplified for fumarase. Enzyme Microb Technol 1991;13:991–996.

156. Peleg Y. Inducible overexpression of the fum1 gene in *Saccharomyces cerevisiae*: Localization of fumarase and efficient fumaric acid bioconversion to l-malic acid. Appl Environ Microbiol 1990;56:2777–2783.

157. Vrsalovic P. Production of l-malic acid by permeabilized cells of commercial *Saccharomyces* sp. Strains. Biotechnol Lett 2005;27:1835–1839.

158. Yamamoto K. Continuous production of l-malic acid by immobilized brevi *Bacteririum ammoniagenes* cells. Eur J Appl Microbiol 1976;3:169–183.

159. Giorno L. Study of an enzyme membrane reactor with immobilized fumarase for production of l-malic acid. Biotechnol Bioeng 2001;72:77–84.

160. Scott A. BASF launches anti-limescale polymer. Chem Week 2002;164:28.

161. Battat E, Peleg Y, Bercovitz A, Rokem JS, Goldberg I. Optimization of l-malic acid production by *Aspergillus flavus* in a stirred fermentor. Biotechnol Bioeng 1991;37:1108–1116.

162. Geiser DM. Cryptic speciation and recombination in the aflatoxin-production fungus *Aspergillus flavus*. Proc Natl Acad Sci USA 1998;95:399–393.

163. Schwartz H. Formation of l(-)malate by *Saccharomyces cerivisae* during fermentation. Appl Microbiol Biotechnol 1988;27:553–560.

164. Taing O. Production of malic and succinic acids by sugar-tolerant yeast *Zygosaccharomyces rouxii*. Eur Food Res Technol 2007;224:343–347.

165. Rhodes R. Production of fumaric acid by *Rhizopus arrhizus*. Appl Microbiol 1959;7:74–80.

166. Rhodes R. Production of fumaric acid in 20-l fermentors. Appl Microbiol 1962;10:9–15.

167. Foster J. Aerobic formation of fumaric acid in the mold *Rhizopus nigricans*: Syntheses by direct 2 c2 condensations. Proc Natl Acad Sci USA 1949;35:663–671.

168. Kenealy W. Biochemical aspects of fumaric acid accumulation by *Rhizopus arrhizus*. Appl Environ Microbiol 1986;52:128–133.

169. Peleg Y. Malic acid accumulation by *Aspergillus flavus*. ^{13}C-NMR and isoenzyme analysis. Appl Microbiol Biotechnol 1989;32:334–339.

170. Zelle R. Malic acid production by *Saccharomyces cerevisiae*: Engineering of pyruvate carboxylation, oxaloacetate reduction, and malate export. Appl Environ Microbiol 2008;74:2766–2777.

171. Pines O. Overexpression of cytosolic malate dehydrogenase (mdh2) causes overproduction of specific organic acids in *Saccharomyces cerevisiae*. Appl Microbiol Biotechnol 1997;48:248–255.

172. Moon SY. Metabolic engineering of *Escherichia coli* for the production of malic acid. Biochem Eng J 2008;40:312–320.

173. Mcginn S. Methane emissions from beef cattle: Effects of monensin, sunflower oil, enzymes, and fumaric acid. J Anim Sci 2004;82:3346–3356.

174. Kane J. Production of Fumaric acid. US2327191. 1939.

175. Waksman S. Process for the production of Fumaric acid. US2326986. 1943.

176. Ng T. Production of tetrahydrofuran/1,4-butanediol by a combined biological and chemical route. Biotechnol Bioeng Symp 1986;17:344–363.

177. Ling L. Fermentation process for carboxylic acid. US4877731. 1989.

178. Foster J. The production of fumaric acid by molds belonging to the genus *Rhizopus*. J Am Chem Soc 1938;61:127–127.

179. Cao N, Du J, Gong C, Tsao G. Simultaneous production and recovery of fumaric acid from immobilized *Rhizopus oryzae* with a rotary biofilm contractor and an adsorption column. Appl Environ Microbiol 1996;62:2926–2931.

180. Zhou Y. Comparison of fumaric acid production by *Rhizopus oryzae* using different neutralizing agents. Bioprocess Biosyst Eng 2002;25:179–181.

181. Gangl I. Economic comparison of calcium fumarate and sodium fumarate production by *Rhixopus arrhizus*. Appl Biochem Biotechnol 1990;24–25(1):663–677.

182. Riscaldati E. Direct ammonium fumarate production by *Rhizopus arrhizus* under phosphorus limitation. Biotechnol Lett 2000;22:1043–1047.

183. Otsuka K. Cis-trans isomerase-isomerization from maleic acid to fumaric acid. Agric Biol Chem 1961;25:726.

184. Takamura Y. Studies on induced synthesis of maleate cis-trans isomerase by malonate. Purification and properties of maleate cis-trans isomerase induced by malonate. Agric Biol Chem 1969;33:718.

185. Scher M. Maleate isomerase. J Biol Chem 1969;244:1878–1882.

186. Goto M. Method for producing fumaric acid. US5677156. 1998.

187. Nakajimakambe T. Bioconversion of maleic acid to fumaric acid by *Pseudomonas alcaligenes* strain xd-1. J Ferment Bioeng 1997;84:165–168.

188. Ichikawa S. Improvement of production rate and yield of fumaric acid from maleic acid by heat treatment of *Pseudomonas alcaligenes* strain xd-1. Biochem Eng J 2003;13:7–13.

189. Fall RR, Kuzma J, Nemecek-Marshall M. Materials and methods for the bacterial production of isoprene. US patent 5849970. The Regents of the University of Colorado; 1998.

190. Zanzig DJ, Sandstrom PH, Hubbell JK, Hsu W-L, Halasa AF, Verthe JJA. Styrene-isoprene rubber for tire tread compounds. The Goodyear Tire & Rubber Company United States; 1996.

191. Cervin M, Whited GM, Chotani GK, Valle F, Fioresi C, Sanford KJ, Mcalaffe JC, Feher FJ, Puhala AS, Miasnikov A, Aldor IS. Composition and methods for producing isoprene. 20090203102. 2009.

192. Genencor International, I.N.C.. Bio-based process for production of isoprene from renewable resources. Renewable resources and biorefineries 5; 2009. [cited June Accessed 2009 Dec 10.

INDEX

Note: Italicized *f* and *t* refer to figures and tables

Biocatalysis for Green Chemistry and Chemical Process Development, First Edition.
Edited by Junhua (Alex) Tao and Romas Kazlauskas.

Printed and bound by CPI Group (UK) Ltd, Croydon, CR0 4YY
16/09/2026
14963415-0002